SEMICONDUCTORS AND SEMIMETALS

VOLUME 27

Highly Conducting Quasi-One-Dimensional Organic Crystals

Semiconductors and Semimetals

A Treatise

Edited by R. K. Willardson
ENICHEM AMERICAS INC.
PHOENIX, ARIZONA

Albert C. Beer
BATTELLE COLUMBUS LABORATORIES
COLUMBUS, OHIO

SEMICONDUCTORS AND SEMIMETALS

VOLUME 27

Highly Conducting Quasi-One-Dimensional Organic Crystals

Volume Editor

ESTHER CONWELL

XEROX CORPORATION
WEBSTER RESEARCH CENTER
WEBSTER, NEW YORK

ACADEMIC PRESS, INC.
(Harcourt Brace Jovanovich, Publishers)

Boston   San Diego   New York
Berkeley   London   Sydney
Tokyo   Toronto

COPYRIGHT © 1988 BY ACADEMIC PRESS, INC.
ALL RIGHTS RESERVED.
NO PART OF THIS PUBLICATION MAY BE REPRODUCED OR
TRANSMITTED IN ANY FORM OR BY ANY MEANS, ELECTRONIC
OR MECHANICAL, INCLUDING PHOTOCOPY, RECORDING, OR
ANY INFORMATION STORAGE AND RETRIEVAL SYSTEM, WITHOUT
PERMISSION IN WRITING FROM THE PUBLISHER.

ACADEMIC PRESS, INC.
1250 Sixth Avenue, San Diego, CA 92101

United Kingdom Edition published by
ACADEMIC PRESS, INC. (LONDON) LTD.
24-28 Oval Road, London NW1 7DX

The Library of Congress has cataloged this serial publication as follows:

Semiconductors and semimetals.—Vol. 1—New York: Academic Press, 1966-

v.: ill.; 24 cm.

Irregular.
Each vol. has also a distinctive title.
Edited by R. K. Willardson and Albert C. Beer.
ISSN 0080-8784 = Semiconductors and semimetals

1. Semiconductors—Collected works.   2. Semimetals—Collected works.
I. Willardson, Robert K.   II. Beer, Albert C.
QC610.9.s48   621.3815′2—dc19   85-642319
ISBN 0-12-752127-5 (vol. 27)

PRINTED IN THE UNITED STATES OF AMERICA

88 89 90 91   9 8 7 6 5 4 3 2 1

Contents

LIST OF CONTRIBUTORS vii
PREFACE ix

Chapter 1 Introduction to Highly Conducting Quasi-One-Dimensional Organic Crystals
E. M. Conwell

I. History 1
II. "Metal-Insulator" Transition 6
III. Properties in the Metallic Range 10
IV. Properties in the Semiconducting Range 19
References 25

Chapter 2 A Reference Guide to the Conducting Quasi-One-Dimensional Organic Molecular Crystals
I. A. Howard

I. Introduction 29
II. Chart of Crystals 31

Chapter 3 Structural Instabilities
J. P. Pouget

I. Introduction 88
II. Basic Structural Properties 89
III. Charge Transfer 98
IV. $2k_F$ and $4k_F$ Structural Fluctuations 110
V. Three-Dimensional CDW Ordering Phase Transitions . . . 156
VI. General Conclusion 193
Appendix 194
References 204

Chapter 4 Transport Properties
E. M. Conwell

I.	The Metallic Range	215
II.	The Peierls Transition: Implications for Transport	254
III.	The Semiconducting Range	274
	References	288

Chapter 5 Optical Properties
C. S. Jacobsen

I.	Introduction	293
II.	Methods	304
III.	General Optical Properties	304
IV.	Infrared Properties	345
V.	Conclusion	376
	List of Symbols and Conversion Factors	377
	References	

Chapter 6 Magnetic Properties
J. C. Scott

I.	Magnetic Measurements	386
II.	The Metallic State	398
III.	The Peierls Transition and the Charge-Density Wave State	412
IV.	The Spin-Density Wave State	418
V.	Summary by Material	425
	References	431

Chapter 7 Irradiation Effects: Perfect Crystals and Real Crystals
L. Zuppiroli

I.	Introduction	437
II.	Experimental Results	439
III.	Discussion	460
IV.	Irradiation by Ionizing Particles: A Local Chemistry of Excited Molecules	473
V.	Conclusion: Perfect Crystals and Real Crystals	477
	References	479

INDEX	483
CONTENTS OF PREVIOUS VOLUMES	493

List of Contributors

Numbers in parentheses indicate the pages on which the authors' contributions begin.

E. M. CONWELL, *Xerox Corporation, Webster Research Center, 800 Phillips Road, Webster, New York 14580* (1, 215)

I. A. HOWARD*, *Xerox Corporation, Webster Research Center, 800 Phillips Road, Webster, New York 14580* (29)

CLAUS S. JACOBSEN, *Physics Laboratory 3, Technical University of Denmark, DK-2800 Lyngby, Denmark* (293)

JEAN PAUL POUGET, *Laboratoire de Physique des Solides, Associe au CNRS, Universite de Paris-Sud, F-91405 Orsay, France* (88)

J. C. SCOTT, *IBM Almaden Research Center K46/803, 650 Harry Road, San Jose, California 95120-6099* (386)

L. ZUPPIROLI, *Section d'Etudes des Solides Irradies, Centre d'Etudes Nucleaire de Fontenay-aux-Roses, 92260 Fontenay-aux-Roses, France* (437)

---

*Present address: *Department of Physics, Queen Mary College, University of London, Mile End Road, London, E1 4NS.*

Preface

A quasi-one-dimensional material has a structure that is so anisotropic that overlap of electron wavefunctions is much larger in one crystallographic direction than the others. There are two main classes of such materials: polymers, which by definition are long-chain molecules, and organic molecular crystals. Members of the latter class have as at least one constituent a large planar molecule such as TCNQ (tetracyanoquinodimethane), TTF (tetrathiafulvalene), or TMTSF (tetramethyltetraselenafulvalene). The quasi-one-dimensional nature of the crystals arises from the fact that these molecules are stacked "like poker chips," with much smaller distances between them along the stacks or chains than in any other direction. They are usually metallic in their properties along the chain direction at room temperature, but at some lower temperature undergo a phase transition to a semiconducting state.

Quasi-one-dimensional materials have aroused a great deal of interest in the past decade, both because of their novel physics and the prospects for applications. One fascinating aspect has been the unexpected overlap with other exciting areas of physics. One-dimensional states with fractional charge found in relativistic field theory have strong similarities to those predicted for polyacetylene and other quasi-one-dimensional conductors. Fractional charge and its ramifications have, of course, been eagerly researched since the discovery of the fractional quantum Hall effect. Most recently, it has been found that the new high T_c superconductors show photoinduced infrared activity of a number of vibrational modes, a well-known property of polyacetylene and other highly conducting polymers that results from the electron-phonon interaction.

Another intriguing feature of quasi-one-dimensional materials is the wide range of phenomena they display. Apart from exhibiting a variety of phases and phase transitions, to charge-density wave, spin-density wave, and superconducting states, they are particularly sensitive to electron–electron correlations and to disorder. Carrier mobilities in the semiconducting phase range from very low values, characteristic of hopping, to one of the highest values found in semiconductors—greater than 10^5 cm^2 V sec in tetramethyltetraselenafulvalene hexafluorophosphate, (TMTSF)$_2$PF$_6$, at 4 K. There are several varieties of nonlinear one-dimensional excitations, solitons, that have been studied extensively by both theorists and experimentalists. As an additional incentive, polyacetylene and some of the other polymers have commercial prospects as conducting plastics, rechargeable battery electrodes, sensors of various kinds, and even as a replacement for copper, because recent advances in synthesis have produced polyacetylene with higher conductivity than copper per unit mass.

This volume is devoted to the organic molecular crystals, in particular to TTF–TCNQ and its close relatives. It begins with a chapter introducing these materials in simple nonmathematical terms. The chapter starts with a brief account of their history, now some 15 years in duration, and goes on to discuss the various types of phase transitions that occur in these materials and their transport and optical properties in the metallic and semiconducting states. The treatment includes only a relatively small number of references; it is left to the chapters on special topics to be more thorough.

The second chapter is a list of the currently known conducting quasi-one-dimensional organic molecular crystals in alphabetical order. Arbitrarily, the lower limit on the room-temperature conductivity that qualified a crystal for inclusion was taken as 10^{-4} ohm^{-1} cm^{-1}; this left ~400 entries at the time of writing! The list includes the structure of the molecular constituents, room-temperature conductivity, and some references where further information may be obtained. Following this there is a chapter by J. P. Pouget covering more than 10 years of experimental and theoretical study of the phase transitions in TTF–TCNQ and its close relatives. The chapter highlights the close relation between the structural instabilities and the electronic instabilities of the one-dimensional electron gas, "the lattice being the fingerprint of the electronic instability."

The next chapter, on transport properties of these materials, covers the coupling of the electrons to the various types of phonons and phonon scattering in both the metallic and semiconducting ranges, with theory compared to experiment. In order to have wavefunctions and the dispersion relation for the semiconducting range, the theory of the Peierls-distorted state is derived for the non-half-filled band case. This chapter is followed by a very thorough treatment by C. Jacobsen of optical properties of these materials from the far infrared through the ultraviolet and how these properties relate to the other properties of the materials. The approach taken is to consider what happens to the optical absorption spectra in passing from the isolated molecule to molecular dimers and then to stacks of molecules.

Measurement of magnetic properties has contributed a great deal to the overall understanding of this family of materials, particularly on interactions of the Fermi gas to which these properties are particularly sensitive. The chapter on these properties by J. C. Scott summarizes magnetic susceptibility, electron spin resonance, and nuclear magnetic resonance data and theoretical interpretations. The concluding chapter, by L. Zuppiroli, deals with the effects of disorder introduced by irradiation on many aspects of these materials—charge-density waves, spin-density waves, single-electron transport, collective modes, phase transitions, etc.

The nature of these materials requires that there be more discussion of what is usually considered chemistry than has been usual in this series. Indeed, close collaboration between physicists and chemists has been and will continue to be vital for progress in this field. It will be demonstrated that the range of phenomena in these quasi-one-dimensional materials is so large as to provide a microcosm of condensed matter physics.

E. M. CONWELL

CHAPTER 1

Introduction to Highly Conducting Quasi-One-Dimensional Organic Crystals

E.M. Conwell

XEROX WEBSTER RESEARCH CENTER
WEBSTER, NEW YORK

I. HISTORY	1
1. *TCNQ*	1
2. *Bechgaard Salts*	4
II. "METAL-INSULATOR" TRANSITION	6
3. *Charge-Density and Spin-Density Waves*	6
4. *Peierls Transition*	8
III. PROPERTIES IN THE METALLIC RANGE	10
5. *Carrier Concentration*	10
6. *Transport*	11
7. *Optical Properties*	15
IV. PROPERTIES IN THE SEMICONDUCTING RANGE	19
8. *Transport*	19
9. *Optical Properties*	23
REFERENCES	25

I. History

1. TCNQ

The term *quasi-one-dimensional* is used to describe materials with structure so anisotropic that overlap of electron wavefunctions, and, as a consequence, conductivity, is much larger in one crystallographic direction than the others. An early landmark in the study of this type of material, which might be taken as its birth, was the synthesis of the large planar molecule tetracyanoquinodimethane by researchers at DuPont (Acker *et al.*, 1960). The structure of this molecule, abbreviated TCNQ, is shown in Figure 1. This synthesis was followed in a short time by the introduction of nearly 100 new organic crystals based on TCNQ (Melby *et al.*, 1962). Although most of the new crystals were highly insulating, as had been found traditionally for organic solids, there were some with sizable room-temperature conductivities, the highest being 10^2 ohm^{-1} cm^{-1} for quinolinium Qn(TCNQ)$_2$.

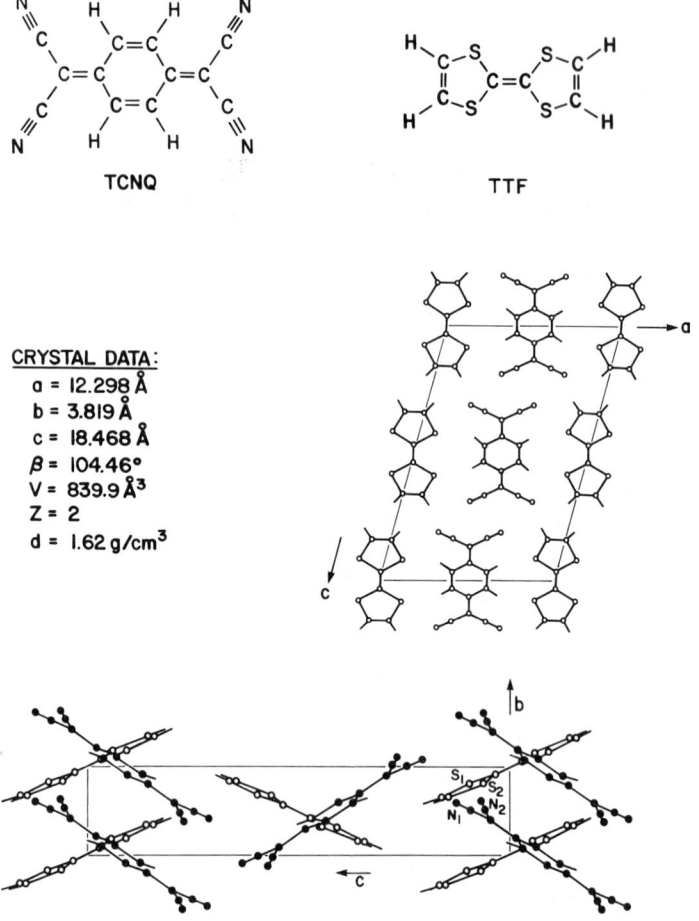

FIG. 1. Molecular constituents and views of the crystal structure of TTF–TCNQ normal to the ac plane and down the (100) direction. V is the volume of the unit cell, Z the number of molecules/unit cell, and d the density of the crystal. [From Kistenmacher et al. (1974).]

These discoveries were followed in short order by those of the highly conducting quasi-one-dimensional (1D) compound of TCNQ with N-methylphenazinium (NMP) and of the molecule tetrathiafulvalene (TTF), synthesized in 1970 by F. Wudl. TTF, also a large planar molecule, is shown in Fig. 1. In 1972 groups from Monsanto (M. G. Miles, J. D. Wilson, and M. Cohen), Johns Hopkins (Pearlstein et al., 1972), and the University of Pennsylvania (A. J. Heeger and A. F. Garito, 1972) independently combined TTF and TCNQ to form TTF–TCNQ, a salt with more strongly metallic properties. This salt is prototypical of what subsequently became a large family of quasi-one-dimensional salts. Its study dominated the field until 1980. In the crystal the TTF and TCNQ molecules are stacked on top of each other, as shown in Fig. 1, with relatively close spacing in the stack or chain direction and much larger spacing between molecules in the other directions. At room temperature the conductivity of TTF-TCNQ along the chain (b) direction is 600 to 900 ohm^{-1} cm^{-1}. The ratio of the b-axis conductivity to that along the a direction, where the current goes between unlike molecules, is 10^3 at 300 K and increases to a maximum close to 6×10^3 near 60 K (Cohen *et al.*, 1974).

The two-chain structure, with molecules of one type segregated in each chain, is characteristic for the class of highly conducting quasi-one-dimensional organic crystals. Generally, the molecules of each chain are closed shell when isolated. In the crystal, however, there is a transfer of electrons from one type of molecule to the other, leaving at least one set of chains with partially filled electron energy bands. In compounds containing TTF or TCNQ, TTF acts as a donor of electrons, while TCNQ is an acceptor.

Typically the quasi-one-dimensional compounds are metallic in their properties along the chain direction at high temperatures, in many cases to temperatures below 100 K. Despite the metallic properties, the bandwidths along the chains are not large, 0.5 eV being a typical value for TCNQ chains. The small bandwidths result in Coulomb interactions being more important in these metals than in the usual ones, as will be discussed later. When temperature falls below 300 K, conductivity parallel to the chains rises rapidly in most cases. For TTF-TCNQ, early measurements of one group gave σ at 60 K about 500 times the room-temperature value, leading them to suggest that superconducting fluctuations were being observed (Coleman *et al.*, 1973). The measurements and resulting suggestion were questioned in a subsequent publication citing much lower σ values at 60 K, typically 10 to 50 times the room-temperature value, obtained by 18 laboratories; the very high σ was attributed to contact difficulties compounded by the anisotropy and small size of the crystals (Thomas *et al.*, 1976). The excitement and the unusual properties of the materials brought physicists and chemists into the field in large numbers. The latter proceeded to make many new quasi-one-dimensional compounds, a number of which are based on molecules with small chemical changes from TTF or TCNQ. Of notable interest were tetraselenafulvalene TSeF-TCNQ (TSeF differing from TTF by the replacement of S by Se), tetramethyltetraselenafulvalene TMTSF-TCNQ (TMTSF differing from TSF by the replacement of four H's by four CH$_3$'s), hexamethylene-tetraselenafulvalene HMTSF-TCNQ, and TMTSF-DMTCNQ, where DMTCNQ stands for dimethyltetracyanoquinodimethane. The structural formulas for these materials, as well as references dealing with their properties, are listed in Chapter 2.

2. BECHGAARD SALTS

The compounds discussed so far were made by single-crystal growth from solution. The year 1979 saw the start of a new group, made electrochemically by K. Bechgaard. They consisted of chains of TMTSF cations alternating with chains of inorganic anions. The chemical formula for these Bechgaard salts, as they came to be called, is $(TMTSF)_2X$. In the earliest versions of these salts, X was PF_6, AsF_6, SbF_6, BF_4, or NO_3 (Bechgaard *et al.*, 1980). Salts with many different anions, including organic ones, have been made since (see Chapter 2). Also, a large number of $(TMTTF)_2X$ salts, in which Se is replaced by S, have been made. The lattice structure of $(TMTSF)_2ClO_4$, which is typical of the group, is shown in Fig. 2.

In the crystal, each TMTSF or TMTTF molecule transfers half an electron to the X chain, resulting in $\frac{3}{4}$-filled bands on the cation chains and singly charged anions. Conduction is entirely on the cation chains and is therefore by holes. Typical room temperature conductivities for the $(TMTSF)_2X$ salts are 400 to 800 $ohm^{-1} cm^{-1}$ (Bechgaard *et al.*, 1980). The TMTSF salts with PF_6, AsF_6, and NO_3 were found, in the initial studies, to remain metallic down to much lower temperatures than the TCNQ salts and others mentioned earlier, reaching conductivities as high as $10^5 ohm^{-1} cm^{-1}$ (Bechgaard *et al.*, 1980). Only at ~ 12 K did the PF_6 salt, for example, undergo the transition to a semiconducting state.

The low transition temperatures of the TMTSF compounds spurred the hunt for superconductivity. Using pressure in the attempt to move the transition temperature even lower, Jerome and his group at Orsay found $(TMTSF)_2PF_6$ to be superconducting at ~ 10 kbar (Jerome *et al.*, 1980). Subsequently other Bechgaard salts were found to be superconducting under similar pressures. Taking the pressure requirement as a hint that superconductivity required a smaller spacing along the chain, Bechgaard made $(TMTSF)_2ClO_4$, for which smaller spacing was predicted, and indeed this salt was found to be superconducting at ambient pressure (Bechgaard *et al.*, 1981).

The superconducting transition temperature in all these cases was ~ 1 K. The suggestion has been made that these materials show superconducting fluctuations to temperatures as high as 30 or 40 K (Jérome and Schulz, 1982). This suggestion is quite controversial, however (Chaikin *et al.*, 1983). It should be noted that, by any reasonable index of what constitutes a 1D material, these materials are at least 2D rather than 1D at the superconducting transition temperature and, in fact, well above it. $(TMTSF)_2PF_6$, for example, shows a plasma edge for light polarized perpendicular to the chains, as well as for light polarized parallel to the chains, at temperatures of ~ 100 K and below (Jacobsen *et al.*, 1981). New organic superconductors

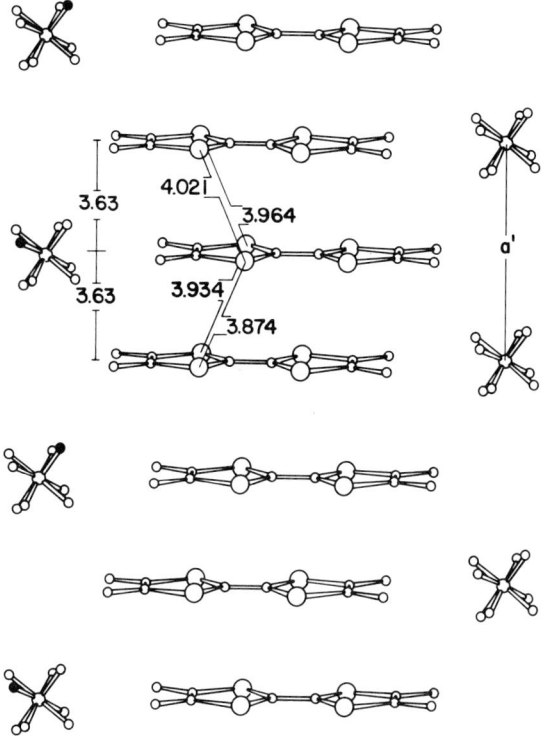

FIG. 2. Side-view of the stacking in (TMTSF)$_2$ClO$_4$ (note that the three H's on each of the four C's at the ends of the molecule have not been shown.) The slight dimerization is visible in the Se–Se intermolecular contacts. Because the ClO$_4^-$ ions are noncentrosymmetric they can, as shown in the anion chain at the right, occupy two equivalent positions related by inversion. At high temperatures they are distributed randomly in these two positions, the lattice constant being a'. They may order at low temperatures, as shown in the anion chain at the left, doubling the lattice constant (Bechgaard, 1982).

continue to be found, the latest at the time of writing being β-(BEDT-TTF)$_2$I$_3$, where BEDT stands for bis-ethylenedithiolo, reported to have a critical temperature of 8 K (Murata *et al.*, 1985) under a pressure of 1.3 kbar.

The Bechgaard salts have many other interesting properties. In addition to superconductivity, they display a remarkable variety of other phase transitions, depending on pressure, temperature, or magnetic field. These will be discussed briefly in the next section. Of particular interest is the phase transition of (TMTSF)$_2$ClO$_4$ at low temperatures from a metallic, nonmagnetic state to a semiconducting magnetic state (Chaikin *et al.*, 1985).

II. "Metal-Insulator" Transition

3. Charge-Density and Spin-Density Waves

With the exception of the few materials that become superconducting, the conductivity of the quasi-one-dimensional materials we are discussing is very small at low temperatures unless the material is quite impure or disordered. In general, the loss of conductivity occurs through one or more phase transitions that culminate in an insulating or, more accurately, semiconducting state. At low temperatures, where almost all the electrons are in states below the Fermi energy ε_F, the total energy of a metallic chain can generally be lowered by the creation of a periodic distortion in the electron gas that opens an energy gap at ε_F, pushing down the states at and close to that energy. When Coulomb effects are small, the periodicity required to do this corresponds to a wavevector $2k_F$, k_F being the magnitude of the wavevector at ε_F. In a half-filled band case, for example, where the spatial period of the undistorted state is b, $2k_F = \pi/b$. The spatial periodicity corresponding to $2k_F$ is then $2b$; thus, the period is doubled by the $2k_F$ distortion, reducing the Brillouin zone by a factor 2 and thereby opening the gap at k_F and ε_F.

When Coulomb effects, specifically electron–electron correlations, are large, some of the statements of the previous paragraph must be modified. Inclusion of electron–electron correlation effects in a 1D case is conveniently, if approximately, done by means of the Hubbard model (Hubbard, 1963). According to this model, the two important parameters determining the behavior of electrons on a chain are the bandwidth W (four times the transfer integral t, in terms of which the model is usually couched) and the repulsion U for a second electron on the same site, with antiparallel spin, of course, due to the Pauli Principle. It is clear that the existence of finite U will result in enhanced magnetic susceptibility, which has indeed been observed in these materials (Torrance *et al.*, 1977) (see Chapter 6). In the limit $U \gg W$ no two electrons may occupy the same site, which is equivalent to saying that no two electrons may be in the same energy state. This has some immediate consequences (Torrance, 1977). First, the conduction band is split into bands of singly and doubly occupied states, the latter having much higher energy than the former. Second, to accommodate the same number of electrons in the conduction band as in the case of negligible U, k_F must be twice as large. Thus the periodicity of the wavevector required to open a gap at ε_F in a large U case would be $4k_F$, rather than $2k_F$, when k_F is calculated in the usual way.

Estimates of U based on isolated molecules lead to values much larger than the bandwidths of these materials, which we have noted are small. These estimates neglect screening by the conduction electrons, however. Experiments to be discussed subsequently show that some TCNQ salts are in the small U limit, while others are in the large U limit. This situation was rationalized by Mazumdar and Bloch (1983), who showed that, due to screening, the importance of electron–electron correlations varies greatly with η, the number of conduction electrons per molecule. They found the correlations to be important for $\eta = 1$, the half-filled band, and $\eta = \frac{1}{2}$, the quarter-filled band, while they are relatively unimportant in the range $0.63 \leq \eta \leq 0.8$. (For further discussion see Chapter 4, Section 3.)

There are several ways of introducing a periodic distortion of the electron gas in these materials. In the one that has been most studied, the new periodicity of the electron gas is accompanied by a distortion of the underlying lattice, the Peierls distortion. It was Peierls who first noted that the distorted state is stable in the low-temperature limit, the cost in energy of distorting the lattice being offset by the decrease in electronic energy due to lowering the states in the gap formation (Peierls, 1955). The periodic distortion of the electron gas in this case is called a *charge-density wave* (CDW). As will be discussed further below, this type of distortion may be thought of as arising from the electron–phonon interaction. It is this type of distortion that is responsible for the metal-to-semiconductor transitions in the TCNQ family.

Another type of gap-opening distortion that is found in quasi-one-dimensional materials involves the charge density for each spin state of the electron varying with period $2k_F$. The charge densities for the two spin states are 180° out of phase with each other, as shown in Fig. 3, so that the total electronic charge density is constant, independent of position. Thus no motion of the ions occurs in this case. There is a net spin polarization, however, that varies with position, and the distortion is known as a

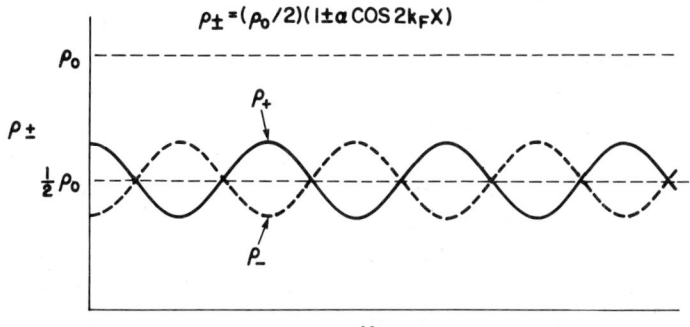

FIG. 3. Electronic charge density vs. distance of *up*-spin (ρ_+) and *down*-spin (ρ_-) electrons in a material with a spin-density wave. The total charge density is a constant, ρ_0. α is a constant less than unity. In a charge-density wave ρ_+ and ρ_- are in phase. [From A. W. Overhauser (1978).]

spin-density wave (SDW). Spin-density-wave distortion is favored by repulsive Coulomb interactions, i.e., $U > 0$, among the electrons. SDWs rather than CDWs are the common form of distortion at low temperatures in the TMTSF salts. Another source of $2k_F$ distortions that give rise to the "metal-insulator" transition in the TMTSF salts is the occurrence of a $2k_F$ periodicity on the anion chains, due usually to an orientational ordering of the anions (Ravy *et al.*, 1986). An example of this is shown in Fig. 2.

The discussion of this section has been based on mean-field theory, that is, it has neglected fluctuations in the values of the various quantities involved. One must be careful about applying such theories, particularly to 1D systems, because fluctuations may have a strong effect on the behavior of the system. For example, fluctuations have been proven to prevent the occurrence of long-range order in a one-dimensional system whose elements interact through finite-range forces (White and Geballe, 1979). In the case of the $(TMTSF)_2X$ compounds, it has been noted earlier that they can be considered two-dimensional above the transition temperatures. The TCNQ compounds, in general, have less interchain coupling than the $(TMTSF)_2X$ compounds. Here, too, however, there is sufficient coupling between chains that the lattice tends to order three-dimensionally near the mean-field transition temperature, leading to a true phase transition. Thus, although based on the Peierls distortion, the actual phase transitions in the TCNQ family are quite complex; they will be discussed in Chapter 3. The only material in which sizable fluctuations persist above the transition temperature is TTF–TCNQ, as will be discussed in the next section. In any case, mean-field theory is widely used in calculations on quasi-one-dimensional materials, and its predictions are at least qualitatively correct.

4. Peierls Transition

It is instructive to study further the Peierls transition, viewing it as originally done by Fröhlich (1954). This will be done for the case of a small electron–electron interaction or small U. Considerations are similar for the large U case, provided $2k_F$ is replaced by $4k_F$. Although an acoustic phonon has very little energy compared to that of an electron at the Fermi surface, it may have comparable momentum and can therefore readily scatter electrons form $\pm k_F$ to $\mp k_F$, i.e., from one side of the Fermi surface to the other. There is thus a strong interaction of the electrons with acoustic phonons whose wavevector is $2k_F$. This produces a softening, or dip in frequency (called a Kohn anomaly), of the $2k_F$ phonons, an effect that has been seen by neutron scattering (Shirane *et al.*, 1976) for temperatures not far above the Peierls transition temperature, T_P, in TTF–TCNQ, for example (see Chapter 3). Significantly, the soft phonons, i.e., the phonons with wavevector $2k_F$ ($4k_F$ in the large-U case), can also be seen by diffuse x-ray scattering, because the x-ray intensity due to phonon-induced vibrations is proportional to the mean square amplitude of those vibrations, in turn proportional to the reciprocal of the square of the phonon frequency (see Chapter 3). It is interesting that, according to x-ray diffuse scattering, there are both $2k_F$ and $4k_F$ charge-density fluctuations for TTF-TCNQ (Pouget *et al.*, 1976; Kagoshima *et al.*, 1976). The $2k_F$ spots are barely seen above ~150 K, whereas the $4k_F$ spots persist to room temperature. Experiments of Kagoshima (1980) established that the $4k_F$ spots are due to the TTF chain. As discussed earlier, this indicates $U \gg W$ on the TTF chain, but not on the TCNQ chain, suggesting a smaller bandwidth for the TTF chain. Additional evidence for the smaller bandwidth for TTF comes from band structure calculations (Herman *et al.*, 1977) and comparison of TTF-TCNQ with TSeF-TCNQ (Schultz and Craven, 1979).

As the temperature of the material decreases to T_P, the frequency of the $2k_F$ phonons goes to zero. The lattice distortion at $2k_F$ thus becomes time invarient—it is "frozen in." If the wavelength associated with the $2k_F$ lattice distortion, π/k_F, is mb, an integer multiple of the lattice spacing, the spatial period of the Peierls distortion is commensurate with the original lattice spacing. This is, for example, the case for the half-filled band for which $m = 2$. For the $m = 2$ case, the dependence of the gap, 2Δ, on the electron-phonon coupling is readily calculated by minimizing with respect to Δ the total free energy—the sum of that due to the chain distortion and that of the electrons in the states on the distorted chain. As will be derived in Chapter 4, that dependence is given by

$$\Delta = We^{-1/\lambda},$$

where λ is the dimensionless electron-phonon coupling constant. It is clear from this equation that the gap increases with the strength of the electron-phonon coupling. For the case we are discussing, λ has been found to be (Takayama *et al.*, 1980)

$$\lambda = \frac{(\partial W/\partial u)^2}{4\pi K t},$$

where K is a spring constant for the lattice distortion and $\partial W/\partial u$ the rate of change of the bandwidth with spacing between ions on the chain. It is the modulation of the latter spacing by the phonons that results in the electron-phonon coupling.

As will be seen in Chapter 4, the dependence of energy on wavevector in the $m = 2$ Peierls distortion case and the equation for the Peierls gap are quite similar to the corresponding equations for a superconductor. The similarities are not unexpected in that superconductivity also results from an electron-phonon coupling that pairs charge carriers on opposite sides of the Fermi surface and having opposite spin. The Peierls instability results, however, from the pairing of electrons with holes rather than with other electrons.

The idea of the Peierls distortion must be generalized for quasi-one-dimensional crystals made up of large molecules such as TTF or TCNQ. Because such molecules have many phonon bands arising from internal vibrations, the coupling of the electrons to these internal modes is generally stronger than that to the acoustic, or external, modes. As a result, the Peierls distortion in these materials may arise from freezing in some of the internal modes. For the case of a half-filled band with small electron–electron interactions, the Peierls distorted state—and the resulting charge-density wave—is then due to variations, with period $2b$, in the shape of the molecules, rather than in their spacing. This would also be true for a quarter-filled band in the large U limit, a case that is realized in $Qn(TCNQ)_2$ and a number of other salts (McCall et al., 1985). Striking evidence of this effect is the appearance in the Peierls distorted state of strong absorption at the frequency of the totally symmetric internal modes for infrared radiation polarized parallel to the chains. Although the vibrations of these modes are essentially perpendicular to the chains, and they would be infrared inactive because of their symmetry (a_g), they receive a large dipole moment parallel to the chains from the phase oscillations (phonons) of the charge-density wave (Rice, 1976).

III. Properties in the Metallic Range

5. Carrier Concentration

The fact that $2k_F$ or $4k_F$ can be determined, as just discussed, by neutron or x-ray scattering means that one can determine the density of conduction electrons or holes, i.e., the charge transfer. For TTF–TCNQ, for example, $2k_F$ is found at ambient pressure to be $0.59\pi/b$, which, incidentally, makes this an incommensurate case. A transfer of one electron/molecule would give $2k_F = \pi/b$. Thus the charge transfer η at ambient pressure in TTF–TCNQ is 0.59 electrons/molecule. With $b = 3.819$ Å, and the number of chains/cm^2 in TTF–TCNQ 8.79×10^{13}/cm^2, $\eta = 0.59$ leads to room-temperature concentrations at ambient pressure of 1.36×10^{21}/cm^3 electrons or holes.

It is of interest that, according to neutron scattering experiments, under pressure η for TTF–TCNQ increases (Megtert et al., 1976), reaching 0.66 electrons/molecule in the range 14.5–17.5 kbar (Megtert et al., 1981). In this pressure range then the charge-density waves and the Peierls distortion are commensurate with the underlying lattice, corresponding to $m = 3$. Similar behavior is found for TSeF–TCNQ, for example. In that case η at ambient pressure is 0.63 electrons/molecule and reaches 0.66 at ~6 kbar (Thomas and Jérome, 1980).

6. Transport Properties

Typically in the metallic range the dc conductivity σ_{dc} parallel to the chains increases with decreasing temperature as $T^{-\alpha}$, where $\alpha = 2.3$ for TTF-TCNQ samples. This dependence holds even for samples with different ratios of 300 to 60 K conductivity (Groff et al., 1974). A similar temperature dependence, with $\alpha \simeq 2$, is found for other quasi-one-dimensional conductors. A typical plot of σ_{dc} vs. T for TTF-TCNQ is shown in Fig. 4. The metallic range may be considered to end at ~60 K, below which the steep

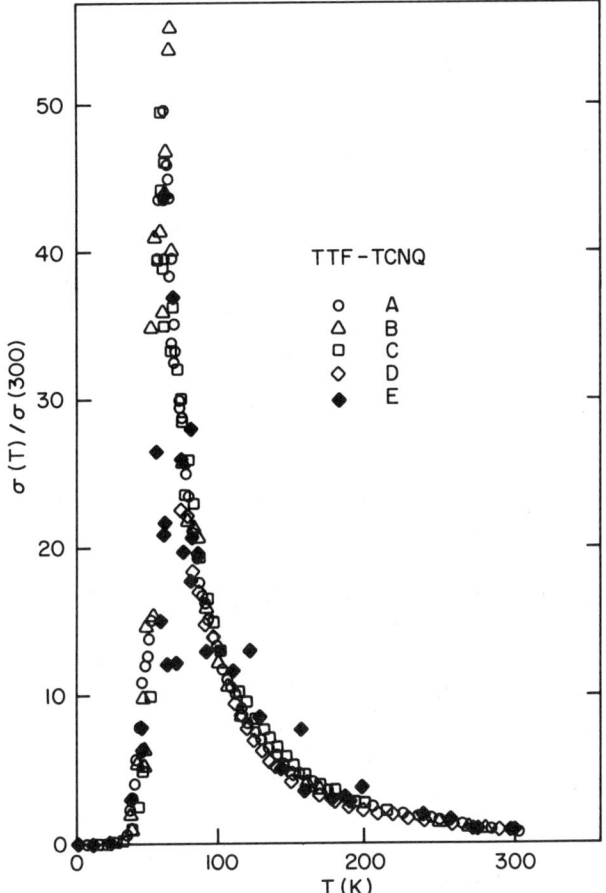

FIG. 4. Conductivity parallel to the chains vs. temperature for several TTF-TCNQ samples (A-D) and the microwave conductivity E for sample D. As noted in the text, the ratio of the peak σ to σ at 300 K is sample dependent, with values of 10 to 15 quoted frequently. [From Ferraris and Finnegan (1976).]

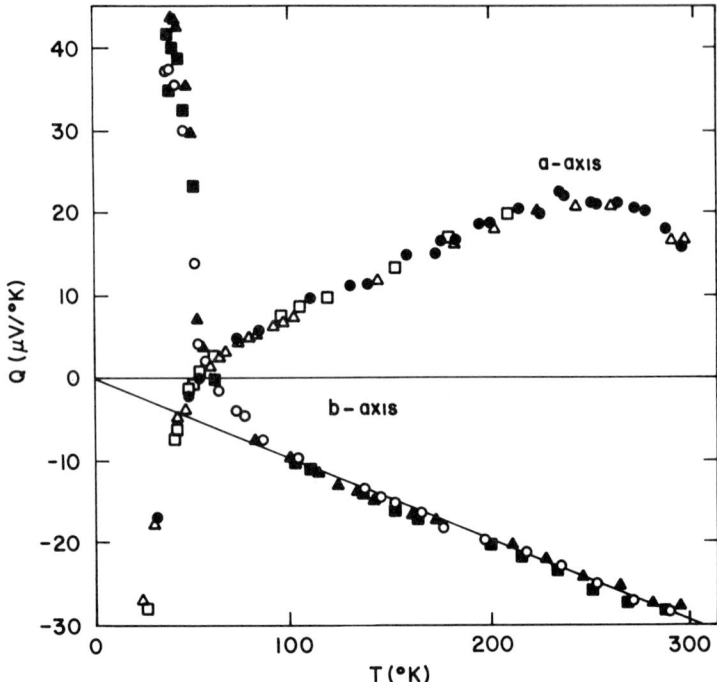

FIG. 5. Thermoelectric power vs. temperature along the *a*- and *b*-axes for single crystals of TTF-TCNQ. [From Kwak *et al.* (1975).]

descent in σ_{dc}, leading to the Peierls transition at 54 K for the TCNQ chains, begins (Tomkiewicz *et al.*, 1976). The dramatic drop in σ_{dc} due to the gap opening on the TCNQ chains, although the Peierls transition does not occur until lower temperature on the TTF chains, indicates that the contribution of the TCNQ chains dominates transport in the metallic range. Further evidence for this is provided by the fact that the thermoelectric power of TTF–TCNQ along the chain direction is negative down to ~60 K, as shown in Fig. 5. Also, the measured Hall constant over a wide temperature range is close to what one would expect if only the TCNQ chain were conducting (Cooper *et al.*, 1977). The dominance of the TCNQ chain in transport appears to result from its bandwidth being larger, a result obtained from band structure calculations (Herman *et al.*, 1977), as well as other types of evidence mentioned earlier.

Impressed by the strong electron–phonon coupling indicated by the Kohn anomaly and Peierls transition, some early investigators suggested that conductivity in the metallic range is due to collective transport of coupled electrons and phonons, or fluctuating charge-density waves (Bardeen, 1973;

Allender et al., 1974; Heeger, 1979). Indeed, as indicated earlier, there is evidence that such fluctuations contribute to current in TTF-TCNQ from T_P up to perhaps 150 K. In an incommensurate case, which we have noted TTF-TCNQ to be, the energy of a charge-density wave is independent of its position. It can therefore translate in an electric field and contribute to current. Andrieux et al. (1979) have done an experiment that strongly suggests this is occurring. Measuring the conductivity of TTF-TCNQ as a function of pressure, they found a sizable drop in σ_{dc} around 20 kbar, about where the charge-density wave was found to be commensurate (Megtert et al., 1981), for temperatures between ~150 and 80 K. With other contributions to σ_{dc} unlikely to be affected by commensurability, the drop in σ_{dc} is reasonably interpreted as indicating that at other pressures, where the charge-density waves should be free to move, thay are indeed making some contribution to σ_{dc}. Above 150 K, however, there is very little drop, if any, in dc conductivity with pressure. This is consistent with the fact mentioned earlier that on the TCNQ chain, which dominates σ_{dc} at these temperatures, the fluctuations are barely visible to x-rays above 150 K. Similar experiments on TSeF-TCNQ (Thomas and Jerome, 1980) show relatively little drop in σ_{dc} at the pressure for which it becomes commensurate, indicating that conductivity due to fluctuating charge density waves is much less important in this material. The reason for the smaller effect may be that with the transition temperature the same for TSeF and TCNQ chains, there are comparable numbers of charge-density waves on both chains. Due to their opposite signs, the CDWs exert a drag on each other. TTF-TCNQ is unique in having a much lower transition temperature on the TTF chain, so that above T_P for the TCNQ chain it has a much higher CDW population than the TTF chain. From the above considerations we conclude that dc conductivity for TTF-TCNQ above 150 K and for TSeF-TCNQ, and probably all other members of this family in the metallic range, is essentially all due to one-electron processes. Even for TTF-TCNQ one-electron processes should make the dominant contribution to σ_{dc}, except for temperatures close to 60 K.

The observed near-proportionality of σ to T^{-2} suggested electron–electron scattering as the source of resistance (Seiden and Cabib, 1976). This suggestion could be definitively eliminated, however, for TTF-TCNQ and other salts based on TCNQ because, the bands being less than half-filled, Umklapp processes are not permitted; as a result, electron–electron collisions do not remove momentum from the electron distribution. The latter statement is not true for the $(TMTSF)_2X$ compounds, however, and it has been shown that electron–electron collisions do make an important contribution to the resistivity in that case (Lyo, 1983). This leaves phonon scattering as the source of resistivity. Scattering by single acoustic phonons, as in ordinary metals, would lead to resistivity ρ increasing linearly with T, because the number of phonons does so in this range. The strong temperature dependence, therefore, led to sugestions that two-phonon processes (Gutfreund and Weger, 1977) or high-energy internal modes (Conwell, 1977) are responsible for the resistivity.

Not long after these suggestions were made, reexamination became necessary in the light of an important experimental fact: The conductivity of these materials is quite pressure dependent. This is shown in Fig. 6. Coupled with the fact that the thermal expansion of these materials is large at the temperatures concerned, as is characteristic for molecular crystals, the pressure dependence indicates that a significant portion of the temperature dependence of σ is actually due to volume dependence (Cooper, 1979). When a correction is made for the volume dependence of ρ, it is found that,

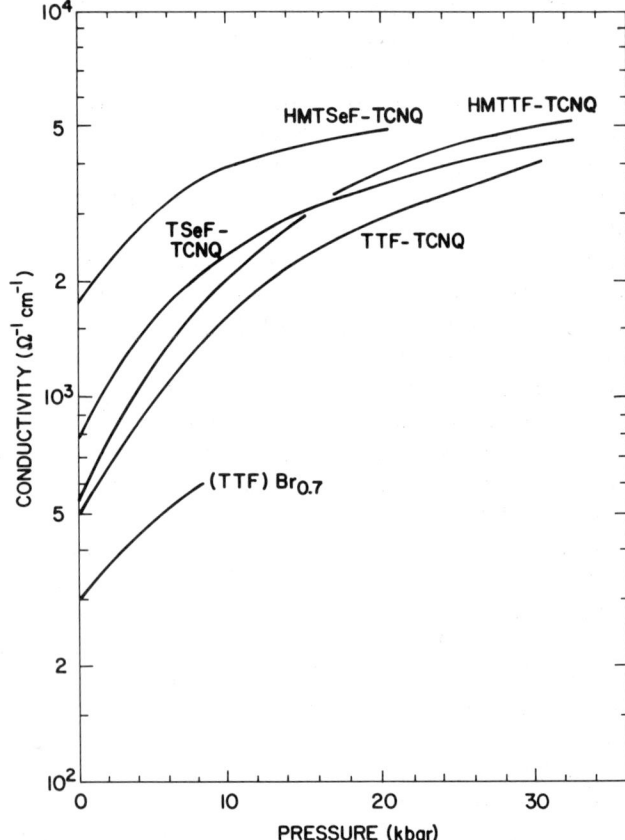

FIG. 6. Room-temperature conductivity parallel to the chains vs. pressure for some organic conductors. [From Cooper (1979).].

in fact, the resistivity at constant volume, ρ_v, is quite close to linear, rather than quadratic (Friend *et al.*, 1978). The near-linear behavior indicates that scattering by internal modes, which have energies high compared to kT in this temperature range, and, in any case, have little pressure dependence, is not of primary importance. Acoustic one-phonon scattering, on the other hand, should be important. Two-phonon processes, however, could not be immediately ruled out, because it was in principle possible that the temperature dependence of other quantities in the expression for ρ, e.g., the frequencies of the modes concerned, could partially cancel the T^2 factor that arises from phonon abundance. To maintain that two-phonon scattering processes are predominant required that one explain why the coupling to these processes is stronger than the coupling to one-phonon processes; many ingenious attempts were made to do this. Recently, however, the coupling constants for one- and two-phonon processes have been calculated for TTF-TCNQ (Van Smaalen *et al.*, 1986), as will be described in Chapter 4, and there is a clear conclusion that the resistivity parallel to the chains of TTF-TCNQ in the metallic range, above ~ 100 K at least, is dominated by one-phonon scattering.

With a room temperature conductivity of ~ 800 cm^2/V sec, carrier concentration of 1.36×10^{21}/cm^3 electrons, and equally many holes, the average mobility parallel to the chains in TTF-TCNQ is ~ 2 cm^2/V sec. It increases as T decreases. The transverse conductivity, which, as noted earlier, is smaller by a factor 10^2 to 10^3, must then be due to hopping.

Thermoelectric power Q of these compounds is also highly anisotropic, as is seen in Fig. 5 for TTF-TCNQ. In the direction parallel to the chains, Q increases linearly with temperature in the metallic range (except for T close to T_P), as expected. We have noted earlier that the fact that $Q < 0$ in the metallic range indicates that the TCNQ chain dominates the conductivity there. It is interesting to note that Q in the a direction, where transport is across both TCNQ and TTF chains, is positive.

7. Optical Properties

The anisotropy of a quasi-one-dimensional material is displayed very clearly in the 300 K reflectance data for single crystal TTF-TCNQ shown in Fig. 7. The reflectance at low frequencies is quite large for the electric vector parallel to the b or chain axis direction, as expected for a metal, but quite small along the a (transverse) direction, as expected for a poor conductor. Along the b direction the reflectance shows what appears to be a well-defined plasma edge at ~ 6500 cm^{-1} or 0.8 eV.

The significance of the reflectance data is better shown by deconvoluting them into real and imaginary parts with the Kramers–Kronig analysis, to be

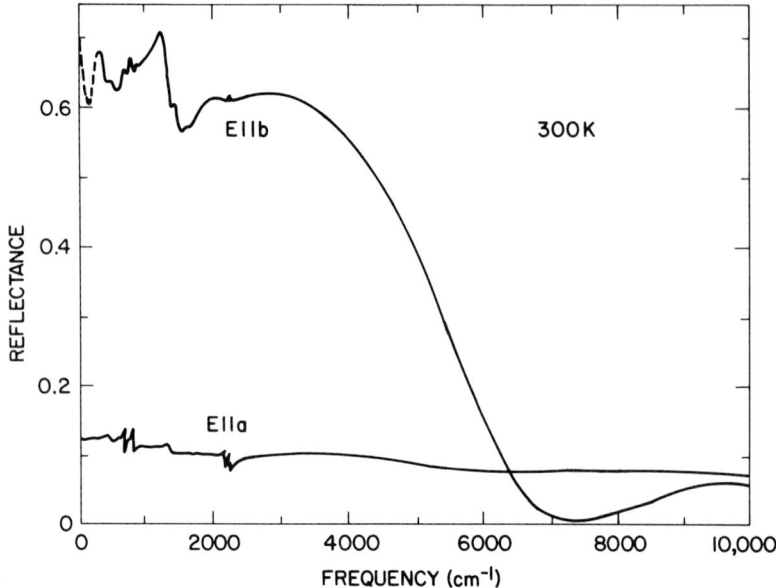

FIG. 7. Polarized reflectance of TTF-TCNQ single crystals vs. frequency at room temperature. [From Tanner et al. (1976).]

discussed in Chapter 5. The resulting real part of the conductivity, $\sigma(\omega)$, and real part of the dielectric constant, $\kappa(\omega)$, for two temperatures in the metallic range are shown in Figs. 8 and 9. It is seen that at high frequencies κ is negative and approaching zero. It actually does go through zero around 6000 cm^{-1} (Tanner et al., 1976), confirming that what looks like a plasma edge in the reflectance, at ~6500 cm^{-1}, is indeed that. However, although the behavior of $\sigma(\omega)$ and $\kappa(\omega)$ above ~2000 cm^{-1} resembles that of an ordinary metal, the behavior below 2000 cm^{-1} is quite different. The fact that $\sigma(\omega)$, which in the low-frequency limit is of the order of σ_{dc}, decreases sharply as the frequency increases from zero, and then increases again, has been attributed to conduction by moving charge-density waves (Tanner et al., 1976). Although a gap does not exist for $T > T_P$, the existence of charge-density waves should result in a thinning of the density of states or a pseudogap. Suggesting that moving charge-density waves would dominate the conduction, Tanner et al. (1976) predicted that $\sigma(\omega)$ would show a narrow peak at $\omega = 0$ with a width $\sim 1/\tau_c$, where τ_c is the collective mode lifetime. If there were truly a gap, ε_G, absorption would vanish as ω increased beyond $\sim 1/\tau_c$, starting again for $\omega \simeq \varepsilon_G/\hbar$ due to single-particle excitation across the gap. In the presence of a pseudogap $\sigma(\omega)$ would not actually vanish beyond $1/\tau_c$ but was predicted to be small compared to σ_{dc}

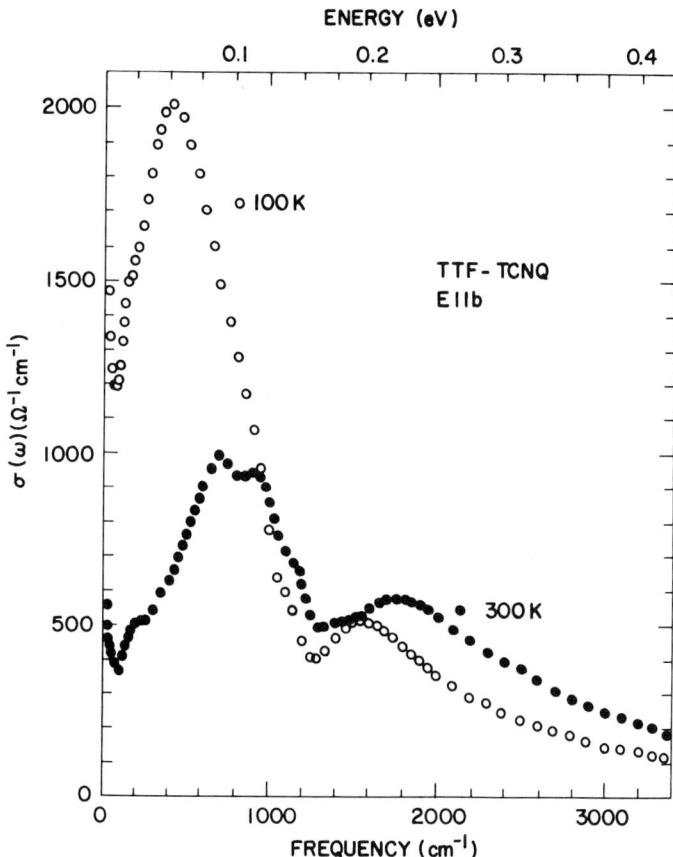

FIG. 8. Frequency-dependent conductivity for $E \| b$ at 300 and 100 K from 20–3400 cm^{-1}. [From Jacobsen (1979).]

on the grounds that electrons in states in the pseudogap would have small mobility, perhaps even being localized in the presence of highly developed fluctuations (Heeger, 1979). It is seen, however, in Fig. 8 that at 300 K, and even at 100 K, the dip in $\sigma(\omega)$ at low frequencies is relatively small. This is consistent with our earlier conclusion that σ is dominated by single-particle conduction rather than moving charge-density waves in the temperature range of these optical data.

It was considered in the early work of Tanner et al. (1976) that the charge-density wave responsible for the initial fall in $\sigma(\omega)$ was the $2k_F$ charge-density wave on the TCNQ chain. This could be true at 100 K where, as noted earlier, the $2k_F$ x-ray spots are well defined. It is not reasonable, however, at 300 K. Later Jacobsen et al. (1984) suggested that at 300 K the initial drop in $\sigma(\omega)$

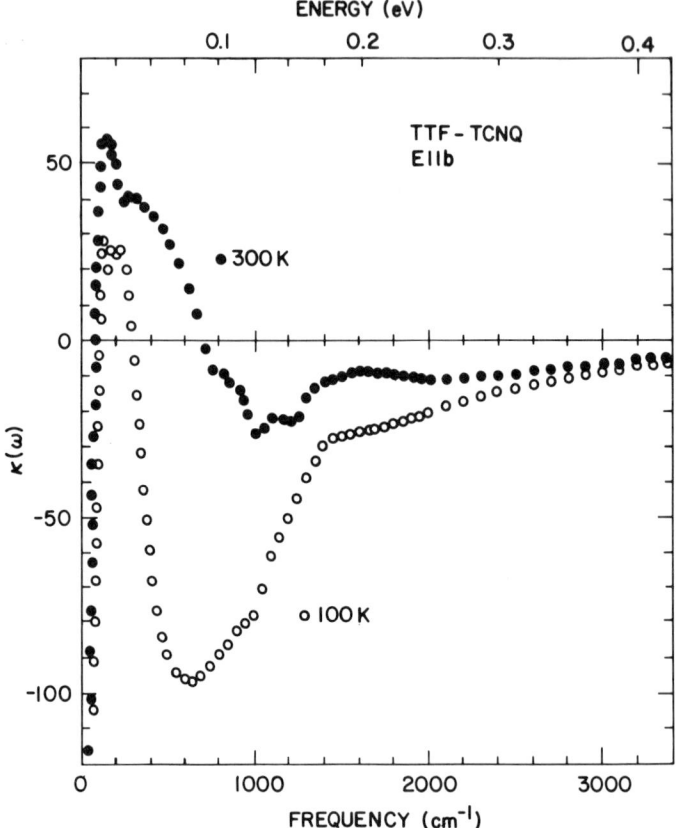

FIG. 9. Real part of dielectric function for $E \| b$, at 300 K and 100 K from 20–3400 cm^{-1}. [From Jacobsen (1979).]

and the rise to a maximum at ~800 cm^{-1} are due rather to the $4k_F$ charge-density wave on the TTF chain, which does show well-developed spots still at room temperature. Consistent with this, Jacobsen et al. (1984) suggested further that at 300 K the TCNQ chain absorption is near Drude-like up to the plasma frequency; this is consistent with the treatment of conductivity in TTF–TCNQ presented earlier.

The dielectric constant also exhibits complicated behavior at low frequencies. The negative value at very low frequencies, ≤ 100 cm^{-1}, is expected for single-particle conductivity. It might also include a negative contribution from moving charge-density waves. It should be noted, however, that at still lower frequencies, i.e., microwave frequencies (~1 cm^{-1}), the room-temperature value of $\kappa(\omega)$ is again positive, $\simeq +2500$ (Gunning et al., 1978).

The suggestion has been made that the positive value at microwave frequencies is due to chain breaks, which in a 1D material can lead to very high polarizability.

IV. Properties in the Semiconducting Range

8. Transport

For either a charge-density-wave or a spin-density-wave transition, theory (to be given in Chapter 4) indicates a rapid increase of the gap as the temperature decreases below the critical temperature for the transition. The type of variation predicted was found in TSeF-TCNQ, for example, as shown in Fig. 10. It was obtained from σ vs. T, as will be described. A similar gap variation was found for TTF-TCNQ and other conducting quasi-one-dimensional salts from infrared absorption data (Bozio and Pecile, 1980; Etemad, 1981), to be described in the next section.

Below T_P it is expected that the charge-density waves be pinned. The pinning may result from charged impurities that happen to be present, or, more likely for the charge-transfer crystals we consider, the electrostatic attraction of charge-density waves of the opposite sign on adjacent chains. The pinning of the charge density waves in TTF-TCNQ below T_P has been demonstrated by infrared absorption, as will be discussed in the next section. Consistent with the pinning, it is expected that the charge-density waves do not contribute to current below T_P. However, in NbSe$_3$, for example,

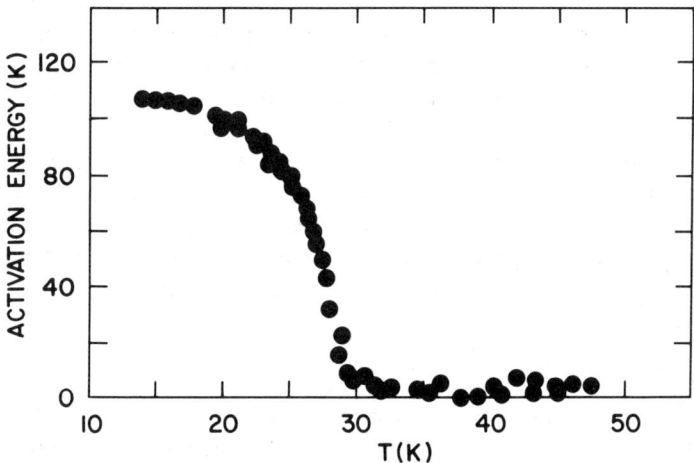

FIG. 10. Variation with temperature of the activation energy for conductivity, or the half gap, for TSeF-TCNQ. [From Etemad (1976).]

charge-density waves can be depinned by an electric field as low as a few mV/cm, and above this threshold field they do contribute to current (Fleming and Grimes, 1979). This has the result that above the threshold field current increases more then linearly with field, or conductivity increases with field. Recently, a similar effect has been found in TTF–TCNQ. For temperatures between ~54 K, where the Peierls transition takes place on the TCNQ chains, and ~38 K, below which ordering takes place and a gap forms on TTF also, there are charge-density waves on the TCNQ chains only. Thus the only pinning in the temperature range between the two transitions is by impurities. Lacoe et al. (1985) have found in this temperature range non-ohmic conductivity above a threshold field E_T, just as in $NbSe_3$. E_T has a minimum value of 0.25 V cm just below the 54 K Peierls transition and increases to 6.4 V cm at 34 K, where the charge-density wave on the TTF chain should be well developed. Below 34 K depinning would require much higher fields, because the charge-density waves on TTF and TCNQ would have to move in opposite directions. Nonlinear conduction with rather different properties has also been found for TTF–TCNQ (Cohen and Heeger, 1977) and $(TMTSF)_2PF_6$ (Chaikin et al., 1980) at much lower temperatures, where it cannot be due to moving charge-density waves. As will be seen later in this section, carrier mobility is quite high at low temperatures, suggesting nonlinearity due to carrier heating. This will be discussed in Chapter 4. In this chapter we confine ourselves to transport in low electric fields, i.e., fields for which σ is constant or Ohm's law is obeyed.

To emphasize the low temperature region, σ_{dc} is plotted vs. $1/T$ for TTF–TCNQ in Fig. 11. Sharp drops are seen at 54 K and at 38 K, where the gaps open on TCNQ and TTF, respectively. (See Chapter 3 for a discussion of the phase transitions in TTF–TCNQ.) The sharp drops are consistent with the rapid opening of the gap below T_P discussed above. There are two sections on the plot of Fig. 11 marked with straight lines to suggest $\log \sigma \propto -K_1(1/T)$ or $\sigma = K_2 e^{-K_1/T}$ where K_1 and K_2 are constants in these sections. For the higher temperature section, $25 \geq T \geq 10$ K, K_1 is found to be ~200 K. Because charge-density waves cannot conduct at low fields in this temperature range, conduction must be due to electrons (or holes) freed from an impurity or defect level or electrons crossing the gap, presumably leaving behind mobile holes. For the latter case, the intrinsic semiconductor, it is well known that

$$\sigma = \sigma_0 e^{-\varepsilon_G/2k_B T},$$

where σ_0 is essentially constant and ε_G is the gap, 2Δ, for the Peierls transition. If this case were applicable here, the gap would be ~400 K. There is good evidence that this is indeed the situation here. As will be shown in the next section, at low temperatures there is strong infrared absorption

1. CONDUCTING QUASI-ONE-DIMENSIONAL ORGANIC CRYSTALS 21

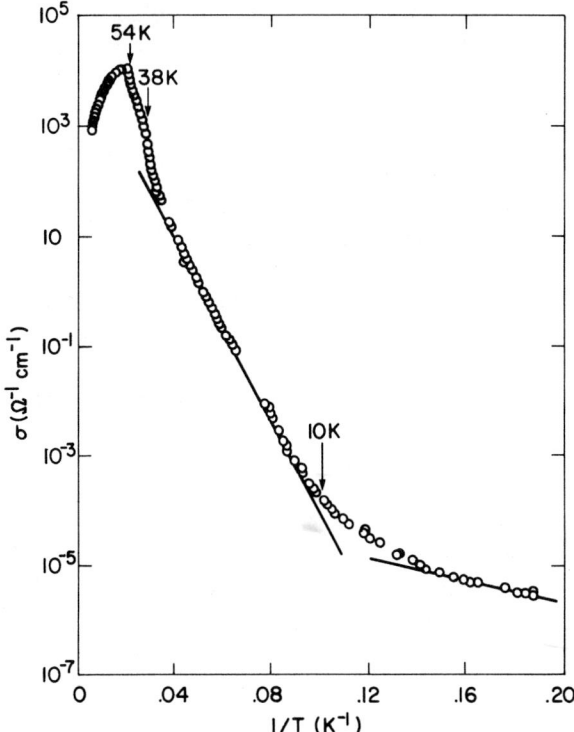

FIG. 11. Low-field dc conductivity parallel to the chains vs. $1/T$ for TTF-TCNQ. [From Cohen and Heeger (1977).]

beginning at ~400 K (Tanner et al., 1981). Also, it has been found that photoconductivity results from absorption of radiation of ~400 K and higher (Eldridge, 1977). Thus in this temperature range, at least down to 10 K, TTF-TCNQ is an intrinsic semiconductor. The relation above between σ and ε_G can be used to determine the gap variation with T for all T's below T_P. It was this relation that yielded the results shown in Fig. 10.

The interpretation of the lower temperature straight line region of ln σ vs. $1/T$ is not as straightforward. It was suggested that it is due to thermal activation of defects, specifically phase kinks corresponding to compressions (Φ-particles) or rarefactions (anti Φ-particles), of the charge-density wave (Cohen and Heeger, 1977). These defects would be charged and mobile, thus capable of contributing to current. The existence of such defects was predicted by Rice et al. (1976), although the expected activation energy was considerably greater than the 14 K obtained from the low-temperature slope. It should be noted, however, that, although such defects should be universal, other researchers do not find a 14 K activation energy for TTF–TCNQ, nor in fact any constant slope below ~10 K. Rather, it is found that the curvature decreases constantly with decreasing temperature in a sample-dependent manner (Banik et al., 1981). An alternative suggestion to account for the smaller and continuously varying slopes seen at low temperatures is that they are due to regions of the material with smaller gaps, these regions dominating σ at low temperatures because the larger gap regions are frozen out. Smaller gaps could result from the presence of disorder or structural defects such as dislocations or strains (Banik et al., 1981), which are known to be present in the crystals. It is well known for $(TMTSF)_2PF_6$, for example, that the gap decreases strongly with pressure, thus with strain, disappearing at about 10 kbar (Jerome et al., 1980).

With the conductivity in the range $125 \geq T \geq 10$ K established as due to electrons crossing the Peierls gap, it is possible to determine the mobility of the carriers parallel to the chains in this range. The carrier concentration may be deduced from the density of states $\rho(k)$ in k space and the known gap. The quantity $\rho(k)$ is obtained from the dependence of the energy on k (see Chapter 4). The resulting electron or hole concentration for a 1D intrinsic semiconductor with gap 2Δ is (Banik et al., 1981)

$$n = p = 2N_c \left(\frac{2\Delta k_B T}{\pi} \right)^{1/2} \frac{e^{-\Delta/k_B T}}{Wb \sin k_F b},$$

where N_c is the number of chains/cm². With $N_c = 8.79 \times 10^{13}$/cm², W the bandwidth 0.5 eV, $b = 3.82$ Å, $k_F b = 0.295\pi$, and $\Delta = 215$ K, this equation leads to $n = p = 10^9$/cm³ for TTF–TCNQ at 10 K. At that temperature σ according to Fig. 11 is 10^{-5} ohm^{-1} cm^{-1}. Thus, the mobility $\mu = 2 \times 10^4$ cm² V sec at 10 K in TTF–TCNQ. This high value is not atypical for quasi-one-dimensional crystalline materials. For TSeF–TCNQ, with a Peierls gap of ~260 K, a similar calculation gives $\mu = 10^4$ cm²/V sec at 20 K. The Hall mobility measured for $(TMTSF)_2PF_6$ at 4 K is 10^5 cm²/V sec (Haen et al., 1981), suggesting electron and hole mobilities of ~10^6/cm² V sec (Conwell and Banik, 1981b). It is these high mobility values that result in carrier heating at modest electric fields.

There are several reasons for these high low-temperature mobilities. One of these is the small size of the Peierls gap. As will be shown in Chapter 4, the effective mass in the Peierls distorted state is essentially $2\Delta/W$ times the effective mass at the bottom of the band in the undistorted state, which is of the order of the free electron mass in these materials. Also, as will be discussed in Chapter 4, the quasi-one-dimensional nature makes scattering by charged defects less effective (Conwell and Banik, 1981a). These high mobilities are perhaps surprising, particularly in light of the expectation of localization of carriers in a 1D system with defects. Of course, all these materials are literally 2D or 3D at these low temperatures, because at least one of the bandwidths transverse to the chains, which can be estimated from various types of measurements, is greater than $k_B T$.

9. Optical Properties

For the semiconducting range, the optical properties that give the most information are in the infrared. A plot of $\sigma(\omega)$ vs. ω in this frequency range for the b-axis (chain) direction in TTF–TCNQ is shown in Fig. 12. Of the four temperatures for which data are given, two are in the metallic range and two in the semiconducting range. The absorption just below 300 cm^{-1} ($\sim$400 K) seen at 25 and 35 K, but not in the metallic range, corresponds to electronic transitions across the semiconducting gap; as indicated earlier, the edge for this absorption is in good agreement with the gap value found from σ vs. T. The other strong feature found in the semiconducting range, but not in the metallic range, is the absorption peak at 40 cm^{-1} for 25 K, a little higher for 35 K. This peak has been identified as absorption due to the pinned

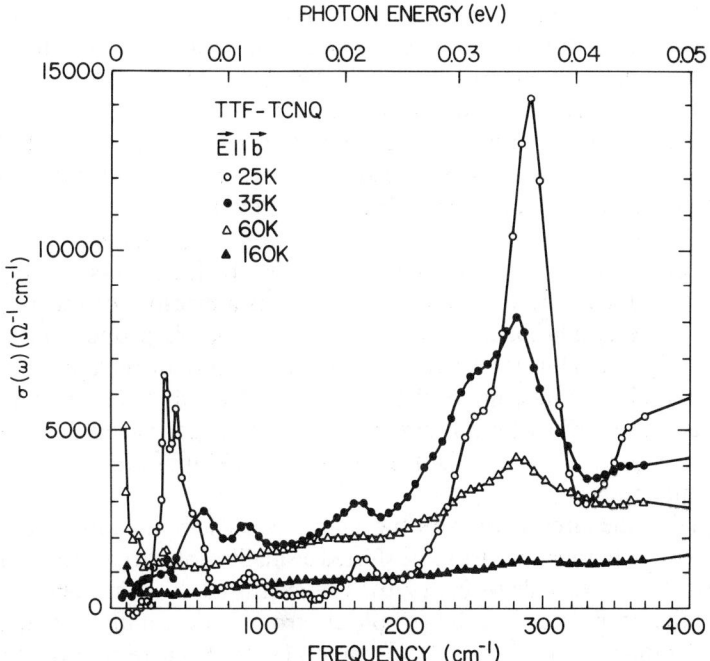

FIG. 12. Frequency-dependent b-axis conductivity of TTF–TCNQ at four temperatures: 25, 35, 60, and 160 K. [From Tanner et al. (1981).]

charge-density wave (Tanner et al., 1981). The pinning provides a restoring force to the charge-density wave motion, shifting the oscillator strength to a finite pinning frequency (Lee et al., 1974). With the pinning frequency $\omega_T = 40 \text{ cm}^{-1}$, the effective mass M^* of the pinned charge-density wave is estimated from the oscillator strength sum rule to be 60 free electron masses (Tanner et al., 1981). A check on these values of M^* and ω_T can be made by calculating from them the contribution of the pinned charge-density wave to the static dielectric constant, $4\pi n e^2/M^*\omega_T^2$ (Lee et al., 1974). With M^* and ω_T given above, this contribution is found to be 2700, which, added to the single-particle contribution of ~600, leads to good agreement with the experimental value, 3000 to 3500, of the low-temperature microwave dielectric constant (Tanner et al., 1981).

Another feature of the infrared data at temperatures below T_P is the appearance of many absorption lines for light with electric vector parallel to the chain direction that are not seen in the metallic range. These features are better revealed at low temperatures by an indirect measurement technique in which the sample is used as a bolometer to yield a measurement of 1-R, R being the reflectivity, rather than R itself (Bates et al., 1981; Eldridge and Bates, 1983). The lines observed in this way have been identified as coming from the totally symmetric internal vibrations (a_g modes) of the molecules referred to earlier and various other modes normally infrared inactive or weak for light with electric vector parallel to the chains. Their giving rise to absorption below T_P indicates that they are involved in the Peierls distortion. The complicated molecular and lattice structure of TTF–TCNQ causes mode mixing, allowing many different kinds of modes to be involved. For those modes included in the Peierls distortion, as indicated earlier, the dipole moment parallal to the chains is greatly enhanced by the phase oscillations of the charge-density wave. In fact, the strength of the absorption due to these modes may be used as a measure of the amplitude of the phase oscillations. Below T_P this amplitude is proportional to the Peierls gap, 2Δ. Thus the variation with temperature of the absorption due to these modes is a measure of the variation of 2Δ with T. This fact has been used, as mentioned earlier, to directly ascertain the gap variation with T for TTF-TCNQ, (TTF) Br$_{0.76}$ (Bozio and Pecile, 1980) and other incommensurate materials.

Actually, the interaction with the charge-density wave causes the absorption lines due to these modes to be shifted a small amount from their location in the isolated molecule (Rice, 1976). The measured shifts have been used to deduce the strength of coupling of the electrons to the totally symmetric (a_g) internal modes in TTF–TCNQ (Etemad, 1981). Such results confirm the statement made in Section 4 that the Peierls distortion in materials of the TTF–TCNQ family is mainly due to electron coupling to the internal modes.

The dimensionless electron–phonon coupling constant λ for the TCNQ chain deduced by Etemad (1981) from this study on TTF–TCNQ has the value $\lambda = 0.35$.

In TCNQ compounds where the lattice is dimerized, another mechanism may operate to activate normally infrared-inactive modes. Optical lattice modes, which cause the two molecules in the dimer to vibrate relative to each other, result in a transfer of charge from one molecule to the other. Again, due to the coupling between electrons and internal modes, this will have the effect of giving the symmetric internal modes a large increase in dipole moment parallel to the chains (Rice et al., 1977). The resulting absorption has been used to deduce the coupling between electrons and internal modes for K–TCNQ (Rice et al., 1977) and MEM(TCNQ)$_2$ (Rice et al., 1980).

A complete discussion of the optical properties of these materials, in the metallic as well as the semiconducting range, will be found in Chapter 5.

References

Acker, D. S., Harder, R. J., Hertler, W. R., Mahler, W., Melby, L. R., Benson, R. E., and Mochel, W. E. (1960). *J. Am. Chem. Soc.* **82**, 6408.
Allender, D., Bray, J. W., and Bardeen, J. (1974). *Phys. Rev. B* **9**, 119.
Andrieux, A., Schulz, H. J., Jerome, D., and Bechgaard, K. (1979). *Phys. Rev. Lett.* **43**, 227.
Banik, N. C., Conwell, E. M., and Jacobsen, C. S. (1981). *Solid State Commun.* **38**, 267.
Bardeen, J. (1973). *Solid State Commun.* **13**, 357.
Bates, F. E., Eldridge, J. E., and Bryce, M. R. (1981). *Can. J. Phys.* **59**, 339.
Bechgaard, K. (1982). *Mol. Cryst. Liq. Cryst.* **79**, 1.
Bechgaard, K., Jacobsen, C. S., Mortensen, K., Pedersen, H. J., and Thorup, N. (1980).*Solid State Commun.* **33**, 1119.
Bechgaard, K., Carneiro, K., Olsen, M., Rasmussen, F. B., and Jacobsen, C. S. (1981). *Phys. Rev. Lett.* **46**, 852.
Bozio, R., and Pecile, C. (1980). In "The Physics and Chemistry of Low Dimensional Solids" (L. Alcacer, ed.), p. 165. Reidel, Dordrecht.
Chaikin, P. M., Gruner, G., Engler, E. M., and Greene, R. L. (1980). *Phys. Rev. Lett.* **45**, 1874.
Chaikin, P. M., Choi, M.-Y., and Greene, R. L. (1983). *J. de Physique* **C3**, 783.
Chaikin, P. M., Choi, M.-Y., Kwak, J. F., Brooks, J. S., Martin, K. P., Naughton, M. J., Engler, E. M., and Greene, R. L. (1985). *Mol. Cryst. Liq. Cryst.* **121**, 79.
Cohen, M. J., Coleman, L. B., Garito, A. F., and Heeger, A. J. (1974). *Phys. Rev. B* **10**, 1288.
Cohen, M. J., and Heeger, A. J. (1977). *Phys. Rev. B* **16**, 688.
Coleman, L. B., Cohen, M. J., Sandman, D. J., Yamagishi, F. G., Garito, A. F., and Heeger, A. J. (1973). *Solid State Commun.* **12**, 1125.
Conwell, E. M. (1977). *Phys. Rev. Lett.* **39**, 777.
Conwell, E. M., and Banik, N. C. (1981a). *Solid State Commun.* **39**, 411.
Conwell, E. M., and Banik, N. C. (1981b). *Phys. Rev. B* **24**, 4883.
Cooper, J. R. (1979). *Phys. Rev. B* **19**, 2404.
Cooper, J. R., Miljak, M., Delplanque, G., Jerome, D., Weger, M., Fabre, J. M., and Giral,, L. (1977). *J. de Physique* **38**, 1097.
Eldridge, J. E. (1977). *Solid State Commun.* **21**, 737.
Eldridge, J. E., and Bates, F. E. (1983). *Phys. Rev. B* **28**, 6972.

Etemad, S. (1976). *Phys. Rev. B* **13**, 2254.
Etemad, S. (1981). *Phys. Rev. B* **24**, 4959.
Ferraris, J. P., and Finnegan, T. F. (1976). *Solid State Commun.* **18**, 1169.
Fleming, R. M., and Grimes, C. C. (1979). *Phys. Rev. Lett.* **42**, 1423.
Friend, R. H., Miljak, M., Jerome, D., Decker, D. L., and Debray, D. (1978). *J. Physique Lett.* **39**, L-134.
Fröhlich, H. (1954). *Proc. Roy Soc.* **A223**, 296.
Groff, R. P., Suna, A., and Merrifield, R. E. (1974). *Phys. Rev. Lett.* **33**, 418.
Gunning, W. J., Heeger, A. J., Shchegolev, I. F., and Zolotukhin, S. P. (1978). *Solid State Commun.* **25**, 981.
Gutfreund, H., and Weger, M. (1977). *Phys. Rev. B* **16**, 1753.
Haen, P., Engler, E. M., Greene, R. L., and Chaikin, P. M. (1981). *Bull. Am. Phys. Soc.* **26**, 213.
Heeger, A. J. (1979). *In* "Highly Conducting One-Dimensional Solids" (DeVreese *et al.*, eds.), p. 69. Plenum, New York.
Heeger, A. J., and Garito, A. F. (1972). *AIP Conf. Proc.* **10**, 1476.
Herman, F., Salahub, D. R., and Messmer, R. P. (1977). *Phys. Rev. B* **16**, 2453.
Hubbard, J. (1963). *Proc. Roy. Soc. A* **276**, 238.
Jacobsen, C. S. (1979). *Lecture Notes in Physics* **95**, 223.
Jacobsen, C. S., Tanner, D. B., and Bechgaard, K. (1981). *Phys. Rev. Lett.* **46**, 1142.
Jacobsen, C. S., Johansen, I., and Bechgaard, K. (1984). *Phys. Rev. Lett.* **53**, 194.
Jérome, D., Mazaud, A., Ribault, M., and Bechgaard, K. (1980). *J. Physique Lett.* **41**, L95.
Jérome, D., and Schulz, H. J. (1982). *Adv. in Phys.* **31**, 299.
Kagoshima, S. (1980). *J. Phys. Soc. Japan* **49**, Suppl. A, 857.
Kagoshima, S., Ishiguro, T., and Anzai, H. (1976). *J. Phys. Soc. Japan* **41**, 2061.
Kistenmacher, T. J., Phillips, T. E., and Cowan, D. O. (1974). *Acta Cryst. B* **30**, 763.
Kwak, J. F., Chaikin, P. M., Russel, A. A., Garito, A. F., and Heeger, A. J. (1975). *Solid State Commun.* **16**, 729.
Lacoe, R. C., Schulz, H. J., Jerome, D., Bechgaard, K., and Johannsen, I. (1985). *Phys. Rev. Lett.* **55**, 2351.
Lee, P. A., Rice, T. M., and Anderson, P. W. (1974). *Solid State Commun.* **14**, 703.
Lyo, S. K. (1983). *Solid State Commun.* **47**, 247.
Mazumdar, S., and Bloch, A. N. (1983). *Phys. Rev. Lett.* **50**, 207.
McCall, R. P., Tanner, D. B., Miller, J. S., Epstein, A. J., Howard, I. A., and Conwell, E. M. (1985). *Synth. Met.* **11**, 231.
Megtert, S., Comès, R., Vettier, C., Pynn, R., and Garito, A. F. (1976). *Solid State Commun.* **19**, 925.
Megtert, S., Comès, R., Vettier, C., Pynn, R., and Garito, A. F. (1981). *Solid State Commun.* **37**, 875.
Melby, L. R., Harder, R. J., Hertler, W. R., Mahler, W., Benson, R. E., and Mochel, W. E. (1962). *J. Am. Chem. Soc.* **84**, 3374.
Murata, K., Tokumoto, M., Anzai, H., Bando, H., Saito, G., Kajimura, K., and Ishiguro, T. (1985). *J. Phys. Soc. Japan* **54**, 1236.
Overhauser, A. W. (1978). *Adv. in Phys.* **27**, 343.
Pearlstein, J. H., Ferraris, J. P., Walatka, V. V., and Cowan, D. O. (1972). *AIP Conf. Proc.* **10**, 1494.
Peierls, R. E. (1955). "Quantum Theory of Solids." Oxford University Press, London.
Pouget, J. P., Khanna, S. K., Denoyer, F., Comès, R., Garito, A. F., and Heeger, A. J. (1976). *Phys. Rev. Lett.* **37**, 437.
Ravy, S., Moret, R., Pouget, J.-P., and Comès, R. (1986). *Synth. Met.* **13**, 63.

Rice, M. J. (1976). *Phys. Rev. Lett.* **37**, 36.
Rice, M. J., Bishop, A. R., Krumhansl, J. A., and Trullinger, S. E. (1976). *Phys. Rev. Lett.* **36**, 432.
Rice, M. J., Lipari, N. O., and Strassler, S. (1977). *Phys. Rev. Lett.* **39**, 1359.
Rice, M. J., Yartsev, V. M., and Jacobsen, C. S. (1980). *Phys. Rev. B* **21**, 3437.
Schultz, T. D., and Craven, R. A. (1979). *In* "Highly Conducting One-Dimensional Solids" (DeVreese *et al.*, eds.), p. 147. Plenum, New York.
Seiden, P. E., and Cabib, D. (1976). *Phys. Rev. B* **13**, 1846.
Shirane, G., Shapiro, S. M., Comès, R., Garito, A. F., and Heeger, A. J. (1976). *Phys. Rev. B* **14**, 2325.
Takayama, H., Lin-Liu, Y. R., and Maki, K. (1980). *Phys. Rev. B* **21**, 2388.
Tanner, D. B., Jacobsen, C. S., Garito, A. F., and Heeger, A. J. (1976). *Phys. Rev. B* **13**, 3381.
Tanner, D. B., Cummings, K. D., and Jacobsen, C. S. (1981). *Phys. Rev. Lett.* **47**, 597.
Thomas, G. A., Schafer, D. E., Wudl, F., Horn, P. M., Rimai, D., Cook, J. W., Glocker, D. A., Skove, M. J., Chu, C. W., Groff, R. P., Gillson, J. L., Wheland, R. C., Melby, L. R., Salamon, M. B., Craven, R. A., DePasquali, G., Bloch, A. N., Cowan, D. O., Walatka, V. V., Pyle, R. E., Gemmer, R., Poehler, T. O., Johnson, G. R., Miles, M. G., Wilson, J. D., Ferraris, J. P., Finnegan, T. F., Warmack, R. J., Raaen, V. F., and Jérome, D. (1976). *Phys. Rev. B* **13**, 5105.
Thomas, J. F., and Jérome, D. (1980). *Solid State Commun.* **36**, 813.
Tomkiewicz, Y., Taranko, A. R., and Torrance, J. B. (1976). *Phys. Rev. Lett.* **36**, 751.
Torrance, J. B. (1977). *In* "Chemistry and Physics of One-Dimensional Metals" (H. J. Keller, ed.), p. 137. Plenum, New York.
Torrance, J. B., Tomkiewicz, Y., and Silverman, B. D. (1977). *Phys. Rev. B* **15**, 4738.
Van Smaalen, S., Kommandeur, J., and Conwell, E. (1986). *Phys. Rev. B* **33**, 5378.
White, R. M., and Geballe, T. H. (1979). "Long Range Order in Solids." Academic Press, New York.

CHAPTER 2

A Reference Guide to the Conducting Quasi-One-Dimensional Organic Molecular Crystals

I. A. Howard*

XEROX WEBSTER RESEARCH LABORATORY
WEBSTER, NEW YORK

I. INTRODUCTION 29
II. CHART OF CRYSTALS 31

I. Introduction

The following is a chart that describes the name, molecular structure, conductivity, and citations in the literature of over 400 conducting organic quasi-one-dimensional molecular crystals. The criterion we have used for the term *conducting* is that a crystal have a room-temperature dc conductivity of $\geq 10^{-4}\,\Omega^{-1}\,\text{cm}^{-1}$. Given the vast literature on such crystals and the rate at which new species are still being synthesized, this chart cannot be claimed to be complete, but is rather a compilation that aims to include the more well-known crystals (e.g., TTF-TCNQ) as well as a sizable selection of less familiar crystals. Crystals that have some two-dimensional character are included if they are of substantial interest.

The chart contains five columns. The first lists the common "shorthand" or abbreviation for the crystal name, if one exists, and the chemical name if no abbreviation is commonly used. All entries are alphabetical; numbers, Greek letters, and the prefixes *cis*- and *trans*- occurring in chemical names are discounted with regard to alphabetization. In the second column the full name of either the cation, the anion, or both is listed if it has not previously appeared in the chart.

The third column depicts the molecular structure of the component(s) listed in the second column. Thus, if only the cation is listed in column two, only its structure is shown in column three. Room-temperature dc conductivities, in $\Omega^{-1}\,\text{cm}^{-1}$, are listed in column four. The conductivity is

---

*Present address: Department of Physics, Queen Mary College, University of London, Mile End Road, London E1 4NS.

distinguished as measured from a single crystal (sc) or a powder (pw). Other notation is used as seems appropriate in some cases. Single-crystal conductivities are those along the highly conducting axis.

The final column lists references. No attempt has been made to list *every* reference pertaining to a given crystal, but in every case the reference from which the listed conductivity was taken is given. Several references, especially to volumes of conference proceedings, are used often, and the following notation has been used:

Reference	Chart Notation
Mol. Cryst. Liq. Cryst. **119** (part C) and **120** (part D), 1985. (Proceedings of the International Conference on the Physics and Chemistry of Low-Dimensional Synthetic Metals (ICSM 84), Abano Terme, Italy, June 17–22, 1984; eds. C. Pecile, G. Zerbi, R. Bozio and A. Girlando).	Abano Terme
Mol. Cryst. Liq. Cryst. **79** (part B), **81** (part C), **85** (part E), and **86** part F), 1982. (Proceedings of the International Conference on Low-Dimensional Conductors, Boulder, CO. (USA), August 9–14, 1981; Eds. A. J. Epstein and E. M. Conwell.)	Boulder
Lecture Notes in Physics **95** (Quasi-One-Dimensional Conductors I) and **96** (Quasi-One-Dimensional Conductors II) (Springer-Verlag, Berlin) 1979. (Proceedings of the International Conference on Quasi-One-Dimensional Conductors, Dubrovnik, Yugoslavia, September, 1978; eds. S. Barišić, A. Bjeliš, J. R. Cooper and B. Leontić).	Dubrovnik I and Dubrovnik II
Chem. Scripta **17**, 1981. (Proceedings of the International Conference on Low-Dimensional Synthetic Metals, Helsingor, Denmark, August 10–15, 1980).	Helsingor
Journal of the American Chemical Society.	JACS
J. de Phys. **44**, Coll. C3, 1983. (Proceedings of the International Conference on the Physics and Chemistry of Conducting Polymers and the International Colloquium of the CNRS on the Physics and Chemistry of Synthetic and Organic Metals, Les Arcs, France, December 11–18, 1982).	Les Arcs
L. R. Melby, R. J. Harder, W. R. Hertler, W. Mahler, R. E. Benson and W. E. Mochel, *JACS* **84**, 3374 (1962).	Melby
I. F. Shchegolev, *Phys. Stat. Sol.* (a) **12**, 9 (1972).	Shchegolev
Solid State Communications	SSC

In order to keep the length of the references to a minimum, only the surname of the first author of a multiple-author article is listed; thus, "Jones, Les Arcs, 100" refers to an article by Jones and coauthor(s) on page 100 of the Les Arcs Proceedings.

Acknowledgments

The author would like to thank Ms. S. Kuhn for her assistance in preparing the manuscript, and D. J. Haack for checking the formulas and nomenclature.

2. QUASI-ONE-DIMENSIONAL ORGANIC MOLECULAR CRYSTALS

Name	Description	Structure	Conductivity	References
Ad (TCNQ)$_2$	Ad: acridinium		75 (sc)	Shchegolev
	TCNQ: 7,7,8,8-tetracyano-p-quinodimethane			
Adz (TCNQ)$_2$	Adz: acridizinium		130 (sc)	Gogolin, *JETP Lett.* **22**, 278 (1975)
8-alkoxyquinolinium (TCNQ)$_m$ R m CH$_3$ 1.94 C$_2$H$_5$ 3.33, 3.27 n-C$_3$H$_7$ 2.11	8-alkoxyquinolinium		R CH$_3$ 500 (pw) C$_2$H$_5$ 390 (pw) n-C$_3$H$_7$ 490 (pw)	Lindner, *Synth. Met.* **9**, 71 (1984)
4-amino-N,N-diethyl-anilinium (TCNQ)$_2$	4-amino-N,N-diethyl-anilinium		6×10^{-3} (pw)	Melby

Name	Description	Structure	Conductivity	References
4-amino-2,3,5,6-tetra-methylanilinium (TCNQ)$_2$	4-amino-2,3,5,6-tetramethylanilinium		0.13 (pw)	Melby
2AP (TCNQ)$_2$	2AP: 2-aminopyridinium		3.4 (pw)	Chyia, Dubrovnik I, 165
3AP (TCNQ)$_2$	3AP: 3-aminopyridinium		0.2 (sc)	Chyia, Dubrovnik I, 165
2,3-BAd (TCNQ)$_2$	2,3-BAd: 2,3-benzacridinium		~60 (needles) 0.7 (plates)	Mihaly, SSC **21**, 1115 (1977)
BDTA-TCNQ	BDTA: benzo-1,3,2-dithiazol-2-yl		1.1 (pw)	Wolmershauser, Abano Terme D, 323

2. QUASI-ONE-DIMENSIONAL ORGANIC MOLECULAR CRYSTALS

Compound	Name/Structure	Conductivity	Reference
(BEDSe-TSeF) ClO$_4$, (BEDSe-TSeF) ReO$_4$	BEDSe-TSeF: bis(ethylenediseleno)-tetraselenafulvalene	~0.3 (sc); ~10^{-2} (sc)	Schumaker, IBM Research Report RJ 3783 (43480) 2/10/83; Schumaker, Les Arcs 1139
(BEDT-TTF)$_2$ AsF$_6$	BEDT-TTF: bis(ethylenedithia)-tetrathiafulvalene	$\sigma_\parallel = 0.1$–0.2, $\sigma_\perp = 2$–3 (sc)	Laversanne, Abano Terme C, 405
(BEDT-TTF)$_3X_2$ $X = $ BF$_4$, Br, ClO$_4$, FSO$_3$, IO$_4$		~100 (sc)	Parkin, Abano Terme C, 375; Parkin, *Phys. Rev. B* **34**, 1485 (1986)
(BEDT-TTF)$_2$ ClO$_4$(TCE)$_{0.5}$	TCE: 1,1,2-trichloroethane	26 (sc)	Saito, Les Arcs, 1215; Shibaeva, Abano Terme C, 361
α-(BEDT-TTF)$_2$I$_3$, β-(BEDT-TTF)$_2$I$_3$		~35 (sc, both)	Henning, Abano Terme C, 337; Shibaeva, Abano Terme C, 361
(BEDT-TTF)PF$_6$		~10^{-3} (sc)	Schumaker, IBM Research Report RJ 3783 (43480), 2/10/83; Schumaker, Les Arcs, 1139

Name	Description	Structure	Conductivity	References
(BEDT-TTF)Q(CN)$_2$	Q(CN)$_2$: dicyano-p-benzoquinone		4.4×10^{-3} (pw)	Saito, Abano Terme D, 341
(BEDT-TTF)$_x$(ReO$_4$)$_y$ (1) $x = 2$, $y = 1$ (2) $x = 3$, $y = 2$			(1) ~20 (sc) (2) ~50 (sc)	Schumaker, IBM Research Report RJ 3783 (43480), 2/10/83; Schumaker, Les Arc, 1139
(BEDT-TTF)$_2$SbF$_6$			$\sigma_\parallel = 0.2$, $\sigma_\perp = 4\text{-}6$ (sc)	Laversanne, Abano Terme C, 405
(BEDT-TTF) TCNQ			0.16 (pw)	Saito, Abano Terme D, 341
[BFDCo(III, III)] (TBDQ)$_3$	BFDCo(III, III): 1,1-bicobaltocene TBDQ: tetrabromo-diphenoquinone		5.3×10^{-3} (pw)	Lau, Boulder F, 131

Compound	Structure	Value	Reference
[BFDCo(III, III)] (TCNQ)$_3$		3×10^{-2} (pw)	Lau, Boulder F, 131
BIPA (TCNQ)$_2$	BIPA: 2,2′-bipyridylium amine	170 (sc)	Strzelecka, Helsingor, 95
2,2′BIP (TCNQ)$_2$	2,2′-BIP: 2,2′-bipyridinium	2 (pw)	Melby; Inoue, Boulder F, 139
4,4′BIP (TCNQ)$_2$	4,4′-BIP: 4,4′-bipyridinium	30 (sc)	Ashwell, Les Arcs, 1261
bis(diphenyl-2,6-telluro-pyrannylidene)-4,4′-TCNQ$_{1.5}$	bis(diphenyl-2,6-telluro-pyrannylidene)4,4′	0.18 (pw)	Amzil, Les Arcs 1249

Name	Description	Structure	Conductivity	References
bis(dithiole-1,2-ylidene)-3,3'-TCNQF$_4$	bis(dithiole-1,2-ylidene)-3,3'; TCNQF$_4$: tetrafluoro–TCNQ		10^{-3} (pw)	Amzil, Les Arcs 1249
bis(8-quinoliniumoxy) alkane (TCNQ)$_m$ n m 5 4.8 6 4.4	bis(8-quinoliniumoxy) alkane		n 5 490 (pw) 6 500 (pw)	Lindner, *Synth. Met.* **9**, 71 (1984)
bis(thieno [3,2-c] dithiole-1,2-ylidene)-3,3'-TCNQ	bis(thieno[3,2-c] dithiole-1,2-ylidene)-3,3'		10^{-3} (pw)	Amzil, Les Arcs 1249
bis(thiopyrannylidenes)-2,2'-TCNQ	bis(thiopyrannylidenes)-2,2' R = C$_6$H$_5$, p-CH$_3$–C$_6$H$_4$, p-Cl–C$_6$H$_4$		0.016 (pw) 0.035 (pw) 0.23 (pw)	Amzil, Les Arcs 1249

2. QUASI-ONE-DIMENSIONAL ORGANIC MOLECULAR CRYSTALS   37

bis(thiopyrannyli-denes)-4,4'-TCNQ	bis(thiopyrannylidenes)-4,4' R = p-Cl-C_6H_4, SCH_3		0.3 (pw) 1.3 (pw)	Amzil, Les Arcs, 1249
(bMDODBF)$_2$AsF$_6$	bMDODBF: 2,3,6,7-bis-methylene-dioxydibenzofuran		2×10^{-4} (sc)	Hellberg, Abano Terme D, 273
(BMDT–TTF)ClO$_4$	BMDT–TTF: bis(methylenedithio)-TTF		$\sim 2 \times 10^{-2}$ (sc)	Schumaker, IBM Research Report RJ 3783 (43480), 2/10/83; Schumaker, Les Arcs 1139
(BMDT–TTF)PF$_6$			~ 2.0 (sc)	
(BMDT–TTF) HCBD	HCBD: hexacyanobutadiene		0.11 (pw)	Saito, Abano Terme D, 341
(BMDT–TTF) TCNE	TCNE: tetracyanoethylene		0.30 (plates)	Saito, Abano Terme D 341

Name	Description	Structure	Conductivity	References
(BMDT–TTF) TCNQ			0.29 (needles)	Saito, Abano Terme D, 341
(BMDT–TTF) TCNQF$_4$	TCNQF$_4$: see bis(dithiole-1,2-ylidene)-3,3'-TCNQF$_4$		4.8×10^{-3} (pw)	Saito, Abano Terme D, 341
(BPDT–TTF) TCNQ	BPDT–TTF: bis(propylenedithio)-TTF		0.06 (pw)	Saito, Abano Terme D, 341
BSePPh$_4$–DDQ	BSePPh$_4$: 4,4'-bis(2,6-diphenyl-selenopyranylidene); DDQ: 2,3-dichloro-5,6-dicyano-p-benzoquinone		8×10^{-3} (pw)	Es-Seddiki, Boulder F, 71; Es-Seddiki, *Tetrahedron Lett.* **22**, 2771 (1981)

Compound	Structure	Value	Reference
BSePPh₄-3.141		3.6×10^{-2} (pw)	Es-Seddiki, Boulder F, 71; Es-Seddiki, *Tetrahedron Lett.* **22**, 2771 (1981)
BSePPh₄-TCNQ		0.5 (pw)	Es-Seddiki, Boulder F, 71; Es-Seddiki, *Tetrahedron Lett.* **22**, 2771 (1981)
(BVDT-TTF)AsF₆	BVDT-TTF: bis(vinylenedithio)-TTF	~70 (sc)	Schumaker, IBM Research Report RJ 3783 (43480), 2/10/83; Schumaker, Les Arcs, 1139
(BVDT-TTF)ClO₄		~30 (sc)	
cobalticinium(TCNQ)₂	cobalticene	0.15 (pw)	Melby
(CoPc)I$_x$ $x = 0.60, 1.0$	Pc: phthalocyanine	0.1 (pw), 0.06 (pw)	Petersen, *JACS* **99**, 286 (1977)

Name	Description	Structure	Conductivity	References
[Co(taa)]$I_{1.9}$	taa: dihydrodibenzo [b, i] [1,4,8,11]tetra-azacyclotetradecine		1.1×10^{-2} (pw)	Lin, *JCS. Chem. Comm.* 954 (1980)
$Cs_2(TCNQ)_3$			10^{-3} (sc)	Shchegolev; Murgich, Boulder E, 285
Cu(bpy)$_2$ (TCNQ)$_2$	bpy: 2,2'-bipyridine (see 2,2'-BIP (TCNQ)$_2$)		0.04 (pw)	Inoue, Boulder F, 139
Cu(5Cl-phen) (TCNQ)$_2$	(5Cl-phen): 5-chloro-1,10-phen-anthroline		4.7×10^{-2} (pw)	Inoue, Boulder F, 139
Cu(2,2'-dipyridylamine) (TCNQ)$_2$	(2,2'-dipyridylamine):		2.5×10^{-2} (pw)	Melby
Cu(4,4'DMe-bpy) (TCNQ)$_2$	(4,4'DMe-bpy): 4,4'-dimethyl-2,2'-bi-pyridinium		2.8×10^{-2} (pw)	Inoue, Boulder F, 139
Cu(6,6'DMe-bpy) (TCNQ)$_2$	(6,6'DMe-bpy): 6,6'-dimethyl2,2'-bi-pyridium		9×10^{-3} (pw)	Inoue, Boulder F, 139

Cu(2,9-DMe-phen)(TCNQ)$_2$	(2,9-DMe-phen): 2,9-dimethyl-1,10-phenanthroline	[structure: 2,9-dimethyl-1,10-phenanthroline with CH$_3$ groups]	3.8×10^{-3} (pw)	Inoue, Boulder F, 139
Cu(4,7-DMe-phen)(TCNQ)$_2$	(4,7-DMe-phen): 4,7-dimethyl-1,10-phenanthroline	[structure: 4,7-dimethyl-1,10-phenanthroline with CH$_3$ groups]	2.3×10^{-2} (pw)	Inoue, Boulder F, 139
Cu(en) (TCNQ)$_2$	en: ethylenediamine	H$_2$NCH$_2$CH$_2$NH$_2$	0.026 (pw)	Inoue, Boulder F, 139
Cu(5Me-phen)(TCNQ)$_2$	(5Me-phen): 5-methyl-1,10-phenanthroline	[structure: 5-methyl-1,10-phenanthroline with CH$_3$]	3.2×10^{-2} (pw)	Inoue, Boulder F, 139
Cu(NH$_3$)$_2$(TCNQ)$_2$	Cu(NH$_3$)$_2$	NH$_3$–Cu–NH$_3$	1.43×10^{-3} (pw)	Melby
Cu(5NO$_2$-phen)(TCNQ)$_2$	(5NO$_2$-phen): 5-NO$_2$-1,10-phenanthroline	[structure: 5-nitro-1,10-phenanthroline with NO$_2$]	6×10^{-3} (pw)	Inoue, Boulder F, 139

Name	Description	Structure	Conductivity	References
$Cu(OEP)(I)_x$ $x = 3.5$ 2.9 1.3 1.2	OEP: octaethylporphinato	(octaethylporphinato copper structure)	6×10^{-3} (pw) 4×10^{-3} (pw) 4.4×10^{-3} (pw) 3×10^{-3} (pw)	Wright, *Synth. Met.* **1**, 43 (1979/80)
$(CuPc)I_{1.71}$	Pc: see $(CoPc)I_x$		4.2 (pw)	Petersen, *JACS* **99**, 286 (1977)
$Cu(phen)(TCNQ)_2$ $Cu(phen)_2(TCNQ)_2$	phen: 1,10-phenanthroline	(1,10-phenanthroline structure)	1.9 (pw) 0.23 (pw)	Inoue, Boulder F, 139
$[Cu(taa)]I_{1.8}$	taa: see $[Co(taa)]I_{1.9}$		1.1×10^{-3} (pw)	Lin, *JCS Chem. Comm.* 954 (1980)
$Cu(TCNQ)$ $Cu(TCNQ)_2$	copper	Cu^+, Cu^{2+}	5×10^{-3} (pw, both)	Melby

4-cyano-N-methyl pyridinium (TCNQ)$_2$	4-cyano-N-methyl pyridinium	(structure of 4-cyano-N-methyl pyridinium cation)	0.02 (pw)	Melby
trans-cyc Co^{3+}(TCNQ$^-$)$_2$Cl^{2-}	cyc: cyclam (tetraaza-cyclotetradecane)	(structure of Co(III) cyclam complex)	5.6×10^{-4} (pw)	Matsuoka, Boulder F, 155
Cyc-2H$^+$(TCNQ$^-$)$_2$ (TCNQ)$_2$			3.2×10^{-2} (pw)	Matsuoka, Boulder F, 155
(Cyc)$_2$La^{3+}(TCNQ$^-$)$_3$ TCNQ			4.2×10^{-4} (pw)	Matsuoka, Boulder F, 155
(Cyc)$_2$Mn TCNQ$_6$			2×10^{-3} (pw)	Matsuoka, Boulder F, 155
Cyc Pd TCNQ$_4$			1.3×10^{-2} (pw)	Matsuoka, Boulder F, 155

Name	Description	Structure	Conductivity	References
(DBTSF)X X = (TaF$_6$)$_{0.6}$ (AsF$_6$)$_{0.65}$ (SbF$_6$)$_{0.7}$	DBTSF: dibenztetra- selenafulvalene		~60 (sc) ~5 (sc) ~0.6 (sc)	Johannsen, Les Arcs, 1361
(DBTTF)$_2$ClO$_4$	DBTTF: dibenztetrathiafulvalene		2 (sc)	Johannsen, Les Arcs, 1361
DBTTF-I$_3$			5×10^{-2} (sc)	Shibaeva, Dubrovnik II, 167
(DBTTF)$_2$ PF$_6$(THF)$_{1/3}$	THF: tetrahydrofuran		~10 (sc)	Johannsen, Les Arcs, 1361
(DBTTF) [PtBr$_5$S(CH$_3$)$_2$]			10 (sc)	Shibaeva, Dubrovnik II, 167
(DBTTF)$_8$(PtCl$_6$)$_3$ (DBTTF)$_8$(SnCl$_6$)$_3$			500 (sc, both) 0.5 (pw, both)	Shibaeva, Dubrovnik II, 167
DBTTF-TCNQ Cl$_2$	TCNQ Cl$_2$: dichloro-TCNQ		40 (sc)	Mortenson, Les Arcs, 1349; Jacobsen, *J. Phys. C.* **13**, 3411 (1980)

DClDCNQI-TTF	DClDCNQI: 2,5-dichloro-N,N'-dicyano-1,4-benzo-quinonediimine		5×10^{-3} (pw)	Andreetti, Abano Terme D, 309
DCNQI-TTF	DCNQI: N,N'-dicyano-1,4-benzoquinonediimine		4.2×10^{-4} (pw)	Andreetti, Abano Terme D, 309
cis/trans-DEDMTSeF-TCNQ	cis/trans-DEDMTSeF: cis/trans-diethyldimethyl-tetraselenafulvalene (cis and trans forms presumably mixed in crystal)			Jacobsen, Phys. Rev. B **18**, 905 (1978)

Name	Description	Structure	Conductivity	References
DEM (TCNQ)$_2$	DEM: diethylmorpholinium	[structure: morpholinium ring with N$^+$ bearing two CH$_2$CH$_3$ groups]	8×10^{-2} (sc)	Visser, Boulder E, 265; Kuindersma, *J. Phys. C* **8**, 3005 (1975)
DEPA (TCNQ)$_{4.5}$	DEPA: 1,2-bis(1-ethyl-4-pyridinium) ethane	CH$_3$–CH$_2$–N$^+$(pyridine)–CH$_2$–CH$_2$–(pyridine)N$^+$–CH$_2$–CH$_3$	100–500 (sc)	Ashwell, Boulder F, 147
DEPE (TCNQ)$_{4.53}$	DEPE: 1,2-bis(1-ethyl-4-pyridinium) ethylene	CH$_3$–CH$_2$–N$^+$(pyridine)–CH=CH–(pyridine)N$^+$–CH$_2$–CH$_3$	10–500 (sc)	Ashwell, Boulder F, 147
DEPP (TCNQ)$_5$	DEPP: 1,3-bis(1-ethyl-4-pyridinium) propane	CH$_3$–CH$_2$–N$^+$(pyridine)–CH$_2$–CH$_2$–CH$_2$–(pyridine)N$^+$–CH$_2$–CH$_3$	1–10 (sc)	Ashwell, Boulder F, 147

Compound	Structure	Conductivity	Reference
DHPA (TCNQ)$_3$	DHPA: 1,2-bis(4-pyridinium) ethane	0.3 (sc)	Ashwell, Boulder F, 147
DHPE (TCNQ)$_3$	DHPE: 1,2-bis(4-pyridinium) ethylene	0.5 (pw)	Ashwell, Boulder F, 147
DHPP (TCNQ)$_4$	DHPP: 1,3-bis(4-pyridinium) propane	1–10 (sc)	Ashwell, Boulder F, 147
diethylmethylamine (TCNQ)$_2$	diethylmethylamine	1.67×10^{-3} (pw)	Melby
5,8-dihydroxy-quinolinium(TCNQ)$_2$	5,8-dihydroxy-quinolinium	6.7×10^{-2} (pw)	Melby
dimethylferricinium-(TCNQ)$_2$	dimethylferrocene	0.03 (pw)	Melby

Name	Description	Structure	Conductivity	References
(2,6-diphenyl 4-phenyliminopyrone) (TCNQ)$_2$	2,6-diphenyl 4-phenyliminopyrone		0.04 (pw)	Strzelecka, Dubrovnik II, 340
(DIPSΦ$_4$)I$_{2.28}$	DIPSΦ$_4$: tetraphenyldithia-pyranylidene		250 (sc)	Faulques, Boulder F, 63; Daoben, Boulder F, 57
(DIPSΦ$_4$)I$_{\sim 3.3-3.45}$ (DIPSΦ$_4$)I$_{2.41}$ (DIPSΦ$_4$) TCNQ (DIPSΦ$_4$) TCNQ$_2$			2 (sc) 10 (sc) 0.9 (pw) 0.2 (pw)	
(DMeFc)$_2$(TCNQ)$_2$	DMeFc: decamethylferrocene		~0.1 (pw)	Miller, Dubrovnik II, 313
DMM (TCNQ)$_2$ (monoclinic)	DMM: N,N-dimethyl-morpholinium		~2×10^{-3} (sc)	Visser, Boulder E, 265

2. QUASI-ONE-DIMENSIONAL ORGANIC MOLECULAR CRYSTALS

DMN_3X_2 $X = PF_6, AsF_6$	DMN: dimethylnaphthalene	~10^3 (sc)	Enkelmann, Les Arcs, 1147
$DMPA (TCNQ)_4$	DMPA: 1,2-bis(1-methyl-4-pyridinium) ethane	1–50 (sc)	Ashwell, Boulder F, 147
$DMPE (TCNQ)_4$	DMPE: 1,2-bis(1-methyl-4-pyridinium) ethylene	1 (pw)	Ashwell, Boulder F, 147
$DMPP (TCNQ)_{4.44}$	DMPP: 1,3-bis(1-methyl-4-pyridinium) propane	10–100 (sc)	Ashwell, Boulder F, 147
$DMtiM (TCNQ)_2$ $(DMTM (TCNQ)_2)$	DMtiM (DMTM): dimethylthio-morpholinium	~4×10^{-3} (sc)	Visser, Boulder E, 265; Visser, Abano Terme D, 167

Name	Description	Structure	Conductivity	References
(DMTTF)₂-X X = Br, Cl, I ClO₄, BF₄, IO₄ PF₆ SCN	DMTTF: dimethyl-4,5-TTF	(structure)	150, 50–100, 100 50–100, 100–150, 50–100 250–300 2–10 (all, sc)	Abderraba, Les Arcs 1243
(DMtTSeF)₂X X = ClO₄ AsF₆ PF₆ ReO₄ SbF₆	DMtTSeF: dimethyltrimethylene- tetraselenafulvalene	(structure)	 1000–1300 (sc) 1200–1500 (sc) 500–600 (sc) 300–700 (sc) 300–500 (sc)	Delhaes, Abano Terme C, 269
(DMTTTF)₂ ClO₄ BF₄ PF₆	DMTTTF: dimethyltri- methylene–TTF	(structure)	60–100 (all, sc) 80 (ClO₄, sc)	Fabre, Les Arcs, 1153; Chasseau, Les Arcs, 1223; Delhaes, Les Arcs 1239
DPB (TCNQ)₄	DPB: dipyridylbutane	(structure)	~0.1 (sc)	Drew, Boulder F, 123

DPPA (TCNQ)₅	DPPA: 1,2-bis(1-n-propyl-4-pyridinium) ethane	0.2 (pw)	Ashwell, Boulder F, 147
DPPE (TCNQ)₅	DPPE: 1,2-bis(1-n-propyl-4-pyridinium) ethylene	0.05 (pw)	Ashwell, Boulder F, 147
DPPP (TCNQ)₅	DPPP: 1,3-bis(1-n-propyl-4-pyridinium) propane	0.05 (sc)	Ashwell, Boulder F, 147
cis/trans-DSeDTF-TCNQ	*cis/trans*-DSeDTF: *cis/trans*-diselena-dithiafulvalene (*cis* and *trans* forms presumably mixed in crystal)	550 (sc)	Etemad, *Phys. Rev. B* **13**, 2254 (1976); Chaikin, *Phys. Rev. B* **13**, 1627 (1976); Etemad, *Phys. Rev. Lett.* **34**, 741 (1975)

Name	Description	Structure	Conductivity	References
(DTB)I$_{1.17}$	DTB: 4,4′,5,5′-tetramethoxy-2,2′-dithiobiphenyl	[structure: tetramethoxy dithiobiphenyl with H$_3$CO, H$_3$CO, OCH$_3$, OCH$_3$ groups and S-S linkage]	5 (sc)	Stender, Abano Terme, D, 277
DTP–TCNQ	DTP: 1,6-dithiapyrene	[structure: 1,6-dithiapyrene]	130 (sc)	Thorup, Abano Terme D, 349
DTTTF (PF$_6$)	DTTTF: dithiopheno–TTF	[structure: dithiopheno-TTF]	10 (sc)	Cowan, Boulder F, 1
cis/trans-α-DTTTF (TCNQ)	α-DTTTF: dit(4,5-dihydrothieno)-[2,3-b;2′,3′-h]-1,4,5,8-tetrathiafulvalene	[structures: cis and trans α-DTTTF]	30 (pw)	Engler, *JACS* **100**, 3769 (1978)

2. QUASI-ONE-DIMENSIONAL ORGANIC MOLECULAR CRYSTALS

Compound	Abbreviation: Name	Structure	Conductivity	Reference
$DZ_3\,(AsF_6)_2$	DZ: diacenaphtho-[1,2-j;1',2'-1]fluoranthene		$\sim 10^3$ (sc)	Enkelmann, Les Arcs, 1147
EiPM (TCNQ)$_2$	EiPM: N-ethyl N-isopropyl-morpholinium		$\sim 2 \times 10^{-3}$ (sc)	Visser, Boulder E, 265
[en-OP]: [H$_3$NCH$_2$CH$_2$NH$_3$]$_{0.82}$ [Pt(C$_2$O$_4$)$_2$]·2H$_2$O	en-OP: ethylenediamine-bis(oxalato)platinate		~ 4 (sc)	Underhill. Les Arcs, 1331
4,5-(ethylenedithio)-3-methylthio-1,2-dithiolium iodide-(TCNQ)$_2$	4,5-(ethylenedithio)-3-methylthio-1,2-dithiolium iodide		0.1–1.0 (pw)	Papavassiliou, Abano Terme D, 315
ethyltriphenylarsonium (TCNQ)$_2$	ethyltriphenylarsonium		0.5 (sc)	Melby

Name	Description	Structure	Conductivity	References
ethyltriphenyl-phosphonium-(TCNQ)$_2$	ethyltriphenyl-phosphonium	(structure: Ph$_3$P$^+$–CH$_2$–CH$_3$)	>0.1 (sc)	Melby
Et$_2$MeS$^+$(TCNQ)$_2^-$	Et$_2$MeS$^+$: diethylmethylsulfonium	(structure)	4×10^{-3} (sc)	Lequan, Abano Terme D, 353
EtMe$_2$S$^+$(TCNQ-I)$^-$	EtMe$_2$S$^+$: ethyldimethylsulfonium	(structure)	90 (sc)	Lequan, Abano Terme D, 353
Et$_3$S$^+$(TCNQ)$_2^-$	Et$_3$S$^+$: triethylsulfonium	(structure)	1.5×10^{-4} (sc)	Lequan, Abano Terme D, 353
FA$_2$X X = ClO$_4$, PF$_6$, AsF$_6$, SbF$_6$, SbCl$_6$	FA: fluoranthene	(fluoranthene structure)	$\sim 10^3$ (sc)	Enkelmann, Les Arcs, 1147

2. QUASI-ONE-DIMENSIONAL ORGANIC MOLECULAR CRYSTALS

Compound	Description/Structure	Value	Reference
Fe (dmit)$_2$ TTF	dmit: dimercaptoisotrithione	10^{-2} (pw)	Kubel, Les Arcs, 1265
Fe (dmit)$_2$ (CH$_3$)$_2$ CO·PYZ (PYZ = pyrazine)		3×10^{-3} (pw)	
Fe OCl (TTF)$_{0.10}$	TTN: tetrathianaphthalene TTT: see (TTT) (AsF$_6$)$_{0.52}$	3.5×10^{-3} (pw)	Averill, Les Arcs 1373
Fe OCl (TTT)$_{1/7}$		2.2×10^{-4} (pw)	
Fe OCl (TTN)$_{1/7}$		5.5×10^{-4} (pw)	
Fe Pc I$_x$ x = 1.93, 2.74	Pc: see (CoPc)I$_x$	4×10^{-3} (pw) 2×10^{-3} (pw)	Petersen, *JACS* **99**, 286 (1977)
ferricinium (TCNQ)$_2$	ferrocene	4.2 (sc)	Melby; Miller, Dubrovnik II, 313
HEM-TCNQ (rhombic)	HEM: N-hydrogen N-ethyl-morpholinium	$\sim 2 \times 10^{-3}$ (sc)	Visser, Boulder E, 265
HEM (TCNQ)$_2$		~ 0.06 (sc)	Visser, Boulder E, 265

Name	Description	Structure	Conductivity	References
HiPM (TCNQ)$_2$	HiPM: N-hydrogen N-isopropyl-morpholinium	[structure: N-isopropyl morpholinium cation]	$\sim 8 \times 10^{-3}$ (sc)	Visser, Boulder E, 265
HMM (TCNQ)$_2$	N-hydrogen N-methylmorpholinium	[structure: N-methyl morpholinium cation]	~ 0.2 (sc)	Oostra, Les Arcs, 1381; Visser, Boulder E, 265
(HMM)$_3$(TCNQ)$_4$			~ 0.1 (sc)	Visser, Boulder E, 265
HMTSF–TNAP	HMTSF: hexamethylene-tetraselenafulvalene TNAP: 11,11′,12,12′-tetracyano-2,6-naphthoquinodimethane	[structures: HMTSF and TNAP]	1800–3000 (sc)	Bechgaard, SSC **25**, 875 (1978)
HMTSF–TCNQ			50 (pw) 2000 (sc)	Engler, JACS **100**, 3769 (1978)

2. QUASI-ONE-DIMENSIONAL ORGANIC MOLECULAR CRYSTALS

HMTTeF		HMTTeF: hexamethylene-tetratellurafulvalene	2×10^{-3} (pw) Saito, Abano Terme D, 337
HMTTeF DBDQ			
DCNQ			8.3×10^{-3} (pw)
DDQ		DBDQ: 2,3-dibromo-5,6-dicyano-p-benzoquinone	1.2×10^{-2} (pw)
$(DMTCNQ)_{0.5}$			4.5 (pw)
FTCNQ		DCNQ: dicyanoquino-dimethane	0.37 (pw)
		DDQ: see BSePPhy-DDQ	
TCNQ		DMTCNQ: 2,5-dimethyl-TCNQ	7.1 (pw), 1400 (sc)
TCNQF$_4$		FTCNQ: fluoro–TCNQ	0.3 (pw)
		TCNQF$_4$: see bis(dithiole-7,2-ylidene)-3,3′-TCNQF$_4$	

Name	Description	Structure	Conductivity	References
HMTTF-TCNQ	HMTTF: hexamethyl-enetetrathiafulvalene		5 (pw) 500 (sc)	Engler, *JACS* **100**, 3769 (1978)
HnBM (TCNQ)$_2$	HnBM: N-hydrogen N-n-butylmorpholinium		~0.5 (sc)	Visser, Boulder E, 265
4-hydroxy-N-benzyl-anilinium-(TCNQ)$_2$	4-hydroxy-N-benzyl-anilinium		0.02 (pw)	Melby
4-hydroxy-2,3,5,6-tetramethylanilinium-TCNQ	4-hydroxy-2,3,5,6-tetramethylaniline		6.25×10^{-3} (pw)	Melby
IPDMA (TCNQ)$_2$	IPDMA: hydrogen isopropyl-dimethylammonium		10^{-3} (pw) 10^{-2} (sc)	Bandrauk, *J. Phys. Chem.* **89**, 434 and 1478 (1985)
K-TCNQ			2×10^{-4} (pw)	Melby

Li$_{0.8}$Pt(mnt)$_2$·2H$_2$O	Pt(mnt): bis(maleonitrile-dithiolato) platinate	[structure]	30–200 (sc)	Cooper, Les Arcs, 1391
MBDTA–TCNQ	MBDTA: methylbenzo-1,3,2-dithiazol-2-yl	[structure]	3 (pw)	Wolmershauser, Abano Terme D, 323
MBTM (TCNQ)$_2$	MBTM: methyl-butyl-thiomorpholinium	[structure]	~1 (sc)	Oostra, Les Arcs, 1387
Me CNQn–TCNQ$_2$	Me CNQn: 1-methyl-4-cyano-quinolinium	[structure]	0.1–1.0 (sc)	Ashwell, Abano Terme D, 137; Shchegolev
α-Me-HMTTF (TCNQ)	α-Me-HMTTF: di(3-methylcyclopenteno)-[1,2-b; 1′,2′-h]-1,4,5,8-tetrathiafulvalene	[structure]	0.1 (sc)	Engler, *JACS* **100**, 3769 (1978)

Name	Description	Structure	Conductivity	References
β-Me-HMTTF (TCNQ)	β-Me-HMTTF: di(4-methylcyclopenteno)-[1,2-b; 1',2'-h]-1,4,5,8-tetrathiafulvalene		80 (sc)	Engler, JACS **100**, 3769 (1978)
MEM (TCNQ)$_2$	MEM: N-methyl-N-ethylmorpholinium		~10^{-2} (sc)	Kuindersma, J. Phys. C **8**, 3005 (1975); Visser, Boulder, E, 265
Me$_3$S$^+$(TCNQ-I)$^-$	Me$_3$S$^+$: trimethylsulfonium		40 (sc)	Lequan, Abano Terme D, 353
4,5-(methylenedithio)-3-methylthio-1,2-dithiolium-iodide (TCNQ)$_2$	4,5-(methylenedithio)-3-methylthio-1,2-dithiolium iodide		0.1-1 (pw)	Papavassiliou, Abano Terme D, 315
MEtiM (TCNQ)$_2$	MEtiM: methylethylthio-morpholinium		~2 (sc)	Visser, Boulder E, 265

M-E-t-TTF–TCNQ	M-E-t-TTF: methyl-ethyl-trimethylene-tetrathiafulvalene	80 (sc)	Keryer, Dubrovnik I, 65
MnBM (TCNQ)$_2$	MnBM: N-methyl N-n-butyl morpholinium	~1 (sc)	Visser, Boulder E, 265
MNEB (TCNQ)$_2$ [(Me-1 N-Et-Bz) (TCNQ)$_2$]	MNEB: methyl-1 N-ethyl-benzimidazolium	0.6 (pw) 6.7 (sc)	Chasseau, C.R. Acad. Sci. (Paris) C **276**, 661 (1973)
MnPM (TCNQ)$_2$	MnPM: N-methyl N-n-propyl-morpholinium	~0.2 (sc)	Visser, Boulder E, 265
morpholinium$_2$–(TCNQ)$_3$	morpholinium	2×10^{-4} (sc)	Melby

Name	Description	Structure	Conductivity	References
(MPht) (TCNQ)$_2$	MPht: N-methyl-phthalazinium		120 (sc)	Graja, Les Arcs, 1365
MTPA (TCNQ)$_2$ [Φ$_3$MeAs (TCNQ)$_2$]	MTPA: triphenylmethyl-arsonium		2×10^{-2} (sc)	Melby; Shchegolev; Devreux, SSC **16**, 275 (1975); Krause, Les Arcs 1429
MTPP (TCNQ)$_2$ [Φ$_3$MeP(TCNQ)$_2$]	MTPP: triphenylmethyl-phosphonium		2×10^{-2} (sc)	Melby; Shchegolev; Swietlik, Les Arcs 1457
Napth$_2$X X = PF$_6$, AsF$_6$, SbF$_6$	Napth: naphthalene		10^3 (sc)	Enkelmann, Les Arcs 1147
N-butyl-N,N-dimethyl-ammonium-(TCNQ)$_2$	N-butyl-N,N-dimethyl-ammonium		0.11 (pw)	Melby

NDTe-TCNQ	NDTe: naphtho[1,8-c,d]-1,2-ditellurole	0.02 (pw)	Meinwald, *JACS* **99**, 255 (1977)
N-ethyl-quinolinium (TCNQ)$_2$	N-ethylquinolinium	0.13 (pw)	Melby
Ni(dpg)$_2$I	Ni(dpg)$_2$: nickel bisdiphenyl glyoximate	2.3–11×10^{-3} (sc)	Cowie, *JACS* **101**, 2921 (1979)
[Ni(oaoH)(oaoH$_2$)] TCNQ	Ni(oaoH)(oaoH$_2$): bis(oxamide oximato) nickel complex	15 (sc)	Endres, Abano Terme D, 365

Name	Description	Structure	Conductivity	References
Ni(OEP)(I)$_x$	OEP: see Cu(OEP)I$_x$			Wright, *Synth. Met.* **1**, 43 (1979/80)
$x = 5.7$			2.8×10^{-2} (pw)	
3.8			1.8×10^{-2} (pw)	
1.7			2.8×10^{-2} (pw)	
1.4			1.4×10^{-2} (pw)	
1.2			6×10^{-3} (pw)	
0.4			1.5×10^{-4} (pw)	
Ni(omtbp)I$_{1.08}$	Ni(omtbp): octamethyltetrabenz-porphyrinato-nickel	(structure of octamethyltetrabenzporphyrinato-nickel)	16 (sc)	Hoffman, *Acc. Chem. Res.* **16**, 15 (1983)
NiPcI$_{1.0}$	Pc: see (CoPc)I$_x$		250–650 (sc)	Schramm, *Science* **200**, 47 (1978); Schramm, *JACS* **102**, 6702 (1980)
(NiPc)I$_x$	Pc: see (CoPc)I$_x$			Petersen, *JACS* **99**, 286 (1977)
$x = 0.56$			0.7 (pw)	
1.0			0.7 (pw)	
1.74			0.8 (pw)	

Compound	Values	Reference
(NiPc)I$_x$ $x = 0.58$ 1.07, 1.44 1.68, 3.81	2.2 7.7, 7.7 0.8, 0.6 (pw)	Schramm, JACS **102**, 6702 (1980)
[Ni(taa)]I$_x$ $x = 0.8$ 1.0 1.8 2.6 7.0	0.21 (pw) 0.45 (pw) 1–50 (sc) 8.1×10^{-2} (pw) 0.23 (pw)	Lin, JCS. Chem. Comm. 954 (1980)
taa: see [Co(taa)]I$_{1.9}$		
Ni(tatbp)I Ni(tatbp): triaza-tetrabenz-porphyrinato-nickel	110–200 (sc)	Hoffmann, Acc. Chem. Res. **16**, 15 (1983)
Ni(tbp)I Ni(tbp): tetrabenzporphyrinato-nickel	180–330 (sc)	Hoffman, Acc. Chem. Res. **16**, 15 (1983)

Name	Description	Structure	Conductivity	References
Ni(tmp)I	Ni(tmp): nickel mesotetra-methylporphyrin		40–280 (sc)	Hoffman, *Acc. Chem. Res.* **16**, 15 (1983)
[Ni(tmtaa)]I$_x$ x = 1.7 2.44 2.9	Ni(tmtaa): nickel tetramethyl-taa		1.4×10^{-2} (pw) 1–20 (sc) 3.8×10^{-2} (pw)	Lin, *JCS. Chem Comm.* 954 (1980)
NMeAd (TCNQ)$_2$	NMeAd: N-methylacridinium		~100 (sc)	Holczer, *SSC* **26**, 689 (1978)
NMe-4-MePy (TCNQ)$_2$	NMe-4-MePy: N-methyl 4-methyl pyridinium		~10^{-3}–10^{-4} (sc)	Graja, Boulder E, 257

2. QUASI-ONE-DIMENSIONAL ORGANIC MOLECULAR CRYSTALS

Compound	Structure	Conductivity	Reference
NMe-2,4-MePy (TCNQ)$_2$	NMe-2,4-MePy: N-methyl 2,4-methyl-pyridinium	$\sim 10^{-3}$–10^{-4} (sc)	Graja, Boulder E, 257
NMe-2,6-MePy (TCNQ)$_2$	NMe-2,6-MePy: N-methyl 2,6-methyl-pyridinium	$\sim 10^{-3}$–10^{-4} (sc)	Graja, Boulder E, 257
NMe-3,5-MePy (TCNQ)$_2$	NMe-3,5-MePy: N-methyl 3,5-methyl-pyridinium	$\sim 10^{-3}$–10^{-4} (sc)	Graja, Boulder E, 257
NMe-2,4,6-MePy (TCNQ)$_2$	NMe-2,4,6-MePy: N-methyl 2,4,6-methyl-pyridinium	$\sim 10^{-3}$–10^{-4} (sc)	Graja, Boulder E, 257
(NMeMeTz) TCNQ$_2$	NMeMeTz: N-methyl-2-methyl-tiazolinium	$\gtrsim 10^{-3}$ (sc)	Schegolev

Name	Description	Structure	Conductivity	References
NMePy (TCNQ)$_2$	NMePy: N-methylpyridinium		~10^{-3}–10^{-4} (sc)	Graja, Boulder E, 257
NMeQn (TCNQ)$_2$	NMeQn: N-methylquinolinium		~50 (sc)	Shchegolev; Mihaly, SSC **17**, 1007 (1975); Mihaly, SSC **19**, 1091 (1976)
			0.33 (pw)	Melby
N-methyl-2-(4-dimethyl-aminophenylazo)-pyridinium-(TCNQ)$_2$	N-methyl-2-(4-dimethyl-aminophenylazo)-pyridinium		0.08 (pw)	Melby
NMP-TCNQ	NMP: N-methylphenazinium		170 (sc)	Shchegolev; Melby, Can. J. Chem. **43**, 1448 (1965)
N-(2-phenethyl)-quinolinium-(TCNQ)$_2$	N-(2-phenethyl)-quinolinium		0.33 (pw)	Melby

Nph₂(TCNQ)₃	Nph: 1,8-naphthyridine	1.4 (pw)	Strzelecka, Dubrovnik II, 340
NPQn(TCNQ)₂	NPQn: N-propylquinolinium	~1 (sc) 0.5 (pw)	Janossy, Boulder E, 233; Melby
(PcH₂)I₂.₂	Pc: see (CoPc)I$_x$	2.3 (pw)	Petersen, *JACS* **99**, 286 (1977)
[Pd(taa)]I₀.₈	taa: see [Co(taa)]I₁.₉	0.13 (pw)	Lin, *JCS Chem. Comm.* 954 (1980)
Pe₂X₁.₁(CH₂Cl)₀.₈	Pe: perylene (THF = tetrahydrofuran)	~10³ (sc, all)	Enkelmann, Les Arcs, 1147
Pe₂X₁.₁(THF)₀.₇ X = AsF₆, PF₆			
Pe₂(AsF₆)₁.₄₃			
2Pe(3I₂) [PeI₂.₉₂]		0.1 (pw) 5–20 (sc)	Teitelbaum, *JACS* **101**, 7568 (1979)

Name	Description	Structure	Conductivity	References
ΦPh (TCNQ)$_2$	ΦPh: Φ-phenanthrolinium		75 (sc)	Strzelecka, Helsingor, 95
p-phenylenediamine-TCNQ	p-phenylenediamine		3.3×10^{-4} (pw)	Melby
(PPN)$_2$(TCNQ)$_3$·(CH$_3$CN)$_2$	PPN: bis-(triphenylphosphor-anylidene) cation		4.6×10^{-4} (max) (sc)	Ahmad, Abano Terme D, 361
PPy-TCNQ [Py-(Py$^+$)$_n$-X·n TCNQ] $X = $ Cl, Br $n \cong 30$–100	Py-(Py$^+$)$_n$-X: polypyridinium halides		$X = $ Cl, $n = 29 \sim 30$: 0.054 (pw) $X = $ Cl, $n = 45$-46: 0.12 (pw) $X = $ Br, $n = 97$-98: 0.41 (pw)	Anzai, Les Arcs 1229

2. QUASI-ONE-DIMENSIONAL ORGANIC MOLECULAR CRYSTALS

Compound	Name / Structure	Value	Reference
4,5-(propylenedithio)-3-methylthio-1,2-dithiolium iodide (TCNQ)$_2$	4,5-(propylenedithio)-3-methylthio-1,2-dithiolium iodide	0.1–1 (pw)	Papavassiliou, Abano Terme D, 315
[Pt(oaoH)(oaoH$_2$)] TCNQ	Pt(oaoH)(oaoH$_2$): bis-(oxamide oximato) platinum complex	5 (sc)	Endres, Abano Terme D, 365
(PtPc)I$_{0.93}$	Pc: see (CoPc)I$_x$	2.4 (pw)	Petersen, JACS **99**, 286 (1977)
Py$_2$(ClO$_4$)(ClO$_4$)-(THF)$_{0.5}$ Py$_7$Py$_4$(PF$_6$)$_4$(CH$_2$Cl$_2$)$_4$ Py$_{12}$(SbF$_6$)$_7$ Py$_2$AsF$_6$	Py: pyrene	~10^3 (sc)	Enkelmann, Les Arcs, 1147
pyridinium (TCNQ)$_2$	pyridinium	0.03 (pw)	Melby
Qn (TCNQ)$_2$	Qn: quinolinium	100 (sc) 2 (pw)	Melby; Shchegolev

Name	Description	Structure	Conductivity	References
QP$_2$SbF$_6$(CH$_2$Cl$_2$)	QP: quaterphenyl	(quaterphenyl structure)	~10^3 (sc)	Enkelmann, Les Arcs, 1147
(R$_1$R$_2$R$_3$-1,2-dithiolylium)(TCNQ)	R$_1$R$_2$R$_3$-1,2-dithiole	(1,2-dithiole structure with R$_1$, R$_2$, R$_3$)		Amzil, Helsingor, 65
R$_1$ = C$_6$H$_5$, R$_2$ = H, R$_3$ = H			6×10^{-2} (pw)	
R$_1$ = pCH$_3$OC$_6$H$_4$, R$_2$ = H, R$_3$ = H			0.1 (pw)	
R$_1$ = H, R$_2$ = C$_6$H$_5$, R$_3$ = H			0.1 (pw)	
R$_1$ = C$_6$H$_5$, R$_2$ = H, R$_3$ = C$_6$H$_5$			0.1 (pw)	
(R$_1$R$_2$R$_3$-1,2-dithiolylium)(TCNQ)$_2$				Amzil, Helsingor, 65; Amzil, Abano Terme D, 357
R$_1$ = C$_6$H$_5$, R$_2$ = H, R$_3$ = H			3.3 (pw)	
R$_1$ = pCH$_3$OC$_6$H$_4$, R$_2$ = H, R$_3$ = H			0.7 (pw)	
R$_1$ = (CH$_3$)$_3$C, R$_2$ = H, R$_3$ = H			2×10^{-3} (pw)	
R$_1$ = H, R$_2$ = C$_6$H$_5$, R$_3$ = H			0.6 (pw)	
R$_1$ = C$_6$H$_5$, R$_2$ = H, R$_3$ = C$_6$H$_5$			0.5 (pw)	
R$_1$ = pCH$_3$OC$_6$H$_4$, R$_2$ = H, R$_3$ = pCH$_3$OC$_6$H$_4$			5×10^{-2} (pw)	
R$_1$ = (CH$_3$)$_3$C, R$_2$ = H, R$_3$ = (CH$_3$)$_3$C			7×10^{-2} (pw)	
R$_1$ = CH$_3$, R$_2$ = H, R$_3$ = CH$_3$			8×10^{-3} (pw)	
R$_1$ = C$_6$H$_5$, R$_2$ = H, R$_3$ = SCH$_3$			0.5 (pw)	
R$_1$ = pCH$_3$OC$_6$H$_4$, R$_2$ = H, R$_3$ = SCH$_3$			0.4 (pw)	
R$_1$ = (CH$_3$)$_3$C, R$_2$ = H, R$_3$ = SCH$_3$			0.12 (pw), 4.0 (sc)	
R$_1$ = SCH$_3$, R$_2$ = C$_6$H$_5$, R$_3$ = SCH$_3$			4×10^{-3} (pw)	
R$_1$ = H, R$_2$ = CH$_3$, R$_3$ = SCH$_3$			7×10^{-3} (pw)	
R$_1$ = SCH$_3$, R$_2$ = SCH$_3$, R$_3$ = SCH$_3$			0.4 (pw)	

Compound	Structure	R	R'	R''	Conductivity	Reference
[R-S-C(NR'R'')₂] (TCNQ)₂	R-S-C(NR'R'')₂: S-alkylchalcogenourium cations $R-S-C{\oplus}(N R'R'')_2$	CH₃	CH₃	H	4×10^{-2} (pw)	Amzil, Abano Terme D, 357
		CH₃	CH₃	CH₃	3.2×10^{-4} (pw)	
		CH₃	CH₂⁻	H	8×10^{-3} (pw)	
		C₂H₅	CH₃	H	0.13 (pw)	
		C₂H₅	CH₃	CH₃	0.35 (pw), 70 (sc)	
		n-C₃H₇	CH₃	CH₃	5×10^{-3} (pw)	
		n-C₄H₉	CH₃	CH₃	4×10^{-2} (pw)	
[R-Se-C(NCH₃CH₃)₂] (TCNQ)₂	R-Se-C(NCH₃CH₃)₂: Se-alkylchalcogenourium cations $R-Se-C{\oplus}(N(CH_3)_2)_2$	CH₃			3.7×10^{-4} (pw)	Amzil, Abano Terme D, 357
		C₂H₅			0.22 (pw), 20 (sc)	
TEA-TCNQ₂ [(Et₃NH) TCNQ₂]	TEA: triethylammonium $CH_3-CH_2-\overset{H\ \oplus}{\underset{\underset{CH_3}{CH_2}}{N}}-CH_2-CH_3$				4 (sc)	Melby; Shchegolev
(TEDA)₂ (TCNQ)₃	TEDA: triethylenediamine				8×10^{-3} (pw)	Melby; Truong, Abano Terme D, 105

Name	Description	Structure	Conductivity	References
(TEDA)$_2$(TCNQ)			10^{-2} (pw)	Bandrauk, *J. Phys. Chem.* **89**, 434 and 1478 (1985)
TEDA CH$_2$Cl (TCNQ)$_2$			10^{-2} (sc)	Bandrauk, *J. Phys. Chem.* **89**, 434 and 1478 (1985)
(1,2,7,8-tetrahydrocyclopentaperylene)$_2$ PF$_6$CH$_2$Cl$_2$	1,2,7,8-tetrahydrocyclopentaperylene		30–70 (sc)	Lapouyade, Les Arcs 1235
tetraphenylphosphonium (TCNQ)$_2$	tetraphenylphosphonium		10^{-3} (sc)	Melby
tetraphenylstibonium (TCNQ)$_2$	tetraphenylstibonium		0.08 (sc)	Melby

TMA-1-TCNQ [(NMe₃H)(I)(TCNQ)]	TMA: trimethylammonium	$CH_3-\overset{H}{\underset{CH_3}{\overset{\oplus}{N}}}-CH_3$	20–40 (sc)	Chaikin, Dubrovnik II, 335
(TMDTDSF)₁₂PF₆	TMDTDSF: tetramethyldithio-diselenofulvalene	[structure of tetramethyldithiodiselenofulvalene with CH₃, S, Se groups] (one S and one Se on each side of C=C bond; presumably disordered)	~40 (sc)	Lacoe, Abano Terme C, 283
(TMHDA)(I)(TCNQ)₂	TMHDA: tetramethylhexa-methylenediammonium	$H^+N(CH_3)_2\text{-}(CH_2)_6\text{-}(CH_3)_2N^+H$	20–50 (sc)	Flandrois, Dubrovnik II, 188
TMPD-Chloranil	TMPD: N,N,N',N'-tetramethyl-p-phenylenediamine Chloranil: see TMTTF-chloranil	[structure of TMPD with two N(CH₃)₂ groups on benzene ring]	2×10^{-3} (sc)	Pott, *Mol. Phys.* **13**, 373 (1967)
TMPD (TCNQ)₂			$\sim 2 \times 10^{-2}$ (pw)	Somoano, *J. Chem. Phys.* **62**, 1061 (1975)

Name	Description	Structure	Conductivity	References
(TMTSF)$_2$-X	TMTSF: tetramethyltetra-selenafulvalene	[structure: tetramethyltetraselenafulvalene with CH$_3$ and Se groups]		
X = AsF$_6$			430 (sc)	Bechgaard, SSC **33**, 1119 (1980)
BF$_4$			540 (sc)	Bechgaard, SSC **33**, 1119 (1980)
ClO$_4$			430 (sc)	Parkin, *J. Phys. C* **14**, L445 (1981)
H$_2$F$_3$			100 (sc)	Mortensen, preprint
NO$_3$			780 (sc)	Bechgaard, SSC **33**, 1119 (1980)
PF$_6$			540 (sc)	Bechgaard, SSC **33**, 1119 (1980)
ReO$_4$			300 (sc)	Jacobsen, *J. Phys. C* **15**, 26 (1982)
SbF$_6$			500 (sc)	Bechgaard, SSC **33**, 1119 (1980)
TMTSF-DMTCNQ	DMTCNQ: see HMTTeF (DMTCNQ)$_{0.5}$		400–600 (sc)	Jacobsen, *Phys. Rev. B* **18**, 905 (1978)
TMTSF-TCNQ			10^3 (sc)	Jacobsen, *Phys. Rev. B* **18**, 905 (1978)

Compound	Structure/Notes	Conductivity	Reference
(TMTTF)$_2$– BF$_4$ Br ClO$_4$ PF$_6$ SCN	TMTTF: tetramethyl- tetrathiafulvalene	50 (sc) 260 (sc) 30 (sc) 20 (sc) 60 (sc)	Delhaes, *Mol. Cryst.* *Liq. Cryst.* **50**, 43 (1979); Mortensen, *Synth. Met.* **9**, 63 (1984); Bozio, Abano Terme C, 211
TMTTF–bromanil chloranil fluoranil	bromanil: $X = Br$ chloranil: $X = Cl$ fluoranil: $X = F$	1 (pw) 20 (pw) 7 (pw)	Kagoshima, Les Arcs, 1289; Torrance, *JACS* **101**, 4747 (1979); Wolfe, Boulder E, 337
(TMTTF)$_3$–DETCNQ	DETCNQ: 2,5-diethyl-TCNQ	130 (sc)	Fabre, *Tetrahedron Lett.* **21**, 607 (1980)
TMTTF–DMTCNQ	DMTCNQ: see HMTTeF (DMTCNQ)$_{0.5}$	120 (sc)	Jacobsen, *Phys. Rev. B* **18**, 905 (1978)
(TMTTF)$_2$(HCBD)	HCBD: see (BMDT–TTF) HCBD	3×10^{-4} (sc)	Saito, Abano Terme D, 345

Name	Description	Structure	Conductivity	References
TMTTF-TCNQ			67 ± 10 (sc)	Tomkiewicz, SSC **20**, 767 (1976); Tohumoto, Boulder E, 195; Ishiguro, *J. Phys. Jpn.* **48**, 456 (1980)
TP$_2$SbF$_6$(CH$_2$Cl$_2$)	TP: terphenyl		~10^3 (sc)	Enkelmann, Les Arcs, 1147
2,4,6-triphenyl-pyrylium-(TCNQ)$_2$	2,4,6-triphenyl-pyrylium		0.06 (pw)	Melby
triphenylselenonium (TCNQ)$_2$	triphenylselenonium		2.5 × 10^{-3} (pw)	Melby

triphenylsulfonium-(TCNQ)$_2$	triphenylsulfonium	[structure: triphenylsulfonium cation]	1 (sc)	Melby
tris-(dimethylamino)-sulfonium-(TCNQ)$_2$	tris-(dimethylamino)-sulfonium	[structure]	0.03 (pw)	Melby
3,4,5-tris(methylthio)-1,2-dithiolium iodide-(TCNQ)$_2$	3,4,5-tris(methylthio)-1,2-dithiolium iodide	[structure]	0.1–1 (pw)	Papavassiliou, Abano Terme D, 315
(TSA)Br$_{2,3}$ Cl·H$_2$O I$_{1,2}$ TCNQF$_4$	TSA: tetraselenaanthracene	[structure]	200–400 (sc, all) 0.6–0.03 (pw, all)	Endres, Boulder F, 111
TSA-TCNE	TCNE: see (BMDT–TTF) TCNE		0.6–0.03 (pw)	Endres, Boulder F, 111

Name	Description	Structure	Conductivity	References
TSeA–TCNQ	TSeA: 1,4,9,10-tetraselena-anthracene		1.0 (pw)	Nogami, Les Arcs 1253
(TSeF)Br$_{0.8}$	TSeF: tetraselenafulvalene		10 (pw)	Kaufman, *JACS* **98**, 1596 (1976)
TSeF–DETCNQ	DETCNQ: see (TMTTF)$_3$–DETCNQ		800 (sc)	Andersen, *J. Chem. Soc., Chem. Comm.* 526 (1977)
TSeF–TCNQ			800 ± 100 (sc)	Kistenmacher, *Acta Cryst. B* **30**, 763 (1974)
(TSeT)$_2$Br	TSeT: tetraselenotetracene		2100 (sc)	Shibaeva, Dubrovnik II, 167

2. QUASI-ONE-DIMENSIONAL ORGANIC MOLECULAR CRYSTALS

Compound	Value	Reference
(TSeT)AsF$_6$	50 (sc)	Shibaeva, Dubrovnik II, 167; Delhais, Helsingor, 41; Kaminsky, Les Arcs, 1167
(TSeT)$_2$Cl	2.1×10^3 (sc)	
(TSeT)(CuBr$_2$)	30 (sc)	
(TSeT)$_3$(Hg$_2$Br$_6$)	30 (sc)	
(TSeT)$_2$(Hg$_2$I$_6$)	10^{-3} (sc)	
(TSeT)$_3$(Hg$_2$I$_5$)	200–400 (sc)	
(TSeT)$_4$(Hg$_4$I$_9$)	900–1500 (sc)	
(TSeT)$_2$I	3.9×10^3 (sc)	
(TSeT)I$_{0.71}$(TSeT)I$_{0.77}$	~80 (mixture, sc)	
(TSeT)(PF$_6$)$_{0.4}$	50 (sc)	
(TSeT)$_2$PF$_6$	10 (sc)	Shibaeva, Dubrovnik II, 167; Delhais, Helsingor, 41
(TSeT)$_2$ SCN	20 (sc)	
TSeT (TCNQ)$_2$	10^2 (sc)	Shibaeva, Dubrovnik II, 167
TTF–Br$_{0.71-0.76}$	200–500 (sc)	La Placa, SSC **17**, 635 (1975)
TTF–Br$_{0.79}$	~400 (sc)	Torrance, IBM Research Report RJ2314 (31256) 8/11/78
TTF–Br$_n$ $n \approx 0.7$	100–550 (sc)	Warmack, *Phys. Rev. B* **12**, 3336 (1975)

tetrathiafulvalene

Name	Description	Structure	Conductivity	References
TTF$_2$ [C$_6$(CN)$_6$]	[C$_6$(CN)$_6$]$^{2-}$: hexacyanotrimeth- ylenecyclopropanediide	(structure of [C$_6$(CN)$_6$]$^{2-}$ shown)	1 (pw)	Fukunaga, *JACS* **98**, 610 and 611 (1976)
TTF-I$_n$ $n \approx 0.7$			100–550 (sc)	Warmack, *Phys. Rev.* B **12**, 3336 (1975)
(TTF)$_7$I$_5$ [TTF-I$_{0.71}$]			360 ± 50 (sc)	Tomkiewicz, *Phys. Rev.* B **18**, 733 (1978)
TTF-MTCNQ	MTCNQ: methyl-TCNQ	(structure of MTCNQ shown)	200–500 (sc) 4–5 (pw)	Jacobsen, *SSC* **19**, 1209 (1976)
(TTF)$_{12}$(SeCN)$_7$ [TTF-SeCN$_{0.58}$]			750 ± 150 (sc)	Somoano, *Phys. Rev.* B **15**, 595 (1977)
(TTF)$_{12}$(SCN)$_7$ [TTF-SCN$_{0.58}$]			550 ± 250 (sc)	Tomkiewicz, *Phys. Rev.* B **18**, 733 (1978); Somoano, *Phys. Rev.* B **15**, 595 (1977)

TTF-TCNQ		192–652 (sc)	Ferraris, *JACS* **95**, 948 (1973); Kistenmacher, *Acta Cryst.* **B 30**, 763 (1974)
TTF-TNAP	TNAP: see HMTSF-TNAP	40 (sc)	Berger, *Phys. Rev. B* **12**, 4085 (1975)
TTN-TCNQ	TTN: dehydrotetrathia-naphthazarin	40 (sc)	Wudl, *JACS* **98**, 252 (1976)
(TTT)(AsF$_6$)$_{0.52}$	TTT: tetrathiatetracene	0.13 (sc)	Nogami, Les Arcs, 1253
(BF$_4$)$_{0.25}$(TCE)$_{0.20}$ Br$_{0.70}$(TCE)$_{0.16}$	TCE: 1,1,2-trichloroethane	2.5×10^{-2} (sc) 45 (sc)	
(ClO$_4$)$_{0.56}$ Cl$_{0.75}$(TCE)$_{0.52}$		6×10^{-3} (sc) 1.47 (sc)	
I$_{0.90}$(TCE)$_{0.20}$ (PF$_6$)$_{0.61}$ (SbF$_6$)$_{0.64}$		158 (sc) 100–760 (sc) 40 (sc)	

Name	Description	Structure	Conductivity	References
(TTT)Br			10^{-4} (sc)	Kaminsky, Les Arcs, 1167; Shibaeva, Dubrovnik II, 167
TTT$_{1.25}$(Br$_2$I$_5$)			900 (sc)	
(TTT)(CuBr$_2$)			6.5×10^{-2} (sc)	
TTT$_3$(Hg$_2$Br$_6$)			10^{-2} (sc)	
(TTT)$_4$I$_4$			50 (sc)	
TTT$_{1.2}$(NiS$_4$C$_4$H$_4$)			30 (sc)	
(TTTF)$_2X$ X = AsF$_6$ BF$_4$ PF$_6$	TTTF: trimethylenetetra-thiafulvalene		40 (sc) 0.3 (sc) 10–20 (sc)	Fabre, Les Arcs 1153; Chasseau, Les Arcs 1223
α-TTTF (TCNQ)	α-TTTF: dithieno [2,3-b; 2',3'-h]-1,4,5,8-tetrathiafulvalene (probably mixture of *cis* and *trans*)		10^{-3} (sc)	Engler, *JACS* **100**, 3769 (1978)

t-TTF-TCNQ	t-TTF: see (TTTF)$_2X$		
(TTT)$_2$I$_{3.0-3.1}$		200–400 (sc)	Keryer, SSC **26**, 541 (1978); Keryer, Dubrovnik I, 65
TTT(TCNQ)$_2$		~10^3 (sc)	Somoano, *Phys. Rev. B* **17**, 2853 (1978); Abrahams, SSC **25**, 521 (1978); Shibaeva, Dubrovnik II, 167
		10^2 (sc)	Shibaeva, Dubrovnik II, 167

CHAPTER 3

Structural Instabilities

Jean Paul Pouget

LABORATOIRE DE PHYSIQUE DES SOLIDES
ASSOCIÉ AU CNRS — UNIVERSITÉ PARIS SUD
F-91405 ORSAY, FRANCE

I.	INTRODUCTION	88
II.	BASIC STRUCTURAL PROPERTIES	89
	1. Stacking	89
	2. Crystal Structures	91
	3. Structural Disorder	94
	4. Elastic Properties	95
III.	CHARGE TRANSFER	98
	5. Definition	98
	6. Structural Instabilities	98
	7. Experimental Values	100
IV.	$2k_F$ AND $4k_F$ STRUCTURAL FLUCTUATIONS	110
	8. Introduction	110
	9. Dimensionality of the Structural Fluctuations	111
	10. Disorder	116
	11. The 1D Fluctuation Regime in Nondisordered Salts	120
	12. Lattice Fluctuations	128
	13. Physical Analysis of the $2k_F$ and $4k_F$ Instabilities	135
	14. Conclusion	155
V.	THREE-DIMENSIONAL CDW ORDERING PHASE TRANSITIONS	156
	15. Introduction	156
	16. Interchain Couplings	158
	17. Phase Diagrams	161
	18. Theoretical Analysis	175
	19. Conclusion	193
VI.	GENERAL CONCLUSION	194
VII.	APPENDIX	195
	20. X-ray Diffuse Scattering	195
	21. Neutron Scattering	201
	22. Satellite Reflections	204
	23. Resolution Corrections in X-ray Diffuse Scattering Experiments	204
	REFERENCES	205

I. Introduction

Since the pioneering observation, more than 10 years ago, by x-ray diffuse scattering techniques of the structural features of the Peierls instability in the one-dimensional conductors $K_2Pt(CN)_40.3Br, xH_2O$ (Comès et al., 1973), and TTF-TCNQ (Dénoyer et al., 1975; Kagoshima et al., 1975), a great number of experimental and theoretical studies have been devoted to the physics of charge density waves (CDW) (see Chapter 1, Section 3) in quasi-one-dimensional (1D) conductors. Among them, x-ray, electron, and neutron scattering studies, which directly probe the lattice counterpart of the CDW instability, have played a key role in the understanding of the unusual properties of these materials.

Donor (D) and acceptor (A) molecules can form either mixed or segregated stacks in organic charge-transfer salts. Mixed (ADAD...) stack salts are insulating. They can undergo spin-Peierls [TTF-CuBDT (Bray et al., 1983)] or neutral-ionic [TTF-Chloranil (Torrance, 1985)] phase transitions, which are not the subject of this review. The segregated D and A stack salts are generally 1D conductors. Metallic salts of the TTF-TCNQ family show low-temperature CDW instabilities. The experimental study of this structural instability is the purpose of this chapter. It is intentionally written at a more specialized level than earlier reviews, which have more generally treated the lattice instabilities of 1D inorganic and organic conductors (Comés and Shirane, 1979; Kagoshima, 1982; Moret and Pouget, 1986; Pouget, 1987). Of course, this chapter does not address the newly discovered classes of inorganic 1D conductors like the transition metal tri- and tetrachalcogenides (Monceau, 1985) and the blue bronzes (Schlenker and Dumas, 1986), which also exhibit CDW instabilities, with nonlinear conductivity attributed to sliding CDW transport. (For discussion of sliding CDW transport in TTF-TCNQ see Chapter 1, Section 6 and Chapter 4, Section 17.) It will not cover the single-chain conductors of the $(TMTSF)_2X$ and $(TMTTF)_2X$ families, which show low-temperature instabilities towards the formation of superconducting, spin-density-wave or spin-Peierls ground state (Jérome and Schulz, 1982a) and (or) a new kind of structural transition associated with the ordering of the anions X, when they are noncentrosymmetric ($X = ClO_4^-, ReO_4^-, NO_3^-, ...$) (Moret and Pouget, 1986; Pouget, 1987).

The format of this review is as follows: in Part II, we shall describe the basic structural properties of organic conductors. The charge transfer between donor and acceptor stacks will be the subject of Part III. Part IV will be concerned with the study of structural fluctuations above the Peierls transition temperature (T_P). The modulated phases occurring below T_P will be analyzed in Part V.

II. Basic Structural Properties

1. STACKING

The organic charge-transfer salts of the TTF–TCNQ family are composed of segregated stacks of donor (D) and acceptor (A) molecules, as schematically shown in Fig. 1. Typical molecules involved in the materials considered in this review are shown in Fig. 2. They have a planar geometry.

The presence of segregated stacks, where a sequence of either A or D planar molecules are piled face to face along a given direction of the crystal favors the delocalization of pπ electrons in this direction and is a key structural feature to understanding the anisotropy and the conducting properties of these materials. Depending on the D and A molecules, several stacking modes have been observed (Khidekel and Zhilyaeva, 1981; Williams et al., 1983). The geometry of the stack has important consequences:

(1) For the 1D electronic properties, by fixing the sign and the value of the transfer integrals (Silverman, 1981; Berlinsky et al., 1974; Herman et al., 1977; Herman, 1977).

(2) For the electron phonon coupling, by fixing the symmetry of the intermolecular vibrational modes coupled to the electrons.

The stacking mode may depend on the degree of charge transfer (Silverman, 1981; Berlinsky et al., 1976). The slipped overlap, in which adjacent molecules present a ring-double-bond overlap (as shown for the

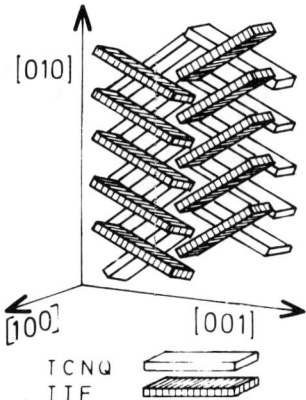

FIG. 1. Schematic representation of the TTF–TCNQ structure. The molecular plane is not perpendicular to the stacking axis b ([010]) but tilted about the a axis ([100]). There are two identical molecules with opposite tilts for one repeat unit along the c direction ([001]). TCNQ and TTF stacks alternate in the a direction.

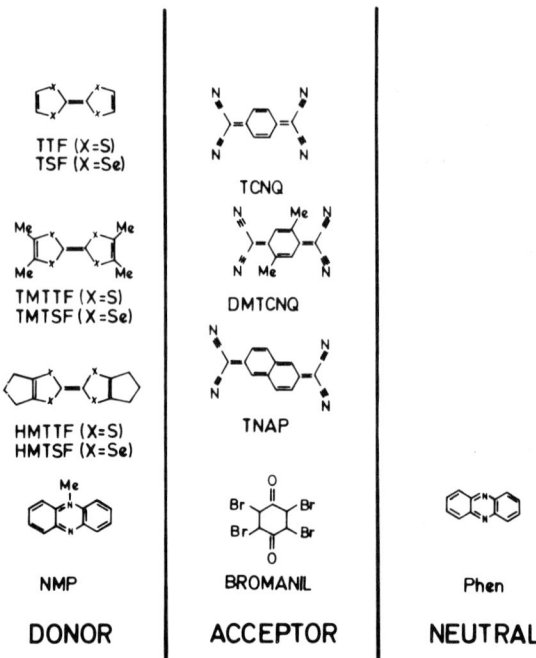

FIG. 2. Molecular structure of typical donor, acceptor, and neutral molecules.

TMTSF and TCNQ dimers, respectively, in Figs. 3a and 3b), is frequently observed in highly conducting crystals. With this overlap two kinds of stacking are usually observed:

- "Skew" stacking, where the molecular planes are tilted with respect to the stacking axis (Fig. 3c), which is found for the D and A stacks in the two chain compounds of the TTF–TCNQ family.
- "Zig-zag" stacking, where the molecular planes are roughly perpendicular to the stacking axis (Fig. 3d), which is observed in the D or A one-chain compounds like $(TMTSF)_2X$ (Thorup *et al.*, 1981) or TMA-I-TCNQ (Filhol *et al.*, 1979).

More complex stacking sequences can be found in other compounds, with, often, a direct influence on their electronic properties. For example, $TEA(TCNQ)_2$ has a repeat sequence that contains four TCNQ (and two TEA) molecules (Kobayashi *et al.*, 1970). This lattice periodicity coincides with the Peierls wavelength $2\pi/2k_F$ deduced from the full transfer of one electron from the TEA to two TCNQ molecules. Consequently, the Fermi level is situated in the gap opened in the "conduction" band at the Brillouin zone boundaries.

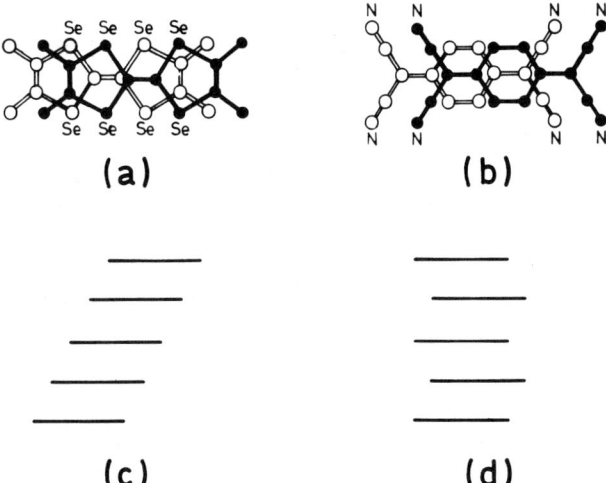

FIG. 3. Molecular overlap of TMTSF molecules (a) and TCNQ molecules (b), skew (c) and zig-zag (d) stackings of molecules.

2. CRYSTAL STRUCTURES

Table I gives the space group and the room temperature lattice parameters for the two-chain conductors, which will be considered in this review. These salts differ mainly in the lateral arrangement between the acceptor and donor stacks. In some of them, built with molecules of high symmetry, and shown in Fig. 4, stacks of donors and of acceptors have a parallel alignment of the long axes of the molecules. In other salts, where one or two CH_3 groups are branched from the central cycle of the donor (case of NMP) or acceptor (case of DMTCNQ) molecule, stacks of donors and acceptors are differently oriented, probably for steric reasons due to the presence of these CH_3 groups.

In the group of salts considered in Fig. 4, sulfur and selenium analogues adopt the same array. Only the pseudo orthorhombic compounds HMTSF-TCNQ and HMTTF-TCNQ adopt the chessboard array (Fig. 4a) where donor and acceptor stacks alternate in the two lateral directions. The other arrays can be viewed as resulting from a shear deformation (leading to a lattice of pseudo monoclinic symmetry) of this arrangement, either along the long axis of the molecules (TMTTF-TCNQ, TMTSF-TCNQ, HMTSF-TNAP, TMTTF-Bromanil—Fig. 4b) or along its short axis (TTF-TCNQ, TSF-TCNQ—Fig. 4c). Other important structural features, not shown in Fig. 4, are the relative shift of the stacks with respect to each other in the chain direction, and the tilt angle of the molecules. These parameters determine the shortest interstack distances, along which a preferential electrostatic coupling or interchain bonding could be established. Generally,

TABLE I

Crystallographic Structures of Some Two-Chain Compounds. The Setting of the Crystallographic Directions is That Used in the References. * Denotes the Stacking Direction, Z is the Number of Formula Units Per Unit Cell. The Lattice Parameters are Those Measured at Room Temperature.

	Space Group	Z	a	b	c	α	β	γ	References
TTF-TCNQ	$P\,2_1/c$	2	12.298	3.819*	18.468	90	104.46	90	(a)
TSF-TCNQ	$P\,2_1/c$	2	12.514	3.876*	18.511	90	104.2	90	(b)
HMTTF-TCNQ	$Pmna$	2	12.462	3.901*	21.597	90	90	90	(c)
HMTSF-TCNQ	$C\,2/m$	2	21.999	12.573	3.890*	90	90.29	90	(d)
TMTTF-TCNQ	$P\,2/c$	2	18.82	3.850*	15.08	90	103.70	90	(e)
TMTSF-TCNQ	$P\bar{1}$	1	3.883*	7.645	18.846	77.34	89.67	94.63	(f)
TMTSF-DMTCNQ	$P\bar{1}$	1	3.938*	8.085	18.956	97.31	98.12	91.37	(g)
TMTTF-DMTCNQ	$P\bar{1}$	1	?	?	?	?	?	?	(h)
HMTSF-TNAP	$P\bar{1}$	1	7.678	20.46	3.919*	90.29	96.36	82.60	(i)
TMTTF-Bromanil	$C\,2/c$	4	31.316	3.950*	19.300	90	121.19	90	(j)
NMP-TCNQ	$P\bar{1}$	1	3.868*	7.781	15.735	91.67	92.71	95.38	(k)
Qn(TCNQ)$_2$	$C\,2/c$	4	28.468	3.838	25.704	90	113.64	90	(l)

(a): Kistenmacher et al. (1974); Blessing and Coppens (1974); Schultz et al. (1976); Filhol et al. (1982). (b): Etemad et al. (1975). (c): Greene et al. (1976); Chasseau et al. (1978). (d): Phillips et al. (1976). (e): Phillips et al. (1977). (f): Bechgaard et al. (1977). (g): Andersen et al. (1978). (h): Result quoted by Jacobsen et al. (1978). (i): Kistenmacher, private communication. (j): Mayerle and Torrance (1981). (k): Fritchie (1966). (l): Kobayashi et al. (1971).

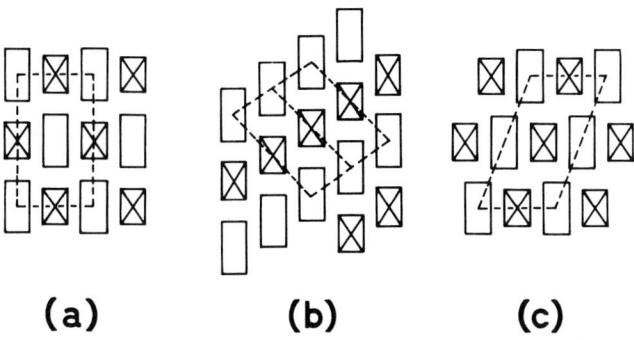

FIG. 4. Schematic arrays of acceptor and donor stacks in (a) HMTSF-TCNQ and HMTTF-TCNQ; (b) TMTTF-TCNQ, TMTSF-TCNQ, HMTSF-TNAP, and TMTTF-Bromanil; and (c) TTF-TCNQ and TSF-TCNQ. The dotted lines show the unit cell defined in Table I [in (b) only the cells of formula unit $Z = 1$ and $Z = 2$ are represented].

short distances are observed between the Se (or S) atom of the donor and the N atom of the acceptor. Among all the compounds listed in Table I, remarkable short Se-N contact distances have been found in the (b, c) plane of HMTSF-TCNQ (Phillips et al., 1976; Kistenmacher, 1978).

Among all the sulfur and selenium analogues of Table I, only TTF-TCNQ and TSF-TCNQ (and perhaps TMTSF-DMTCNQ and TMTTF-DMTCNQ) are isostructural. HMTSF-TCNQ and HMTTF-TCNQ differ mostly by the relative orientation of the tilt angle of the molecules in the long direction of the unit cell, TMTSF-TCNQ and TMTTF-TCNQ differ in a similar way in the two transverse directions. Weak interstack coupling seems to prevail in these directions, where a substantial disorder is observed (see Section 3).

The relative stability of the different arrays, like those considered in Fig. 4, is believed to be the result of a competition between the Madelung energy of charged chains and the Van der Waals energy of metallic chains composed of highly polarizable molecules: the former interaction favoring the alternation of D and A stacks in the two lateral directions and the latter interaction leading to the formation of sheets of D or A stacks (Debray et al., 1977). However, in the light of recent theoretical estimates of these energies (Barisic and Bjelis, 1983 and 1985; Zupanovic et al., 1985) some questions regarding the cohesion of the two-chain organic stacks remain open. In principle the problem of cohesion cannot be disconnected from that of the determination of the partial charge transfer ρ between donor and acceptor stacks (see Part III). In this respect, it appears after inspection of Table III, that, for decreasing ρ, the chain array changes from that shown in Fig. 4a to that of Fig. 4c, then to that of Fig. 4b. The synthesis of the Mott-Hubbard organic insulators HMTTF-TCNQF$_4$ and HMTTF-TCNQF$_4$, for which $\rho = 1$, and which are isostructural to HMTTF-TCNQ and HMTSF-TCNQ, respectively (Torrance et al., 1980; Hawley et al., 1978), strengthened this observation. It shows that large charge transfers stabilize the structure, shown in Fig. 4a, which minimizes the Madelung Coulomb energy term.

3. STRUCTURAL DISORDER

Not too much attention has been paid to the characterization of structural disorder in organic conductors and solid solutions of organic conductors. In some cases, x-ray patterns clearly reveal diffuse streaks or diffuse sheets passing through the layers of Bragg reflections perpendicular to the stacking direction, which probably originate from faults in the relative angle of tilt of molecules belonging to adjacent stacks. In HMTSF-TCNQ (Phillips et al., 1976; Pouget, unpublished results) and HMTTF-TCNQ (Chasseau et al., 1978; Megtert et al., 1978a), streaks along the a^* and b^* reciprocal directions, respectively, are observed. They indicate a disorder between (b, c) and (a, c) planes of stacks, respectively (see Table I for the definition of these directions); i.e., in the direction along which there is the weakest interstack coupling. Diffuse sheets perpendicular to the stacking direction are observed in TMTTF-TCNQ, TMTSF-TCNQ, and HMTSF-TNAP (Pouget, unpublished results) and are indicative of a weak interstack coupling in the two transverse directions. Such disorder is not observed in TTF-TCNQ and TSF-TCNQ.

Orientational disorder may arise in materials containing noncentrosymmetric molecules like quinolinium (Qn) or NMP (see Figure 2). In Qn(TCNQ)$_2$, the Qn cations are disordered, taking with equal probability two orientations related by a center of inversion (Kobayashi et al., 1971). In NMP-TCNQ, there is local order of the methyl group of the NMP molecules in the conducting direction, long-range order in the b direction, and no order at all in the c direction (Kobayashi, 1975; Pouget et al., 1979a and 1980). This observation has been interpreted on the basis of electrostatic dipole interactions between the NMP molecules. The extent of the local order of the NMP dipoles in the chain direction depends upon the sample: the shortest range order being observed in the most conducting crystals (Pouget et al., 1979a). This observation, and the increase in the order of donor molecules along the chain direction when the number of electron per donor chain decreases (i.e., for x decreasing in the NMP$_x$Phen$_{1-x}$TCNQ solid solution) indicate that the coupling between NMP dipoles is certainly screened by the 1D electron gas (Pouget et al., 1982).

Additional questions concern the eventual ordering between substituted molecules that may occur in solid solutions of organic conductors. In the substituted solution $NMP_xPhen_{1-x}TCNQ$ (Miller and Epstein, 1978) ordering between NMP and Phen molecules occurs in the (a, b) planes of donor molecules for $0.5 < x < 0.57$ (Pouget, 1981; Epstein et al., 1981; Pouget et al., 1982). Because of the triclinic environment of the TCNQ stack, the alternate order of NMP and Phen creates (site and bond) potentials of period 2a in the chain direction. The alternate order of NMP dipoles gives rise to (site and bond) potentials of period 4a. Their Fourier components coincide respectively with the $4k_F$ and $2k_F$ wavevectors of the TCNQ 1D electron gas for $x = 0.5$. It can induce a commensurate structural distortion of the TCNQ stack, which may help in the formation of charged solitons for x close to 0.5. Such charged defects has been detected in the electronic properties of these salts (Epstein et al., 1982). However, optical measurements have shown that the electrical gap results mainly from a temperature dependent Peierls distortion of the acceptor stack (see Part IV) and that the contribution of the NMP-Phen ordering potential at this gap is weak at low temperature (Epstein et al., 1985).† The alloy series $TTF_{1-x}TSF_xTCNQ$ has also been prepared (Etemad et al., 1978) without significant ordering in the position of the substituted TSF (or TTF) molecules or their segregation (Kagoshima et al., 1980). For low values of x, the observation of sizable x-ray Laue scattering (see Fig. 18c), due to the difference between the TTF and TSF molecular structure factors, is the signature of a random distribution of TTF and TSF molecules on the donor stack (result quoted by Forro et al., 1984).

Finally, it has been observed that x-rays can severely damage the organic conductors. Very dramatic effects have been observed on the phase transitions of TMTTF-TCNQ (Kagoshima et al., 1983a) and TMTSF-DMTCNQ (Forro et al., 1983). At low doses ($\leq 10^{-2}$ mol defects), enough to suppress the Peierls transition, it has been shown (Zuppiroli et al., 1986) that the damage is characterized by individual point defects on the molecular scale. They act as independent strain sources and induce volume and lattice parameter changes of about one molecular volume per defect. Radiation damage introduces much stronger disorder than the chemical disorder previously considered in this section (see Chapter 7).

4. ELASTIC PROPERTIES

Most of the work reported in this part has been performed on TTF-TCNQ. The thermal expansion of TTF-TCNQ has been studied by x-ray diffraction

---

† However, as demonstrated for the $(TMTTF)_2X$ series (Barisic and Brazovskii, 1981; Emery et al., 1982), the commensurability effect provided by the $4k_F$ NMP-Phen ordering potential can allow additional Umklapp electron-electron (g_3) interaction processes whose effects are considered in Part IV, Section 13.

(Schultz et al., 1976), neutron diffraction (Pouget et al., 1979b; Filhol et al., 1981), and capacitance dilatometry (Schafer et al., 1975). All the determinations agree with the following thermal expansion coefficients at room temperature (in K^{-1}):

$$\alpha_a = 0.37\ 10^{-4}, \quad \alpha_b = 1.47\ 10^{-4}, \quad \alpha_c = 0.27\ 10^{-4}, \quad \alpha_{vol} = 2.09\ 10^{-4}.$$

The largest coefficient is α_b, and Fig. 5 shows that the stacking axis undergoes a significant contraction when the temperature decreases: $\Delta b/b = 2.4\%$ between 300 K and 10 K. A similar degree of contraction in the chain direction, $\Delta a/a = 2.5\%$, has been measured in TMTSF–DMTCNQ between 300 K and 10 K (Gay et al., 1982). The lattice parameters do not present any sizable anomaly at the Peierls transition.

The effect of pressure on the structure of TTF–TCNQ has been studied by neutron diffraction at 4.6 kbar (Filhol et al., 1981), and the compressibility coefficients have been obtained up to about 20 kbar (Debray et al., 1977; Megtert et al., 1979a; Filhol et al., 1981). At ambient pressure and room temperature, the most reliable experimental determination (Filhol et al., 1981) gives: $K_a = 0.18\%$ kbar^{-1}, $K_b = 0.59\%$ kbar^{-1}, $K_c = 0.13\%$ kbar^{-1}, and $K_{vol} = 0.88\%$ kbar^{-1}. The transverse compressibilities are only weakly temperature and pressure dependent, whereas the compressiblity in the chain direction decreases significantly for decreasing temperature (by a factor 2 between 300 and 10 K) and increasing pressure (Megtert et al., 1979a; Filhol et al., 1981). Above 10 kbar, the compressibility of TTF–TCNQ is nearly isotropic.

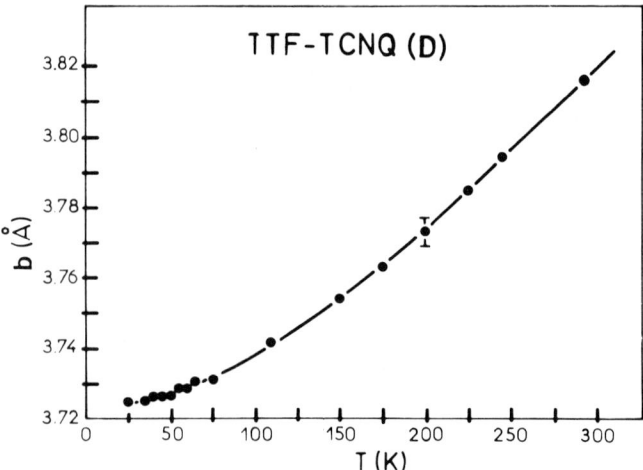

FIG. 5. Temperature variation of the b parameter of deuterated TTF–TCNQ. [From Pouget et al. (1979a).]

Finally, let us remark that the contraction of the stacking axis parameter of TTF-TCNQ between 300 and 10 K, at ambient pressure, is equivalent to the application of a pressure of 4 kbar at 300 K. This effect may significantly alter the physical quantities that depend strongly on the intrachain distance, such as the overlap integral between neighboring molecules, which determines the 1D electronic bandwidth. It also causes a change in the charge transfer as a function of temperature (see Part III).

Compressibility measurements show that TTF-TCNQ exhibits sizable anistropy in its elastic properties at room temperature. However, such anisotropy is not found in the slope of the longitudinal acoustic branches measured by inelastic neutron scattering experiments (Pouget et al., 1979b), which yield nearly isotropic values for the elastic constants: $C_{11} \sim C_{22} \sim C_{33}$. Within experimental error, ultrasonic measurements (Tiedje et al., 1977) agree roughly with the neutron measurements. Table II presents the elastic constants obtained by several experimental techniques operating at different frequencies. Among them, the ultrasonic techniques (Tiedje et al., 1977) and the vibrating reed techniques (Barmatz et al., 1974) show that several acoustic modes exhibit an increase in sound velocity below the 52 K Peierls transition of TTF-TCNQ when the electronic screening of the force constants is reduced by the opening of energy gaps in the band structure.

TABLE II

Comparison of the Room Temperature Elastic Constant of TTF-TCNQ Obtained from the Slope of the Acoustic Branches Measured by Neutron Scattering (Pouget et al., 1979a), the Sound Velocity Measured by Ultrasonic Techniques (Tiedje et al., 1977) and Vibrating Reed Techniques (Barmatz et al., 1974), and from the Inverse of the Compressibility Coefficients Measured on the a, b, c Directions, Respectively (Filhol et al., 1981). Most of the Elastic Constants (C_{eff}) in this Table Have Been Derived by Neglecting the Off-Diagonal Elastic Constants (C_{ij} with $i \neq j$) in the Analysis of the Data. The Order of Magnitude of the Frequency ν Used in Each Type of Measurement is Indicated. The Tensor of Elastic Constants is Expressed in the (a, b, c*) Frame.

C_{eff} ($\times 10^{11}$ dynes/cm^2)	Neutron measurements ($\nu \sim$ 1 THz)	Ultrasonic measurements ($\nu \sim$ 1 MHz)	Vibrating reed measurements ($\nu \sim$ 1 KHz)	Compressibility measurements
C_{11}	$\simeq 2.0 \pm 0.4$	3.4 ± 1		~ 5.55
C_{22}	$= 1.7 \pm 0.3$	1.3 ± 0.15	0.3	1.7
C_{33}	$\simeq 2.2 \pm 0.4$			~ 7.70
C_{44}	$= 0.88 \pm 0.22$			
C_{66}	$= 0.42 \pm 0.03$	$\begin{cases} 0.66 \pm 0.2 \\ 0.62 \pm 0.06 \end{cases}$		
C_{46}	$= 0.04 \pm 0.60$			

III. Charge Transfer

5. Definition

One of the most important physical parameters characterizing the 1D organic metals is the charge transfer, ρ, that occurs between donor (D) and acceptor (A) molecules constituting the solid according to the reaction:

$$A + D \rightarrow A^{-\rho} + D^{+\rho}$$

This quantity is *a priori* unknown and generally cannot be guessed from the stoichiometry of the salts. In principle, it can be obtained by minimizing the part of the cohesive energy that depends on ρ (Klimenko *et al.*, 1976; Torrance and Silverman, 1977; Torrance, 1979; Metzger, 1981 and 1982; Bloch and Mazumdar, 1983; Barisic and Bjelis, 1984 and 1985; Noguera, 1985). With respect to the neutral DA system, there are four primary contributions to the cohesive energy and, therefore, to ρ: (1) the cost of ionization (affinity) energy of the constituent molecules; (2) the first-order Coulomb interactions between the charges (Madelung energy); (3) the gain of energy due to the delocalization of the charges along the stacks, and (4) the second-order Coulomb (Van der Waals) interactions between the stacks, involving the intramolecular and metallic (intrachain) polarizabilities. Practically, these contributions cannot be calculated with enough accuracy to get reliable values of the charge transfer because of an insufficient knowledge of many-body effects. Experimentally, ρ has been determined in a great variety of salts (see, for example, Table III). Two kinds of determination have been used:

1. Direct methods: measuring either directly the charge per molecule [very accurate structural determination, x-ray photo-electron spectroscopy (XPS)] or the relative change of a intramolecular parameter sensitive to this charge (bond length, intramolecular vibration frequency). These methods give ρ with a relatively poor accuracy ($\gtrsim 10\%$).

2. "Structural" methods: measuring the critical wave vector of the structural instability accompanying the $2k_F$ or $4k_F$ CDW instability of the 1D electron gas as described below. This method has been used for all the charge-transfer salts considered in this review and yields ρ within an accuracy of a few percent.

6. Structural Instabilities

Let us first consider a stack composed of identical molecules having the repeat periodicity b. The highest occupied electronic states overlap in this direction to form an energy band. Let us first neglect the electron–electron interactions and assume that there is only one energy band, which is partly filled

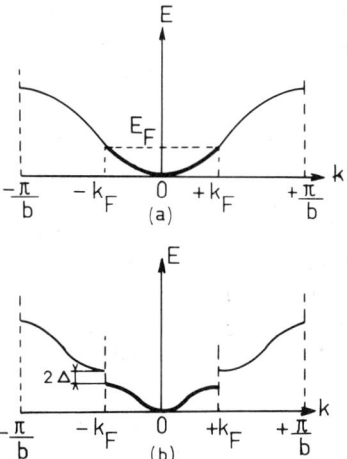

FIG. 6. (a) Electronic band dispersion $E(k)$ of an electron gas in a 1D lattice of periodicity b. The Fermi surface is composed of two points at $\pm k_F$. (b) Effect of a periodic lattice distortion of wavevector $2k_F$ opening an energy gap 2Δ, at the Fermi level.

with the ρ charges provided by each molecule. Such a band is represented in Fig. 6a. Its Fermi surface is composed of two points at $\pm k_F$. The band can accommodate two electrons per repeat unit (per molecule); the factor of 2 is due to the spin degeneracy. Because of the 1D nature of the electron band, there is a very simple relationship between k_F and the band-filling ρ

$$\rho = 2 \times 2k_F. \tag{1}$$

It was Peierls who first pointed out, in 1955, that such a 1D metallic chain is not stable at $T = 0$ K against a periodic lattice distortion of wavevector $2k_F$, opening an energy gap 2Δ at the Fermi level (Fig. 6b); the gain in electronic energy always overcomes the lattice energy lost by the distortion. More refined treatments (Rice and Strassler, 1973; and the reviews of Berlinsky, 1976; Tombs, 1978; Jérome and Schulz, 1982a; and see Chapter 1, Section 4 and Chapter 4, Section 11) have shown that such structural instability comes from the $2k_F$ instability of a 1D electron gas towards the formation of CDW via the electron-phonon coupling. Thus, in this picture of independent electrons, the measurement of the wavevector of the structural instability in reciprocal wavevector units ($2\pi/b$ unit in Fig. 6) gives $2k_F$. Twice this value gives ρ, according to the Eq. (1).

Let us now consider the case in which the electron gas experiences very strong electron–electron repulsion. In the limit where U, the intrasite Coulomb repulsion, is infinite, a very simple description can be obtained (Ovchinnikov, 1973; Bernasconi et al., 1975; Emery, 1977; Tombs, 1978;

Schultz, 1980; see Chapter 1, Section 3). Since a given lattice site cannot be occupied by more than one electron (we still assume here that there is no orbital degeneracy), we are faced with the problem of a spinless 1D Fermi gas. The energy dispersion for this system is similar to that shown in Fig. 6. However, since the spin degree of freedom is lost, the same number of electrons now fill the "conduction band" from $-2k_F+$ to $+2k_F$ (here we still define k_F for non-interating electrons). Similar to the $2k_F$ Peierls instability, the strongly correlated electron gas is also unstable at $T = 0$ K against a periodic lattice distortion of the lattice. However, this instability now occurs at the vector $4k_F$, directly related to ρ from Eq. (1).

The electron–electron interactions experienced by the 1D electron gas of organic conductors are of course, situated between these two extreme cases. It has been shown (for reviews see Solyom, 1979; Emery, 1979) that, depending on the value and on the range of the electron–electron interactions, the 1D electron gas can still present CDW instabilities at the $2k_F$ and/or $4k_F$ wavevectors defined above. However, in order to determine ρ one must know whether the observed structural instability occurs at the $2k_F$ or at the $4k_F$ electronic wavevector.

7. Experimental Values

The electronic structure of charge-transfer salts is more complex than that described in part 6 of this section, because the donor and acceptor stacks must be considered together. If there is an incomplete charge transfer from the donors to the acceptors, we are faced with two conducting subsystems. In an independent electron description, two bands cut the Fermi level. Two cases have to be considered, depending on whether or not the donor has a closed-shell electronic configuration. In the first case, relevant for TTF and its symmetric derivatives, the DA molecular pair share two electrons. In the second case, relevant for the nonsymmetric NMP molecule, there is only one electron to share per DA pair of molecules. These two examples are illustrated by the 1D tight binding band schemes shown in Fig. 7.

In Fig. 7a, relevant to the case of TTF–TCNQ and its family, there are two electrons to share per chemical formula, and the number of electrons transferred to the acceptor chain is equal to the number of holes created on the donor chain. Thus, assuming normal and inverted 1D bands, which hold for TTF–TCNQ (Berlinsky et al., 1974; Herman et al., 1977; Herman, 1977), $\pm k_F$ corresponds to the crossing points of the two bands. The two conducting subsystems have the same critical wavevector: $k_F = k_F^A = k_F^D \, (=\rho/4)$. If the two bands have the same curvature, k_F^A is different from k_F^D, but $2k_F^A$ differs from $2k_F^D$ only by a reciprocal lattice wavevector and $2k_F^A = 1 - 2k_F^D$ because of charge conservation. The critical wavevector of the structural instability remains the same.

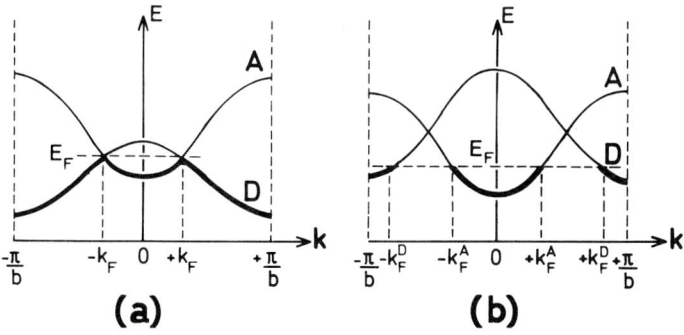

FIG. 7. Schematic representation of the one-electron (conduction) band structure of charge-transfer salts with: (a) two electrons to share between donor and acceptor molecules; (b) one electron to share between donor and acceptor molecules.

In Fig. 7b, relevant to the case of NMP–TCNQ, where there is only one electron to share per chemical formula, the Fermi wavevector differs on the acceptor and donor bands, whatever their curvature. They are, however, related by charge conservation. For NMP–TCNQ this relationship is $\rho^D + \rho^A = 1$, with $2 - 4k_F^D = \rho^D$, and $4k_F^A = \rho^A$, assuming that the band curvatures shown by the Fig. 7b hold for this salt. We now discuss the values of the charge transfer found in these two categories of materials.

a. TTF–TCNQ and its Family

(1) *Low temperature and atmospheric pressure.* In a great number of organic salts the observation of structural instabilities at the $2k_F$ and $4k_F$ wavevectors allows us to determine the charge transfer, ρ, per molecule without ambiguity. In the other salts, containing Se-based donors, there are reasons (see Part IV) to believe that only the $2k_F$ instability remains. The most convincing experimental evidence of such an assertion is obtained from a study of the solid solution $TTF_{1-x}TSF_xTCNQ$. In TTF–TCNQ the $2k_F$ and $4k_F$ instabilities are observed at the wavevectors $0.295b^*$ and $0.59b^*$, respectively (Pouget *et al.*, 1976; Kagoshima *et al.*, 1976; Khanna *et al.*, 1977). With increasing x the two structural instabilities are still observed with only a slight increase in the value of k_F; at $x \simeq 0.9$, the values $2k_F = 0.31b^x$ and $4k_F = 0.62b^x$ are obtained (Sakai *et al.*, 1983). For $x = 0.99$ and $x = 1$, only the structural instability of wavevector $0.315b^x$ remains (Sakai *et al.*, 1983; Weyl *et al.*, 1976; Kagoshima *et al.*, 1978; Megtert *et al.*, 1978b and 1979b). Continuity in the band filling requires that it corresponds to the $2k_F$ instability.

Table III gives the charge transfer, ρ, obtained from the $2k_F$ and $4k_F$ wavevectors of the structural instabilities observed at low temperature

TABLE III

Value of the Charge Transfer ρ in Organic Conductors Having Two Conduction Electrons to Share Per Formula Unit

	"Structural" charge transfer $\rho(T = 25$ K and $P = 1$ atm)	Instabilities $2k_F$	Instabilities $4k_F$	References	Other determinations of ρ	Experimental technique	References
TTF-TCNQ	0.590 ± 0.003	Yes	Yes	1–6	$0.47 \pm 0.15^{(a)}$ $0.55 - 0.67^{(b)}$ $0.8 \pm 0.2^{(a)}, 0.56 - 0.67^{(b)}$ $0.56 \pm 0.05^{(c)}$ $\sim 0.55^{(a)}, 0.59 \pm 0.05^{(b,c)}$	X-ray diffraction a: charge integration b: bond lengths XPS Raman scattering	(a) 22, 23, (b) 24 (a) 25, (b) 26, (c) 27 (a) 28, (b) 29, (c) 30
TSF-TCNQ	0.63 ± 0.01	Yes	No	7–10	—		
DSTSF-TCNQ	0.62 ± 0.02	Yes	No	11	—		
HMTTF-TCNQ	0.72 ± 0.01	Yes	Yes	9, 12, 13	0.6 ± 0.07	X-ray diffraction (bond lengths)	31
HMTSF-TCNQ	0.74 ± 0.01	Yes	No	7, 13	—		
TMTTF-TCNQ	0.57 ± 0.01	Yes	Yes	14	$0.65 \pm 0.1; 0.61 \pm 0.1; 0.58 \pm 0.05$ $0.50 \pm 0.05^{(a)}; 0.55 \pm 0.05^{(a,b)}$ $0.4 \pm 0.1^{(b)}$ 0.65	XPS Raman scattering Infrared absorption	30 (a) 30, (b) 32 33
TMTSF-TCNQ	0.57 ± 0.02	Yes	No	15	—		
HMTSF-TNAP	0.58 ± 0.02	Yes	No	16	—		
TMTTF-DMTCNQ	0.50 ± 0.02	Yes	Yes	17	—		
TMTSF-DMTCNQ	0.50 ± 0.01	Yes	Yes	15, 18, 19	—		
TMTTF-Bromanil	0.52 ± 0.01	Yes	Yes	20	0.50 ± 0.05	Raman Scattering	32
DBTTF-TCNQCl$_2$	0.52	Yes	Yes	21	—		

(1) R. Comes et al. (1975); (2) R. Comes et al. (1976); (3) J. P. Pouget et al. (1976); (4) S. Kagoshima et al. (1976); (5) S. K. Khanna et al. (1977); (6) J. P. Pouget, S. M. Shapiro, and G. Shirane, Unpublished results; (7) C. Weil et al. (1976); (8) S. Kagoshima et al. (1978); (9) S. Megtert et al. (1978b); (10) S. Megtert et al. (1979b); (11) M. Sakai et al. (1983); (12) S. Megtert et al. (1978a); (13) J. P. Pouget et al. (1979a); (14) S. Kagoshima et al. (1983a); (15) J. P. Pouget (1981); (16) M. Denoziere, DEA Report, Orsay (1983), unpublished; (17) J. P. Pouget, unpublished results; (18) P. A. Albouy, DEA Report, Orsay (1980), unpublished; (19) J. P. Pouget et al. (1980b); (20) S. Kagoshima et al. (1983b); (21) K. Mortensen et al. (1983); (22) P. Coppens (1975); (23) P. Coppens and T. N. Gururow (1978); (24) S. Flandrois and D. Chasseau (1977); (25) R. S. Swingle et al. (1976); (26) W. D. Grobman and B. D. Silverman (1976); (27) I. Ikemoto et al. (1977); (28) M. Kuzmany and B. Kundu (1979); (29) M. Matsuzaki et al. (1980); (30) M. Tokumoto et al. (1982); (31) Results quoted by Y. Tomkiewicz et al. (1977); (32) J. B. Torrance et al. (1981); (33) J. S. Chappel et al. (1981).

($T \simeq 25$ K) and ambient pressure in all the two-chain compounds of the TTF–TCNQ family that have been investigated up to now. With the exception of TTF–TCNQ, where neutron scattering investigations have been performed with great accuracy (see Ref. 6 in Table III), all the values quoted in this table have been obtained by x-ray scattering. This table also gives the charge transfer obtained by direct methods. Within experimental error, there is an overall agreement on the value of the charge transfer obtained by both kinds of determination. This attests to the reliability of the "structural" method and proves the electronic origin of the structural instability.

A close inspection of Table III shows that:

A. The substitution S → Se in the donor molecule does not change ρ appreciably, except perhaps for the smallest donor, where a weak increase of ρ is observed when TTF is substituted into TSF.

B. The substitution H → CH$_3$ at the periphery of the molecule (TTF → TMTTF) decreases ρ weakly.

C. The addition of a cycle on the donor molecule (TTF → HMTTF) increases ρ significantly.

D. The addition of a cycle on the acceptor molecules (TCNQ → TNAP) or the substitution H → CH$_3$ on the central cycle of the TCNQ molecule (TCNQ → DMTCNQ) decrease ρ significantly.

The most drastic change in ρ is observed for the substitution of methyl groups or the addition of cycles in the central part of the molecule. The reason is that these chemical modifications affect the part of the molecule where the electronic cloud, which has an important overlap with that of the neighboring molecules, originates from. Also, these modifications noticeably change the intramolecular (energy of the highest occupied level, molecular polarizability, Coulomb repulsion, etc.) and intermolecular (overlap integrals, etc.) contributions to the cohesion energy.

Table III shows that in all the two-chain organic conductors of 1:1 stoichiometry investigated until now, the charge transfer lies in a quite narrow range of values: $\frac{1}{2}$–$\frac{3}{4}$. In other salts of 1:1 stoichiometry composed of segregated stacks, a charge transfer of 1 has been observed, but it concerns

- one-chain conductors: Na, K, Rb-TCNQ (Terauchi, 1978)
- Mott Mubbard insulators: HMTTF-TCNQF$_4$ and HMTSF-TCNQF$_4$ (Torrance *et al.*, 1980; Hawley *et al.*, 1978).

The order of magnitude of the various terms that contribute to the ρ-dependent part of cohesion energy has been derived recently (Noguera, 1985) in order to account for the observed values of the charge transfers. This phenomenological study shows that, contrary to direct calculations (Klimenko *et al.*, 1976; Torrance and Silverman, 1977; Torrance, 1979;

Metzger, 1981, 1982), the delocalization energy cannot be neglected. As we shall see below, this term can account quantatively for the relative variation of 20% of ρ when the chain parameter b of TTF-TCNQ is changed.

(2) *Temperature and pressure dependence.* With the present accuracy in the determination of the value of ρ, it has been possible to show that the charge transfer varies with the temperature or under pressure in several salts. For decreasing temperature (between room temperature and 25 K) at ambient pressure, the following has been observed:

- an increase of ρ from 0.55 to 0.59 in TTF-TCNQ (Kagoshima et al., 1976; Khanna et al., 1977) and from 0.48 to 0.52 in TMTTF-Bromanil (Kagoshima et al., 1983b)
- a decrease of ρ from 0.56 to 0.52 in DBTTF-TCNQCl$_2$ (Mortensen et al., 1983).

No significant change of ρ as a function of temperature has been detected for HMTSF-TCNQ and TMTSF-DMTCNQ within the experimental error given in Table III. However, in these two last compounds ρ is very close or equal to the commensurate values $\frac{3}{4}$ and $\frac{1}{2}$, respectively. It is also interesting to note that a small jump of the charge transfer from 0.586 to 0.590 (not due to a change in the b parameter—see below and Fig. 5), has been detected for decreasing temperatures at the 38 K first-order phase transition of TTF-TCNQ (Pouget, Shapiro, and Shirane, unpublished results).

A continuous increase of ρ under pressure has been observed in TTF-TCNQ (Megtert et al., 1979a and 1981). Figure 8 shows the pressure dependence of $2k_F$ ($\equiv \rho/2$), measured in the low-temperature (10-20 K) modulated phases of TTF-TCNQ. In this figure two regimes can be distinguished

- a continuous increase of ρ at low pressure ($P < 14$ kbar)
- a lock in of ρ, at the commensurate value $\frac{2}{3}$, in a sizable pressure range (at least 14-18 kbar).

Let us first discuss the low-pressure regime. It is easy to see that the change of the chain parameter b is mainly responsible for the variation of the charge transfer in TTF-TCNQ, because very similar values of $\partial \log \rho / \partial \log b$ are obtained from the change of ρ by varying temperature (-3, using the thermal contraction of the b parameter at ambient pressure), and from the change of ρ under pressure (-3.5, using the compressibility of the b parameter at low temperature). Among the various contributions to the cohesion energy, delocalization is very sensitive to a change of the b parameter, through the exponential dependence of the overlap integrals. Thus, the change in ρ can be qualitatively understood (Friend et al., 1978;

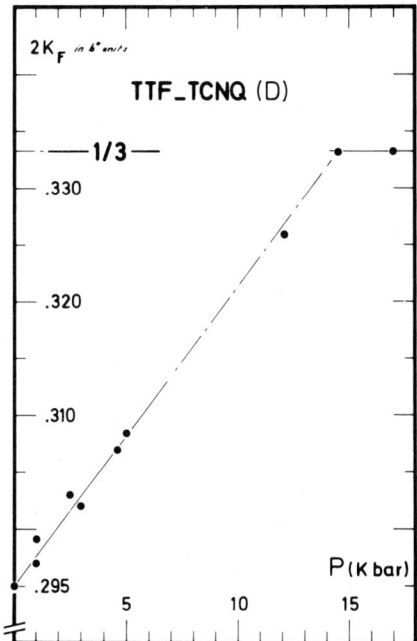

FIG. 8. Pressure dependence of the $2k_F$ component (in b^* units) of the satellite reflections of deuterated TTF-TCNQ. Note the continuous increase of $2k_F$ with pressure until about 14 kbar, where it locks at the commensurate value $\frac{1}{3}$. The charge transfer ρ is twice $2k_F$. [From Megtert et al. (1981).]

Jérome, 1978; Conwell, 1980a; Pouget et al., 1981; see Chapter 4, Section 9a) in the framework of a 1D tight-binding band description. With the inverted band structure shown in Fig. 7a, the following dispersion relation for the TCNQ (Q) and the TTF (F) stacks hold

$$E_Q(k) = E^Q - 2t^Q \cos 2\pi k$$
$$E_F(k) = E^F + 2t^F \cos 2\pi k, \tag{2}$$

where t^Q and t^F are the (positive) intrastack transfer integrals, E^Q and E^F are the midband molecular orbital energy, and k is defined in reciprocal wave vector units $(2\pi/b)$. The crossing point of the two bands, obtained from the equality $E_Q(k) = E_F(k)$, gives k_F. Then the charge transfer is, according to Eq. (1),

$$\rho = \frac{2}{\pi} \mathrm{Arc} \cos \left[\frac{E^Q - E^F}{2(t^F + t^Q)} \right]. \tag{3}$$

Assuming that only t^F and t^Q vary significantly with b, the logarithmic derivative of (3) gives

$$\frac{\partial \log \rho}{\partial \log b} = \left(\frac{2}{\pi\rho} \cot g \frac{\pi\rho}{2}\right) \frac{\partial \log(t^F + t^Q)}{\partial \log b}. \quad (4)$$

From Eq. (4) one obtains: $\partial \log(t^F + t^Q)/\partial \log b \simeq -4$, using the experimental values of $\rho = 0.59$ and $\partial \log \rho/\partial \log b \simeq [-3, -3.5]$. This value is in very good agreement with $\partial \log(t^F + t^Q)/\partial \log b = -4.1$, from the band parameters given by Herman (1977)†.

Let us now turn our attention to the high-pressure regime. Structural determination, based on the value of the $2k_F$ component of satellite reflections observed below the Peierls transition, gives a commensurate charge transfer $\rho = \frac{2}{3}$ between 14 and 18 kbar, at least. A pressure domain of about 4 kbar centered around 19 kbar, where the $2k_F$ value is assumed to be pinned to the commensurate value $\frac{1}{3}$, has also been deduced from conductivity measurements (Andrieux et al., 1979) above the Peierls transition of TTF-TCNQ. In this domain a decrease of electrical conductivity is observed. It has been interpreted as resulting from the suppression of the fluctuative CDW Fröhlich conductivity due to the $2k_F$ commensurability pinning (for reviews, see Jérome and Schulz, 1982a and 1982b). The understanding of the commensurability pinning of the charge transfer go beyond the simple consideration of the delocalization energy which gives a continuous variation of ρ with the chain parameter b. It certainly involve the other contributions at the cohesion energy, such as terms dealing explicitly with the Coulomb interactions between charges.

At pressure greater than 21 kbar, the recovery of a substantial part of the conductivity (Andrieux et al., 1979) indicates that charge transfers higher than $\frac{2}{3}$ are probably achieved in TTF-TCNQ. In a similar way, conductivity measurements suggest that a small commensurability domain with $\rho = \frac{2}{3}$ might be achieved around 6 kbar in TSF-TCNQ (Thomas and Jérome, 1980). Starting at a value $\rho = 0.63$ at ambient pressure (identical to that observed in TTF-TCNQ under 8 kbar, see Fig. 8), it is easily seen that, with a charge transfer of $\frac{2}{3}$ occurring at 6 kbar, TSF-TCNQ has the same rate of increase of ρ under pressure as TTF-TCNQ.

† For $b = 3.73$ Å:

$$\begin{cases} \dfrac{\partial t^Q}{\partial b} = -0.225 \text{ eV/Å}^{-1}, & t^Q = 0.16 \text{ eV} \\ \dfrac{\partial t^F}{\partial b} = -0.095 \text{ eV/Å}^{-1}, & t^F = 0.13 \text{ eV}. \end{cases}$$

These slopes can be also obtained from Fig. 5 in Welber et al. (1978). See also Berlinsky et al. (1974).

b. NMP-TCNQ and the Solid Solution $NMP_xPhen_{1-x}TCNQ$

There has been a long controversy concerning the exact fraction of charge transferred from NMP to TCNQ. Early workers assumed complete charge transfer of one electron (Epstein et al., 1972), while later contributors proposed incomplete charge transfer with ρ close to 1 (Soos, 1974; Torrance et al., 1975; Butler et al., 1975; Kwak et al., 1976; Devreux et al., 1978). X-ray diffuse scattering at reduced wave vectors $\pm 0.094 na^*$ (n integer, varying from 0 to 5; a is the chain direction of NMP–TCNQ—see Table I) was observed at room temperature and proposed as evidence for k_F, $2k_F$, $4k_F$, etc. instabilities, in agreement with $\rho \sim 1$ (Ukei and Shirotani, 1977). However, a later study (Pouget et al., 1979a) showed that this scattering, observed only in a fraction of the samples investigated, is not critical (the intensity decreases with decreasing temperature, as for normal x-ray thermal scattering). In contrast, in all the batches of NMP–TCNQ investigated so far, real quasi-one-dimensional critical scatterings were observed, but at the wavevectors $q_1 = \frac{1}{6}a^*$ and $q_2 = \frac{1}{3}a^*$ (Megtert et al., 1978b; Pouget et al., 1979a and 1980a). However, because of the special values of q_1 and q_2, these studies could not ascribe unambiguously the q_1 and q_2 scattering to $4k_F$ and $2k_F$ instabilities. Also they could not attribute them directly to a given stack and could not decide clearly whether the charge transfer is $\frac{2}{3}$ or $\frac{1}{3}$ (14 assignments were possible—see Table I in Pouget et al., 1980a). The situation was clarified by the study of the $NMP_xPhen_{1-x}TCNQ$ solid solution, whose main results (and final assignments) are summarized in Fig. 9.

The alloy $NMP_xPhen_{1-x}TCNQ$ is achieved by substituting neutral phenazine, phen, for up to 50% of the non-totally-symmetric NMP molecule. The phen is of similar size, shape, and polarizability as NMP^+ but is neutral, closed shell, and symmetric (see Fig. 2). Detailed analysis has shown that the overall (NMP) (TCNQ) crystal structure remains unchanged by the replacement of NMP by phen (Miller and Epstein, 1978). Consequently, the number of conduction electrons per unit cell to be shared among the donors and TCNQ is equal to the fraction x of NMP in the alloy. Figure 9 shows that the scattering of $NMP_xPhen_{1-x}TCNQ$, for low phen concentration, presents two interesting features (Pouget, 1981; Pouget et al., 1980b and 1983; A. J. Epstein et al., 1981):

a. The q_1 scattering is no longer observed.
b. The q_2 scattering occurs at a smaller wave vector.

The suppression of the q_1 CDW instability has been interpreted as being a consequence of the NMP/Phen disorder created on donor stacks upon alloying. This assigns the q_1 scattering to the NMP stack and, consequently, the q_2 scattering to the TCNQ stack. With one electron to share between

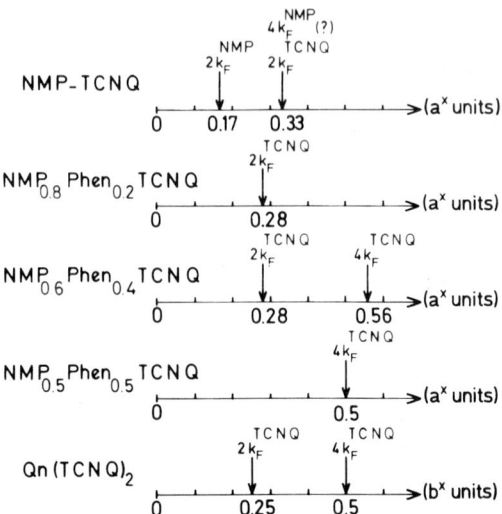

FIG. 9. Wavevectors in the chain direction of the quasi-one-dimensional x-ray scattering observed from NMP-TCNQ, NMP$_x$Phen$_{1-x}$TCNQ, and Qn(TCNQ)$_2$. This scattering has been assigned with respect to the $2k_F$ and $4k_F$ instabilities and the NMP and TCNQ stacks. Note that in NMP-TCNQ: $2k_F^{TCNQ} \equiv 4k_F^{TCNQ}$, because of the special value, $\frac{1}{3}a^*$, of the $2k_F$ wavevector.

the NMP and TCNQ molecules, the value of q_2 gives a charge transfer $\rho^A \simeq 0.67 \pm 0.02$ in pristine NMP-TCNQ. Thus, with $\rho^D \simeq 0.33$, the q_1 scattering corresponds at a $2k_F$ instability on the NMP stack. However, as $q_2 \simeq 2q_1$, one cannot exclude that a $4k_F$ instability on the NMP stack may also contribute to the q_2 scattering. In addition because of the commensurate value $\frac{1}{3}$, the q_2 scattering of pristine NMP-TCNQ may be due either to a $2k_F$ or to a $4k_F$ instability of the TCNQ stack, because $\frac{1}{3}a^*$ is equivalent to $\frac{2}{3}a^*$, within one reciprocal wavevector a^*. The decrease of q_2 when x decreases (i.e., when the amount of available electrons decreases) means that the q_2 scattering corresponds at a $2k_F$ instability of the TCNQ chain (for a $4k_F$ instability occurring at $1 - q_2$, an increase of q_2 is expected for x decreasing). A similar value of the charge transfer $\rho = 0.63 \pm 0.05$ has been obtained by Raman spectroscopy of NMP-TCNQ (Kuzmany and Elbert, 1980).

The charge transfer and the assignment of the instabilities of NMP-TCNQ being clarified, the measurement of the value of q_2, in reciprocal wavevector units, directly yields the charge transferred to the TCNQ chains ($\rho = 2 \times q_2$) in the NMP$_x$Phen$_{1-x}$TCNQ series. The observation, for $x = 0.6$, of an additional $4k_F$ scattering at twice the q_2 wavevector (Fig. 9), confirms its $2k_F$ assignment. For $x = 0.50$, only the $4k_F$ scattering is observed (Fig. 9). Figure 10 gives, as a function of x, the average number of electrons per

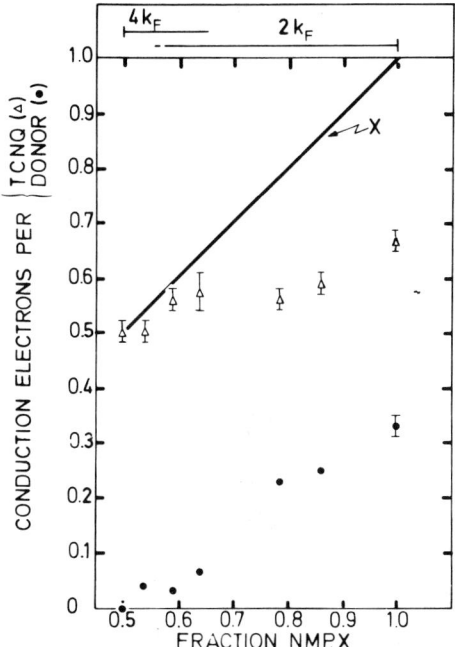

FIG. 10. Number of conduction electrons per TCNQ (△) and per donor (●) vs. NMP concentration x. The solid line shows the total conduction electron concentration available. The experimental determinations of the number of conduction electrons per TCNQ (△) deduced from the value of the wavevector in chain direction of the $2k_F$ and/or $4k_F$ scattering is given with their error bars; the number of conduction electrons per donor (●), which is obtained by subtraction from the total available electrons x, is given without error bars. In the upper part of the figure the concentration ranges where the $2k_F$ and/or $4k_F$ instabilities of the TCNQ stack are observed, are indicated. [From Epstein et al. (1981) and Pouget et al. (1982).]

TCNQ molecule. The difference between this value and the total conduction electron concentration available, x, yields the averge number of electrons per donor molecule. Figure 10 shows that the number of electrons per TCNQ decreases only slightly for decreasing x, reaching 0.56 ± 0.02 for $x \sim 0.6$, a value also in relatively good agreement with the Raman scattering determination (0.48 ± 0.05 by Kuzmany and Elbert, 1980). The number of conduction electrons on the donor stack decreases rapidly around $x_c \sim \frac{2}{3}$, leading to a nearly complete charge transfer for x smaller than x_c. This has serious consequences for several electronic properties like the thermoelectric power and the plasma frequency, which show a sudden change when x crosses the value x_c (Epstein and Miller, 1980). For $x = 0.5$, a commensurate and complete charge transfer is observed. By the number of electron per TCNQ molecule (and also by its electronic transport and magnetic behavior)

NMP$_{0.5}$Phen$_{0.5}$TCNQ resembles Qn(TCNQ)$_2$. The x-ray observation of $2k_F$ and $4k_F$ scatterings, at the wavevectors $\frac{1}{4}b^*$ and $\frac{1}{2}b^*$, respectively, in Qn(TCNQ)$_2$ (see Fig. 9) confirms the complete charge transfer of one electron from the Qn to the TCNQ stacks (Pouget, 1981). This complete charge transfer has been previously inferred from electron-spin-resonance (Clarke et al., 1979) and Raman scattering (Bozio and Pecile, 1980) studies of Qn(TCNQ)$_2$. With an average charge of $\frac{1}{2}$ electron per TCNQ molecule, NMP$_{0.5}$Phen$_{0.5}$TCNQ and Qn(TCNQ)$_2$ belong to the large class of single-chain salts with a quarter-filled TCNQ-band. Among them, MEM(TCNQ)$_2$ has been extensively studied. It shows also $2k_F$ and $4k_F$ structural instabilities (Huizinga et al., 1979; Van Bodegom et al., 1981; Viser et al., 1983).

IV. $2k_F$ and $4k_F$ Structural Fluctuations

8. Introduction

In a mean-field description, the intrachain CDW instability leads to a periodic lattice distortion at finite temperature, T_P^{MF} (Rice and Strassler, 1973). However, it is well known (see, for example, Landau and Lifshitz, 1980) that a 1D system cannot undergo a phase transition at a finite temperature. The mean-field transition is suppressed when the fluctuations of the 1D system are properly taken into account. Therefore, the CDW susceptibility χ and the intrachain correlation length, $\xi^{\parallel}$, diverge at 0 K. In 1D metals, CDW fluctuations are due to the electron-electron interactions and/or to the electron-phonon coupling. In the case of the electron-phonon coupling, low-temperature thermal fluctuations of the phase of the incommensurate CDW (for a review see, for example, Dieterich, 1976) lead to the divergences

$$\chi_P \sim T^{-2}$$
$$\xi_P^{\parallel} \sim T^{-1}. \tag{5}'$$

In the case of electron-electron interactions, the divergence of the CDW electronic response function has been expressed in the form of power laws (see, for example, Solyom, 1979; Emery, 1979)

$$\chi_e \sim T^{-\gamma}$$
$$\xi_e^{\parallel} \sim T^{-\nu}, \tag{5}''$$

where the exponents γ and ν depend implicitly on the electron-electron coupling constants.

However, in real materials, the chains are weakly coupled and, because of this coupling, a 3D phase transition occurs at a finite temperature, T_P, much lower than T_P^{MF}. But the phase transition is still anticipated over a large temperature range by 1D structural fluctuations related to the $2k_F$ and/or $4k_F$ intrachain CDW instabilities. The temperature range where the $2k_F$ and $4k_F$ precursor structural fluctuations are observed is given in Table IV for several members of the TTF-TCNQ family that undergo a Peierls transition. In other salts, like NMP$_x$Phen$_{1-x}$TCNQ, and Qn(TCNQ)$_2$, probably because of disorder on donor stacks, a structural phase transition does not occur, but quasi-one-dimensional $2k_F$ and $4k_F$ structural modulations are still observed on a large temperature range.

The format of this part is as follows. In Section 9 we define the dimensionality of structural fluctuations. Disorder effects are analyzed in Section 10. Spatial and dynamical correlations in the regime of 1D fluctuations are the object of Section 11. The lattice modes involved in these fluctuations are considered in Section 12. Finally, the connection between experimental findings and relevant theoretical interpretations of the $2k_F$ and $4k_F$ scatterings is done in Section 13.

9. Dimensionality of the Structural Fluctuations

The spatial anisotropy of the CDW fluctuations observed above T_p can be best measured by x-ray diffuse scattering experiments. It is given by the measurement of the anisotropy of the correlation lengths—defined in (A.6) in the Appendix—obtained by the q dependence of the CDW susceptibility χ_ρ, directly measured by x-ray diffuse scattering (see Section 20 in the Appendix). Generally, 1D conductors show a large regime of 1D (or quasi-one-dimensional—see below) fluctuations at high temperature. Then, below a cross-over temperature, T_{co}, interchain coupling becomes relevant and a regime of 3D (or 2D) anisotropic fluctuations take place over a temperature range, the extent of which depends on the compounds.

Above T_{co}, in the 1D regime, χ_ρ and ξ_ρ behave as quantities that tend to diverge towards 0 K—see Eq. (5)—while below T_{co}, in the 3D regime, χ_ρ and ξ_ρ increase more rapidly in order to really diverge at T_P, according to Eq. (A.8) in the Appendix. Experimental measurements of these two quantities are shown in Fig. 11 for the $2k_F$ scattering of TMTSF-DMTCNQ (a), TMTSF-TCNQ (b), and HMTSF-TNAP (c). $\chi_{2k_F}^{-1}$ and $\Delta_{/\!/}$ clearly show a break in their temperature dependence at T_{co}. Furthermore, Figs. 11a and 11c show, for TMTSF-DMTCNQ and HMTSF-TNAP, respectively, that at about the same temperature the interchain correlation lengths are such that $\xi_\perp \simeq d_\perp$ ($d_\perp$ is the distance between first neighbor stacks in the direction along which $\xi_\perp$ is measured). This last criterion is used (see, for example, Dieterich, 1976) to define the crossover temperature T_{co}.

In TMTSF-DMTCNQ, TMTSF-TCNQ, and HMTSF-TNAP, T_{co} occurs at about $1.5 T_P$. Above T_{co}, some interchain correlations still persist until

TABLE IV

Temperature Range of $2k_F$ and $4k_F$ Fluctuations, Peierls Critical Temperature and Wavevector of Modulation of the Charge Transfer Salts of the TTF–TCNQ Family Undergoing Well-Defined Structural Phase Transitions. The Wavevector of Modulation is Expressed in the Reciprocal Space Frame Defined by the Lattice Parameters of Table I. References to Structural Work are Those of Table III. The Data Reported Here are Those Obtained at Ambient Pressure. (The Asterisk, *, Means that 3D Long-range Order is Not Completely Achieved Below T_P)

	Temperature range of structural fluctuations		T_P (K)	Modulation Wave vectors	Remarks
	$2k_F$	$4k_F$			
TTF–TCNQ	$T \lesssim 150$ K	$T <$ above R.T.	$\begin{cases} 54(T_H)^* \\ 49(T_M)^* \\ 38(T_L) \end{cases}$	$(\frac{1}{2}, 0.295, 0)$ $[q_a(T), 0.295, 0] + [2q_a(T), 0.41, 0]$ $(\frac{1}{4}, 0.295, 0) + (\frac{1}{2}, 0.41, 0)$	T_P measured in fully deuterated samples
TSF–TCNQ	$T \lesssim 250$ K	—	29	$(\frac{1}{2}, 0.315, 0)$	+ additional (0, 0.41, 0) reflections
DSDTF–TCNQ	$T \lesssim 150$ K	—	~45*	$(\frac{1}{4}, 0.31, 0)$	+ (0, 0.315, 0) 3D short-range order below 50 K
HMTTF–TCNQ	$T \lesssim 300$ K	$T \lesssim 100$ K	$\begin{cases} 49(T_H)^* \\ 42(T_L) \end{cases}$	$(0.42, 0.36, 0)$ $(0.42, 0.36, 0) + (0.16, 0.28, 0)$	
HMTSF–TCNQ	$T <$ above R.T.		~24	$(0, 0, 0.37)$	
TMTTF–TCNQ	$T \lesssim 180$ K	60 K $\lesssim T < 180$ K	~40	$(0, 0.285, 0)$	+ additional (0, 0.43, 0) reflections
TMTSF–TCNQ	$T <$ above R.T.	—	57	$(0.285, \frac{1}{2}, ?)$	
HMTSF–TNAP	$T \lesssim 300$ K	—	41	$(?, ?, 0.29)$	
TMTTF–DMTCNQ	$T \lesssim 100$ K	$T <$ above R.T.	~40	$(\frac{1}{2}, ?, ?) + (\frac{1}{4}, ?, ?)$	It has not been determined if the critical temperature is the same for the $2k_F$ and $4k_F$ satellite reflections
TMTSF–DMTCNQ	$T \lesssim 225$ K	60 K $\lesssim T <$ above R.T.	42	$[\frac{1}{4}, (\pm)^? \frac{1}{3}, 0]$	Only $2k_F$ reflections are present below 42 K
DBTTF–TCNQCl$_2$	No precursors	$T <$ above R.T.	180 38	$(0, 0.52, ?, ?)$ $(0.52, ?, ?) + (0.26, ?, ?)$	

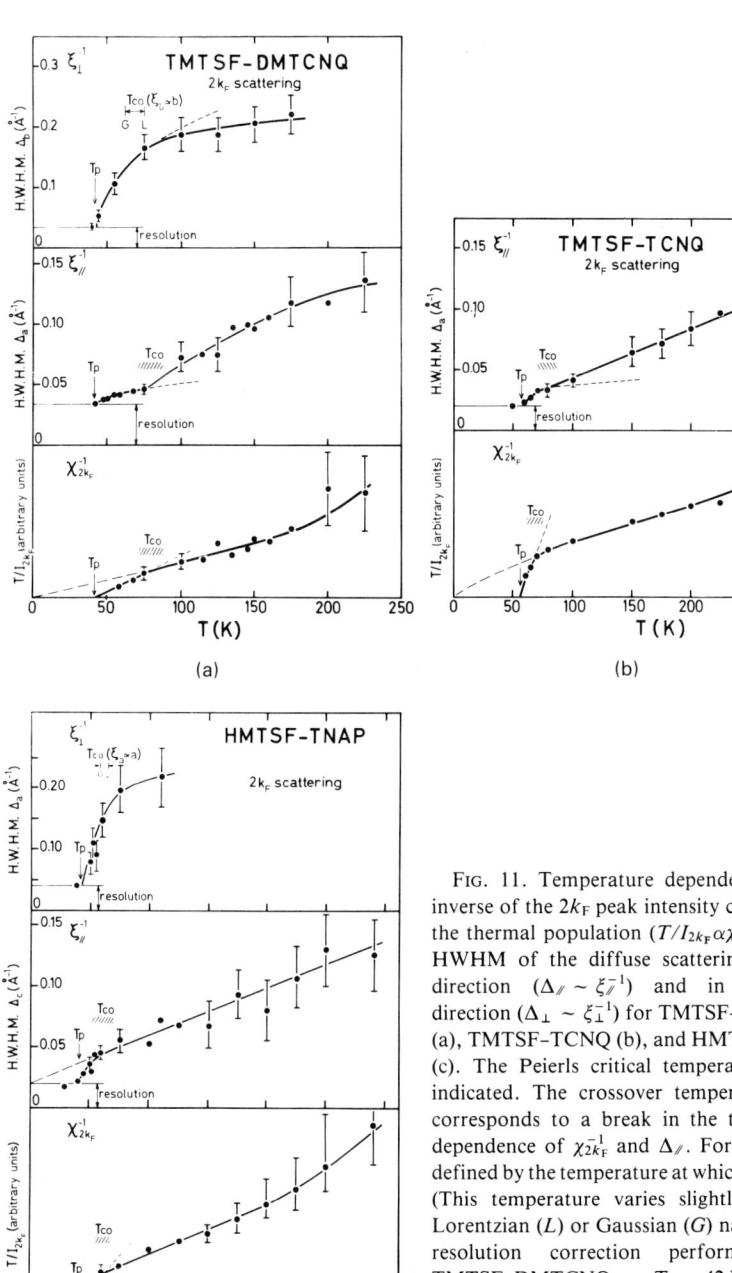

FIG. 11. Temperature dependence of the inverse of the $2k_F$ peak intensity corrected by the thermal population ($T/I_{2k_F} \alpha \chi_{2k_F}^{-1}$), of the HWHM of the diffuse scattering in chain direction ($\Delta_{\parallel} \sim \xi_{\parallel}^{-1}$) and in transverse direction ($\Delta_{\perp} \sim \xi_{\perp}^{-1}$) for TMTSF-DMTCNQ (a), TMTSF-TCNQ (b), and HMTSF-TNAP (c). The Peierls critical temperature, T_P is indicated. The crossover temperature, T_{co}, corresponds to a break in the temperature dependence of $\chi_{2k_F}^{-1}$ and $\Delta_{\parallel}$. For $\Delta_{\perp}$, T_{co} is defined by the temperature at which $\xi_{\perp} \simeq d_{\perp}$. (This temperature varies slightly with the Lorentzian (L) or Gaussian (G) nature of the resolution correction performed.) For TMTSF-DMTCNQ: $T_P = 42$ K, $T_{co} \sim 65$–85 K; for TMTSF-TCNQ: $T_P = 57$ K, $T_{co} \sim 70$–80 K; and for HMTSF-TNAP: $T_P = 41$ K, $T_{co} \sim 60$ K.

about 200 K for TMTSF-DMTCNQ (Fig. 12), 225 K for TMTSF-TCNQ, and 100 K for HMTSF-TNAP. However, having $\xi_\perp < d_\perp$ in the two transverse directions, we shall thus speak of a regime of quasi-one-dimensional fluctuations in that temperature range. A regime of quasi-one-dimensional fluctuations at $2k_F$ can coexist with a regime of true 1D fluctuations at $4k_F$, as observed in TMTSF-DMTCNQ (Pouget, 1981).

T_{co} is observed at about 60 K in the $2k_F$ scattering of TTF-TCNQ (Khanna et al., 1977), only a few degrees above the highest phase transition, which occurs at $T_H = 54$ K in deuterated samples (Comès et al., 1975 and 1976). In contrast to the compounds considered above, TTF-TCNQ has a very narrow temperature range ($\sim 10\% T_P$) of 3D fluctuations at $2k_F$. Three-dimensional fluctuation effects have been observed in the specific heat (Craven et al., 1974; Djurek et al., 1977), the electrical resistivity (Horn and Rimai, 1976; Horn and Guidotti, 1977), and the magnetic susceptibility (Horn et al., 1977) of TTF-TCNQ. The conductivity of TTF-TCNQ in its metallic phase has been interpreted as being due to the sum of a fluctuative CDW Fröhlich contribution and of a single-particle contribution, the first term representing about 80% of the longitudinal conductivity at 60 K

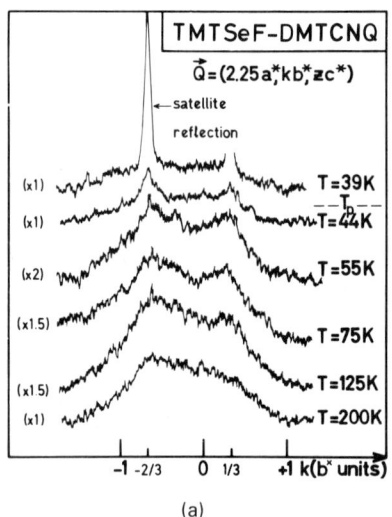
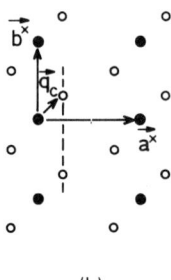

FIG. 12. (a) Microdensitometer readings of x-ray patterns from TMTSF-DMTCNQ in the $2 + 2k_F$ diffuse sheet along the direction shown in b. They reveal an important regime of interchain short-range order above $T_P = 42$ K. The smooth variation of the x-ray diffuse intensity at 200 K is mostly due to the change with k of the TMTSF molecular structure factor. The scan performed is shown in (b) by the interrupted line (● main Bragg reflections, ○ satellite reflections). In (a), as in (b), the triclinic deformation of the lattice has not been considered, thus the wavevector scanned can be either $\mathbf{k}$ or $-\mathbf{k}$.

3. STRUCTURAL INSTABILITIES

(Jérome and Schulz, 1982b). In this picture, the loss of conductivity below T_{co} could be caused by a drastic reduction of the collective contribution, due to pinning of the CDW on the impurities and the defects of the lattice. This pinning mechanism becomes very efficient below T_{co} due to the rapid development of the 3D coupling between CDWs. Within this temperature range, the $4k_F$ fluctuations keep their 1D character.

A somewhat different behavior is shown by HMTTF-TCNQ, where interchain correlations begin to grow below the temperature $T_H = 49 \pm 1$ K, (see Fig. 41) at which specific heat (Biljakovic-Franulovic et al., 1979), resistivity (Greene et al., 1976; Delhaes et al., 1977), and g-factor (Tomkiewicz et al., 1977b) anomalies suggest that a phase transition occurs. From the structural point of view, T_H more likely corresponds to the onset of a 3D ordering between chains. The interchain correlation length increases progressively upon cooling and, at the scale of the experimental resolution of Fig. 41, the long-range periodic lattice distortion is established only below $T_L = 43 \pm 1$ K (Megtert et al., 1978a, 1978b; Pouget et al., 1979a), a temperature at which additional anomalies in electronic properties are observed.

Among all the 1D organic conductors investigated, TSF-TCNQ shows the most intricate regime of $2k_F$ fluctuations. Starting from a high-temperature regime of quasi-one-dimensional fluctuations, the $2k_F$ CDW already manifests near 100 K a 2D coupling within the (b, c) plane of donors (Megtert et al., 1979b; Kagoshima et al., 1978). Around 50–70 K, the HWHM of the corresponding diffuse rods along c gives a correlation length on the order of the distance between first-neighbor stacks (7.5 Å from Yamaji et al., 1981; 12 Å from Kagoshima et al., 1978a). Then at 50 K short-range order in the third direction, a, begins to develop (Kagoshima et al., 1978; see also Fig. 13). At 45 K, the correlation length along the a axis reaches 8 Å, slightly more than the distance between TSF and TCNQ first neighbor stacks. However, as shown in Fig. 13, this short-range order, with wavevector $(0, 2k_F, 0)$, does not become critical when the temperature decreases. Below about 40 K, another short-range order, characterized by a different wavevector component in the a^x direction, $q_a = a^x/2$, grows critically and drives the 29 K Peierls transition of TSF-TCNQ (Weyl et al., 1976; Kagoshima et al., 1978a). Below T_P, the $(0, 2k_F, 0)$ short-range order still coexists with the $(\frac{1}{2}, 2k_F, 0)$ long-range order (Fig. 13). This result differs from that of Kagoshima et al. (1978a), who reports a continuous shift of q_a from 0 to $a^x/2$ above T_P. A change in the rate of decrease of the magnetic susceptibility at about 45 K (Scott et al., 1978) and a decrease of the electrical conductivity below about 40 K (Etemad, 1976), also characterize the onset of critical 3D fluctuations. Below this crossover temperature a faster rate of increase of the in-chain correlation length of the $2k_F$ scattering of TSF-TCNQ can be observed (see Fig. 21).

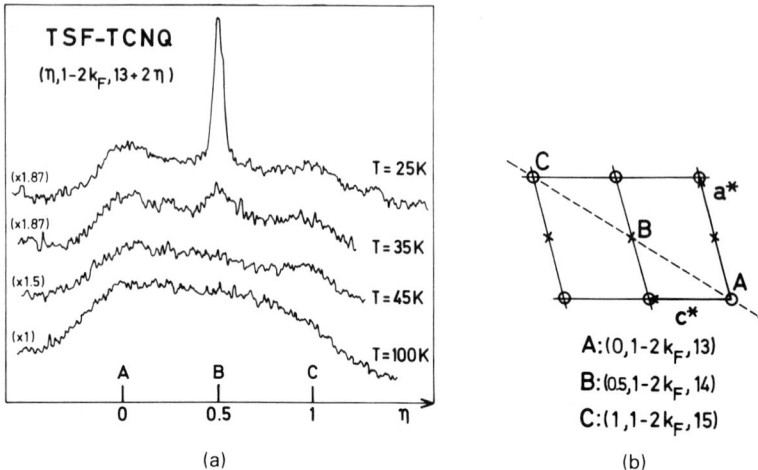

FIG. 13. (a) Microdensitometer readings of x-ray patterns from TSF–TCNQ in the $1-2k_F$ diffuse sheet, along the direction shown by the interrupted line in (b). They reveal the competition between the $(0, 2k_F, 0)$ (points A and C) and the $(\frac{1}{2}, 2k_F, 0)$ (point B) CDW-ordering wavevectors. At 100 K, the smooth variation of the diffuse intensity with η is due to the TSF molecular structure factor.

Finally, we want to stress the puzzling behavior of HMTSF–TCNQ, where x-ray diffuse scattering data show a well-defined regime of 1D fluctuations at $2k_F$ down to about 25 K (Yamaji et al., 1983), while the low-temperature electronic properties (diamagnetism, Hall constant, magnetoresistance) suggest the presence of 3D (or 2D) portions of Fermi surface below 100 K (see, for example, Jérome and Schulz, 1982a). A change in the dimensionality of the Fermi surface provides a mechanism of interchain coupling (Horovitz et al., 1975), which must be accompanied by a crossover in the dimensionality of structural fluctuations. Since such a structural crossover is not observed around 100 K, a possible explanation might be that HMTSF–TCNQ has, at low temperature, a Fermi surface composed of several contours: small pockets of small area, determining the electronic properties, and planar portions, containing most of the electrons driving the 1D CDW instability.

10. Disorder

In the presence of disorder, the Peierls transition to a 3D long-range ordered CDW state is suppressed, as shown by the structural studies of:

- substitutional alloys: $TTF_{1-x}TSF_xTCNQ$ (Kagoshima et al., 1980) and $NMP_xPhen_{1-x}TCNQ$ (Epstein et al., 1981; Pouget et al., 1982)

- salts containing orientationally disordered molecules: NMP-TCNQ (Pouget et al., 1980a) and Qn(TCNQ)$_2$ (Pouget, 1981)
- salt disordered in a controlled way by irradiation: TMTSF-DMTCNQ and TTF-TCNQ (Forro et al., 1983 and 1984).

One of the major issues of these studies was to show, as expected by theory (see, for example, the reviews of Sham, 1979; Jérome and Schulz, 1982a), that disorder

1. enhances the intensity of the $2k_F$ or $4k_F$ diffuse scattering, especially at low temperature. This reflects a local increase of the amplitude of the order parameter due to the formation of Friedel oscillations in the 1D electron gas by the disorder potential. This effect is all the more important when the CDW response function is strong [see expression (A.25) in the Appendix].
2. limits the intrachain and interchain correlation lengths to finite values. Such a decorrelation effect is due to large-amplitude fluctuations resulting from the superimposition of Friedel oscillations and, at low temperature, to pinning effects of the phase of the CDW on disorder centers randomly distributed. This prevents the establishment of long-range order.

These two aspects are clearly illustrated by Fig. 14, which compares the temperature dependence of the peak intensity and the HWHM of the $4k_F$ scattering in the chain direction for TTF-TCNQ and the substitutionally disordered alloy TTF$_{0.97}$TSF$_{0.03}$TCNQ.

Disorder produced by irradiation damage, which affects both the donor and acceptor stacks, perturbs the $2k_F$ and $4k_F$ scattering differently, because the divergence of their respective CDW response functions differs. In particular, in salts exhibiting the $2k_F$ and $4k_F$ scattering, it has been shown that a few mol% defects suppress the $4k_F$ scattering of TMTSF-DMTCNQ (Forro et al., 1983) and the $2k_F$ scattering of TTF-TCNQ (Forro et al., 1984). Also, defects induce mostly $2k_F$ Friedel oscillations in TMTSF-DMTCNQ and $4k_F$ Friedel oscillations in TTF-TCNQ, in agreement with a strongest $2k_F$ CDW response function in TMTSF-DMTCNQ (the only one that diverges at low temperature) and a strongest $4k_F$ CDW response function in TTF-TCNQ (the dominant one at high temperature).

Another observation is that for strong enough disorder, the anisotropy in the correlation lengths at low temperature bears some similarity to that observed at high temperature in the nondisordered compound. This result can be easily understood since disorder suppresses, as do the thermal fluctuations, the weakest CDW correlations, which occur in the directions of smallest coupling. The spatial correlations that remain are those generally established by the electronic response functions. In this respect, the observation of 1D or quasi-one-dimensional $4k_F$ scattering in TMTTF-Bromanil,

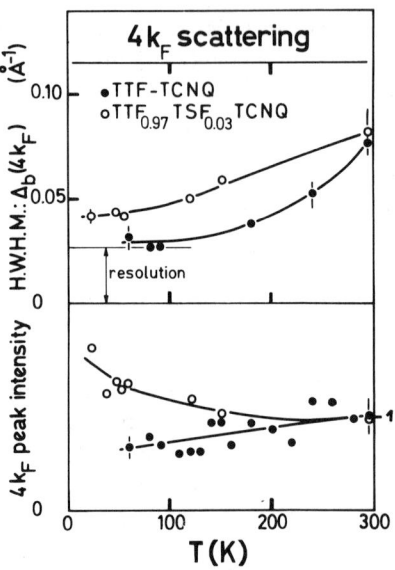

FIG. 14. Temperature dependence of the peak intensity and HWHM in the chain direction for the $4k_F$ scattering of TTF-TCNQ and TTF$_{0.97}$TSF$_{0.03}$TCNQ. The $4k_F$ peak intensities have been normalized at room temperature.

NMP$_x$Phen$_{1-x}$TCNQ ($0.5 < x < \frac{2}{3}$) and Qn(TCNQ)$_2$, shows that the 1D electronic gas is responsible for the $4k_F$ local modulation. However, the observation of nearly isotropic $2k_F$ scattering in the first salt and quasi-one-dimensional $2k_F$ scattering in the last two salts show that forces of different anisotropy seem to be at the origin of the $2k_F$ instability.

These three salts are also interesting because of the different temperature dependence of their $2k_F$ and $4k_F$ scattering. The $4k_F$ correlation length in chain direction remains constant, and the intensity of this scattering increases only slightly upon cooling [this last effect could be due to a slight increase of the CDW response function (or distortion) or to a decrease of the Debye Waller factor in the diffuse structure factor]. A typical behavior is illustrated by Fig. 15 for NMP$_{0.59}$Phen$_{0.41}$TCNQ, which shows that the $4k_F$ mean square amplitude of fluctuations, relatively well established at room temperature, remains nearly constant in temperature, probably because of the disorder. This behavior contrasts with that of the $2k_F$ scattering which appears below room temperature, grows in intensity when the temperature decreases (see Fig. 16 for NMP$_{0.59}$Phen$_{0.41}$TCNQ). This shows that, in spite of the disorder, these salts can develop a low-temperature $2k_F$ instability. This relatively well-decoupled behavior of the $2k_F$ and $4k_F$ scattering is reminiscent of that of MEM(TCNQ)$_2$. In this last salt, x-ray and neutron

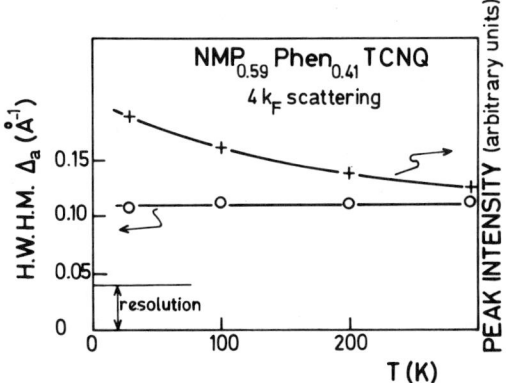

FIG. 15. Peak intensity and HWHM in chain direction of the $4k_F$ scattering of $NMP_{0.59}$-$Phen_{0.41}TCNQ$.

diffraction studies have shown that the TCNQ stack undergoes a high-temperature (335 K) $4k_F$ instability (Van Bodegom, 1981), then a low-temperature (17.5 K) $2k_F$ instability (Visser et al., 1983), which have been associated, respectively, with instabilities in the charge and spin degrees of freedom. The spin Peierls instability has been found to be enhanced by disorder effects (Hirsch and Kariotis, 1985). In these salts disorder certainly helps repulsive Coulomb interactions in the $4k_F$ charge localization process.

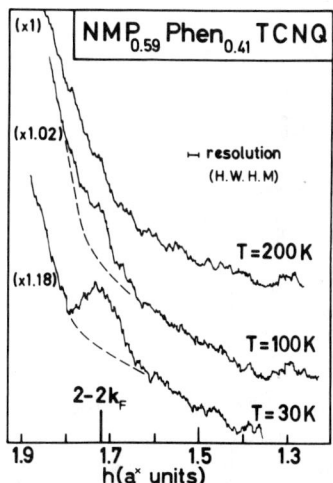

FIG. 16. Microdensitometer readings from x-ray patterns of $NMP_{0.59}Phen_{0.41}TCNQ$ showing the temperature-dependent $2k_F$ scattering. [From Pouget et al. (1982a).]

11. THE ONE-DIMENSIONAL FLUCTUATION REGIME IN NONDISORDERED SALTS

a. Spatial Correlations

These correlations have been studied by x-ray diffuse scattering experiments using either the counter method at the University of Tokyo or the photographic method at the University of Paris-Sud. The 1D spatial fluctuations appear in the form of diffuse sheets in the reciprocal space. The intersection of these diffuse sheets with the Ewald sphere gives rise to diffuse lines on x-ray patterns. Figures 17a and 18a show such lines for HMTSF–TCNQ and TTF–TCNQ, respectively. In HMTSF–TCNQ, only one set of lines occurs at the reduced wavevectors $\pm 0.37c^*$ from the layers of main Bragg reflections, perpendicular to the chain direction. It corresponds at a $2k_F$ scattering. In TTF–TCNQ, two sets of lines occur at the reduced wave vectors $\pm 0.295b^*$ and $\pm 0.41b^*$ (or $\pm 1 \mp 0.59b^*$) from the layers of main Bragg reflections. They correspond, respectively, to $2k_F$ and $4k_F$ scatterings. The x-ray intensity is not continuous along a diffuse line. The intensity distribution, which will be explicitly considered later in Section 12, is due to the $\mathbf{Q}$ dependence of the diffuse structure factor $F_d(\mathbf{Q})$—see (A.2) in the Appendix. The $2k_F$ and $4k_F$ diffuse scattering of HMTSF–TCNQ and TTF–TCNQ are more clearly

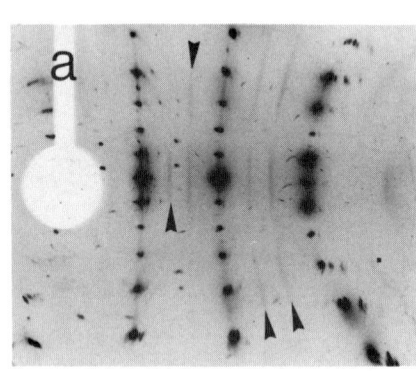
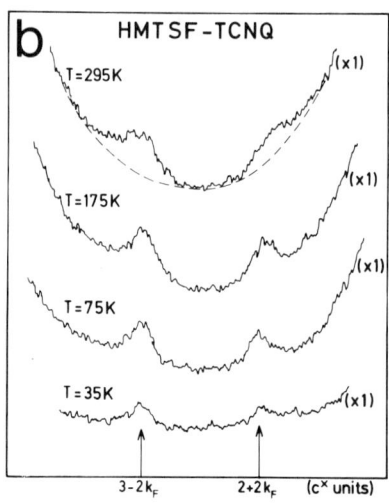

FIG. 17. (a) X-ray pattern from HMTSF–TCNQ at 40 K showing the $2k_F$ diffuse lines (black arrows). The chain axis c is horizontal and the b direction is vertical. Note the presence of well-defined diffuse lines at a temperature lower than 100 K (temperature at which a 1D to a 2D or 3D crossover of the Fermi surface seems to occur, according to electronic properties). (b) Microdensitometer scans along c^* of x-ray patterns similar to (a), showing the temperature dependence of the $2k_F$ diffuse lines.

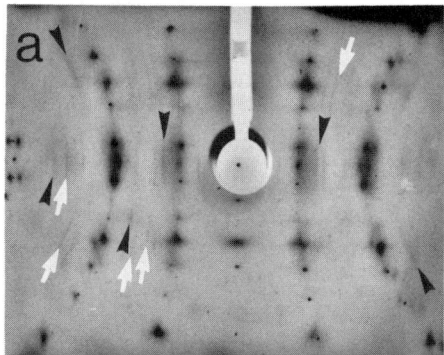
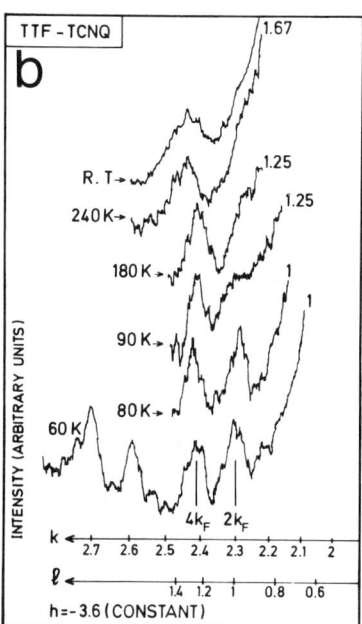
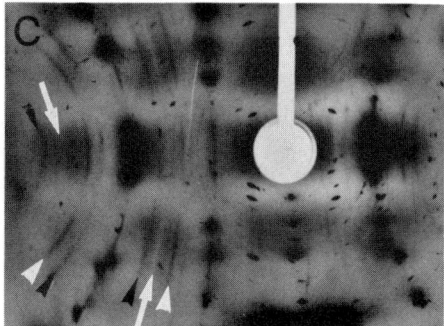

FIG. 18. (a) X-ray pattern from TTF–TCNQ at 80 K showing the $2k_F$ (black arrows) and $4k_F$ (white arrows) diffuse lines. The chain axis b is horizontal and the a direction is vertical. This x-ray pattern has been taken with the synchrotron radiation of LURE (Orsay). [From S. Megtert (1984).] (b) Microdensitometer scans of x-ray patterns from TTF–TCNQ showing the temperature dependence of the $2k_F$ and $4k_F$ scatterings. [From Khanna et al. (1977).] Note the slight shift of the wavevector of the $4k_F$ scattering as a function of the temperature. The right side of the scans gives the scale factor by which the intensity has to be multiplied in order to scale with the lower temperature readings. (c) X-ray pattern from TTF$_{0.97}$TSF$_{0.03}$TCNQ at 50 K showing the $2k_F$ (black arrows) and $4k_F$ (white arrows) diffuse lines. The white long arrows point toward "negative" $4k_F$ diffuse lines, resulting from destructive interferences between the $4k_F$ scattering and the Laue scattering. The chain axis b is horizontal, and the a direction is vertical.

revealed by the microdensitometer readings shown in Fig. 17b and Fig. 18b, respectively. Performed at a constant **Q** wave vector, such scans allow a study of the temperature dependence of the $2k_F$ or $4k_F$ CDW fluctuations from the measurement of the x-ray diffuse scattering above the background (see Section 20 in Appendix).

This extra scattering is a small quantity, ~15% of the background for HMTSF–TCNQ at 300 K (Moret and Pouget, 1986) and ~6% of the background for TTF–TCNQ at 300 K (Kagoshima et al., 1976), which represents

10^{-5} to 10^{-6} times the intensity of a typical Bragg reflection. The difficulty in measuring the extra scattering with good statistics using conventional (sealed-tube or rotating-anode) generators has permitted only semiquantitative analysis of the $2k_F$ and $4k_F$ scatterings to be performed until now.

The $2k_F$ and $4k_F$ scatterings have been characterized by the temperature dependence of

- the peak intensity above the background I
- the half width at half maximum (HWHM) of the scattering in chain direction $\Delta_{/\!/}$.

The peak intensity above the background (I_{q_c} here, δI in the Appendix), is proportional to the instantaneous displacement correlation function, $\delta S^D(q_c, t = 0)$ defined in (A.23). The factor of proportionality, $|F_d(\mathbf{Q})|^2$, is, in fact, slightly temperature dependent through the Debye Waller factor $W_j(\mathbf{Q})$ [see (A.2)]. In the classical limit (which will be justified in Subsection b), the intensity divided by the temperature gives a quantity proportional to the CDW susceptibility χ_{q_c} (χ_ρ in the Appendix). Figures 11, 19, and 20 show, respectively, the temperature dependence of χ for the $2k_F$ and $4k_F$ scatterings observed in the 1D fluctuation regime of various materials. $\chi^{1D}_{2k_F}$ always diverges towards low temperatures. This is not always the case for $\chi^{1D}_{4k_F}$. In spite of the uncertainties in the measurement of I/T, and the assumptions done to relate this quantity to χ_ρ, Figs. 19 and 20 show large differences in the temperature dependence of $\chi^{1D}_{2k_F}$ and $\chi^{1D}_{4k_F}$ among the organic salts. These differences will be analyzed in Section 13.

The HWHM, $\Delta_{/\!/}$, corrected by the experimental resolution (see Section 23 in the Appendix), gives the inverse CDW correlation length $\xi_{/\!/}^{-1}$. Figures 11 and 21 give the temperature dependence of $\Delta_{/\!/}$ for the $2k_F$ scattering of various materials, and Fig. 14 for the $4k_F$ scattering of TTF-TCNQ. They show that in the 1D fluctuation regime, $\xi_{/\!/}$ increases when the temperature decreases. A comparison of Figs. 19 and 21 indicates that the rate of increase of $\xi_{/\!/2k_F}$ is related to the divergence of $\chi^{1D}_{2k_F}$. The correlation lengths will be discussed more quantitatively in Section 13.

b. Temporal Fluctuations

Because of the lack of large enough crystals, the organic conductors have not been studied in detail by neutron inelastic scattering. Only the acoustic branches have been measured in TTF-TCNQ (Shirane *et al.*, 1976; Shapiro *et al.*, 1977; Pouget *et al.*, 1979a) for which fully deuterated single crystals of about 10 mm^3 were available. Figure 22 shows for deuterated TTF-TCNQ at room temperature, the dispersion curve of the acoustic modes propagating along the chain direction. The acoustic branches do not exhibit any pronounced anomaly. Only a shallow, but clearly existing, anomaly can

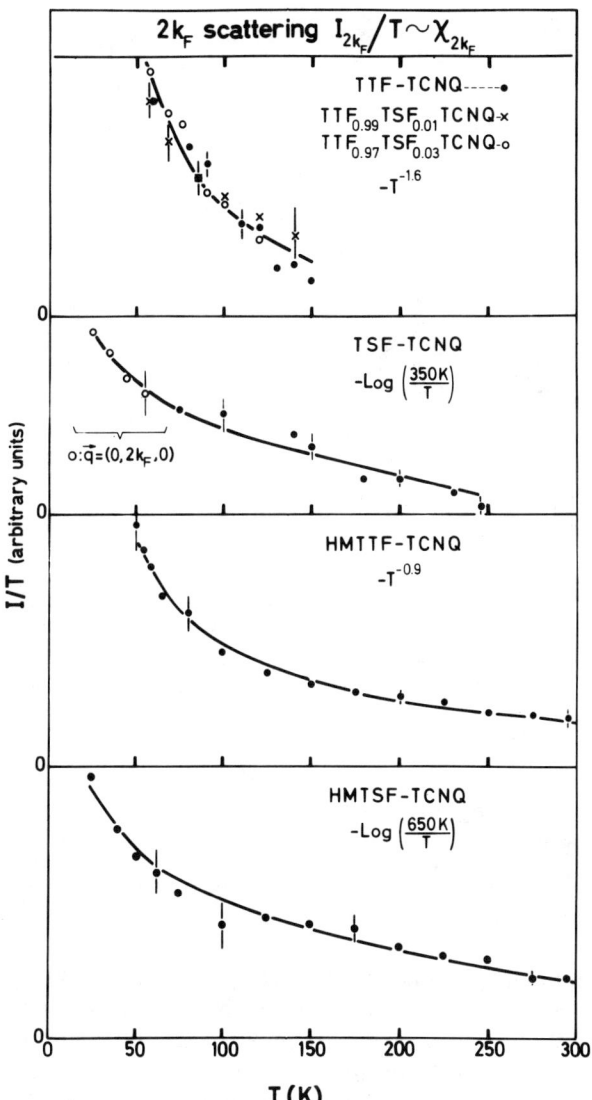

FIG. 19. Temperature dependence of the peak intensity of the $2k_F$ diffuse scattering corrected by the thermal population factor, I/T, in TTF$_{1-x}$TSF$_x$TCNQ ($x = 0, 0.01, 0.03$), TSF–TCNQ, HMTTF–TCNQ, and HMTSF–TCNQ in the regime of 1D fluctuations [except for TSF–TCNQ, where the data have been taken at $\mathbf{q} = (0, 2k_F, 0)$ below 50 K]. The continuous line gives the best fit of the data from a simple expression ($T^{-\alpha}$ or $\log A/T$), which diverges at low temperature. Note that within experimental error the same law accounts for the divergence of $\chi_{2k_F}^{1D}$ in TTF$_{1-x}$TSF$_x$TCNQ with $x = 0, 0.01$, and 0.03.

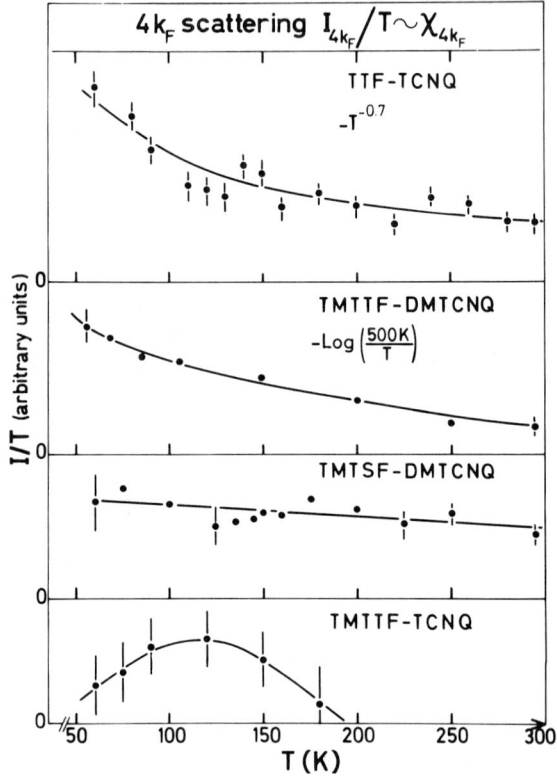

FIG. 20. Temperature dependence of the peak intensity of the $4k_F$ diffuse scattering corrected by the thermal population factor, I/T, in TTF-TCNQ, TMTTF-DMTCNQ, TMTSF-DMTCNQ, and TMTTF-TCNQ in the regime of 1D fluctuations. For the two former compounds, the continuous line gives the best fit of the data by a simple expression ($T^{-\alpha}$ or $\log A/T$), which diverges at low temperature.

be found on the longitudinal branch around $2k_F$. This result contrasts with an earlier report of the formation in this branch, for protonated samples, (Mook and Watson, 1976) of a giant Kohn anomaly extending down to 1.2 meV at room temperature. Repeated neutron inelastic scattering measurements, carried out jointly at Brookhaven and Oak Ridge National Laboratories, on both protonated and deuterated samples have unambiguously shown that no distinct $2k_F$ anomaly can be observed at room temperature in the longitudinal acoustic branch (Mook et al., 1977).

At lower temperature, the shallow longitudinal anomaly does not develop, but another anomaly, also at the wave vector $2k_F$, appears below 150 K in the tranverse acoustic branch mainly polarized along c^* (Shirane et al., 1976; see the insert in Fig. 22). The progressive development of this "transverse"

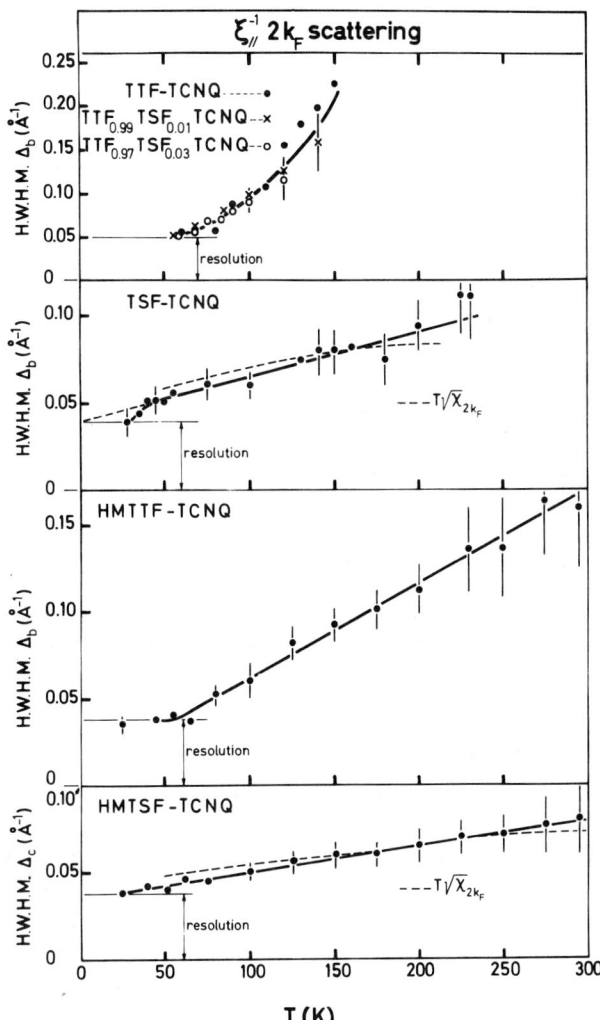

FIG. 21. Temperature dependence of the HWHM of the $2k_F$ diffuse scattering in chain direction in $TTF_{1-x}TSF_xTCNQ$ ($x = 0, 0.01, 0.03$), TSF-TCNQ, HMTTF-TCNQ, and HMTSF-TCNQ. Note that within experimental error Δ_b has the same temperature dependence in $TTF_{1-x}TSF_xTCNQ$ for $x = 0, 0.01$, and 0.03. For TSF-TCNQ the rapid change, around 45 K in the temperature dependence of Δ_b corresponds to a crossover toward a regime of 3D fluctuations. For TSF-TCNQ and HMTSF-TCNQ, the interrupted line gives a fit of the HWHM by law $\xi_\parallel^{-1} \propto T\sqrt{\chi_{2k_F}}$ added at the experimental resolution (Lorentzian resolution correction).

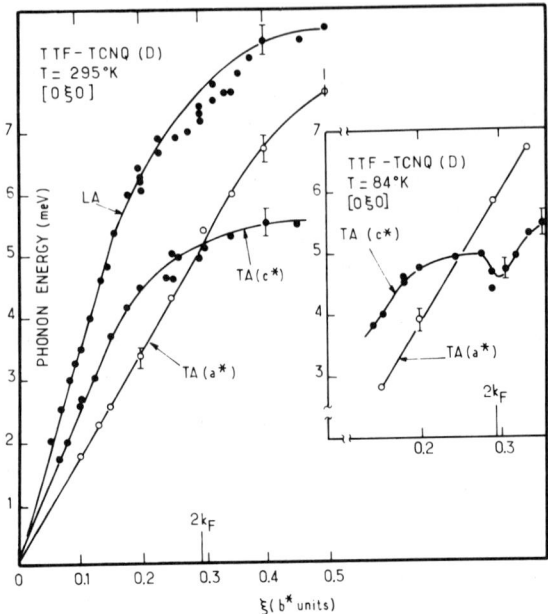

FIG. 22. Dispersion curves for acoustic modes of TTF–TCNQ propagating along the chain direction at 295 K. Note the shallow anomaly around $2k_F$ in the longitudinal branch. The insert shows the transverse branch around $2k_F$ at 84 K. Note that the Kohn anomaly occurs in the branch mainly polarized along c^* and not in the branch polarized along a^*, which has the same symmetry. [From Comès and Shirane (1979).]

anomaly is clearly visible on the intensity contours of Fig. 23. This result is in perfect agreement with x-ray diffuse scattering experiments showing the development of a $2k_F$ scattering below 150 K (Pouget et al., 1976; Kagoshima et al., 1976; Khanna et al., 1977; see also Fig. 19). However, the frequency of the Kohn anomaly decreases only slightly with the temperature, showing a relative dip of only

$$\frac{\omega_0 - \omega_a}{\omega_0} = 15\%$$

at 60 K (6 K above the Peierls transition). In this respect, TTF–TCNQ differs markedly from the inorganic conductors, showing a well-developed Kohn anomaly well above the Peierls transition [$K_2Pt(CN)_4Br_{0.3}xH_2O$, Comès and Shirane, 1979; $K_{0.3}MoO_3$, Sato et al., 1983 and Pouget et al., 1985a]. The neutron results seem to indicate that below 60 K the phase transition of TTF–TCNQ is more likely driven by the critical growth of a central peak in energy, than by the sudden decrease of the frequency ω_a of its Kohn anomaly. In this respect, an elastic neutron scattering study has shown

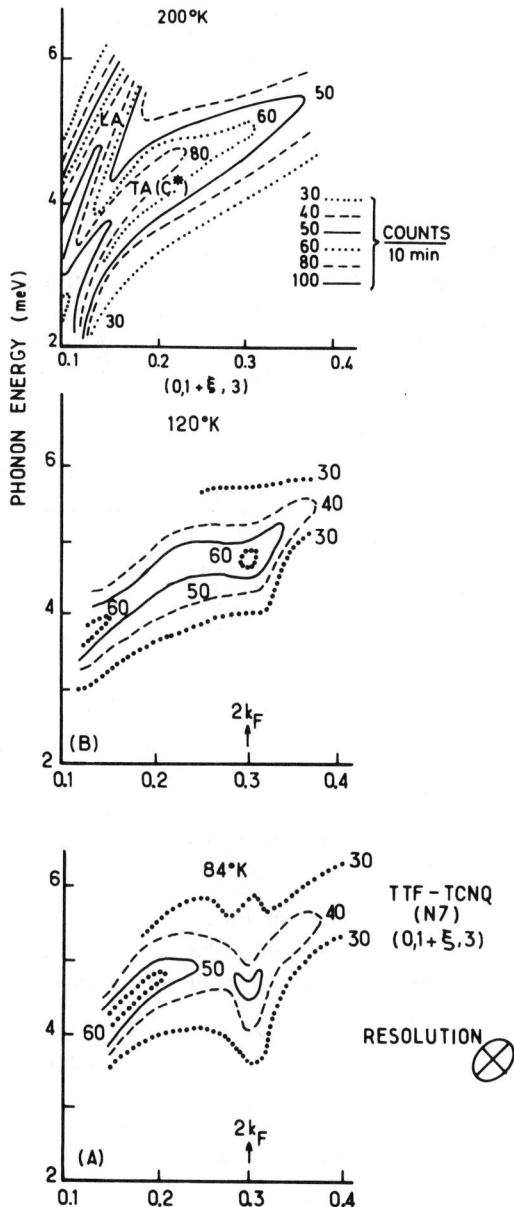

FIG. 23. Normalized ω, q intensity contours of the neutron scattering of TTF-TCNQ from the acoustic excitation of wavevector $(0, \xi, 0)$ mainly polarized in the c^* direction, at 200 K, 120 K, and 84 K. [From Comès and Shirane (1979).]

evidence for critical scattering a few degrees above $T_H = 54$ K, in the regime of 3D fluctuations (Comès et al., 1976). However, the level of detection of an eventual 1D elastic scattering above 60 K is presently out of reach with the size of available crystals. Thus, it is not clear if the 1D scattering observed with x-rays above 60 K is of purely inelastic nature or if a quasi-elastic component also contributes to the scattering. If there is only an inelastic contribution to the x-ray scattering, the neutron data of Shirane et al. (1976), showing a Kohn anomaly at $\hbar\omega_a = 4.4$ meV (i.e., 48 K) at 60 K, prove that the high-temperature approximation of the fluctuation dissipation theorem can be used for the analysis of the x-ray diffuse scattering data [i.e., with this value of ω_a, the replacement of (A.26) by (A.28) in the Appendix introduces an error of 5% at 60 K and this error decreases when the temperature increases]. This approximation is further justified if a quasi-elastic scattering also contribute to the fluctuations.

A neutron investigation has also been performed in order to search for the dynamical origin of the $4k_F$ anomaly of TTF–TCNQ (Pouget et al., 1979a). The $4k_F$ anomaly, shown to be longitudinally polarized from x-ray studies (Kagoshima et al., 1976; Khanna et al., 1977), does not occur in the longitudinal acoustic branch. This leaves the possibility of a $4k_F$ anomaly arising from an optical branch or being quasi-elastic.

12. Lattice Fluctuations

The $2k_F$ and $4k_F$ scatterings are due to a wave of displacement or of deformation of the molecules of the stack with respect to the long-range ordered high-temperature structure. Thus, the nature of the atomic displacements, which enter in the diffuse intensity through the diffuse structure factor $F_d(\mathbf{Q})$—see Eq. (A.2)—can be obtained by an analysis of the $\mathbf{Q}$ dependence of the diffuse intensity $I(\mathbf{Q})$. Such an analysis gives in principle:

1. the organic stack, which undergoes the $2k_F$ or the $4k_F$ instabilities
2. the lattice degrees of freedom, which are coupled, by the electron-phonon interaction, to the CDW electronic instability.

In practice, because of its very weak intensity, the analysis of the diffuse scattering has only been performed in a few salts (Yamaji et al., 1981, 1982, 1983), with a clear conclusion when a single molecular species dominates the scattering. Thus, presently, a general discussion of lattice fluctuations has to include information obtained from other physical measurements.

a. Assignment of the $2k_F$ and $4k_F$ Instabilities

Two situations, which depend on whether the Fermi wave vector is or is not common to the donor and acceptor bands, have already been considered in

Part III. The second case is the simpler one because, there a one to one correspondence with the value of the critical wavevector and the stack bearing the instability. This occurs in the NMP$_x$Phen$_{1-x}$TCNQ solid solution, where x-ray and spectroscopic studies assign the $2k_F$ and $4k_F$ instabilities to the TCNQ stacks (see Section 7b and Fig. 9). However, as already emphasized in Part III, because of the special value $\rho = \frac{2}{3}$ of the charge transfer of NMP-TCNQ, an ambiguity remains concerning the attribution of the $q_2 = \frac{1}{3}a^*$ scattering to the TCNQ ($2k_F$ scattering) and/or to the NMP ($4k_F$ scattering) stacks. An analysis of the **Q** dependence of the intensity of the q_2 scattering was unable to decide satisfactorily between these two attributions (Yamaji et al., 1982). The presence of instabilities on the TCNQ stack in NMP$_{0.55}$Phen$_{0.45}$TCNQ and Qn(TCNQ)$_2$ is corroborated by the observation of CDW-induced infrared resonances at the frequencies of the intramolecular TCNQ Ag modes (Epstein et al., 1985; McCall et al., 1985).

The x-ray studies of Qn(TCNQ)$_2$ (Pouget, 1981) and NMP$_{0.59}$Phen$_{0.41}$TCNQ (Epstein et al., 1981; Pouget et al., 1982) show that the acceptor stack undergoes a high-temperature $4k_F$ instability, then a low-temperature $2k_F$ instability. A similar observation has been made in MEM(TCNQ)$_2$, which shows a high-temperature $4k_F$ distortion, followed by a low-temperature $2k_F$ distortion of the TCNQ stack (Van Bodegom, 1981; Visser et al., 1983).

The assignment of the $2k_F$ and $4k_F$ instabilities in charge transfer salts, where there is a common k_F wavevector between D and A bands, is not straightforward. Let us first consider the compounds showing only the $2k_F$ scattering. In all these materials, the donor stack is composed of molecules containing Se atoms. The analysis performed in TSF-TCNQ (Megtert et al., 1979b; Yamaji et al., 1981) and in HMTSF-TCNQ (Yamaji et al., 1983) shows clearly that the observed $2k_F$ x-ray diffuse intensity originates from the donor stack. But, the donor contribution is so strong that a contribution of the TCNQ stack, of intensity similar to that observed in NMP-TCNQ, for example, could hardly be detected. High-resolution polarized far-infrared spectra of TSF-TCNQ obtained below T_P give strong evidence that the Peierls distortion produces little displacement or rotation of the TCNQ molecules (Bates et al., 1981). These data mean that, at most, a weak contribution of the acceptor stack is expected for the $2k_F$ instability of TSF-TCNQ.

TMTSF-DMTCNQ shows $2k_F$ and $4k_F$ scatterings. A preliminary analysis of the diffuse scattering of TMTSF-DMTCNQ (Albouy, 1980, unpublished results) shows that the observed $2k_F$ and $4k_F$ scattering originates mostly from the TMTSF stack. In TMTSF-TCNQ, the **Q** dependence of the $2k_F$ diffuse scattering strongly resembles that of TMTSF-DMTCNQ. Thus, it certainly originates from the TMTSF stack. However, in HMTSF-TNAP the

Q dependence of the $2k_F$ diffuse scattering differs noticeably from that of HMTSF-TCNQ. This and its weakness suggest that it may originate from the TNAP stack.

Detailed analysis of the diffuse scattering has not been performed in salts in which the donor molecule contains S atoms. Thus, indirect methods have to be used to ascribe the $2k_F$ and $4k_F$ scattering that is observed in these salts. In this respect, there are several pieces of evidence that in TTF-TCNQ the TCNQ stack drives the $2k_F$ instability and that the $4k_F$ anomaly belongs to the TTF stack. Structural evidence comes from the study of the 1D diffuse scattering of the solid solution $TTF_{1-x}TSF_xTCNQ$, for a small value of x (results quoted by Forro et al., 1984). The first evidence comes from an inspection of Figs. 19 and 21, which show that the random substitution of 1 mol% or 3 mol% of TTF molecule by TSF molecules does not appreciably change the temperature dependence of the $2k_F$ fluctuations, while such a substitution has a dramatic effect on the $4k_F$ scattering (Fig. 14). The most direct evidence of the presence of $4k_F$ fluctuations on the donor stacks is given by the x-ray pattern of Fig. 18c, which shows destructive interference ("white" diffuse lines) between the $4k_F$ scattering and the Laue scattering, due to the random substitution of the two molecular species TTF and TSF. By a general principle of diffraction, the two contributions interfere because they originate from the same stack. No such interference occurs for the $2k_F$ scattering. The formation of a $2k_F$ CDW on the TCNQ stacks below about 150 K is corroborated by the slight change in the rate of decrease of the uniform magnetic susceptibility of the TCNQ stack (Takahashi et al., 1984) and the deviation from linearity of the thermoelectric power, which is dominated by the negative contribution of the TCNQ stack (Chaikin et al., 1976). These findings are also in agreement with the anomalous increase of the integrated intensity and inversion of shape of some TCNQ Ag modes (Elridge and Bates, 1982; Jacobsen, 1987). The presence of the $4k_F$ CDW on the TTF stack is corroborated by the presence of nearly temperature-independent dips in the far-infrared reflectance at the position of TTF Ag modes (Jacobsen, 1987). The shift, weakening, and change of shape of these dips below 60 K, have been tentatively assigned to the onset of a $2k_F$ CDW on the TTF stack. Finally, the sharp decrease below 100 K of the negative contribution to the thermoelectric power (due to the DMTCNQ stack) in TMTTF-DMTCNQ (Jacobsen et al., 1978) could be related to the rapid growth of $2k_F$ CDW fluctuations on the acceptor stack.

b. Nature of the Modulation

The interaction between structural and electronic degrees of freedom is established by the electron–phonon coupling. The instability of the 1D electron gas leads to 1D structural fluctuations via linear electron–phonon coupling terms (second-order terms couple the 1D electron gas with several phonon modes, and generally they do not lead to a $2k_F$ or $4k_F$ 1D instability in a particular phonon branch). The linear coupling was considered on phenomenological grounds in the appendix in order to show that the x-ray scattering intensity is proportional to the CDW response function χ_ρ. However, as emphasized by Scalapino and Hirsh (1983), the expression in the response function depends on the electronic parameter coupled to the phonon field. This can be clearly illustrated on the 1D tight-binding extended Hubbard Hamiltonian (Hubbard, 1978):

$$H = \varepsilon \sum_{\ell,\sigma} n_{\ell,\sigma} + t \sum_{\ell,\sigma} (C^+_{\ell+1,\sigma} C_{\ell,\sigma} + C^+_{\ell,\sigma} C_{\ell+1,\sigma})$$
$$+ U \sum_\ell n_{\ell\uparrow} n_{\ell\downarrow} + V \sum_\ell n_{\ell+1} n_\ell, \tag{6}$$

where $C^+_{\ell,\sigma}(C_{\ell,\sigma})$ is the creation (anihilation) operator of an electron of spin σ on the site ℓ, $n_{\ell,\sigma}$ is the occupation operator, and ε, t, U, V are, respectively, the one-electron site energy, the one-electron transfer integral, the one-site Coulomb interaction, and the nearest neighbor Coulomb interaction. In a molecular crystal, each of these energies can be linearly coupled to the lattice mode. For example, it is easy to see that the intrasite energies, ε and U, are affected by a change of the intramolecular coordinates and that the intersite energies, t and V, are affected by a change of intermolecular coordinates, due, respectively, to intramolecular and intermolecular vibration modes. In addition, it has been pointed out (Shultz, 1980) that intermolecular modes, which also affect the environment of a given molecule, may perturb the one-site energy via the crystalline potential. The charge density polarizability, χ_ρ^ε, and the charge transfer polarizability, χ_ρ^t, associated with the one-electron terms, show singular responses at the critical wavevector $2k_F$ and/or $4k_F$, which depend on the strength of the Coulomb interaction parameters (U, V) and on the filling of the band (Hirsh and Scalapino, 1983, 1984). χ_ρ^U and χ_ρ^V associated with the Coulomb terms are related to the two particle–two hole polarizabilities. They show also a singular response at the wave vector $4k_F$ (Emery, 1976; Hirsh and Scalapino, 1983, 1984). These response functions will be further discussed, together with experimental data, in Section 13.

Only the totally symmetric intramolecular Ag modes couple linearily to the electrons for nondegenerate molecular levels (Duke et al., 1975). Intermolecular phonons involve translational and rotational rigid body displacements of the molecules. These last two kinds of degree of freedom could be mixed for phonons of general wavevector (Morawitz, 1981) and even mixed with internal molecular modes of particular low frequencies (Bates et al.,

1981). The external modes, which are linearly coupled to the 1D electron gas in a stack of $2/m$ symmetry [case of TTF(TSF)–TCNQ, HMTTF(HMTSF)–TCNQ], are (Conwell, 1979; Pouget et al., 1981; Van Smaalen et al., 1986; see also Chapter 4, Section 4):

- the molecular translations in the symmetry plane of the stack
- the molecular rotation around the twofold axis perpendicular to this plane.

Figure 24 shows how these displacements induce a first-order change of the transfer integrals, t, or of the crystal potential on molecular sites, ε.

In the following we shall present evidence which shows that translational modes play a key role in the structural instabilities of the organic conductors. Let us first consider the two conducting-chain compounds showing a $2k_F$ instability, which drives the low-temperature Peierls transition. The more extensive study has been performed for the $2k_F$ scattering occurring on the TSF stack in TSF–TCNQ. In a first study, Megtert et al. (1979b) have shown that, contrary to the assertions of Weyl et al. (1976), molecular rotations around the a axis (see Fig. 24) do not contribute significantly to the $2k_F$ diffuse scattering. The analysis of Yamaji et al. (1981) has shown that translation components along the b and c^* directions (T_b and T_{c^x} in Fig. 23) are combined in such a way to produce a resultant translation nearly parallel to the longest dimension of the molecule (Fig. 25). A rigid body translation of the HMTSF molecule, approximately along its long axis, was also found for the $2k_F$ instability of HMTSF–TCNQ (Yamaji et al., 1983). For molecules in close contact along the chain direction, as seems to be the case with molecules containing Se atoms, a translation along the longest molecular axis will certainly involve the least cost of "elastic" energy, while changing to

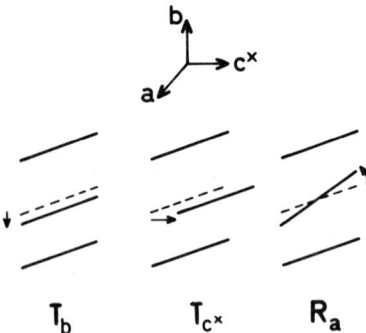

FIG. 24. External vibration modes modulating, to first order, the one-electron site energy or the in-chain transfer integrals of a stack of $2/m$ symmetry. The coordinate system is that of TTF(TSF)–TCNQ. T_b and T_{c^x} are translations along the b and c^* directions respectively. Ra is a rotation around the a axis.

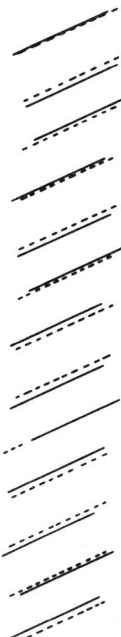

FIG. 25. Schematic pattern showing the $2k_F$ tanslational displacement of TSF molecules in the bc^x symmetry plane (full line). Note that the phasing between longitudinal and transverse components produces a translation nearly along the longest axis of the molecule. (The amplitude of the displacement on this figure is arbitrary.) The interrupted lines give the steady position of the molecules. [From Yamaji et al. (1981).]

first-order the intrastack transfer integrals. In TTF-TCNQ, the molecular displacements contributing to the $2k_F$ scattering were also found to be polarized in the c (or c^*) and b directions (Kagoshima et al., 1976; Khanna et al., 1977). These displacements mainly correspond to translations, as evidenced:

• for the c or c^* polarization by the observation of a Kohn anomaly in the transverse acoustic branch (Shirane et al., 1976)
• for the longitudinal polarization, by satisfactorily agreement between observation and calculation of the intensity of the diffuse scattering, in $[hk\ell]$ zones of the reciprocal space with no sizable ℓ components, using translation of rigid molecules (Pouget et al., 1976).

In agreement with an electron-phonon coupling mechanism for the $2k_F$ structural instability, it is interesting to remark that in TTF-TCNQ the Kohn anomaly occurs in the transverse acoustic branch polarized along c^*, but not in the acoustic branch of the same symmetry (with respect to the 3D crystal space group $P\,2_1/c$) polarized along a^x (or a) for which the atomic displacements are not linearily coupled to the electrons (Shapiro et al., 1977; see the insert in Fig. 22). In contrast to TSF-TCNQ, the transverse and longitudinal components of translation seem to be decoupled in TTF-TCNQ at ambient pressure. Experimental evidence comes from the different temperature dependence of the width and the intensity of the "transverse" (below T_H) and "longitudinal" (below $T^* \simeq T_M$) $2k_F$ satellites (Comès, 1977; Khanna et al., 1977; Pouget et al., 1979a; see Fig. 35). However, the merging together of the T^* and T_H lines (Megtert et al., 1982) (Fig. 36a), suggests that under pressure these two components combine, as in TSF-TCNQ, in order to displace the molecule in its long direction. The symmetry of these two modes has been identified (Megtert et al., 1983 and 1985), and electrical measurements under pressure (Friend et al., 1978) show that they belong to the TCNQ stack.

In relation to these findings, a transverse displacement of dimers of TCNQ molecules along the long molecular axis was mainly found for the (non-symmetry-breaking) $2k_F$ distortion of TEA(TCNQ)$_2$ (Filhol and Thomas, 1984; Farges, 1985)†.

Let us now consider the $4k_F$ diffuse scattering of TTF-TCNQ and its family. Only preliminary calculations of its **Q** dependence have been performed for TTF-TCNQ (Pouget et al., 1976) and TMTSF-DMTCNQ (Albouy, 1980, unpublished results). Both calculations agree with a translational instability. Furthermore, the polarization analysis performed in TTF-TCNQ (Kagoshima et al., 1976; Khanna et al., 1977) and TMTSF-DMTCNQ (Pouget, 1981) mainly gives a longitudinal component of displacement. It has been argued (Sham, 1976; Kondo and Yamaji, 1977) that longitudinal displacements are expected when phonons modulate the Coulomb interaction V_{ij} between neighboring charges in chain direction.

In its "longitudinal" polarization in the displacement of the donor, the $4k_F$ instability of TTF-TCNQ differs from that of MEM(TCNQ)$_2$, which is transverse, as shown by the observation below 335 K of a deformation of the nearly uniform zigzag of acceptors, in a direction mainly perpendicular to the symmetry plane of the TCNQ stack (Van Bodegom and Bosch, 1981). However, the $4k_F$ structural instability of MEM(TCNQ)$_2$ differs from that of TTF-TCNQ by several aspects: (1) the MEM$^+$ cation sublattice provides an external (3D) $4k_F$ potential on the TCNQ stacks and (2) large elastic

† This displacement is defined with respect to the mean benzyl ring plane of the TCNQ. In addition it is found that the TCNQ(B) shows a boat-like conformation whose bending angles of the C≡N terminal groups with respect to the benzyl ring increase for decreasing temperature (Filhol and Thomas, 1984). This kind of deformation is expected when the TCNQ molecule, which is displaced, has one of its C≡N terminal groups strongly linked by short contact forces to a cation (here the TEA$^+$).

deformations (due to a strong coupling with the cationic sublattice ?) are involved in the $4k_F$ distortion of the former salt. MEM(TCNQ)$_2$ shows also at low temperature, 17.4 K, another transversally polarized distortion, but with the wavevector $2k_F$. At this phase transition, which has been interpreted as a spin-Peierls transition, the structural change consists of a transverse shift (in the plane of the TCNQ molecule along a direction nearly parallel to the diagonal linking opposite CN groups) of rigid TCNQ dimers (Visser *et al.*, 1983). Assuming a $4k_F$ Wigner lattice of charges strongly localized on the TCNQ molecules, where several sites are available for their localization, it has been pointed out by Yamaji (1979) that the electric field caused by an antiferroelectric-like ordering of charges in the chain direction may drive a $2k_F$ lattice instability in a transverse acoustic mode. In the case of MEM(TCNQ)$_2$, if the charges are mainly localized on the diagonally opposed CN groups of the TCNQ molecule, this mechanism may explain the transverse nature of the $2k_F$ displacement wave of TCNQ dimers, which is stabilized below 17.4 K.

13. Physical Analysis of the $2k_F$ and $4k_F$ Instabilities

a. Classification of the Salts

Table V gives a tentative classification of the organic charge-transfer salts reviewed here, with respect to the nature and the divergence of their $2k_F$ and $4k_F$ structural instabilities. Depending on the $2k_F$ or $4k_F$ critical wavevector of the dominant instability, two extreme categories of salts can be considered:

  I. Salts containing TCNQ, DMTCNQ, and TNAP as acceptors and TTF, TMTTF, HMTTF, and their Se analogues as donors. They exhibit a second-order Peierls transition around 50 K driven by the $2k_F$ instability, where both charge and spin degrees of freedom are involved, as shown by the observation of an activation energy in transport and magnetic properties. The $4k_F$ instability (except perhaps in TMTTF–DMTCNQ) does not play a dominant role at low temperature. These salts are good conductors, with electrical conductivity increasing for decreasing temperatures (see, for example, Jérome and Schulz, 1982a). $2k_F$ and $4k_F$ instabilities show 1D fluctuations in a wide temperature range.

  II. DBTTF–TCNQCl$_2$ and MEM(TCNQ)$_2$. They show a $4k_F$ high-temperature metal–insulator phase transition freezing the charge degrees of freedom [second-order phase transition at 180 K announced by a regime of quasi-1D fluctuations for DBTTF–TCNQCl$_2$ and strong first-order phase transition at 334 K, without percursors, for MEM(TCNQ)$_2$]. Then they undergo a low-temperature (at 38 K and 17.4 K, respectively) second-order

TABLE V

CLASSIFICATION OF THE ORGANIC TRANSFER SALTS ACCORDING TO
THEIR STRUCTURAL INSTABILITIES

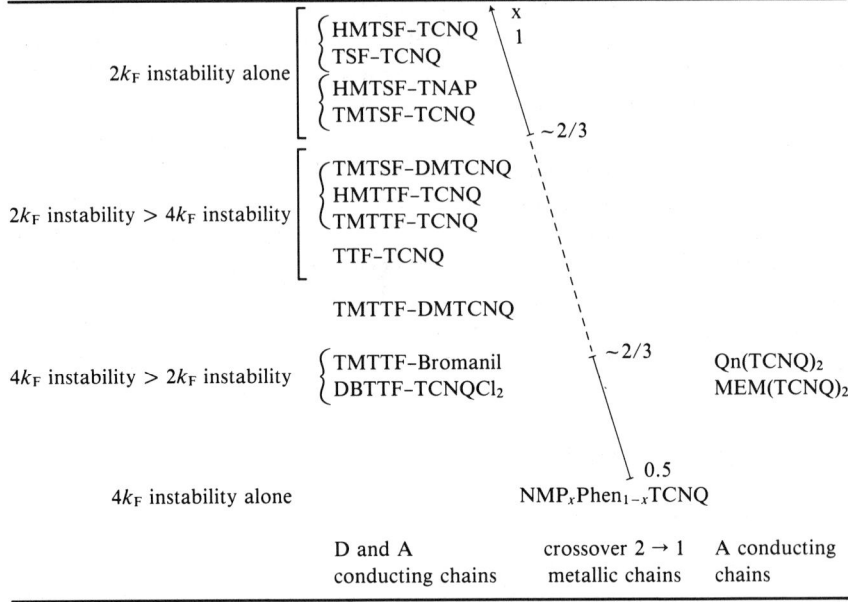

$2k_F$ phase transition, involving the spin degrees of freedom and which has been interpreted as a spin-Peierls transition (Mortensen *et al.*, 1983; Sawatzky *et al.*, 1979).

The other salts considered in Table V have an intermediate behavior:

• TMTTF-Bromanil, where electronic and magnetic properties show an incomplete decoupling of charge and spin degrees of freedom in temperature (Torrance *et al.*, 1981). However, its structural features bear some resemblance with the salts of category II: a $4k_F$ distortion occurs above room temperature and a $2k_F$ phase transition arises around 80 K. This last transition is announced by 3D precursors (Kagoshima *et al.*, 1983b). It has not been studied in detail because of disorder effects, probably due to irradiation damages produced during the x-ray study itself.

• The solid solution $NMP_x Phen_{1-x} TCNQ$ whose instability shifts, as a function of x, from the regime of dominant $2k_F$ to the regime of dominant $4k_F$. These instabilities are observed under the form of quasi-1D (or very anisotropic 3D) scattering (Pouget *et al.*, 1980a, 1982; Epstein *et al.*, 1981). These salts do not exhibit true structural phase transitions, probably because of disorder effects. In the concentration range where both the $2k_F$ and $4k_F$ instabilities exist, the dominant role is played by the $4k_F$ scattering; the $2k_F$ scattering appears only at low temperature. Similar structural features are observed in Qn(TCNQ)$_2$ (Pouget, 1981). This behavior is reminiscent of the salts of category II, with, however, the difference that the $2k_F$ instability develops a well-defined regime of 1D fluctuations. After a broad maximum in conductivity slightly below room temperature, these materials have a low-temperature activated electrical conductivity (Epstein and Miller, 1978), which might correspond to the progressive freezing of the charge degrees of freedom. At low temperature, when the $2k_F$ instability arises, the charge and spin degrees of freedom are relatively well decoupled. The $2k_F$ fluctuations might be viewed as precursors of a spin-Peierls transition, never achieved because of the disorder.

b. Semiquantitative Analysis of One-Dimensional Structural Fluctuations in TTF-TCNQ and its Family

The quantities of interest are the CDW susceptibility (χ_ρ^{1D}—Figs. 19 and 20) and the interchain CDW correlation length ($\xi_\parallel^{1D}$—Figs. 21 and 24) of the $2k_F$ and $4k_F$ instabilities.

(1) *CDW susceptibilities.* In all the compounds investigated, $\chi_{2k_F}^{1D}$, proportional to I_{2k_F}/T in Fig. 19, shows a low-temperature divergence, which drives the Peierls transition at T_P. The $2k_F$ divergence varies from compound to compound, being the strongest in TTF-TCNQ, intermediate in HMTTF-TCNQ, and the weakest in TSF-TCNQ and HMTSF-TCNQ. The temperature dependence of the $4k_F$ scattering changes appreciably among the series of organic charge-transfer salts. From I_{4k_F}/T one gets a vanishing of $\chi_{4k_F}^{1D}$ at low temperature in TMTTF-TCNQ, a constant value of $\chi_{4k_F}^{1D}$ in TMTSF-DMTCNQ, and a weak divergence of $\chi_{4k_F}^{1D}$ in TMTTF-DMTCNQ and in TTF-TCNQ.

In order to analyze more quantitatively the nature of these divergences, $\chi_{2k_F}^{1D}$ and $\chi_{4k_F}^{1D}$ have been fitted by the simplest expression, which shows a divergence at 0 K (as expected by the theory of 1D systems): power law $T^{-\alpha}$ or $\log(A/T)$. The values of α or A that reproduce the temperature dependence of the diffuse intensity, corrected by the thermal population factor, are given in Table VI. The analysis performed is only semiquantitative because

- there are large uncertainties in the measurement of the diffuse scattering
- the classical limit (justified for TTF-TCNQ in Section 11b) is used

● all the temperature dependence of the intensity is assumed to come from $S(q, t = 0)$.†

The link between the observed dependence in temperature of χ_ρ and that calculated by various theories of the 1D electron gas is not straightforward. The first difficulty is that measurements are performed at constant (ambient) pressure. With the lattice contraction occurring in the chain direction when the temperature decreases (Fig. 5), some of the parameters of the 1D electron gas are changed: (1) increase of the intermolecular transfer integral and (2) decrease of the various constants g_i characterizing the electron–electron interactions.‡ Its main effect is to reduce the relative strength of the Coulomb interactions and thus to inhibit the growth of the $4k_F$ CDW anomaly. The second difficulty is that the CDW susceptibility measured in structural experiments involves the electron–phonon coupling explicitly. In the RPA approximation of the electron–phonon coupling, it is shown in the Appendix [Eq. (A.21)] that, in the 1D fluctuation regime,

$$\chi_\rho^{1D} = \frac{\chi_e^{1D}}{1 - |\lambda|^2 \chi_L \chi_e^{1D}}; \qquad (7)$$

where χ_e^{1D} is the electronic polarizability of the 1D electron gas. This quantity presents a low-temperature divergence, characterized by the exponent ν [see Eq. (5)″]. Well above the Peierls mean-field temperature of the lattice instability, T_P^{MF} [corresponding to the vanishing of the denominator of Eq. (7)], where $|\lambda|^2 \chi_L \chi_e^{1D} < 1$, the expression, Eq. (7), can be approximated by

$$\chi_\rho^{1D} \simeq \chi_e^{1D}[1 + |\lambda|^2 \chi_L \chi_e^{1D}] \sim \chi_e^{1D}. \qquad (8)$$

In this limit, the electron–phonon coupling does not appreciably modify the CDW electronic divergence. Near T_P^{MF}, however, structural fluctuations are important because they suppress, as expected for a 1D system, the divergence of Eq. (7). They have to be incorporated explicitly in the calculation of χ_ρ^{1D}. A Hartree Fock treatment of amplitude fluctuations already suppresses the

---

† There is, in fact, a slight decrease of the square of the diffuse structure factor $|F_d|^2$ for increasing temperatures through the Debye Waller factor [see Eq. (A.2)]. This decrease is estimated, for the wavevectors scanned in Figs. 17 and 18, at about 5–10% between 60 K and 300 K, from the Debye Waller data available for TTF-TCNQ (Schultz et al., 1976). Its effect is smaller than the experimental uncertainties given in Table VI.

‡ This is, for example, the case for the density of state at the Fermi level. From Eqs. (10) and (12), defined later, it is easy to show that:

$$\frac{\Delta g_i}{g_i} \simeq -\frac{\Delta v_F}{v_F} = -\frac{\Delta b}{b}\left[1 + \frac{\delta \log t_\parallel}{\delta \log b} + \left(\frac{\pi}{2}\rho \cot g \frac{\pi\rho}{2}\right)\frac{\delta \log \rho}{\delta \log b}\right].$$

This leads to a relative decrease of $\Delta g_i/g_i$ of about 12% between 300 K and 60 K for TTF-TCNQ using the data of Part III.

TABLE VI

Temperature Dependence of the $2k_F$ and $4k_F$ CDW Susceptibilities in the 1D or Quasi-1D Fluctuation Regime + χ_{2k_F} Determined at $\mathbf{q} = (0, 2k_F, 0)$ Between 30 K and 55 K, in the 3D Short Range Fluctuation Regime

	$\chi_{2k_F}^{1D}$	$\chi_{4k_F}^{1D}$
TTF-TCNQ	$T^{-\alpha}$; $\alpha \sim [1.3-2]$ for 60 K $\leq T \leq$ 150 K	$T^{-\alpha}$; $\alpha \sim [0.6-0.9]$ for 60 K $\leq T \leq$ 300 K
TSF-TCNQ	$\log(A/T)$; $A \sim [250$ K-450 K$]^+$ for 30 K $\leq T \leq$ 250 K	—
HMTTF-TCNQ	$T^{-\alpha}$; $\alpha \sim [0.8-1]$ for 50 K $\leq T \leq$ 300 K	Not determined (χ_{4k_F} increases below 100 K)
HMTSF-TCNQ	$\log(A/T)$; $A \sim [500$ K-800 K$]$ for 25 K $\leq T \leq$ 300 K	—
TMTTF-TCNQ	$T^{-\alpha}$ (α cannot be determined because of irradiation damages)	Not determined (χ_{4k_F} increases between 180 K and 120 K, then decreases below 120 K)
TMTSF-TCNQ	$T^{-\alpha}$; $\alpha \sim [0.7-0.9]$ for 75 K $\leq T \leq$ 300 K	—
TMTTF-DMTCNQ	Not determined (χ_{2k_F} increases below 100 K)	$\log(A/T)$; $A \sim [400$ K-600 K$]$ for 55 K $\leq T \leq$ 300 K
TMTSF-DMTCNQ	$T^{-\alpha}$; $\alpha \sim [1-1.5]$ for 75 K $\leq T \leq$ 225 K	Nearly constant for 60 K $\leq T \leq$ 300 K
HMTSF-TNAP	$T^{-\alpha}$; $\alpha \sim [1-1.2]$ for 60 K $\leq T \leq$ 300 K	—

divergence of χ_ρ^{1D} at finite temperature (Jérome and Shulz, 1982a and 1982b). At low temperature, for incommensurate band filling, phase fluctuations have to be included in the calculation of χ_ρ^{1D} (Scalapino et al., 1972; Dieterich, 1976). Strong lattice fluctuations lead to the formation of a pseudogap in the density of states at the Fermi level (Lee et al., 1973).

It has been argued (Barisic, 1985; Barisic and Bjelis, 1985) that in TTF-TCNQ the electron–phonon interaction is a small quantity compared to the electron–electron interactions and that χ_e^{1D} drives the divergence of χ_ρ^{1D}. In particular, the weak decrease of the density of states (estimated from spin susceptibility measurements—Takahashi et al., 1984) observed below 150 K in TTF-TCNQ, when the $2k_F$ fluctuations grow, suggests that lattice fluctuations affect only weakly the electronic properties. In this limit, the expression (8) could be a good approximation for χ_ρ^{1D}, but lattice contraction effects make an accurate comparison of the experimental data with the results of the theory of the 1D electron gas difficult. The observation (see Table VI) for several compounds of a logarithmic divergence of χ_ρ over a large temperature range also suggests that the approximation (8) might be reasonable. A logarithmic divergence of the electronic polarizability at $2k_F$ is obtained for a gas of independent electrons [see Eq. (A.16)]. It is also found at the wavevector $4k_F$ for a gas of electrons in the presence of on-site repulsion in the limit $U \to \infty$ (see, for example, Hirsch and Scalapino, 1983). The $2k_F$ logarithmic divergence is observed in TSF-TCNQ and HMTSF-TCNQ. However, the constant A that is deduced from the fit with the data, 350 K and 650 K, respectively, is 4-5 times less than expected for the electronic cutoff energy ($\sim E_F$, i.e., 1600 K and 2600 K respectively) of the donor chains. This difference might be due to the presence of (logarithmic) correction terms to the expression (8), including the electron-phonon coupling in a perturbative way (similar terms also occur when the electron-electron interaction is treated in perturbation in χ_e^{1D}). It could be due also to a Fermi energy (velocity) renormalization by the electron-electron interactions (Lee et al., 1977; Emery, 1979).

A $4k_F$ logarithmic divergence is observed in TMTTF-DMTCNQ, also with a small value of A (≈ 500 K). TTF-TCNQ has a $4k_F$ divergence, close to a logarithmic law. Further evidence in favor of the measurement of χ_e^{1D} is given below by the calculation of the intrachain CDW correlation length $\xi_\parallel$.

The classification of the salts of the TTF-TCNQ family given in Table V, is partly based on the nature of the divergence of $\chi_{2k_F}^{1D}$. The stronger divergence of $\chi_{2k_F}^{1D}$ occurs in salts showing the $4k_F$ lattice instability.

(2) *Intrachain correlation lengths.* Let us begin with TSF-TCNQ and HMTSF-TCNQ the $2k_F$ instability of which takes place on the donor stack and is close to that shown by a gas of independent electrons. In that case, it is easy to show, after a development in $\delta q = q - 2k_F$ of the $2k_F$ electronic polarizability [see Eq. (A.16)], that the inverse of the CDW correlation length is given by

$$\xi_\parallel^{-1} = 4.34 \frac{k_B T}{\hbar v_F} \sqrt{\ln(A/T)}, \qquad (9)$$

where the Fermi velocity, for a 1D tight-binding band, amounts to

$$\hbar v_F = \frac{W d_\parallel}{2} \sin \frac{\pi \rho}{2}, \qquad (10)$$

with W, $d_\parallel$, ρ being, respectively, the bandwidth, the intrachain lattice spacing, and the charge transfer. For TSF-TCNQ, Fig. 21 gives $\xi_\parallel^{-1}$ between 0.05 Å$^{-1}(L)$ and 0.07 Å$^{-1}(G)$ at 200 K; these two values depend on the nature of the correction of resolution performed (L = lorentzian; G = gaussian). From Eqs. (9) and (10) $\xi_\parallel^{-1}$ leads to a bandwidth of 0.7(L)-0.5(G) eV,

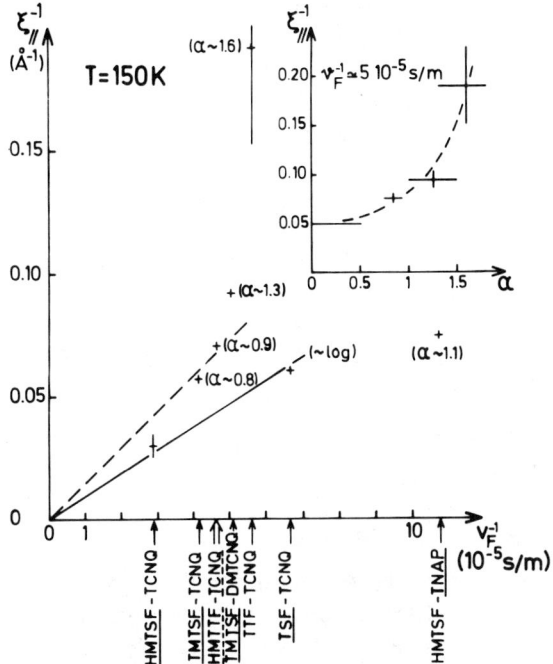

FIG. 26. Inverse correlation length of $2k_F$ CDW in chain direction at 150 K in the regime of 1D fluctuation (obtained after a Gaussian resolution correction) as a function of the inverse Fermi velocity (the stack assumed to bear the $2k_F$ instability is underlined; the Fermi velocity has been calculated from Eq. (10) using the bandwidth obtained from the optical data of Jacobsen, 1985). This figure shows that for a logarithmic divergence of the CDW response function $\xi_\parallel^{-1}$ scales with v_F^{-1}. The exponent α, driving the divergence of the CDW response function, is indicated for each compound (α depends also on v_F—see, for example, the expressions 11-13). The insert shows that at constant v_F, $\xi_\parallel^{-1}$ increases with the exponent α.

in very good agreement with the value, 0.6 eV, calculated by Herman (1977) for the TSF stack and that deduced from optical measurements (Jacobsen, 1985). In HMTSF–TCNQ, the data obtained at room temperature give, with the resolution corrections mentioned above, $\xi_\parallel^{-1}$ between 0.04 Å$^{-1}(L)$ and 0.06 Å$^{-1}(G)$. This leads at a donor bandwidth of 1.3(L)-0.9(G) eV. The former value is identical to that deduced from optical measurements (Jacobsen, 1985).

Figure 26 gives the absolute value of $\xi_\parallel^{-1}$ at 150 K for the $2k_F$ scattering of some charge-transfer salts. According to the independent electron behavior [Eq. (9)], this quantity is plotted as a function of v_F^{-1} [v_F is given by Eq. (10), using the value of W deduced from optical measurements (Jacobsen, 1985), for the stacks assumed to bear the $2k_F$ anomaly, according to Section 12]. Figure 26 shows clearly that $\xi_{\!/\!/}^{-1}$ is generally enhanced from the independent electron value. This enhancement increases with the exponent α, driving the divergence of CDW response function, $\chi_{2k_F}^{1D}$, which exponent is itself a function of v_F. This observation shows that, as expected, the stronger the fluctuations, the smaller is $\xi_{\!/\!/}$.

Now let us consider the temperature dependence of $\xi_{\!/\!/}^{-1}$. Figure 21 shows that expression (9) also reproduces reasonably the temperature dependence of the HWHM of the $2k_F$ scattering of TSF-TCNQ and HMTSF-TCNQ [the experimental data show a slightly greater rate of increase than that predicted by Eq. (9), which could be accounted for by perturbative corrections to the independent electron description, as already mentioned]. In the other salts, the divergence in temperature of $\xi_{\!/\!/ 2k_F}^{1D}$ seems to scale with that of $\chi_{2k_F}^{1D}$. One gets, using a $T^{-\beta}$ dependence to express the divergence of $\xi_{\!/\!/ 2k_F}$:

- $\beta \lesssim 1$ for TSF-TCNQ, HMTSF-TCNQ, HMTSF-TNAP, and TMTSF-TCNQ
- $\beta \sim 1.3$ for HMTTF-TCNQ and TMTSF-DMTCNQ
- $\beta \sim 2.5$ for TTF-TCNQ.

It should be remarked that the observation of a $\xi_{\!/\!/}^{-1} \sim T$ dependence in several salts is not the proof of a phase fluctuation regime, because the CDW susceptibility does not show the T^{-2} divergence expected in this regime [see Eq. (5)'].

c. Electronic Origin of the Structural Instabilities

Theoretical Background

The best evidence for an electronic origin for the $2k_F$ and $4k_F$ structural instabilities comes from the direct observation (see Figs. 17 and 18) of 1D structural fluctuations (i.e., with the anisotropy of the electron gas) in a large temperature range above T_P. Another feature well established experimentally is that in their temperature dependence, polarization, and modification by disorder, the $2k_F$ and $4k_F$ scatterings behave differently. They correspond to two different kinds of structural instabilities. For example the $4k_F$ scattering cannot be considered as a second harmonic of the $2k_F$ scattering in the 1D fluctuation regime. Similarly, the observed divergences of the $4k_F$ CDW response function cannot be obtained by a second-order perturbation theory of the electron gas with respect to Coulomb interactions (see, for example, Sham, 1976; Gasser, 1985). The presence of a $4k_F$ instability is often taken as the best evidence that substantial electron–electron interactions are experienced by the electron gas. However, the relevant physical parameters controlling this instability are not completely clarified (see Subsection d).

Depending on the strength of the electron–electron interactions relative to the kinetic energy (or the one-electron bandwidth), the instability of the 1D electron gas toward the formation of $2k_F/4k_F$ CDW has been discussed in various limits.

(1) *Weak coupling limit.* A vast literature exists on the treatment of the electron–electron interactions in the weak coupling limit (for reviews see Solyom, 1979; Emery, 1979; Barisic and Bjelis, 1985). It predicts a low-temperature divergence of the CDW electronic response function, with, in Eq. (5)″, an exponent γ expressed as a function of the $q = 0$ and $q = 2k_F$ Fourier transforms of the electron–electron interactions, g_2 and g_1 (forward and backward scattering), respectively

$$\chi_{2k_F} \sim \left(\frac{E_C}{k_B T}\right)^{2g_2^*}$$
$$\chi_{4k_F} \sim \left(\frac{E_C}{k_B T}\right)^{2[1 - 2\sqrt{1 - 2g_2^*})/(1 + 2g_2^*)]}, \quad (11)$$

where $g_2^* = g_2 - g_1/2$. The $4k_F$ CDW response function diverges only for $g_2^* > 3/10$.

In the case of an on-site Hubbard model, with RPA-screened 3D Coulomb interactions (Barisic 1983, Schulz, 1983)

$$g_1 \sim n(E_F)U$$
$$g_2 \sim n(E_F)\left[U + \frac{e^2}{d_\parallel}\ln\left(\frac{E_C}{\hbar\omega_p}\right)\right], \quad (12)$$

where $n(E_F) = d_\parallel/\pi v_F h$ is the density of states at the Fermi level per spin, and ω_p is an average plasma frequency involved both in the intrachain and interchain screening effects. The g_1 and g_2 matrix elements of the Coulomb interaction at $q = 2k_F$ and $q = 0$ can be also related to the U and V parameters of the extended Hubbard model, Eq. (6), by the following Fourier transform relationships, where the charge transfer ρ appears explicitly, and where Eq. (10) has been used to express v_F ($W = 4t$).

$$g_1 = \frac{U + 2V\cos\pi\rho}{2\pi t d_\parallel \sin(\pi\rho/2)}$$
$$g_2 = \frac{U + 2V}{2\pi t d_\parallel \sin(\pi\rho/2)}. \quad (13)$$

The $2k_F$ CDW electronic instability is coupled to the lattice via the linear electron–phonon coupling. The $4k_F$ CDW electronic instability is coupled to the lattice either through the linear electron–phonon coupling or through a linear variation of the Coulomb interaction with displacements (Emery, 1976; see also Barisic and Bjelis, 1985). The various coupling mechanisms have already been considered in Section 12. However renormalization-group calculations do not distinguish between site and bond CDW susceptibilities (χ_ρ^ε and χ_ρ^t respectively).

For completeness let us mention that in the half-filled band case ($\rho = 1$) Umklapp processes (g_3 matrix elements of the Coulomb interaction) have to be added. When g_3 is relevant, enhanced $4k_F$ CDW will develop below a certain temperature T_0 (defined by Barisic and Brazovskii, 1981). The $2k_F$ CDW response function remains finite, because a gap appears below T_0 in the charge density spectrum. Coupling to the lattice will produce associated distortion at $4k_F$ and $2k_F$ (spin Peierls transition); there is no ordinary Peierls transition in that case (Emery et al., 1982). The effects of g_3 processes are progressively inhibited by deviation from half-filling (Firsov et al., 1985; Montambaux et al., 1986).

(2) *Intermediate coupling limit.* No analytical result exists for the case that electron–electron interactions are comparable to the electron bandwidth. Only numerical calculations have been performed with the extended Hubbard Hamiltonian, Eq. (6). In particular, Monte Carlo simulations (Chui and Bray, 1980; Hirsh and Scalapino, 1983, 1984) have studied the interplay between $2k_F$ and $4k_F$ instabilities for the electron polarizabilities defined in Section 12 as a function of the Coulomb interactions and the band filling. First, these calculations show that the part of the two particle-two hole correlation functions χ_ρ^U and χ_ρ^V not reducible to a single-particle-hole correlation function does not show any singularity at $4k_F$. Only the single-particle-hole charge-density susceptibility χ_ρ^ε and charge-transfer susceptibility χ_ρ^t show singularities at $4k_F$ (and, of course, at $2k_F$). Secondly, both the on-site (U) and near-neighbor (V) Coulomb interactions are necessary in order to obtain a strong divergence at $4k_F$ (if $V = 0$: χ_ρ^ε and χ_ρ^t do not diverge for any $U < \infty$; there is a logarithmic divergence only for $U = \infty$). The Coulomb interaction U suppresses the $2k_F$ response of χ_ρ^ε and χ_ρ^t, but it has a weaker effect on χ_ρ^t. χ_ρ^t can show both $2k_F$ and $4k_F$ singularities, for values of U and V comparable to the bandwidth. For the half-filled band case ($\rho = 1$), $\chi_\rho^\varepsilon(2k_F)$ and $\chi_\rho^t(2k_F)$ behave differently. U rapidly suppresses $\chi_\rho^\varepsilon(2k_F)$, while $\chi_\rho^t(2k_F)$ is first enhanced by a small U and only suppressed for larger values of U (see also Caron and Bourbonnais, 1984). However, both $\chi_\rho^\varepsilon(2k_F)$ and $\chi_\rho^t(2k_F)$ are enhanced by the presence of a small-neighbor repulsion V.

(3) *Strong coupling limit.* Very strong electron–electron interactions lead to charge localization on molecular sites or between them. The Wigner lattice thus formed has the $4k_F$ wave vector as the first Fourier component of the charge distribution (Kondo and Yamaji, 1977; Hubbard, 1978; Torrance, 1978). In this limit, the $4k_F$ instability can be also viewed as a standard Peierls instability of a strongly correlated electron gas (Ovchinnikov, 1973; Bernasconi et al., 1975). Depending on the site or bond localization of the charges, the electronic $4k_F$ instability arises in the charge-density polarizability (χ_ρ^ε) or charge-transfer polarizability (χ_ρ^t). In the especially important quarter-filled band case ($\rho = \frac{1}{2}$), there is localization of one electron either on one site out of two (accompanied by the internal deformation of one molecule out of two) or between two molecules (which shift towards each other to form a dimer). With the $4k_F$ electron localization and associated lattice distortion, the charge degrees of freedom are lost. At low temperature an additional $2k_F$ lattice instability can occur. It has been suggested that the $2k_F$ instability can be due either to the pairing in a singlet state of the spin degrees of freedom, which have not been affected by the $4k_F$ instability (case of Fig. 27—Sumi, 1979), or to an antiferroelectric ordering of charges on molecules, when two molecular sites are available for the localization (case of Fig. 28—Yamaji, 1979). In the former case, known as a spin Peierls transition or magnetically driven spin-lattice dimerization of an antiferromagnetic chain, the $2k_F$ divergence occurs in a four-spin response function that is proportional to χ_ρ^t (in the strong coupling limit, the antiferromagnetic exchange integral $J = 4t^2/U$ can be modulated by $2k_F$ phonons linearly coupled to the transfer integral t—see, for example,

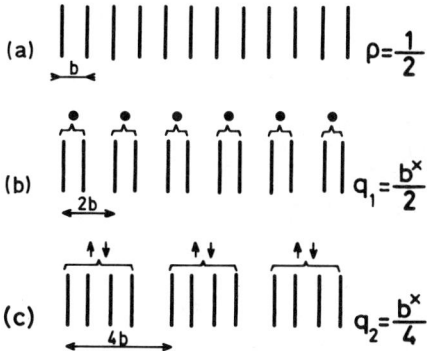

FIG. 27. Instabilities of a strongly correlated 1D electron system, with $\frac{1}{2}$ electron per molecule (a) for lattice distortions of wavevector $q_1 = b^*/2$ (b) and $q_2 = b^*/4$ (c), leading, respectively, to a $4k_F$ charge localization [one charge (●) per dimer] and to a $2k_F$ spin Peierls pairing of the spin degrees of freedom in a singlet (↑↓) state.

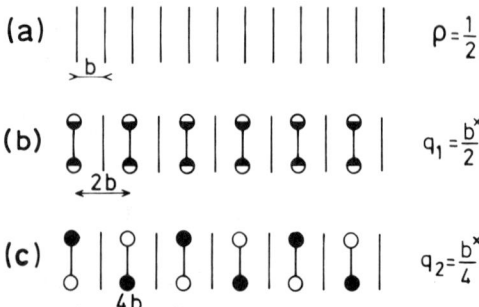

FIG. 28. Same as Fig. 27, with two available sites per molecule for electron localization. In (b) an electron is localized on one molecule out of two, but equally distributed on the two molecular sites (the wavevector of the charge distribution is $q_1 = b^*/2$). In (c), because of the repulsive Coulomb interactions between electrons, an antiferroelectric ordering of charge occurs with the $q_2 = b^*/4$ wavevector.

Hirsch and Scalapino, 1984). Numerical calculations performed by these authors show that the $2k_F$ singularity is absent in the on-site charge density response χ_ρ^ε.

Band filling other than $\rho = \frac{1}{2}$ has been considered in the literature. Let us mention, for completeness, the possibility of solitons with charge $\pm e/2$ for ρ close to $\frac{1}{2}$ (Rice and Mele, 1982) and some exact results concerning the nature of the ground-state broken symmetry for $\rho = 1$ (Dixit and Mazumdar, 1984; Mazumdar and Campbell, 1985).

Comparison with Experiments

It is now interesting to qualitatively compare the experimental findings with these theoretical predictions. The numerical calculations performed in the intermediate and strong coupling regimes seem to indicate that χ_ρ^t can account for the observed $2k_F$ and $4k_F$ instabilities. This is in agreement with the experimental results of Section 12 which show that translational vibration modes, directly coupled to the nearest neighbor transfer integral t, are strongly involved in the lattice instabilities of organic charge transfer salts.

Salts of group I (TTF–TCNQ and its family) seem to be well described in the framework of weak and intermediate coupling theories. The weak-coupling theory certainly applies to salts that do not show the $4k_F$ anomaly. This is the case of TSF–TCNQ and HMTSF–TCNQ, which are close to the independent electron limit. It has been argued (Barisic and Bjelis, 1985) that the intermediate coupling theory should apply to TTF–TCNQ. The observation in TTF–TCNQ and TMTTF–DMTCNQ of a weakly T-dependent (quasilogarithmic) $4k_F$ scattering corroborates the numerical calculation of Hirsch and Scalapino (1984) for intermediate coupling (typically, $U \sim 4t$

and $V \sim 2t$ on the donor stack). A recent analysis of the NMR data of TTF-TCNQ in the framework of the Hubbard model gives $U \sim 3.2t$ for both stacks (Takahashi et al., 1984), and intraband oscillator strength measurements give $U \sim 6t$, assuming $U = 3V$ (Jacobsen, 1986). However, a more detailed analysis is necessary in order to discriminate between the various components of the electron–electron interaction experienced by each stack in order to explain why the $4k_F$ anomaly occurs on the TTF stack and why the TCNQ stack drives the $2k_F$ instability. Weaker Coulomb interactions are certainly present in TMTTF-TCNQ and TMTSF-DMTCNQ, where, respectively, the vanishing at low temperature of the $4k_F$ response and the constancy of the $4k_F$ response can be reproduced with smaller values of V (i.e., for $U \sim 4t$ and $V \sim 0$). These last effects could also result of the thermal contraction of the lattice.

The presence of weak electron–electron interactions in TSF-TCNQ is in accordance with the observation below 29 K of the same energy gap in electrical (charge excitation gap) and magnetic (spin excitation gap) measurements (Etemad, 1976; Scott et al., 1978). In TTF-TCNQ the observation below 38 K of an electrical gap larger than the magnetic gap (especially on the TTF stack, where the $4k_F$ instability is observed)—(Etemad, 1976; Herman et al., 1976; Tomkiewicz et al., 1977; Takahashi et al., 1984) is in accordance with the presence of substantial electron–electron interactions (electron–electron interactions lead to different excitation energies for charge and spin degrees of freedom—see, for example, Emery, 1979; Solyom, 1979). In all these salts, when it occurs, the diverence of the $4k_F$ instability remains weak. The low-temperature Peierls transition is always driven by the $2k_F$ instability, with perhaps an exception for TMTTF-DMTCNQ.

The opposite behavior is exhibited by salts of group II. With a dominant (high temperature) $4k_F$ instability, and a low temperature $2k_F$ one, they could be described in the framework of a strong-coupling (or intermediate-coupling) theory. The $4k_F$ instability gives rise either to a high temperature phase transition in ordered crystals [MEM(TCNQ)$_2$ and DBTTF-TCNQCl$_2$] or to a (weakly temperature-dependent) local distortion in disordered crystals [TMTTF-Bromanil, NMP$_x$Phen$_{1-x}$TCNQ for $\frac{2}{3} > x > \frac{1}{2}$, Qn(TCNQ)$_2$]. With the exception of MEM(TCNQ)$_2$, which shows a first-order phase transition, the $4k_F$ pretransitional fluctuations above the second-order phase transition of DBTTF-TCNQCl$_2$ and the $4k_F$ local distortion in the disordered salts are basically 1D in nature. It means that the observed $4k_F$ scattering can be mostly associated with an in chain instability, acting to localize the charge carriers. Electrical properties agree with a $4k_F$ charge localization, due to the opening of an electron–electron charge correlation gap in MEM(TCNQ)$_2$ (Sawatzky et al., 1979) and DBTTF-TCNQCl$_2$ (Mortensen et al., 1983). The nearly constant value ($\sim -k_B/e \ln 2$) of the high temperature thermoelectric power indicates that a somewhat similar picture could hold for NMP$_{0.59}$Phen$_{0.41}$TCNQ (Epstein et al., 1979) and Qn(TCNQ)$_2$ [Chaikin et al., 1979—see also Conwell (1979) for a critical discussion of the data]. Optical absorption data obtained in these latter salts (McCall et al., 1985) provide evidence for a Peierls-like gap in the temperature range where the $4k_F$ distortion is observed. This gap is, however, small and decays slowly with the temperature (see Conwell and Howard, 1986 for a discussion of its nature and of its temperature dependence). In TMTTF-Bromanil, electrical measurements (Torrance et al., 1981) show also that the charge degrees of freedom are not completely frozen by the $4k_F$ distortion.

Well decoupled from the $4k_F$ instability, a $2k_F$ instability occurs at low temperature in the form either of a second-order phase transition in ordered crystals or of temperature-dependent fluctuations in disordered crystals. The phase transition of MEM(TCNQ)$_2$ and DBTTF-TCNQCl$_2$ is observed in the magnetic susceptibility, but not in the electrical conductivity, which shows, that only the (nonfrozen) spin degrees of freedom are involved. In TMTTF-Bromanil the 78 K transition is observed both in the magnetic susceptibility and the electrical conductivity (Torrance et al., 1981). It means that this $2k_F$ phase transition, which occurs on the TMTTF stack, as evidenced by vibronic absorption spectra, also involves the charge degrees of freedom (a situation more likely decribed by intermediate coupling theories). There is, however, a substantial difference between these different salts on the basis of the anisotropy of $2k_F$ fluctuations:

1. Weak and nearly isotropic fluctuations are observed in MEM(TCNQ)$_2$ (Van Bodegom et al., 1981; Pouget, unpublished results), TMTTF-Bromanil (Kagoshima et al., 1983b) and DBTTF-TCNQCl$_2$ (Mortensen et al., 1983). By the isotropy of the $2k_F$ fluctuations, these salts resemble to the spin Peierls insulator TTF-CuBDT (Bray et al., 1983).

2. Quasi-1D fluctuations are observed in Qn(TCNQ)$_2$ (Pouget, 1981), NMP$_{0.59}$Phen$_{0.41}$TCNQ (Epstein et al., 1981; Pouget et al., 1982a). From the anisotropy of $2k_F$ fluctuations, these salts resemble to the spin Peierls "conductors" (TMTTF)$_2$PF$_6$ (Pouget et al., 1982b) and Per$_2$Pd(mnt)$_2$ (Henriques et al., 1984).

Case 1 is generally observed in salts with well-localized charges, while case 2 is observed in salts showing a progressive increase of charge localization upon lowering the temperature. In the second case the phonon dynamics can still be considered as slow compared to that of the electronic degrees of freedom. This leads to 1D fluctuation effects for a Peierls transition, driven by the instability of the 1D electronic gas. The first case corresponds probably to the opposite situation. The phase transition is uniquely driven by the interchain coupling, giving rise only to 3D fluctuations, which are generally weak (Schulz, 1987).

Finally, let us mention that the study of the NMP$_x$Phen$_{1-x}$TCNQ solid solution furnishes the first example where an external parameter (the Phen concentration) shifts the Coulomb interactions from weak values ($2k_F$ instability alone) to strong values ($2k_F$ and $4k_F$ instabilities, then $4k_F$ instability alone) on the TCNQ stack. This effect is apparently correlated with the emptying of conduction electrons from the donor stack (Epstein *et al.*, 1981; Pouget *et al.*, 1982a).

d. Physical Parameters Controlling the $2k_F$ and $4k_F$ Instabilities

Table VI shows that among the salts with two conducting subsystems, those having donor molecules based on the TTF entity (TTF, TMTTF, HMTTF) present the $4k_F$ instability. The only case among the Se analogues is TMTSF-DMTCNQ, which has a quarter-filled band ($\rho = \tfrac{1}{2}$). However, Fig. 20 shows that for the same value of ρ the $4k_F$ instability of TMTTF-DMTCNQ diverges at low temperature, as distinguished from TMTSF-DMTCNQ. These data, and the more direct evidence derived from the study of the solid solution TTF$_{1-x}$TSF$_x$TCNQ (see Section 12), allow us to ascribe the $4k_F$ scattering to the donor stack. In these salts the $4k_F$ instability has not the dominant divergence at low temperature (except, perhaps, in TMTTF-DMTCNQ). The $2k_F$ instability drives the Peierls transition. In TTF-TCNQ and TMTTF-DMTCNQ (and perhaps in HMTSF-TNAP), it belongs to the acceptor stack, while in TSF-TCNQ, HMTSF-TCNQ, TMTSF-DMTCNQ, and TMTSF-TCNQ it is situated on the donor stack (see Section 12). The $4k_F$ instability can also occur on acceptors (TCNQ), but in salts with only one conducting chain, like MEM(TCNQ)$_2$ and Qn(TCNQ)$_2$. In this respect, the study of the NMP$_x$Phen$_{1-x}$TCNQ solid solution shows clearly that the $4k_F$ instability develops when, by varying x, one passes from partial charge transfer salts to nearly complete charge transfer (to the TCNQ) salts. In these last salts, the $4k_F$ instability has a dominant role. In this respect, they differ from TTF-TCNQ and its family. A dominant $4k_F$ instability has also been observed in salts like TMTTF-Bromanil and DBTTF-TCNQCl$_2$ which have a partial charge transfer. But the stack on which it develops is not known.

(1) $4k_F$ *instability*. Among all the physical parameters that could influence the temperature dependence of the $4k_F$ instability, we have previously mentioned the thermal contraction of the intrachain lattice parameter. A decrease of this parameter certainly inhibits the divergence of the $4k_F$ response function by decreasing the relative value of the electron–electron

interactions with respect to the one-electron delocalization energy. It could explain the nondivergence at low temperature of the $4k_F$ instability of TMTTF-TCNQ and TMTSF-DMTCNQ. In contrast, the tendancy at a $2k_F$ instability, for the stack bearing the $4k_F$ instability, is probably increased by such lattice contraction effects. In a similar way, the decrease of the chain parameter under pressure, in addition to the associated increase of the charge transfer in the case of TTF-TCNQ (see below), disfavors the $4k_F$ instability.

Let us now discuss the physical parameters at the origin of the $4k_F$ instability. First, it has often been argued that, with comparable electron–electron interactions on both stacks, the $4k_F$ scattering arises on the stack of smallest bandwidth. In particular, it was explained that charge transfer salts with TSF-based donor molecules do not exhibit the $4k_F$ instability (on the donor stack) because of their larger donor bandwidth, due to the presence of Se orbitals of larger spatial extent. However, from a recent determination of donor and acceptor bandwidths, based on optical data (Jacobsen, 1985), it is easy to see that bandwidth (or Fermi energy) effects alone cannot explain the $4k_F$ scattering, because

- the $4k_F$ instability appears in TTF-TCNQ but not in HMTSF-TNAP: These two salts have the same charge transfer ($\rho \sim 0.58$) and the same small bandwidth ~ 0.4 eV (but corresponding to the D and A stack, respectively)
- the $4k_F$ instability is observed on the D stack of TMTSF-DMTCNQ, but not in that of TSF-TCNQ, which has a smaller bandwidth (0.62 eV against 0.93 eV). (These two salts differ, however, by the average number of holes per D molecule 0.50 against 0.63—see Table III).

These examples show that the nature of the molecule and the value of the charge transfer are two important parameters to consider. We shall see below that they influence the $4k_F$ CDW instability through electron–electron interactions.

Before, it should be mentioned that numerical calculations, quoted in the last subsection (Chui and Bray, 1980; Hirsch and Scalapino, 1983 and 1984), show that substantial nearest neighbor Coulomb interactions, V, are essential for the development of a $4k_F$ instability. In the strong coupling limit, where $4k_F$ corresponds to the first Fourier component of a (Wigner) lattice of localized charges, it is obvious (for ρ different from 1) that the formation of a periodic lattice is conditioned by the value of the Coulomb interaction terms V_i between the ith neighbors (Hubbard, 1978; Kondo and Yamaji, 1977). A first neighbor Coulomb interaction term V comparable to the on-site term U is not surprising in molecular crystals for which the distance between first-neighbor molecules is comparable to intramolecular distances; various estimates (Hubbard, 1978; Mazumdar and Soos, 1981) give $U \sim 2$–$3V$. It is, however, difficult to evaluate precisely how U and V vary with the nature of the molecular stack and with its surrounding (effect of screening by conduction and polarization of neighboring chains). However, it is expected that Coulomb interactions are reduced for molecules of larger size or higher polarizability [i.e., by adding cycles (TTF → HMTTF, TCNQ → TNAP) or by making the substitution S → Se].

The observation of a $4k_F$ instability in the Se analogue TMTSF-DMTCNQ, which has the smallest charge transfer ($\rho = 0.5$), and the interchange of the temperature range of the 1D fluctuations at $2k_F$ and $4k_F$ between HMTTF-TCNQ and TMTTF-DMTCNQ, which have comparable bandwidth (~ 0.8 eV) on both stacks, but different charge transfer (0.72 and 0.50, respectively), suggest a ρ dependence in the competition between the $2k_F$ and $4k_F$ instabilities. It has been pointed out (Mazumdar and Bloch, 1983; Bloch and Mazumdar, 1983; Mazumdar et al., 1984) that the short-range Coulomb interactions depend on the band-filling ρ through correlation effects that fix the number of statistically available electronic configurations. These authors argue that short-range Coulomb interactions, which they consider to be at the origin of the $4k_F$ scattering, remain large for quarter-filled ($\rho = \frac{1}{2}$) and half-filled ($\rho = 1$) bands and are smaller for intermediate band filling. In the strong coupling limit it can be easily seen that the tendancy at the formation of a $4k_F$ CDW is enhanced near these limits, because a long range Wigner lattice of localized charges requires only U for $\rho = 1$ and U and V for $\rho = \frac{1}{2}$.

In addition to the band-filling effect, other physical parameters may influence the $4k_F$ instability. For example, in the series HMTTF-TCNQ, TTF-TCNQ, TMTTF-TCNQ, and TMTTF-DMTCNQ showing $4k_F$ scattering, the strongest divergence of the $4k_F$ response function, observed for TTF-TCNQ, could be due to the smallest bandwidth of the TTF stack (0.4 eV against 0.8 eV for the other D stacks). Also the presence of a long and highly polarizable HMTTF molecule could act against a strong $4k_F$ instability in HMTTF-TCNQ.

Band-filling effects are unable to explain why the $4k_F$ instability of the $NMP_xPhen_{1-x}TCNQ$ solid solution appears when x becomes lower then $x_c \simeq \frac{2}{3}$, because there is not a sizable change in the electronic density on the TCNQ stack for x around x_c (see Fig. 10). The simultaneous appearance of the $4k_F$ scattering with the emptying of donor stacks of mobile conduction electrons in the solid solution $NMP_xPhen_{1-x}TCNQ$ seems to indicate that reduction of interchain screening effects has a drastic influence on the value of electron–electron interactions. From the expression (12) showing how the screening medium contributes to the reduction of the forward Coulomb interactions g_2, an enhancement of g_2 is expected from the emptying of neighboring chains of mobile electrons. With the increase of g_2, the $4k_F$ CDW response function can have a faster rate of increase than the $2k_F$ one [see Eq. (11), for example].

It has also been proposed (Shaik and Whangbo, 1986) that even in the absence of strong Coulomb interactions, the $4k_F$ instability might be due to intramolecular structural localization. In their model the intrasite electronic localization is favored by the strong tendency for intramolecular relaxation in the sulfur-based donor molecules. However, it is not obvious that this effect produces a charge ordering with a $4k_F$ wave vector. More likely, as shown by Beni *et al.* (1974), the effective (attractive) coupling between neighboring polarons (entities formed by the localized electron and the internal deformation of the molecule) leads to the formation of a lattice of pairs of polarons, where the first Fourier component is $2k_F$. In a general way, the effect of energetic intramolecular phonon modes (i.e., with frequencies higher than $2\pi k_B T$) is to renormalize the Coulomb interaction constants g_i by adding to the repulsive Coulomb terms [considered in Eq. (13)] a negative contribution of virtual exchange of phonons between two electrons (see, for example, Barisic and Brazowskii, 1981). Its effect is thus to reduce the pure Coulomb interaction terms g_1 and g_2.

Finally, for completeness, let us mention briefly two other models that have been proposed to explain the $4k_F$ scattering of TTF-TCNQ, as a second harmonic effect of a primary $2k_F$ instability:

1. Anharmonic motion of $2k_F$ librons that are softened by the Kohn anomaly (Weger and Friedel, 1977). It predicts polarizations of atomic displacements that are not observed.

2. Lattice distortion at the second harmonic of the $2k_F$ wave vector of a spin-density wave (SDW) instability of the electron gas (Weger and Gutfreund, 1979). Until now there is no experimental evidence of a SDW instability in TTF-TCNQ and related compounds that show the $4k_F$ scattering.

(2) *$2k_F$ instability.* Before discussing $2k_F$ instability, it should be pointed out that the same stack may present both $2k_F$ and $4k_F$ instabilities, as shown by the theory. This is true for the TMTSF stack in TMTSF-DMTCNQ. However, with a dominant $2k_F$ instability that drives the Peierls transition, TMTSF-DMTCNQ is different from the one-chain conductors like MEM(TCNQ)$_2$, which first show a high-temperature $4k_F$ instability then a $2k_F$ spin-Peierls instability. This last scenario is well understood in the strong coupling limit.

Experimentally, the $2k_F$ instability of the two-conducting-chain salts seems to be driven by a single stack (TCNQ in TTF-TCNQ, TSF in TSF-TCNQ, for example). The details of the interactions that make the difference between

D and A stacks for the development of the $2k_F$ instability are probably very subtle, as shown by HMTSF-TCNQ and HMTSF-TNAP, where the $2k_F$ instability apparently arises on the D and A stacks, respectively. However, for a particular species of stacks, the divergence of the $2k_F$ instability seems to be controlled by the value of the charge transfer which, according to the previous discussion of the $4k_F$ instability, is an important parameter for determining the strength of the electron–electron interactions. Table VI shows that for the $2k_F$ instability on

- the donor stacks: the passage from a logarithmic-like divergence (HMTSF-TCNQ and TSF-TCNQ) to a power law divergence (with an exponent that increases from TMTSF-TCNQ to TMTSF-DMTCNQ) occurs for decreasing charge transfers (see Table III). The bandwidth of the TMTSF stack ~ 1 eV is intermediate between that of the TSF (0.6 eV) and HMTSF (1.3 eV) stacks (see Jacobsen, 1985)
- the acceptor stacks: the divergence increases from HMTTF-TCNQ to TTF-TCNQ and TMTTF-TCNQ, a sequence along which there is decrease of the charge transfer (without an important change of bandwidth: $W_A \sim 0.8$ eV). With comparable values of charge transfer ($\rho \sim 0.58$) the decrease of the $2k_F$ divergence from TTF-TCNQ to HMTSF-TNAP could be due to the decrease of the Coulomb interactions on the acceptor stack by the substitution TCNQ $\rightarrow$ TNAP, in spite of a decrease of the acceptor bandwidth by a factor 2.

The study of TTF-TCNQ and its family may suggest that the $2k_F$ CDW instability always occurs in 1D conductors. This is, in fact, not true, and theories of 1D electron gas (see, for example, Solyom, 1979; Emery, 1979) show that the $2k_F$ CDW instability requires special values of the electron–electron interactions. In this context it is useful to recall that 1D conductors may also present SDW or superconducting instabilities, as recently observed in the $(TMTSF)_2 X$ series (see, for example, Jérome and Schulz, 1982a). In the salts with $X = PF_6$ and AsF_6 only a very weak $2k_F$ CDW instability, which vanishes at low temperature, has been observed (Pouget et al., 1982). However, because of the commensurate value ($\rho = \frac{1}{2}$) of the charge transfer the periodicity of the anion (X) sublattice coincides with the $4k_F$ wave vector of the 1D electron gas. Umklapp terms (g_3) in the electron–electron interactions have thus to be considered. It is predicted (Barisic and Brazowskii, 1981; Emery et al., 1982), as already mentioned, that a relevant g_3 coupling favors the $2k_F$ SDW and $4k_F$ CDW, at the expense of the $2k_F$ (site) CDW. A g_3 coupling is also expected in $NMP_{0.5}Phen_{0.5}TCNQ$ because the NMP/Pen ordering wave vector coincides with the $4k_F$ wave vector of the TCNQ electron gas (see Part II, Section 3).

Finally, let us remark that although we have only considered the stack where the dominant $2k_F$ instability develops, the study of phase transitions and of the electronic properties below T_P show that both stacks undergo a $2k_F$ distortion. Such a distortion may result either from the coupling between the stacks (in the 3D fluctuation regime) or from the development of another 1D $2k_F$ instability on the other stack. The only direct experimental evidence concerns NMP-TCNQ (Pouget et al., 1980a), where the $2k_F(q_1)$ instability of the NMP stack manifests itself below about 70 K, in the regime of 3D coupling between the $2k_F(q_2)$ CDW on the TCNQ stack. In TTF-TCNQ it has been suggested on the basis of optical data (Jacobsen, 1987) that the $2k_F$ CDW on the TTF stacks develops below about 60 K, the cross over temperature from the 1D to the 3D fluctuation regimes.

(3) *Coulomb interactions.* We have previously discussed in a qualitative way some physical parameters that control the $2k_F$ and $4k_F$ instabilities. Among them are the Coulomb interactions between electrons. However, it must be realized that the components of the Coulomb interactions affect the physical quantities differently. In this respect, it is often argued (Mazumdar and Bloch, 1983) that the $4k_F$ instability occurs on the stack that has the strongest magnetic susceptibility. However, susceptibility decompositions performed in TTF-TCNQ (Takahashi et al., 1984) and TMTSF-DMTCNQ (Tomkiewicz et al., 1978) assign the strongest susceptibility to the acceptor stack and the weakest one to the donor stack, which undergoes the $4k_F$ instability. This difference can be understood in the weak coupling limit where

- the forward Coulomb interaction g_2 favors the $4k_F$ instability [see Eq. (11)],
- the backward Coulomb interaction g_1 favors the uniform magnetic susceptibility (Lee et al., 1977).

Thus, if, in TTF-TCNQ, g_1 dominates on the acceptor stack and g_2 dominates on the donor stack, it can be easily understood that the $4k_F$ instability does not occur on the stack that has the largest magnetic susceptibility (Barisic and Bjelis, 1985; Barisic, 1985). In the framework of the nearest neighbor extended Hubbard model, it is easy to see, with the ratio obtained from Eq. (13),

$$\frac{g_2}{g_1} = \frac{U + 2V}{U + 2V \cos \pi \rho}, \quad (14)$$

that, because of the coefficient $\cos \pi \rho$ for $\frac{1}{2} < \rho < 1$, V favors g_2 and disfavors g_1. Through the combination g_2^*, V enhances the $2k_F$ and $4k_F$ CDW response function, in agreement with numerical calculations. Thus, assuming comparable value of U, the stack with the larger V will show the $4k_F$ instability and the smallest magnetic susceptibility.

Finally, from the overall behavior of the $2k_F$ and $4k_F$ response functions, it is easy to see that Coulomb interactions [or, more likely, the combination entering under the form $g_2^* \sim g_2 - g_1/2$ in Eq. (11)] increases roughly along the sequence

a. Se-based donor stacks (HMTSF, TSF, TMTSF)
b. Acceptor stacks (TCNQ, TNAP, DMTCNQ)
c. S-based donor stacks (HMTTF, TTF, TMTTF).

These three classes are based on the observation that stacks of type b have a stronger $2k_F$ CDW divergence than stacks of type a, while only stacks of type c show the $4k_F$ CDW instability. Within each class of stacks, the values of the charge transfer, the bandwidth, and the intramolecular and intermolecular Coulomb terms influence the value of the appropriate combination of electron–electron interactions, as previously discussed. The difference between the classes a, b, c of stacks is probably due to molecular quantities like the polarizability. This classification holds for salts with two conducting stacks. In salts with one conducting stack, Coulomb interactions are probably enhanced by the reduction of interchain screening effects, as observed for acceptor (TCNQ) based salts. The same effect probably occurs in donor-based salts like $(TMTSF)_2X$ and $(TMTTF)_2X$. However, in these one stack salts one has to also account for the effects of the periodic potential of the anion/cation sublattice (see, for example, Pouget, 1987) and the fact that, in the surroundings of a given stack the anions or cations are less polarizable than the organic molecules.

14. Conclusion

In summary, the spatial correlations of CDW fluctuations observed above the Peierls transition are relatively well characterized by x-ray diffuse scattering studies. However, because of the lack of large enough single crystals, the associated temporal fluctuations have not been studied by neutron scattering; only a preliminary study has been devoted to TTF–TCNQ. Semiquantitative analyses of the spatial fluctuations have been performed in a large number of charge-transfer salts. In particular, they show that the CDW fluctuations vary appreciably among the various salts investigated. The trends are understood qualitatively. Experimental studies and theoretical estimates agree in recognizing the pertinent role of electron–electron interactions in the physics of organic conductors. A typical manifestation of such interactions is the observation of a $4k_F$ CDW instability of the electron gas in a great number of organic salts. In materials like TTF–TCNQ, the electron gas seems to experience electron–electron interactions of intermediate strength. In this coupling range, the effect of electron–electron interactions is only known by numerical simulations. Although the basic experimental features are accounted for by such treatments, analytic calculations are awaited for a more quantitative understanding of the physical properties.

In the detail of the electron–electron interactions, the charge-transfer organic salts differ appreciably from the inorganic 1D conductors like KCP, $NbSe_3$, or $K_{0.3}MoO_3$, where the CDW fluctuations are dominated by the electron–phonon coupling. The differences between the various families of 1D conductors begin to be understood. In particular it is expected that the $4k_F$ instability which requires comparable values of U and V can only be realized in chains composed of large entities molecules. In this respect, some key parameters controlling the $2k_F/4k_F$ CDW electronic instability of organic conductors are identified. However, their microscopic calculation is only poorly assessed.

V. Three-dimensional CDW Ordering Phase Transitions

15. Introduction

In Part IV, we have shown that the intrachain CDW response function, χ_ρ^{1D}, diverges at low temperature. Thus a modest interchain coupling leads, below T_{co}, to a 3D CDW coupling and below T_P to a periodic lattice distortion (PLD). The crossover temperature from the 1D to the 3D regime of fluctuation has already been considered in Section 9. Here we shall analyze the long-range ordering of the CDW below T_P. However, while the wavevector component in the chain direction is determined by the divergence of χ_ρ^{1D}, the wavevector components perpendicular to the chain direction are determined by the interchain interactions and the 3D crystal structure.

Among all the compounds considered in Part IV, long-range order is observed only in the nondisordered salts of the TTF-TCNQ family. The modulation wave vectors of the PLD are given in Table IV (p. 112). Except in TTF-TCNQ (below 38 K), no structural refinement of the lattice modulation has been performed in the Peierls CDW insulating state. The stacks involved in the Peierls transition are mostly known by local measurements (NMR, EPR, and absorption spectra—see table VII) and the analysis of pretransitional fluctuations (Section 12).

The format of this Part is as follows: in Section 16, we give several mechanisms of interchain coupling. In Section 17, we present the experimental phase diagrams. Then theoretical analysis of the phase transitions will be reviewed in Section 18 in the framework of the Landau theory, taking into account the (3D) crystal symmetry.

TABLE VII

STACKS INVOLVED IN THE PEIERLS TRANSITIONS OF CHARGE TRANSFER SALTS WITH TWO CONDUCTING CHAINS

	Critical temperature (K)	Stack involved	Experimental technique	References
TTF–TCNQ	52 (H); 54 (D)	TCNQ	TCNQ (C^{13}) NMR	Rybaczewski et al. (1976)
			g factor decomposition	Tomkiewicz et al. (1976, 1977c)
	49	Start of TTF	Vibronic absorption	Bozio and Pecile (1981)
	38	TTF	TTF (C^{13}) NMR	Takahashi et al. (1984)
			g factor decomposition	Tomkiewicz (1980)
HMTTF–TCNQ	50–49	TCNQ	Vibronic absorption	Bozio et al. (1983)
	43–42	HMTTF		
TMTTF–TCNQ	~40	TMTTF and TCNQ	Vibronic absorption	Bozio et al. (1983)
TMTTF–Bromanil	78	TMTTF	Vibronic absorption	Torrance et al. (1981)
TMTSF–DMTCNQ	42	TMTSF (mostly)	EPR g value	Tomkiewicz et al. (1978)
TSF–TCNQ	29	TSF (mostly)	Far infrared spectra	Bates et al. (1981)

16. Interchain Couplings

The best way to discuss the interchain (bilinear) coupling is to start with a generalized form of the RPA CDW susceptibility [Eq. (A.21)],

$$\chi_\rho(\mathbf{q}) = \frac{\chi_e(T, \mathbf{q})}{1 - [|\lambda(\mathbf{q})/\hbar\omega(\mathbf{q})|^2 - g_1(\mathbf{q}_\perp)]\chi_e(T, \mathbf{q})}, \quad (15)$$

where the $\mathbf{q}$ dependence of the various quantities defined in the appendix is explicitly introduced ($\mathbf{q} = 2k_F \mathbf{b}^* + \mathbf{q}_\perp$, taking b as the chain direction). To be general, we have also considered that, with respect to $k_B T$, the Fermi surface might deviate from a plane, due to the finite value of the interchain transfer integrals ($t_\perp$). This effect is included in the expression of $\chi_e(T, \mathbf{q})$. Among the various quantities considered in Eq. (15), only the electronic response function depends strongly on the temperature. Starting from high temperatures, the phase transition stabilizes, in the RPA approximation, the wavevector $\mathbf{q}_c$ that first leads to the divergence of Eq. (15). Three kinds of coupling may fix the value of $\mathbf{q}_\perp$ (the $q_\parallel$ component is determined by the $2k_F$ divergence of χ_e):

1. 3D band structure effect: $\mathbf{q}_\perp$ is the wavevector for which $\chi_e(2k_F, \mathbf{q}_\perp)$ shows a maximum value, This occurs for the best-nesting wavevector of the Fermi surface (Horowitz et al., 1975).

2. Interchain (direct and mediated) Coulomb interactions between CDWs, described by $g_1(\mathbf{q}_\perp)$ in Eq. (15). Because of the oscillatory nature of the CDW, the direct Coulomb interaction between two chains decreases exponentially with the interchain distance [Saub et al., 1976—see Eq. (16)]. Also, mediated interactions, where the interchain Coulomb coupling is accompanied by a virtual excitation of the electron gas or polarization on a third chain (Weger and Friedel, 1977; Noguera, 1985), as well as Van der Waals interactions, where the coupling results from simultaneous virtual excitations on the two coupled chains (Barisic and Bjelis, 1983) might be important to consider. $\mathbf{q}_\perp$ is thus the wavevector that minimizes the total interchain Coulomb interactions.

3. Elastic interchain coupling: $\mathbf{q}_\perp$ is thus the transverse wavevector for which there is a preexisting minimum in the phonon dispersion relationship $\omega(2k_F, \mathbf{q}_\perp)$ of the lattice mode coupled to the 1D electron gas. For this wavevector, the dimensionless electron phonon-coupling constant, entering in Eq. (15), has a maximum value.

Let us examine the CDW couplings 1 and 2 for the case of a 2D orthorhombic lattice of identical chains:

1. With the array shown in Fig. 29a, a finite transverse overlap integral $t_\perp$ yields a warped Fermi surface, for which the best nesting wavevector

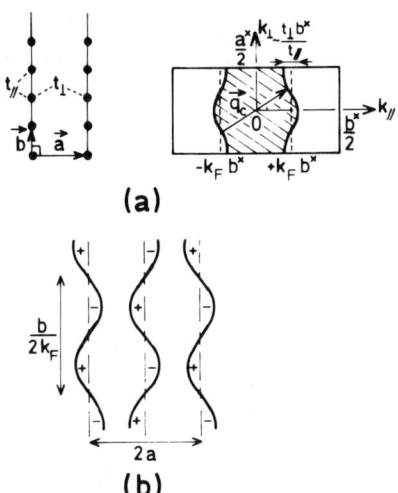

FIG. 29. (a) Warped Fermi surface resulting from finite tunneling coupling ($t_\perp$) between chains. The best nesting wavevector, ($\frac{1}{2}$, $2k_F$), of the Fermi surface is indicated. (b) Out-of-phase coupling between CDW, which minimizes the direct Coulomb interaction. The sign of the CDW modulation with respect to the average charge per chain is indicated.

is ($\frac{1}{2}$, $2k_F$); i.e., $q_\perp = \pi/a$. In a crystal of lower symmetry, the Fermi surface may allow an incommensurate nesting wavevector component $q_\perp$ (Yamaji, 1982). This can occur also if, in the orthorhombic symmetry, $t_\perp$ is a sizable fraction of $t_\parallel$ (Jafarey, 1977; Montambaux, 1985).

2. The direct Coulomb interaction between CDW of amplitude (phase) $\rho_i(\theta_i)$ and $\rho_j(\theta_j)$ belonging to chains i and j, at a distance d_{ij} is given by (Saub et al., 1976)

$$W_{ij} = V_{ij}\rho_i\rho_j \cos(\theta_i - \theta_j), \tag{16}$$

with $V_{ij} = e^2/d_{ij} K_0(2k_F d_{ij})$. $K_0(x)$ is a modified Bessel function, which decays very rapidly for increasing x [$K_0(x) \sim e^{-x}/\sqrt{x}$). For Coulomb interactions between CDW, Eq. (16) is minimum for a phase shift of π between neighboring chains (Fig. 29b). Incommensurate ordering may arise when the (**b**, **a**) angle deviates from $\pi/2$, because the phase shift of π is not strictly aligned with the a direction.

In charge transfer salts, the lateral ordering is more difficult to explain, since each kind (D and A) of stack is metallic and tends to form its own CDW sublattice. The order within each sublattice is governed by the coupling described above, but it is also influenced by the CDW order on the other sublattice. Several situations arise, depending upon the strength of the intersublattice (V_{AD}) and intrasublattice (V_{AA} and V_{DD}) couplings and on the temperature at which each sublattice becomes unstable (see Bjelis and Barisic, 1977, 1978 in the case of Coulomb couplings). We shall come back to them in Section 18. Let us consider here the simplest situation of a dominant intersublattice coupling ($V_{AB} \gg V_{AA}, V_{DD}$). The semimetallic Fermi surface resulting from strong interchain tunneling coupling (hybridization) between like and unlike stacks shown in Fig. 30a could allow (0, $2k_F$) as a nesting wavevector if the $t_\perp$'s are not too different. The same transverse periodicity is also stabilized by dominant Coulomb interactions between oppositely charged CDWs (Fig. 30b).

Coupling by transverse tunneling effects gives the same temperature dependence for the longitudinal ($\xi_\parallel$) and transverse ($\xi_\perp$) correlations lengths in the 3D fluctuating regime between T_{co} and T_P (Horovitz et al., 1975). When one has $(t_\perp)^2/t_\parallel \gtrsim k_B T_P$ the best nesting wavevector does not succeed in completely nesting the Fermi surface into itself at T_P. Small electron-hole pockets remain below T_P (at least in the temperature range where the partial gap opened is such that: $\Delta < (t_\perp)^2/t_\parallel$). This gives a semimetallic character at the phase. The observation of low-temperature semimetallic states suggests that this coupling might be important in salts containing the HMTS(T)F molecules like HMTSF-TNAP (Bechgaard et al., 1978), HMTSF-TCNQ (Jérome and Schulz, 1982 and references therein), and pressurized HMTTF-TCNQ (Friend et al., 1978b).

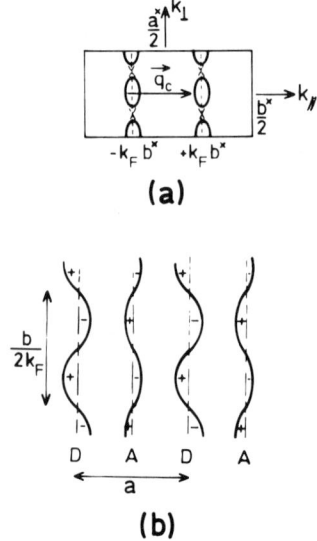

FIG. 30. (a) Semimetallic Fermi surface that results from a strong hybridization between D and A stacks, and its nesting wavevector (0, $2k_F$). (b) Ordering of D and A CDW for a strong direct Coulomb interaction between neighboring stacks.

In TTF–TCNQ, various estimates (Bjelis and Barisic, 1977; Hartzstein et al., 1980a) show that the transverse coupling is probably due to Coulomb interactions. Interchain transfer integrals are of order $t_\perp \sim 1$ meV ($< k_B T_P$), and the transfer integral between nearest neighbors along the a axis vanishes because of the opposite inversion symmetry of the TTF (b_{1u}) and TCNQ (b_{2g}) wavefunctions (Berlinsky et al., 1974; Shitzkovsky et al., 1978).

17. Phase Diagrams

a. *TTF–TCNQ*

(1) *Ambient pressure.* TTF–TCNQ presents three-thermodynamic phase transitions at ambient pressure, detected by specific heat anomalies (Craven et al., 1974; Djurek et al., 1977). The upper phase transition is of second order and has a critical temperature T_H, which differs slightly in protonated and deuterated (TCNQ chain) samples (Cooper and Lukatela, 1979). The lower phase transition, T_L, is first order and also sensitive to deuteration of the TTF chain. These two critical temperatures are easily detected in the temperature dependence of the electrical conductivity (i.e., peaks in $\partial \log \sigma / \partial T$; Coleman et al., 1973; Ferraris et al., 1973; Jérome et al., 1974; Cooper et al., 1975). The intermediate critical temperature, T_M, is more difficult to detect. It was discovered after a reanalysis by Bak and Emery (1976) of the initial report of the temperature dependence of the q_a component of the incommensurate modulation (Comès et al., 1975). Structural data provide the best method to characterize the three low-temperature modulated phases of TTF–TCNQ. The results of several low-temperature x-ray and neutron structural investigations (Comès et al., 1975 and 1976; Kagoshima et al., 1976; Khanna et al., 1977; Ellenson et al., 1976, 1977; Pouget et al., 1979a) can be summarized as follows:

1. Upper modulated phase (49 K < T < 54 K)

The x-ray pattern of Fig. 31a shows, by the observation of well-formed satellite reflections, that there is a 3D ordering of the $2k_F$ CDW with the reduced wavevector $\mathbf{q}_H = (\frac{1}{2}, 2k_F, 0)$ (Figs. 32 and 33). There is no ordering of the $4k_F$ CDW, which stays in the 1D fluctuation regime.

2. Intermediate modulated phase (38 K < T < 49 K)

At 49 K, the q_a component of the $2k_F$ ordering wavevector begins to deviate continuously from the commensurate value $\frac{1}{2}a^*$ (Figs. 32 and 33). The modulation wavevector $\mathbf{q}_M = [q_a(T), 2k_F, 0]$ has thus two incommensurate components. The $q_a(T)$ variation provides evidence for a progressive sliding of the CDW lattice in the a direction (direction along which sheets of TCNQ and TTF stacks alternate). Within the experimental errors, Fig. 33 shows that

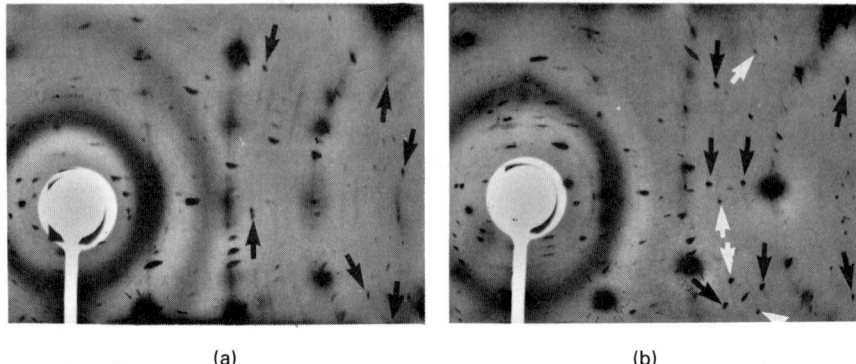

FIG. 31. X-ray diffuse scattering patterns from TTF–TCNQ. (a) At 50 K in the upper modulated phase. Note the presence of well-defined $2k_F$ satellite reflections (black arrows) with the reduced wavevector (0.5, 0.295, 0). In this phase the $4k_F$ scattering keeps its one-dimensional character (no $4k_F$ satellite reflection). (b) At 30 K in the lower modulated phase, $2k_F$ and $4k_F$ satellite reflections at the reduced wavevector (0.25, 0.295, 0) and (0.5, 0.41, 0) are visible and are shown, respectively, by the black and white arrows. [From Khanna et al. (1977).]

$(q_a - \frac{1}{2})^2 = \delta q_a^2$ varies like $(T_M - T)$. The sliding is extremely sensitive to strain effects. Near T_M a continuous ridge of elastic intensity, linking the satellites along a^*, and symmetrically placed with respect to the high-temperature wavevector component $q_a = \frac{1}{2}a^*$, was observed for crystals constrained too rigidly (Ellenson et al., 1976). It has been interpreted as being due to a distribution of q_a components induced by an inhomogeneous distribution of strain in the sample. Linear coupling between external strains and the order parameter has been considered by Bak (1976). Small deviations (<0.04%) from the normal temperature dependence of the β monoclinic angle, which can also be due to strain induced by the CDW sliding, have also been reported between T_M and T_L (Thomas et al., 1979). In addition, data of Figs. 33a and 33b taken, respectively, upon heating and upon cooling, show a global hysteresis of about 1.5–2 K in the temperature dependence of the $q_a(T)$. Furthermore, thermal cycles performed in this intermediate phase (Ellenson et al., 1977; Pouget et al., 1979a) have shown the true character of this hysteresis phenomenon. Weak $4k_F$ satellite reflections, detected by x-ray below 45 K, show that the $4k_F$ CDW are coupled in this phase (because of their weakness it has not been possible to check if the 3D coupling occurs at T_M). Within experimental accuracy the, $4k_F$ satellite reflection position coincides with the $2\mathbf{q}_M$ harmonic wavevector (Fig. 32).

3. Low-temperature modulated phase ($T < 38$ K).

The x-ray pattern of Fig. 31b shows the presence of strong $2k_F$ and $4k_F$ satellite reflections at the reduced wave vectors $\mathbf{q}_L = (\frac{1}{4}, 2k_F, 0)$ and

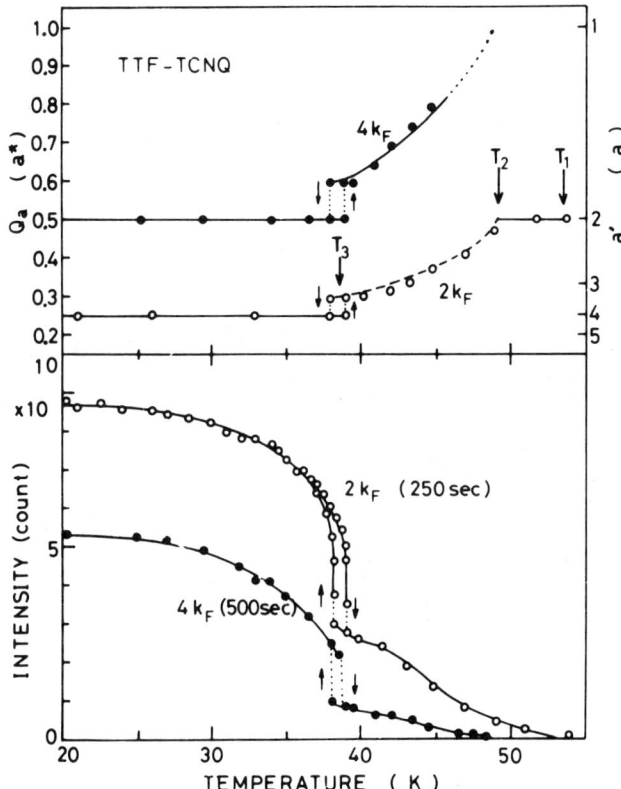

FIG. 32. Temperature dependence of the peak intensity and q_a component of the $2k_F$ and $4k_F$ satellite reflections of TTF–TCNQ in its three modulated phases. The critical temperatures $T_1(T_H)$, $T_2(T_M)$, and $T_3(T_L)$ are indicated. [From Kagoshima (1982).]

$2\mathbf{q}_L = (\frac{1}{2}, 4k_F, 0)$, respectively. The 38 K phase transition is first order, as it can be seen by the jump of the satellite intensity and of the wavevector component q_a (Figs. 32 and 33), and corresponds to the lock in of the q_a component of the $2k_F$ CDW at the commensurate value $\frac{1}{4}a^*$. This phase also corresponds to an antiphase ordering of the $4k_F$ CDW. This first-order phase transition is also accompanied by a slight increase of the $2k_F$ wavevector component from $0.293b^*$ to $0.295b^*$ (Pouget, Shapiro, and Shirane, unpublished results). A new set of $4k_F$ satellite reflections with the reduced wavevector $(0, 4k_F, 0)$ has also been detected below 38 K (Kagoshima et al., 1976). We shall discuss its origin later.

The diffraction from these three modulated phases is schematically represented in Fig. 34, which gives the position of the $2k_F$ and $4k_F$ satellite

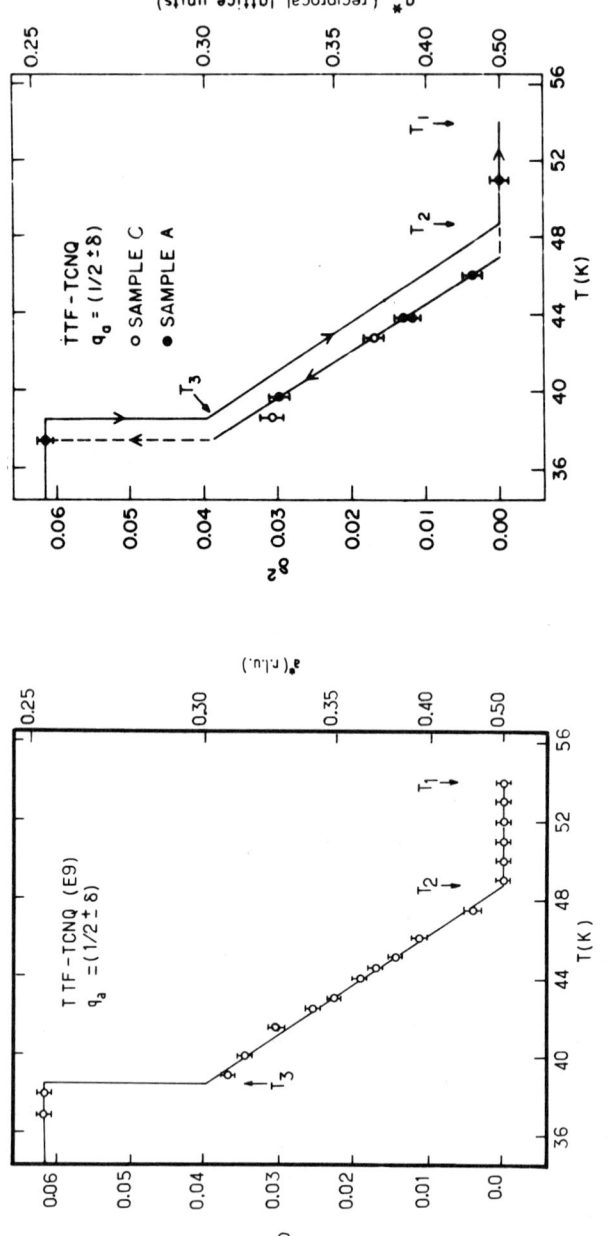

FIG. 33. Temperature dependence of the square of the deviation of the q_a component at the $a^*/2$ value: $\delta^2 = (\frac{1}{2} - q_a)^2$, ($\delta_{q_a}^2$ in the text), as a function of the temperature in TTF–TCNQ. It clearly shows the three phase transitions at $T_1(T_H)$, $T_2(T_M)$, and $T_3(T_L)$. The experimental points shown in (a) and (b) have been taken, respectively, upon heating and cooling. The global hysteresis is shown in (b). [From Ellenson et al. (1976, 1977).]

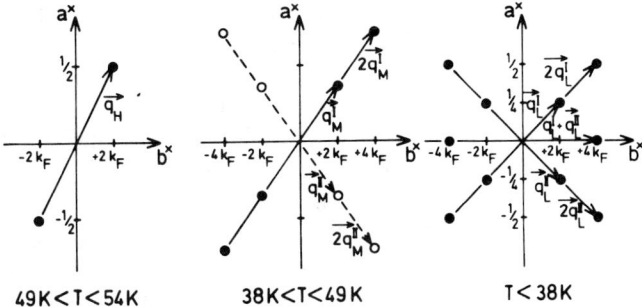

FIG. 34. Position, in the (a^*, b^*) reciprocal plane of the $2k_F$ and $4k_F$ satellite reflections observed in the three modulated phases of TTF-TCNQ at ambient pressure.

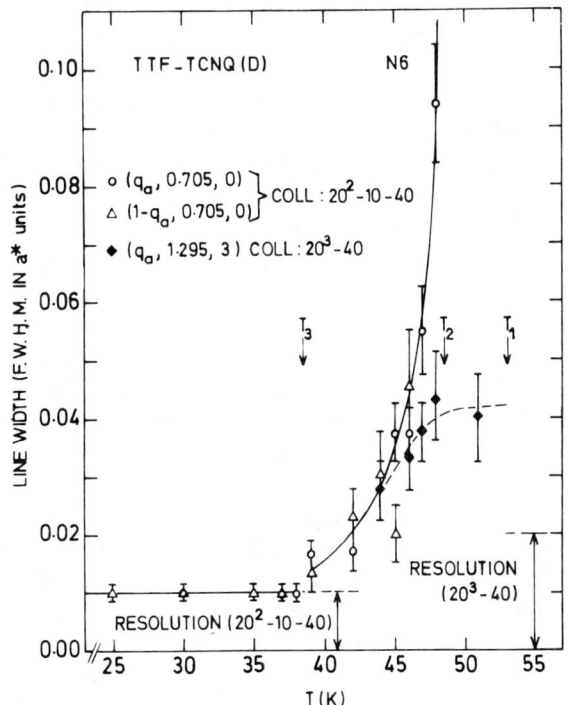

FIG. 35. Temperature dependence of the full width at half maximum (FWHM) along a^* for the "longitudinal" (empty symbols) and "transverse" (full symbols) $2k_F$ satellites of TTF-TCNQ. The experimental resolution is indicated as well as the $T_1(T_H)$, $T_2(T_H)$, and $T_3(T_L)$ critical temperatures. [From Pouget et al. (1979a).]

reflections. The corresponding structures have been classified in terms of four- and five-dimensional superspace groups, and extinction rules for satellite reflections have been predicted (Bak and Janssen, 1978). Finally, Fig. 35 shows that the $2k_F$ CDW order is not longe-range in the a direction (direction along which the TTF and TCNQ stacks alternate) above 38 K. In fact, the short-range order even differs between the $2k_F$ satellite reflections with $\ell \neq 0$ component ("transverse" satellite) and that with $\ell = 0$ component ("longitudinal" satellite). Furthermore, an important difference between these two kinds of satellite reflections is that the "longitudinal" satellites are observed only below 49 K (Pouget et al., 1979a). It seems now fortuitous that at ambient pressure, its onset temperature (T^*), coincides with the T_M critical temperature. A subsequent study of these satellite reflections shows a clear separation of the temperatures T^* and T_M under pressure (Megtert et al., 1982).

According to data quoted in Table VII, the T_H phase transition corresponds to a distortion of the TCNQ stacks. The TTF stacks start to distort at T_M, according to vibronic absorption data of Bozio and Pecile (1981). The TTF distortion increases markedly below the first-order phase transition at T_L, where the TTF gap becomes observable by magnetic measurements (Tomkiewicz, 1980; Takahashi et al., 1984).

A recent determination of the $2k_F$ modulated structure of TTF-TCNQ at 15 K (Coppens et al., 1987; Bouveret and Megtert, 1988) shows that the displacement pattern results in a coherent superposition of two modulation waves of wave vector q_L^I and q_L^{II} (defined by Fig. 34). The modulation consists in a slip of the TTF molecule, with an amplitude of about 0.02 Å, in the direction of its long molecular axis (as does the TSF molecule in TSF-TCNQ, see Section 12) and a displacement of the TCNQ of about 0.05 Å in a direction nearly perpendicular to its molecular plane. A boat-like deformation of the TCNQ molecule, resembling that observed in TEA(TCNQ)$_2$, seems also to occur (Bouveret and Megtert, 1988).

(2) *Pressure dependence.* The phase diagram of TTF-TCNQ has been studied under pressure by neutron elastic scattering experiments, which are most sensitive, due to the values of the neutron scattering lengths, to the distortion on the TCNQ stacks. The resulting phase diagram obtained on deuterated crystals, which combines the studies of Fincher et al. (1980) and Megtert et al. (1979a and 1979c, 1981 and 1982) is shown in Fig. 36a. It is compared in (b) with the phase diagram deduced from electrical measurements (peak anomaly in $\partial \log \sigma/\partial T$) performed on protonated samples (Friend et al., 1978a), which gives the pressure dependence of the temperature at which the electrical gap opens on the TCNQ (upper line) and TTF (lower line) stacks.

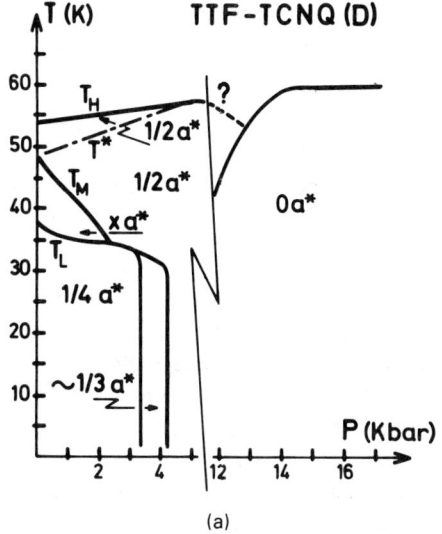

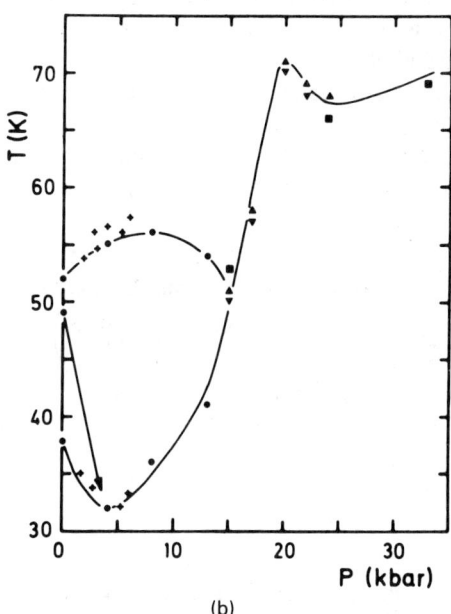

Fig. 36. (a) Phase diagram of TTF-TCNQ deduced from neutron elastic scattering measurement performed on fully deuterated single crystals. [From Megtert (1984).] (b) Phase diagram of TTF-TCNQ derived from electrical measurements performed on protonated single crystals. [From Friend et al. (1978).]

Let us first consider the low-pressure region ($P < 4$ kbar) of the phase diagram. The critical temperatures behave as follows:

- T_H increases slightly at a rate of about 0.5 K/kbar
- T_M decreases rapidly under pressure in such a way that the intermediate sliding phase is suppressed above 2 kbar
- T_L decreases slightly, but at about 3.3 kbar, the phase with $q_a = \frac{1}{4}a^*$ disappears. In a narrow pressure range of about 1 kbar of width, a new phase with $q_a \sim \frac{1}{3}a^*$ is stabilized.

Above 4.2 kbar, the phase with $q_a = \frac{1}{2}a^*$ is stabilized throughout the temperature range (Fig. 37). The temperature T^*, below which the $2k_F$ "longitudinal" satellite reflections are detected, increases strongly under pressure and coincides with T_M above 4–5 kbar. The T^* line cannot be considered to be a true phase boundary, because no change in the symmetry of the distortion or of the ordering wavevector is associated with this line. It corresponds merely to the addition of a new polarization component to the distortion of the TCNQ stacks (Megtert et al., 1983, 1985; see Section 18).

In this low-pressure range, the phase diagram obtained by structural measurements agrees relatively well with that deduced from transport studies; the two anomalies observed in electrical measurements occurring at T_H and "T_L". At 5 kbar, the two phase diagrams are in apparent contradiction: the T_H and T_L anomalies are still observed, around 56 K and 32 K, respectively, in transport measurements, while neutron data show only a single phase transition at 57 K, with a regular increase of the satellite intensity (Fig. 37), which stabilizes over the entire temperature range the $(\frac{1}{2}, 2k_F, 0)$

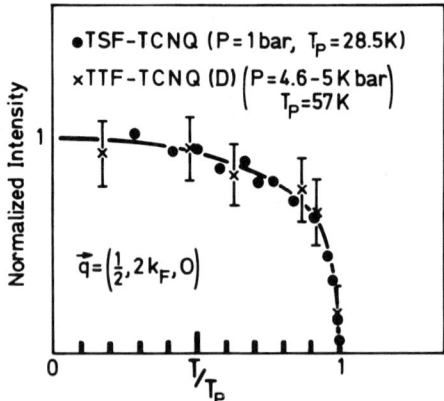

FIG. 37. $(\frac{1}{2}, 2k_F, 0)$ satellite intensities as a function of normalized temperature for TSF–TCNQ [from Kagoshima et al. (1978a)] and TTF–TCNQ at 5 kbar [from Megtert et al. (1979a)], showing the same temperature dependence below the second-order phase transition.

CDW order. Taken together, these data lead to the surprising result that the opening of a gap on the TTF stack below 32 K at 5 kbar and the onset of an associated distortion does not affect the $\frac{1}{2}a^*$ component of the transverse modulation. The distortion of the TTF stack should be detected by an increase of the $2k_F$ x-ray satellite intensity around 32 K, x-rays being in contrast to the case of neutrons, mostly sensitive to the TTF molecule.

At higher pressure, the two lines corresponding to the opening of an electrical gap on the TCNQ and TTF chains merge together (around 15 kbar), and a sharp peaking of the Peierls critical temperature up to about 70 K has been associated with the formation of a commensurability domain (where $2k_F = \frac{1}{3}b^*$) in pressure between 17 and 21 kbar (Friend et al., 1978a; Fig. 36b). These results are not entirely corroborated by the structural measurements of Megtert et al., (1981, 1982) where

● the commensurability domain begins at a pressure as low as 14.5 kbar (Fig. 8)
● no anomalous increase of the value of T_P is observed in the commensurability domain (Figs. 36a and 38).

The reasons why the structural and electrical results differ in this pressure range are not clear. While the deuteration of the TTF-TCNQ samples used in structural (but not in electrical) measurements could be invoked to explain

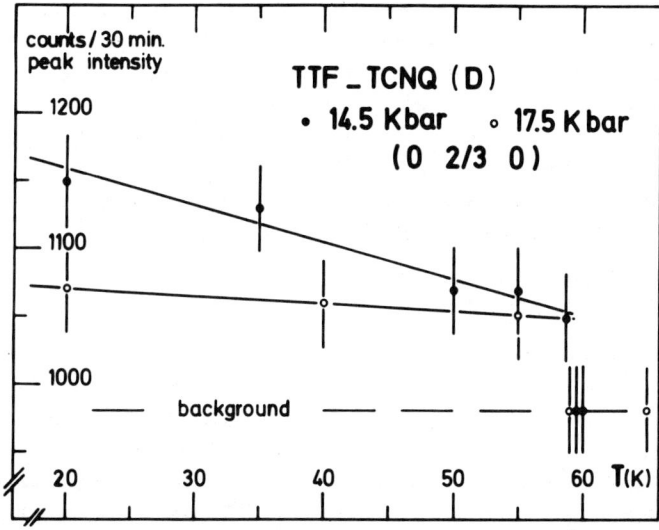

FIG. 38. Temperature dependence of the satellite intensities at two pressures in the commensurate phase $(0, \frac{1}{3}, 0)$ of TTF-TCNQ, showing the first-order nature of the 59 K phase transition. [From Megtert et al. (1982).]

the pressure shift of the commensurability domain, it is more difficult to understand why it suppresses the peaking of T_P.

Between 5 and 12 kbar, the q_a component of the wavevector of the modulation still varies, taking the value $q_a = 0a^*$ in the high pressure region. The exact boundary of this new high-pressure phase is not known, but it certainly does not coincide with the commensurability domain of the $2k_F$ value, since $(0, 2k_F, 0)$ satellite reflections are observed at 12 kbar, when $2k_F$ is still incommensurate. At this pressure, a phase transition around 45 K from the $(0, 2k_F, 0)$ to the $(\frac{1}{2}, 2k_F, 0)$ modulated phases seems to occur (Megtert, unpublished results). Finally, Fig. 38 clearly shows the first-order nature of the 59 K Peierls transition in the commensurability domain, as a jump in the satellite intensity (i.e., discontinuity of the order parameter).

b. TSF-TCNQ and $TTF_{1-x}TSF_xTCNQ$

The phase diagram of TTF-TCNQ can be extended by alloying with the isostructural selenium analog TSF-TCNQ. The physical properties of the resulting solid solution are detailed in the review of Schultz and Craven (1979). The structural studies (Kagoshima et al., 1978b and 1980; Sakai et al., 1983) reveal several new interesting features (Fig. 39). For small concentrations of TSF in TTF-TCNQ the three modulated phases of TTF-TCNQ still persist, but with lower critical temperatures: $T_H \simeq 50$ K, $T_M \sim 45$ K, $T_L \sim 28-30$ K for $x = 3\%$ (Kagoshima et al., 1978b; Forro et al., 1984). The largest decrease occurs for T_L, as expected by the substitutional disorder on the donor stacks. The $4k_F$ fluctuations are not 3D coupled in any of these phases.

The Peierls transition temperature of TTF-TCNQ T_H decreases gently with increasing proportions of TSF until about $x = 0.9$, where it occurs at 40 K. Then a very rapid drop of T_H, distinctly observed by conductivity measurements, occurs with the addition of the last 10% of TSF, the Peierls transition of TSF-TCNQ being at 29 K. This drop of T_H could be correlated with the interchange of the CDW ordering wavevectors:

- $(0, 2k_F, 0)$ order for $x < 0.9$
- $(\frac{1}{2}, 2k_F, 0)$ order for $x \sim 1$.

In addition, competing fluctuations between the $(0, 2k_F, 0)$ and $(\frac{1}{2}, 2k_F, 0)$ CDW ordering are clearly observed in TSF-TCNQ (Kagoshima et al., 1978a). But the $(0, \frac{1}{2}, 0)$ short-range CDW order persists below the $(\frac{1}{2}, 2k_F, 0)$ Peierls transition of TSF-TCNQ (Fig. 13). In TSF-TCNQ T_P increases under pressure, then saturates at around 45 K above 12 kbar (Thomas and Jérome). By analogy with the phase diagram of $TTF_{1-x}TSF_xTCNQ$, this increase could be explained by the stabilization of the $(0, 2k_F, 0)$ CDW ordering wavevector under pressure. In TTF-TCNQ, pressure also shifts the CDW ordering wavevector from $(\frac{1}{2}, 2k_F, 0)$ to $(0, 2k_F, 0)$.

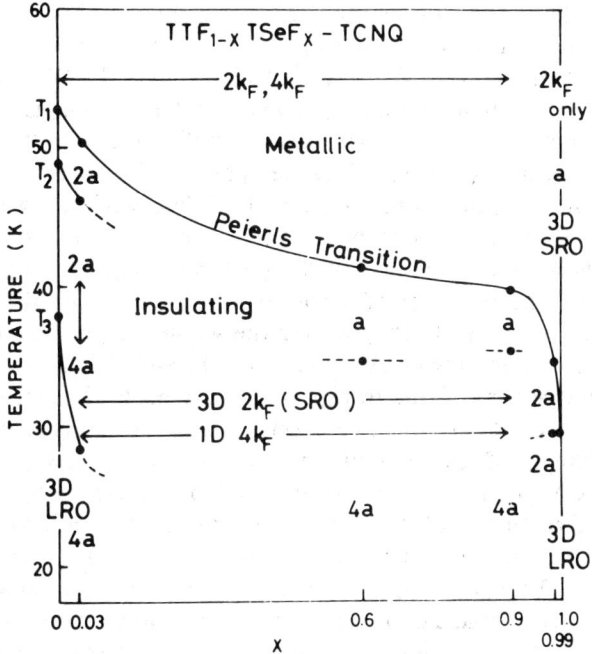

FIG. 39. Phase diagram of the solid solution TTF$_{1-x}$TSF$_x$TCNQ. The various phases are characterized by the wave length of the order in the a direction ($\lambda_a = 2\pi/q_a$). The long-range order (LRO) and short-range order (SRO) nature of the 3D coupling between $2k_F$ CDW, as well as the 1D nature of the $4k_F$ CDW, are indicated. [From Sakai et al. (1983).]

Studies of the solid solutions in the intermediate concentration range ($x = 0.9$, Sakai et al., 1983; $x = 0.6$, Kagoshima et al., 1980) show a transformation at about 35 K from the high-temperature ($0, 2k_F, 0$) short-range order to a low-temperature ($\frac{1}{4}, 2k_F, 0$) short-range order. The ($\frac{1}{4}, 2k_F, 0$) short-range order is also present below 45 K (Sakai et al., 1983) in DSDTF–TCNQ, where the donor molecule is a hybrid of the TTF and TSF molecules. This ordering recalls that stabilized below T_L in TTF–TCNQ.

Electrical conductivity (Etemad et al., 1975) and magnetic susceptibility (Scott et al., 1978) measurements show that a full gap develops below 29 K in TSF–TCNQ, which means that both kinds of stacks undergo the transition. Once more, this shows that the ($\frac{1}{2}, 2k_F, 0$) ordering can be kept after a structural distortion on both stacks. However, from optical measurements, it seems that the most important component of the distortion occurs on the TSF stack (Table VII). This means that the TCNQ stacks can undergo a weak distortion, while keeping the ($\frac{1}{2}, 2k_F, 0$) CDW order (we shall come back to this important point in Section 18). Earlier transport studies (Craven et al.,

1977) showed a splitting of the single-phase transition of TSF-TCNQ by the substitution of small amounts of TTF-TCNQ. Such a behavior was recently confirmed by the structural study of $TTF_{0.01}TSF_{0.99}TCNQ$ (Sakai et al., 1983), in which the temperature dependence of the peak intensity of satellite reflections (Fig. 40) clearly shows two anomalies at about 35 K and 29 K. The upper transition, whose critical temperature increases rapidly with x decreasing, seems to be driven mainly by the TCNQ stack, in agreement with a previous EPR study (Craven et al., 1977). The rapid increase of the x-ray intensity below 29 K (dominated by the TSF scattering), indicates a substantial increase of the distortion of the donor stack. In this salt, the wavevector still stays at the value $(\frac{1}{2}, 2k_F, 0)$ over the whole temperature range. This behavior recalls that previously assumed for TTF-TCNQ at 5 kbar in order to reconciliate structural and transport measurements. A somewhat similar behavior of the satellite intensity, consisting of an increase in two steps without change of the modulation wavevector, is also observed in HMTTF-TCNQ (compare Figs. 40 and 41). However, the latter material has a different structure, and its ordering wavevector has an incommensurate component q_a. The low temperature increase of the $(\frac{1}{2}, 2k_F, 0)$ satellite intensity of $TSF_{0.99}TTF_{0.01}TCNQ$ more likely corresponds to a rapid change in the temperature dependence of the donor order parameter than to a real thermodynamic phase transition associated with a symmetry breaking.

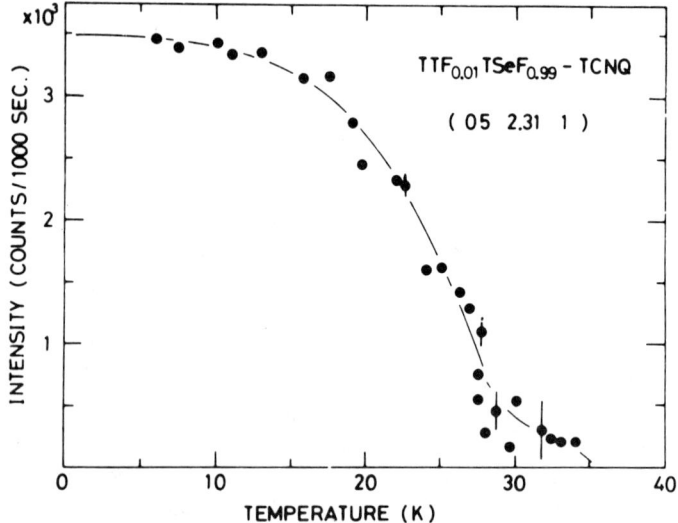

FIG. 40. Temperature dependence of the satellite intensity of $TTF_{0.01}TSF_{0.99}TCNQ$ showing the onset of the $(\frac{1}{2}, 2k_F, 0)$ CDW coupling at about 35 K and a further increase of intensity, below 29 K. [From Sakai et al. (1983).]

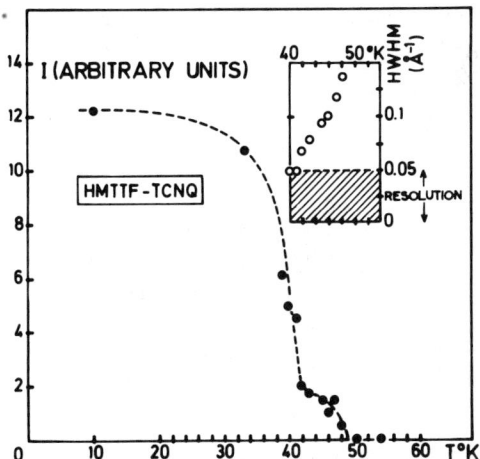

FIG. 41. Temperature dependence of the satellite peak intensity of HMTTF-TCNQ showing the onset of the (0.42, $2k_F$, 0) CDW coupling at 49 K and a further increase of intensity below 42 K. The insert shows the HWHM of the satellite reflection as a function of the temperature. [From Pouget et al. (1979b).]

c. HMTTF-TCNQ and HMTSF-TCNQ

HMTTF-TCNQ also shows very subtle phase transitions at low temperature. The temperature derivative of the electrical conductivity (Greene et al., 1976; Delhaes et al., 1977) and the specific heat (Biljakovic-Franulovic, 1979) present two anomalies at $T_H \simeq 49$ K and $T_L \simeq 43$ K, while x-ray diffuse scattering shows the stabilization of a doubly incommensurate modulation of wavevector $\mathbf{q} = (0.42, 0.36, 0)$ below T_H (Megtert et al., 1978a). Figure 41 shows that T_L corresponds to a net increase of the x-ray intensity. Between T_H and T_L, the CDW ordering is only short range along the a and c directions, where the TMTTF and TCNQ stacks alternate (see the insert of Fig. 41). This behavior is similar to that shown by TTF-TCNQ along a above 38 K (Fig. 35). T_H and T_L correspond, from vibronic absorption spectra (Bozio et al., 1983) and ESR measurements (Tomkiewicz et al., 1977a), to successive distortions of the TCNQ and HMTTF stacks, respectively. Below T_L, very weak $4k_F$ satellite reflections, occurring at the harmonic wave vector $2\mathbf{q}$, are observed (Megtert et al., 1987b; Pouget et al., 1979b). T_H and T_L decrease when pressure is applied: T_L is not observed above 3 kbar, and T_H falls linearly at 1.5 K/kbar (Friend et al., 1978b). Under pressure, the low-temperature ($T < T_L$) insulating behavior observed at ambient pressure is rapidly replaced by a semimetallic behavior.

HMTSF-TCNQ shows very weak satellite reflections at the reduced wave vector (0, 0.37, 0) in the low-temperature range 10–25 K (Megtert et al., 1978b;

Pouget et al., 1979b). This indicates a very weak distortion (of the HMTSF stack?), which can be related to the weak increase of electrical conductivity observed below 24 K (Korin et al., 1981). The electronic state of HMTSF-TCNQ below this temperature is that of a 3D semimetal (Miljak et al., 1978).

d. TMTTF-TCNQ and TMTSF-DMTCNQ

TMTTF-TCNQ shows an anomaly in the temperature dependence of its electrical conductivity at 36 K (Ishiguro et al., 1980). It corresponds to a Peierls transition involving both the TMTTF and TCNQ stacks (Bozio et al., 1983) and to the formation of a 3D modulated structure at the reduced wavevector $\mathbf{q} = (0, 2k_F, 0)$ (Kagoshima et al., 1983a). Very weak $4k_F$ satellite reflections at $2\mathbf{q}$, not due to the divergence of $4k_F$ structural fluctuations (see Fig. 20), are also observed below T_P.

At 42 K, TMTSF-DMTCNQ shows a very well-defined Peierls transition in its transport measurements (Jacobsen et al., 1978) and magnetic properties (Tomkiewicz et al., 1978), which seems to mainly involve the TMTSF-stack. It corresponds, because of the commensurate value of $2k_F$ ($\equiv \frac{1}{4}a^*$), to the formation of a commensurate superstructure characterized by the reduced wavevector $[\frac{1}{4}, (\pm)^?\frac{1}{3}, 0)$, where there is some uncertainty in the sign of the q_b component. Figure 42 shows the temperature dependence of the satellite intensity. Surprisingly, this quantity, proportional to the square of the order parameter, follows the mean field-BCS-like theory of the Peierls transition (Rice and Strassler, 1973), although the BCS relationship between the

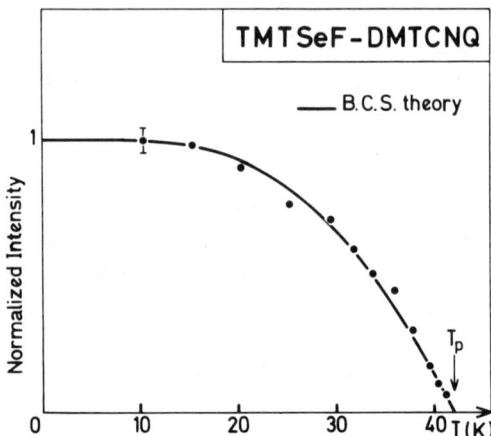

FIG. 42. Temperature dependence of the peak intensity of $2k_F$ satellite reflections of TMTSF-DMTCNQ below 42 K. The solid line gives the square of the order parameter calculated from the BCS theory of the Peierls transition. [From Moret and Pouget (1986).]

magnitude of the Peierls gap and the critical temperature does not hold. In others compounds, like TSF-TCNQ, showing a single Peierls transition, the satellite intensity increases more steeply below T_P than predicted by the mean-field theory (Fig. 37).

The metal-insulator phase transition of TMTSF-DMTCNQ is suddenly suppressed above 10 kbar (Andrieux *et al.*, 1979b). Hysteresis effects indicate some unusual first-order character at the phase boundary under pressure. The structural phase diagram of TMTSF-DMTCNQ under pressure is not established. In particular, it is not known if the CDW 3D order subsists (with another wave vector?) above 10 kbar. There is also some evidence, from transport studies under pressure, of high-temperature anomalies in the metallic state (metal I and metal II phases in Andrieux *et al.*, 1979b), which could be connected with physical features associated with the disappearance of the insulating phase, or with change in the transport conduction process (Conwell, 1980b). Effects that may influence the phase diagram of TMTSF-DMTCNQ under pressure are:

- possible deviations in the charge transfer from commensurate value $\rho = \frac{1}{2}$
- the change of the dimensionality of the Fermi surface. This last feature could also give rise to the metal I-metal II anomalies under pressure. A bad nesting of a warped Fermi surface first leads at a transition towards a semimetallic state then suppresses the PLD for further increase of $t_\perp$. There is experimental evidence that a low-temperature semimetallic state could be achieved by the applying pressure to TMTSF-DMTCNQ (see Hardebusch *et al.*, 1979; Andrieux *et al.*, 1979).

18. THEORETICAL ANALYSIS

The analysis of the PLD that develops below T_P has been performed within the framework of the Landau theory, which explicitly considers the symmetry of the chain array in the 3D crystal, and which differs among the various salts (see Fig. 4). Among them, the TTF(TSF)-TCNQ structures has been mostly investigated, owing to the very rich phase diagram shown by these salts. However, it is outside the scope of this review to justify the description of the 3D CDW ordering within the Landau theory in terms of structural order parameters. This aspect has been recently covered in the review of Barisic and Bjelis (1985), who also give a microscopic derivation of some coefficients entering in the Landau expansion.

In the simplest treatment, each family of chains is described by a displacive mode of a given polarization: the CDW amplitude ρ_A (ρ_D) is assumed to induce only one type of displacement A_A (A_D). Therefore, there is a one-to-one correspondence between ρ and A via the dimensionless electron phonon coupling constant, $\lambda_q/\hbar\omega_q$. A general description of the phase transitions in TTF-TCNQ in the frame work of the Landau theory using this scalar order parameter has been given by Abrahams et al. (1977). However, this simplified approach neglects the vectorial nature of the displacement field. For example, a given CDW may be coupled to several lattice modes through the linear electron phonon coupling. A more complete description using the deformation modes of TTF-TCNQ, like the translational degrees of freedom of the A and D stacks considered in Section 12, has been recently given by Megtert et al. (1985). This description is necessary for a full understanding of the phase diagram of TTF-TCNQ.

a. Description of the Phase Transitions of TTF-TCNQ and TSF-TCNQ

TTF-TCNQ and TSF-TCNQ crystallize in the monoclinic space group $P\,2_1/c$ (Table I), where (bc) sheets of like D or A stacks alternate in the a direction (Fig. 1). Such a simple array shows a great number of phase transitions that are easily distinguished by the transverse ordering of CDW in the a direction in temperature (Figs. 32 and 33), under pressure (Fig. 36), or upon alloying (Fig. 39).

(1) $(0, 2k_F, 0)$ *CDW ordering.* This ordering has already been considered when there is a dominant coupling between first-neighboring stacks (Fig. 30). The wavevector stabilized is easily explained by an antiphase ordering of CDW. This phase ordering, shown in Fig. 43a, is observed in several salts, regardless of the lateral arrangement of stacks (TTF-TCNQ under pressure, HMTSF-TCNQ, and TMTTF-TCNQ, corresponding, respectively, to arrays c, a, and b of Fig. 4).

(2) $(\frac{1}{2}, 2k_F, 0)$ *CDW ordering.* This wavevector can be easily understood if only one kind of CDW orders. It can be easily seen from the array of Fig. 4c that an out-of-phase ordering like CDW retains the c periodicity ($q_c = c^*$) and doubles the a periodicity ($q_a = a^*/2$). This argument was used (Saub et al., 1976; Schultz and Etemad, 1976; Bak and Emery, 1976; Bjelis and Barisic, 1977) to explain the high-temperature phase of TTF-TCNQ, where only the TCNQ stack undergoes a sizable structural distortion. However, the observation of the same ordering wavevector in TSF-TCNQ, where apparently both stacks undergo a distortion at the same temperature, or in TTF-TCNQ under pressure and $TTF_{0.01}TSF_{0.99}TCNQ$, where the temperatures of distortion of D and A stacks are different, suggests that the $(\frac{1}{2}, 2k_F, 0)$ CDW ordering is more subtle. It was found by Abrahams et al. (1977) that, because of the $P\,2_1/c$ symmetry, an out-of-phase ordering of

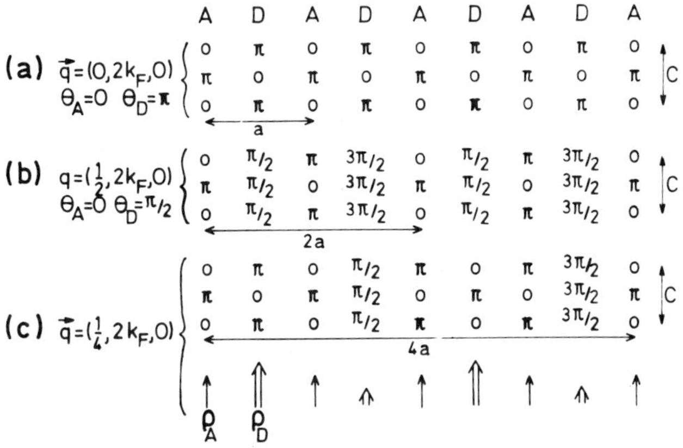

3. STRUCTURAL INSTABILITIES

	A	D	A	D	A	D	A	D	A
(a) $\vec{q}=(0, 2k_F, 0)$, $\theta_A=0$, $\theta_D=\pi$	0	π	0	π	0	π	0	π	0
	π	0	π	0	π	0	π	0	π
	0	π	0	π	0	π	0	π	0

←— a —→

	A	D	A	D	A	D	A	D	A
(b) $q=(\frac{1}{2}, 2k_F, 0)$, $\theta_A=0$, $\theta_D=\pi/2$	0	$\pi/2$	π	$3\pi/2$	0	$\pi/2$	π	$3\pi/2$	0
	π	$\pi/2$	0	$3\pi/2$	π	$\pi/2$	0	$3\pi/2$	π
	0	$\pi/2$	π	$3\pi/2$	0	$\pi/2$	π	$3\pi/2$	0

←—— 2a ——→

(c) $\vec{q}=(\frac{1}{4}, 2k_F, 0)$	0	π	0	$\pi/2$	π	0	π	$3\pi/2$	0
	π	0	π	$\pi/2$	0	π	0	$3\pi/2$	π
	0	π	0	$\pi/2$	π	0	π	$3\pi/2$	0

←——— 4a ———→

↑ ⇑ ↑ ⋏ ↑ ⇑ ↑ ⋏ ↑
  ρ_A ρ_D

FIG. 43. Phase shifts in the (a, c) plane for (a) the $(0, 2k_F, 0)$, (b) the $(\frac{1}{2}, 2k_F, 0)$, and (c) the $(\frac{1}{4}, 2k_F, 0)$ CDW ordering. In (a) and (b) the values of the phases θ_A and θ_D, defined in the text, are given. The modulation of the amplitude of D and A CDW, associated with the ordering (c), is also represented. The phase shifts correspond to $V_{AD} > 0$ along a, V_{AA} and $V_{DD} > 0$ along c.

CDW on one sublattice induces the ordering of CDW on the other sublattice. This is simply illustrated in Fig. 44, which schematically shows the electrostatic forces felt by the molecule a from its neighboring of CDW ordered with the $(\frac{1}{2}, 2k_F, 0)$ wavevector. This figure shows that

1. There is compensation of the forces produced by chains 1 and 2 ($F_{12} = 0$).
2. Because of its tilt, molecule a felt a nonzero resultant force (F_{34} and F_{56}) from the couples of chains (3, 4) and (5, 6).
3. As the symmetry is monoclinic ($d_{56} > d_{34}$), there is no compensation between F_{34} and F_{56}.

As emphasized earlier by Schultz and Craven (1979), a complete decoupling between the D and A sublattices occurs for the $(\frac{1}{2}, 2k_F, 0)$ CDW ordering if one of the following conditions is fulfilled:

1. Only the interactions with the nearest neighboring chains 1 and 2 are considered.
2. The molecules are not tilted out of the ac plane ($F_{34} = F_{56} = 0$, the crystal symmetry is thus P 2/m—Abrahams et al., 1977).
3. The lattice is orthorhombic ($d_{56} = d_{34}$, leading to $F_{56} + F_{34} = 0$—the crystal symmetry is thus Pmcb—Megtert et al., 1985).

The interaction between D and A sublattices can be accounted for by the following quadratic part of the Landau Ginzburg free energy (Bjelis and

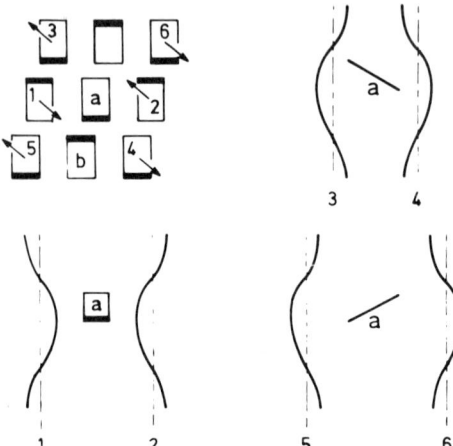

FIG. 44. Schematical representation of the ($\frac{1}{2}$, $2k_F$, 0) CDW neighborhood of a (D) molecule a along the 1–2, 3–4, and 5–6 directions, as indicated on the left upper part of the figure (the arrows show an arbitrary molecular displacements of the A array corresponding at the CDW sublattice shown).

Barisic, 1977; Barisic and Bjelis, 1977, 1985):

$$F^{(2)}(\rho_A, \rho_D, \theta, q_a) = [a_A + V_{AA} \cos q_a a]\rho_A^2 + [a_D + V_{DD} \cos q_a a]\rho_D^2$$
$$+ V_{AD}\left[(1 + \delta)\cos\left(\frac{q_a a}{2} + \theta\right) + (1 - \delta)\cos\left(\frac{q_a a}{2} - \theta\right)\right]\rho_A\rho_D. \quad (17)$$

In this expression only the q_a-dependent part of the free energy has been explicitly considered (the q_c contribution is included in the coefficients a_A, a_D, etc.,). The q_a dependence can be easily derived from Eq. (16), by writing the order parameter of the chain n under the form of a single plane wave,

$$\rho(n)_{A(D)} = \rho_{A(D)} e^{i[q_a n a + \theta_{A(D)}]},$$

which assumes a linear variation of the phase in the a direction. In Eq. (17) V_{AA}, V_{DD}, and V_{AD} are, respectively, the Coulomb interactions between first-neighboring A–A, D–D, and A–D sheets along a. The hypothesis of spatially homogeneous amplitude $\rho_{A(D)}$ and phase difference $\theta = \theta_D - \theta_A$ between D and A CDWs along a (a similar term along c is included in a_A and a_D) has also been made. The coefficients $a_{A(D)}$ contain the intrachain contribution (i.e., the inverse of the CDW response function) as well as the coupling energy between like CDW in the (b, c) plane, as mentioned above. In Eq. (17), δ describes the left and right asymmetry of the D–A coupling between (b, c) sheets of CDW along a. The free energy including δ has been previously discussed by Jérome and Schulz (1982b). The value of θ that minimizes $F^{(2)}$ is given by

$$tg\theta = -\delta tg\frac{q_a a}{2}. \tag{18}$$

If one assumes that the A sublattice is unstable at T_P ($a_A < 0$ below T_P) and that the D sublattice remains stable at this temperature ($a_D > 0$), one obtains after minimization with respect to ρ_D

$$\rho_D = \frac{V_{AD}}{a'_D}\rho_A\left[-\cos\frac{q_a}{2}\cos\theta + \delta\sin\frac{q_a a}{2}\sin\theta\right], \tag{19}$$

and the effective free energy

$$F^{(2)}_{\text{eff}}(\rho_A, q_a) = \left[a_A - \frac{V^2_{AD}(1+\delta^2)}{2a'_D}\right]\rho^2_A + \left[V_{AA} - \frac{V^2_{AD}(1-\delta^2)}{2a'_D}\right]\rho^2_A\cos q_a a, \tag{20}$$

where $a'_D(q_a) = a_D + V_{DD}\cos q_a a$. The ordering at $q_a = \pi/a$ is stable if

$$V^{\text{eff}}_{AA}\left(q_a = \frac{\pi}{a}\right) = V_{AA} - \frac{V^2_{AD}(1-\delta^2)}{2a'_D(q_a = \pi/a)} > 0, \tag{21}$$

and the ordering at $q_a = 0$ is stable if

$$V^{\text{eff}}_{AA}(q_a = 0) = V_{AA} - \frac{V^2_{AD}(1-\delta^2)}{2a'_D(q_a = 0)} < 0. \tag{22}$$

When the contribution of the mediated coupling, is weak enough, condition (21) is obeyed and Eq. (19) yields the amplitude of the induced CDW on the D stack

$$\rho_D = \frac{\rho_A V_{AD}\delta}{a'_D(q_a = \pi/a)}. \tag{23}$$

Expression (18) gives a phase difference $\theta = \pm\pi/2$ [in agreement with the symmetry analysis of Abrahams et al. (1977)]. This analysis also shows that the CDW of weaker amplitude (ρ_D here) orders in phase along c (see Fig, 43b). This result can be easily understood from Fig. 44. With respect to molecule a, molecule b has an environment that is out of phase, but having an opposite tilt with respect to a, b undergoes from its surrounding a resultant force in the same direction. This induces CDW with the same phase on stacks "a" and "b". This unfavorable in-phase coupling of the CDW of sublattice D makes the $(\frac{1}{2}, 2k_F, 0)$ structure unstable if ρ_D is strong enough [the unfavorable coupling energy term is contained in $a'_D(q_a = \pi/a)$].

If the mediated coupling is strong enough, condition (22) is followed. The order at $q_a = 0$ becomes stable and a stronger induced CDW develops on the D stacks. From Eq. (19) one gets

$$\rho_D = \frac{\rho_A V_{AD}}{a'_D(q_a = 0)}, \qquad (24)$$

and from Eq. (18): $\theta = \pi$. Now the CDWs are out of phase on the D sublattice (if $V_{AD} > 0$). This CDW ordering has been previously considered in (1) and is shown in Fig. 43a. The same phase shift between distortions in the $(\frac{1}{2}, 2k_F, 0)$ phase can be obtained using molecular displacements of the TCNQ (Q) or TTF (F) chains either in longitudinal, b (L) or transverse, c^* (T) directions (Megtert et al., 1985); the label acoustic (A) or optic (D) corresponds, respectively, to parallel or antiparallel displacement of neighboring chains in the c direction (Fig. 46). Figure 45 gives a classification of these modes according to the irreductible representations (IR) of the space group $P\,2_1/c$ and its orthorhombic approximation $Pmcb$, for $\mathbf{q}_H = (\frac{1}{2}, 2k_F, 0)$. In $P\,2_1/c$, the representation T^+ corresponds to an out-of-phase displacement on the Q sublattice and an in-phase displacement on the F sublattice. The opposite situation is described by T^-. The coupling between

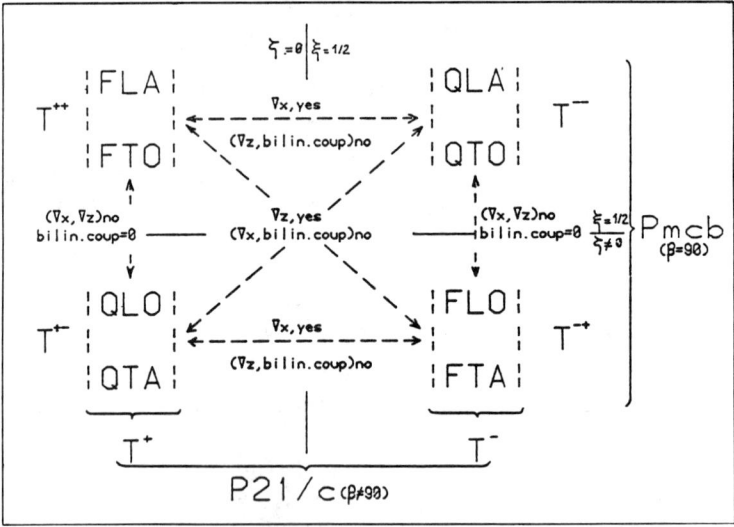

FIG. 45. Classifications of the translational modes of the Q and F sublattices according to the irreductible representations "T" of the two space groups $P\,2_1/c$ and $Pmcb$ for $\mathbf{q} = (\frac{1}{2}, 2k_F, 0)$. The bilinear couplings are indicated. ∇_x and ∇_z denote the Lifshitz couplings between these different modes along the a and c directions, respectively. The modes are more clearly defined by Fig. 46. [From Megtert et al. (1985).]

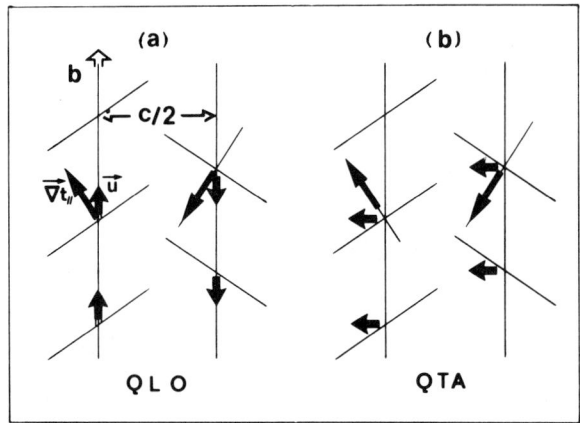

FIG. 46. Translational displacements in the (b, c) plane of Q molecules, involved in the QLO (a) and QTA (b) displacive modes. The optical (O) and acoustic (A) character of the longitudinal (L) and transverse (T) modes, respectively leading to the same out-of-phase ordering of CDW, is clearly shown. [From Megtert et al. (1985).]

these IR is given by Fig. 45. For $\mathbf{q} = (0, 2k_F, 0)$ all these modes belong to the same IR (Megtert, 1984). They are thus bilinearily coupled.

According to the analysis of Yamaji et al. (1981) of the 2D fluctuation regime of TSF-TCNQ, the $2k_F$ instability arises on the IR T^- (out-of-phase ordering of the TSF sublattice).

In TTF-TCNQ, the upper transition driven by the TCNQ stacks corresponds to the instability of the IR T^+ (Megtert et al., 1985). However, from polarization analysis of the satellite reflections (Fig. 35), the dominant contribution comes from the QTA mode. This mode is apparently very weakly coupled with the QLO mode at ambient pressure, because the associated longitudinal distortion is mostly observed below T^* in the sliding phase. No substantial ordering occurs on the TTF sublattice. Under pressure, the bilinear coupling between the two components QTA and QLO increases, and the T_H and T^* lines merge together above 4–5 kbar (Fig. 36a). The onset of the QLO distortion is not associated with a symmetry breaking, thus the T^* line is not a true transition line. The Landau theory for the phase lines T_H and T^* has been developed by Megtert et al. (1983, 1985).

In this picture, the crossover from a dominant instability on the acceptor stack (case of TTF-TCNQ) to a dominant instability on the donor stack (case of TSF-TCNQ) corresponds to a change of the unstable IR from T^+ to T^-. This occurs, with increasing x, in the solid solution $TTF_{1-x}TSF_xTCNQ$, probably during the $(0, 2k_F, 0)$ CDW ordering phase, where all the deformations belong to the same IR. Moreover, the very rapid decrease of T_P for $x > 0.9$ could be due to the fact that for a T^- instability, [associated with a $(\frac{1}{2}, 2k_F, 0)$ CDW ordering] the high-temperature distortion of the TCNQ sublattice occurs in phase, which costs a maximum of Coulomb coupling between the CDW. The stability of the (T^-) ordering is achieved by the out-of-phase ordering of TSF CDWs of larger amplitude. It occurs below 29 K in $TTF_{0.01}TSF_{0.99}TCNQ$ and TSF-TCNQ.

The stabilization of the $(\frac{1}{2}, 2k_F, 0)$ CDW order requires only the divergence of the $2k_F$ CDW response function of one sublattice, which is TCNQ in TTF-TCNQ and TSF in TSF-TCNQ. In the solid solution $TTF_{1-x}TSF_x$-TCNQ, the interchange of the nature of the stack which drives the $2k_F$ instability seems to occur for x between 0.9 and 1. For $x < 0.9$, the donor stack probably still shows a dominant $4k_F$ instability. As $x \to 1$, the divergence of the donor $2k_F$ CDW response function certainly increases with the decreasing amount of TTF/TSF substitutional disorder. Expression (23) shows that a rapid increase of ρ_D, while keeping the $(\frac{1}{2}, 2k_F, 0)$ order, requires either a rapid increase of ρ_A or an anomalous decrease of a'_D (which must, however, remain >0). The latter situation could arise in TTF-TCNQ around 32 K at 5 kbar (ρ_A does not show an anomalous temperature dependence—see Fig. 37).

A quantitative analysis of the successive development of the ρ_A and ρ_D order parameters requires the consideration of higher-order terms in the free energy or the resolution of self consistent gap equations, since the distortion of a given sublattice opens a gap at the Fermi level in its band structure and, by coupling, it also induces a gap at the Fermi level of the band structure of the other sublattice. This removes the divergence of the intrasublattice CDW response functions (see, for example, Sy and Mavroyannis, 1977). To lowest order all these effects contribute to the fourth-order term of the free energy, which thus takes the phenomenological form (Abrahams *et al.*, 1977; Megtert, 1984)

$$F^{(4)}(\rho_A, \rho_D) = B\rho_A^4 + C\rho_D^4 + D\rho_A^2\rho_D^2 + E\rho_A^3\rho_D + F\rho_A\rho_D^3. \quad (25)$$

$F^{(4)}$ fixes the amplitude of the order parameters. Thus it must be explicitly considered to define the domain of stability of the different phases.

The various coefficients entering into Eq. (25) may also depend on the wavevector, q, of the modulation. The q dependence of the coefficient D was explicitly considered by Schultz (1977) to explain the phase diagram of TSF-TCNQ. It is easily shown that the q-dependent part of D considered by Schultz is in fact proportional to δ^2, the square of the asymmetry of the coupling between A–D stacks. This biquadratic q-dependent coupling was used to account for the experimental finding that, in spite of high-temperature $(0, \frac{1}{2}, 0)$ fluctuations [favored by the bilinear coupling term $\rho_A\rho_D$ in Eq. (17), if Eq. (22) is obeyed], the $(\frac{1}{2}, 2k_F, 0)$ CDW order is finally stabilized below 29 K in TSF-TCNQ.

Another explanation of the phase diagram of TSF-TCNQ might be a shift from the condition (22) to condition (21) for decreasing temperatures, resulting either from a decrease of the intersublattice CDW-CDW Coulomb interaction, $|V_{AD}|$ (i.e., increase of V_{AD} from a negative value?—see Hartzstein et al., 1980b) or from an increase of the asymmetry of the coupling $|\delta|$. The opposite behavior of these interactions could explain the change from the $(\frac{1}{2}, 2k_F, 0)$ to the $(0, 2k_F, 0)$ CDW order in TTF-TCNQ under pressure. However, it is not determined experimentally if in TTF-TCNQ the transition between the $(\frac{1}{2}, 2k_F, 0)$ and the $(0, 2k_F, 0)$ CDW orders is direct [by a first-order phase transition, if one passes directly from condition (21) to condition (22)] or if the transition occurs via a succession of phase transitions where the wavevector changes more or less continuously from $q_a = a^*/2$ to $q_a = 0$ [in that case, a temperature range may exist where neither Eqs. (21) nor (22) are obeyed—see Bjelis and Barisic, 1977]. In this respect, it is interesting to remark that intermediate commensurate periodicities can be easily constructed by a periodic succession, along a, of the $q_a = 0$ structure and half of the $q_a = a^*/2$ structure. Figure 43 gives an example with $q_a = a^*/4$. In this respect, let us just mention that on the TSF side of the $TTF_{1-x}TSF_xTCNQ$ phase diagram, the $(\frac{1}{4}, 2k_F, 0)$ phase competes with the $(0, 2k_F, 0)$ and $(\frac{1}{2}, 2k_F, 0)$ phases (see Fig. 39). Surprisingly, the $(\frac{1}{4}, 2k_F, 0)$ phase has greater stability than the $(0, 2k_F, 0)$ phase at low temperature. By analogy with the arguments developed above to explain the stability of the $(\frac{1}{2}, 2k_F, 0)$ CDW order of TSF-TCNQ, the $(\frac{1}{4}, 2k_F, 0)$ order could be favored by q-dependent high-order terms in the free energy or by a decrease of the absolute value of the intersublattice CDW-CDW Coulomb coupling $|V_{AD}|$. Its stabilization also requires Umklapp terms in the free energy [see subsection (4)].

(3) *The transverse sliding phase of TTF-TCNQ.* This phase exists between 49 K and 38 K at ambient pressure. It follows the $(\frac{1}{2}, 2k_F, 0)$ phase, which has established a CDW order on the TCNQ sublattice, and was first analyzed in detail by Bak and Emery (1976) under the form of a two-chain model, where the TTF CDWs begin to order at 49 K, causing a continuous deviation of q_a from the $a^*/2$ value (Figs. 32 and 33). This phase transition has been identified by Megtert et al. (1985) as arising from the instability of a T^- mode, involving primarily the TTF sublattice. Below T_M, the T^- mode is coupled, via a Lifshitz invariant along, a, to the already established T^+ distortion (such coupling exists also between the corresponding T^{+-} and T^{-+} IR of the higher symmetry orthorhombic space group $Pmcb$—see Fig. 45). The sliding of the q_a component is simply explained by the bilinear Lifshitz coupling between T^+ and T^-, which exists only below T_M. A general development of the free energy, with the CDW order parameters, can be found in Abrahams et al. (1977). Here we shall adopt a simpler treatment

due to Bjelis and Barisic (1977), which starts from the free energy Eq. (17) with $\delta = 0$ (such an approximation is reasonable because the TTF sublattice is not substantially distorted above 49 K)

$$F^{(2)}(\rho_A^+, \rho_D^-, \theta, q_a) = [a_A^+ + V_{AA} \cos q_a a](\rho_A^+)^2 + [a_D^- + V_{DD} \cos q_a a](\rho_D^-)^2 + 2V_{AD} \cos \frac{q_a a}{2} \cos \theta^{+-} \rho_A^+ \rho_D^- \quad (17)'$$

In Eq. (17)' the index $+(-)$ refers to the IR $T^+(T^-)$ and the term in $\rho_A^+ \rho_A^-$ is the Lifshitz invariant mentioned above. $F^{(2)}$ is minimum for $\theta^{+-} = \pi$ (if $V_{AD} > 0$) and for

$$\cos \frac{q_a a}{2} = \frac{V_{AD} \rho_D^- \rho_A^+}{2[V_{AA}(\rho_A^+)^2 + V_{DD}(\rho_D^-)^2]}. \quad (26)$$

Expression (26) shows that with the increase of the other parameter ρ_D^- for T decreasing, q_a deviates from $a^*/2$. More quantitatively, assuming the mean-field behavior $(\rho_D^-)^2 \sim (T_M - T)$ and that ρ_A^+ is larger than ρ_D^- and weakly temperature dependent below T_M, expression (26) becomes, for q_a close to $a^*/2$,

$$\left(\frac{a^*}{2} - q_a\right)^2 = \delta q_a^2 \sim T_M - T. \quad (27)$$

This simple behavior is relatively well followed between T_M and T_L (Fig. 33), probably because ρ_D^- remains very small throughout the temperature range. [This is in agreement with the inability to detect an energy gap in the electronic properties of TTF above T_L (Tomkiewicz, 1980; Takahashi et al., 1984)].

Above 38 K, there is only short-range CDW order along a (Fig. 35). First of all this means that above 49 K, where the distortion of the TTF sublattice is negligible, the direct coupling V_{AA} between TCNQ sheets is not strong enough to achieve a long-range order along a (its weakness is in agreement with the numerical estimations of Hartzstein et al., 1980a). Then below 49 K, this order is slightly improved by the distortion of the TTF sublattice, which now provides an additional coupling mechanism between TCNQ sheets, under the form of a mediated interaction of value $|V_{AD}|^2/2a_D'$. However, as the TTF order parameter, ρ_D, remains weak, this interaction is not strong enough to achieve a transverse long-range order, along a. There is only an improvement of the order with the growth of ρ_D when the temperature decreases. The long-range order is observed only below 38 K when a sizable distortion occurs on the TTF sublattice.

In the above description the stabilization of an incommensurate q_a component requires the instability of the IR T^- mode, primarily involving the TTF sublattice. Thus $a_D^-(q_a = \pi/a)$ must vanish at a temperature T_D^-

[if mean-field behavior is assumed: $a_D^-(q_a = \pi/a) = \alpha_D(T - T_D^-)$]. However, because of the coupling between the order parameters ρ_A^+ and ρ_D^-, the real critical temperature T_M is different from T_D^-. Such couplings occur via the bilinear Lifshitz invariant term of the free energy Eq. (17)' (Hartzstein et al., 1980b) and the fourth-order terms of the free energy Eq. (25): $D\rho_A^{+2}\rho_D^{-2}$ (biquadratic interaction) and $F'\delta q_a \rho_A^{+3}\rho_D^-$ (cubic linear Lifshitz invariant-like interaction) (Abrahams et al., 1977; Megtert, 1984). With these terms it is easy to see that

$$T_M = T_D^- + \frac{V_{AD}^2}{2\alpha_D V_{AA}} - \left(D + \frac{V_{AD}F'}{V_{AA}}\right)\frac{\rho_A^{+2}(T_M)}{\alpha_D}, \qquad (28)$$

where $\rho_A^+(T_M)$ is the value at T_M of the TCNQ order parameter that is established below T_H. The first correction, which takes into account the mediated interaction V_{AD}, enhances the critical temperature, while the second one (assuming D and $F'V_{AD} > 0$) depresses it. Under pressure, T_M decreases rapidly (Fig. 36a). Two explanations have been proposed to account for this effect. The first one (Hartzstein et al., 1980) is based on a decrease of the first correction due to the decrease of the modulus of the CDW–CDW Coulomb interaction $|V_{AD}|$ for modest pressure [see also subsection (4)]. The second explanation (Megtert, 1984) is based on a large increase of the second correction, i.e., increase of the biquadratic inter-sublattice coupling D and $V_{AD}F'/V_{AA}$ under pressure. (This effect is also enhanced by the increase of the order parameter ρ_A^+ for T decreasing.)

It has also been predicted, by considering the true space group $P2_1/c$, that a sliding of the q_c component from 0 must accompany the sliding of q_a (Horowitz and Mukamel, 1977; Mukamel, 1977). At the present accuracy of the measurements, the sliding of q_c has not been observed. Figure 45 shows that in the symmetry group $Pmcb$, the Lifshitz invariants along c, responsible for the sliding of the q_c component, require the coupling of T^{+-} and T^{--} or of T^{-+} and T^{++}. In the orthorhombic approximation, only the IR T^{+-} and T^{++} are activated below T_H, thus there is no sliding of q_c. In the real symmetry $P2_1/c$, such a sliding exists if T^{++} or T^{--} is activated, by bilinear coupling, with, respectively, the primary IR T^{+-} and T^{-+}. Experimentally this bilinear coupling appears to be very weak for TTF-TCNQ at ambient pressure. In particular, T^{++} is not activated between T_H and T_M by coupling with T^{+-} since there is not a sizable distortion of the TTF stacks in the $(\frac{1}{2}, 2k_F, 0)$ phase.

Figure 33b shows that the $q_a(T)$ dependence presents a remarkable hysteresis in temperature between T_L and T_M (Ellenson et al., 1976; Pouget et al., 1979a), where the heating and cooling curves are shifted by about 1.5 K, a value comparable to the hysteresis of the first-order transition at T_L. Since this first report global hysteresis effects in the wavevector dependence have been observed in several incommensurate modulated structures. It has been generally proposed that the hysteresis results from the locking of the incommensurate modulation on defects either extrinsic to the structure (crystal imperfections; formation of defect density waves—Lederer et al., 1986) or intrinsic to the lattice itself (i.e., due to its discreteness; formation of devil's staircase—Aubry, 1980). In the case of TTF-TCNQ, a detailed theory of the global hysteresis, based on an intrinsic coupling between the phase and the amplitude of the modulation, has been worked out (Bjelis and Barisic, 1982; Barisic and Bjelis, 1985). The satellites that describe the modulation between T_M and T_L are situated at $\pm \mathbf{q}_I = [\pm q_a(T), \pm 2k_F, 0]$ and $\pm \mathbf{q}_{II} = [\mp q_a(T), \pm 2k_F, 0]$ in the Brillouin zone. They form a star with four arms (Fig. 34). Following the arguments of Abrahams et al. (1977), two cases are important to consider (Fig. 47):

1. Single q modulation. In this case, q_I and q_{II} are activated in different domains of the crystal. In a given domain the activation of two arms of the star describes a modulation of the CDW phase (the CDW phase varies along a). For example, in domain I, one has:

$$\rho = \rho_I \cos(\mathbf{q}_I \mathbf{r} + \theta_I) \tag{29}$$

2. Double q modulation. In this case, the modulations q_I and q_{II} coexist microscopically. Assuming an equal amplitude of modulation, it is easy to see that the superposition of plane waves like Eq. (29) for q_I and q_{II} describes a modulation of the amplitude of the CDW (the phase of the CDW is constant along a, but the amplitude of the CDW varies from chain to chain)

$$\rho = \rho_0[\cos(\mathbf{q}_I \mathbf{r} + \theta_I) + \cos(\mathbf{q}_{II} \mathbf{r} + \theta_{II})] \tag{30}$$

$$\rho = 2\rho_0 \cos\left[2k_F \mathbf{a}^* \mathbf{r} + \frac{(\theta_I + \theta_{II})}{2}\right] \cos\left[q_a(T) \mathbf{a}^* \mathbf{r} + \frac{(\theta_I - \theta_{II})}{2}\right].$$

In this case, one should observe satellite reflections at $\pm \mathbf{q}_I \pm \mathbf{q}_{II}$ and $\pm \mathbf{q}_I \mp \mathbf{q}_{II}$ due to interference between the two primary planes waves. These satellites are different from harmonics of modulation at $\pm 2\mathbf{q}_I$ or $\pm 2\mathbf{q}_{II}$, which can be formed in each single q domain. $(0, 4k_F, 0)$ and $(2q_a, 0, 0)$ satellite reflections have not been reported above 38 K. In addition, it has been pointed out by Abrahams et al. (1977) that in the limit $\delta q_a \to 0$, the single q modulation has a lower free energy than the double q modulation.

Satellite reflections at $2q_I$ and $2q_{II}$ are observed in the sliding phase of TTF-TCNQ until about 45 K (Fig. 32). Their observation at (or slightly below) T_M is in perfect agreement with the presence of the $4k_F$ anomaly on the TTF chain. The formation of $4k_F$ satellite reflections at a harmonic wavevector of the $2k_F$ one can be accounted by linear-quadratic coupling

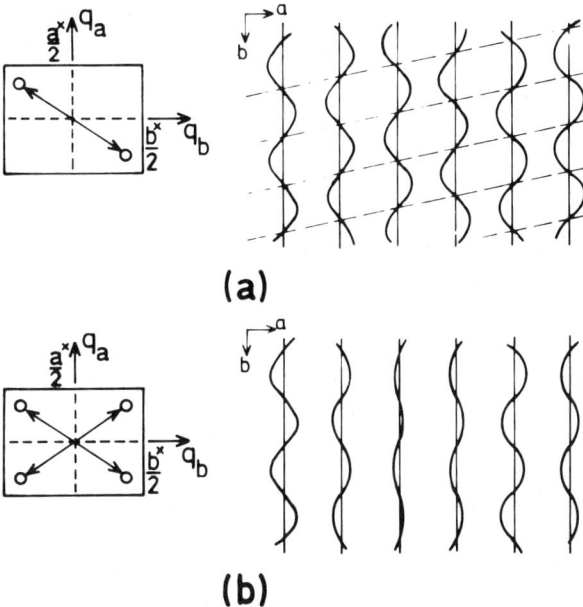

FIG. 47. Ordering of CDW in the case of a phase modulation (a) and of an amplitude modulation (b). The wavevectors involved in the modulations are indicated.

terms in the free energy between, respectively, the $4k_F$ and $2k_F$ CDW order parameters (Sato et al., 1978; Barisic and Bjelis, 1985). This coupling leads to an effective attractive intrachain contribution, which should be added to the fourth-order terms in the free energy previously considered [Eq. (25)]. Such a contribution could decrease under pressure if the $4k_F$ scattering is suppressed. Its effect on the phase transitions of TTF–TCNQ could be tested by the analysis of the phase diagram of TTF$_{1-x}$TSF$_x$TCNQ for low values of x, when the 3D coupling between $4k_F$ CDW is suppressed: T_H and T_M are depressed by the same amount $\Delta T = 4$ K for $x = 3\%$, which means that this coupling has a minor influence on T_M.

(4) *Lock-in phases of TTF–TCNQ.* At ambient pressure TTF–TCNQ undergoes a first-order phase transition at 38 K, where the q_a component of the $2k_F$ satellite reflections lock-in at the value $\frac{1}{4}a^*$. Several theories have been proposed for this phase transition (Bak and Emery, 1976; Bak, 1976; Bjelis and Barisic, 1976; Horowitz and Mukamel, 1977; Weger and Friedel, 1977; Abrahams et al., 1977).

In the limit of dominant $2k_F$ ordering, the $q_a = \frac{1}{4}a^*$ order is a consequence of Umklapp terms in the free energy. When $q_b = 2k_F$ is incommensurate, no Umklapp term of third order is possible. The first Umklapp term is of fourth order. Within a given sublattice, it is given by (Bjelis and Barisic, 1976; Barisic and Bjelis, 1977, 1985; Abrahams et al., 1977)

$$\Delta F_{\text{com}}(4) = G_4 \rho_I^2 \rho_{II}^2 \cos 2(\theta_I + \theta_{II}) \delta_{q_a}, \frac{a^*}{4}. \quad (31)$$

It combines arms I and II of the star of the $2k_F$ wavevector (Fig. 34), which, respectively, have the order parameters $\rho_{I(II)} \exp(i\theta_{I(II)})$. If $\theta_I + \theta_{II} = \pi/2$, this term provides a gain of free energy when ρ_I and ρ_{II} are simultaneously (i.e., microscopically) different from zero. It results a double q modulation similar to that considered in the last subsection. The observation of $(0, 4k_F, 0)$ satellite reflections below T_L (Kagoshima et al., 1976), mixing the wavevectors q_I and q_{II} of the star, was recognized early (Bak, 1976; Schultz and Craven, 1979) as experimental evidence of the importance of Umklapp processes in the 38 K phase transition of TTF-TCNQ.

A theory including Umklapp terms like Eq. (31) for the D and A sublattices has been worked out in the orthorhombic approximation of TTF-TCNQ (Barisic and Bjelis, 1977, 1985). Starting with a dominant order parameter on the A (TCNQ) sublattice, the minimization of the free energy leads to the odering: $\rho_A, -\rho_D, \rho_A, 0, -\rho_A, \rho_D, -\rho_A, 0, \rho_A$, along a (if $V_{AD} > 0$). In this sequence, half of the D (TTF) chains [those placed between two A (TCNQ) chains having an opposite distortion] are not ordered. However, electronic properties below 38 K show that a gap (and thus a stack distortion) appears on all the chains. The ordering of these TTF chains can be explained if the real $P2_1/c$ symmetry of TTF-TCNQ is considered: As shown in Fig. 44, the asymmetry of their coupling with the $(\frac{1}{2}, 2k_F, 0)$ CDW order on the neighboring TCNQ sheets leads to their distortion. By analogy with the results of subsection (2), the induced CDW on these TTF stacks takes the phase shift $\pm \pi/2$, as schematically indicated by Fig. 43c. However, because of their unfavorable environment, these TTF stacks will probably undergo a distortion of smaller amplitude than the TTF stacks placed between two TCNQ planes distorted in phase. This produces the modulation of amplitude previously mentioned and shown by Fig. 43c.†

The 38 K transition could be also interpreted in the opposite limit of a dominant $4k_F$ ordering (Barisic and Bjelis, 1985). In this limit, the $q_a = \frac{1}{4}a^*$ component of the $2k_F$ wavevector could be imposed by the out-of-phase ordering ($q_a = \frac{1}{2}a^*$) of $4k_F$ CDW on one (TTF) sublattice. However, this interpretation seems to be unlikely in TTF-TCNQ, because in the alloy series

† This phase shift has been found in a recent determination of the $2k_F$ modulated structure of TTF-TCNQ below T_L (Bouveret and Megtert, 1988). It leads at an equal (unequal) amplitude of modulation of the TCNQ (TTF) stacks for the arms I and II of the star of the $2k_F$ wave vector.

TTF$_{0.97}$TSF$_{0.03}$TCNQ, the ($\frac{1}{4}$, $2k_F$, 0) satellite reflections are observed, although the $4k_F$ CDWs never couple three dimensionally (Kagoshima et al., 1978b; Forro et al., 1984). However, the observation of the ($\frac{1}{4}$, $2k_F$, 0) order in the concentration range $x < 0.9$ in TTF$_{1-x}$TSF$_x$TCNQ, where the $4k_F$ CDW fluctuations are observed (Fig. 39), suggests that a very subtle link could exist between these two structural features. But, it has been shown in DSDTF-TCNQ (Sakai et al., 1983) that the ($\frac{1}{4}$, $2k_F$, 0) order can appear without the presence of a $4k_F$ instability.

The behavior under pressure of the ($\frac{1}{4}$, $2k_F$, 0) phase and its transformation into a phase with $q_a \sim \frac{1}{3}a^*$ over a very short pressure interval (Fig. 36a) are not clearly explained. It is interesting to remark that pressure stabilizes order with a wavevector $q_a \sim \frac{1}{3}a^*$, very close to that reached in the sliding phase of TTF-TCNQ at ambient pressure just before the first-order transition at T_L (Fig. 33a). However, it is not obvious that the $q_a \sim \frac{1}{3}a^*$ phase is structurally similar to the ambient pressure sliding phase. A $3a$ commensurate phase could be obtained by an alternate ordering of $2a$ (Fig. 43b) and a (Fig. 43a) unit cells in the a direction. The "$3a$" superstructure can be derived from the "$4a$" superstructure of Fig. 43c by the suppression of one subcell, with the (0, $2k_F$, 0) ordering, out of two. The suppression of the other half (0, $2k_F$, 0) subcells gives the ($\frac{1}{2}$, $2k_F$, 0) ordering that is stabilized above 4 kbar. It is somewhat surprising that pressure does not favor the (0, $2k_F$, 0) CDW ordering in the low-pressure range because it is generally expected that the coupling between A and D stacks will increase under pressure. However, a realistic calculation of the CDW-CDW Coulomb interactions (and of its asymmetry S) is very difficult to perform because it requires an accurate knowledge of the charge distribution on the molecules and of the relative distances and orientations between them. The coupling of the CDW to additional degrees of freedom could also be considered (Hartzstein et al., 1980a). All of these quantities can change in a nontrivial way under pressure. Only the pressure dependence of the charge transfer (see Part III) and of the crystal packing (change in the stacking distances, tilt angles, shortening of S-N distances, etc.; see Fihlol et al., 1981) is known experimentally. In addition, it has been pointed out (Weger and Friedel, 1976; Hartzstein et al., 1980a) that, because of the tilting of the molecules and of the almost quarter filling of the band the parameter, V_{AD} is very small. In a model where the excess charges are concentrated on the TCNQ nitrogens and on the TTF sulphurs, a negative value of V_{AD} has even been obtained for the lattice parameters at ambient pressure (Hartzstein et al., 1980b). In that case, the increase under pressure of V_{AD}, leads first to a decrease of $|V_{AD}|$ in the pressure range where $V_{AD} < 0$, then to an increase for higher pressures when $V_{AD} > 0$. It is not known if such a situation corresponds to TTF-TCNQ; however, the nonmonotonic pressure dependence of the temperature

"T_L" at which a sizable gap develops on the TTF chains (Fig. 36b), could indicate an anomalous behavior of $|V_{AD}|$ under pressure. Around 5 kbar, where "T_L" (and thus ρ_D) is minimum (see Fig. 36b), the A–D coupling could reach a minimum, leading to the stabilization of the $(\frac{1}{2}, 2k_F, 0)$ CDW ordering. The increase of V_{AD} for higher pressures leads to the $(0, 2k_F, 0)$ ordering, as considered in subsection (2). In addition, the various fourth-order contributions to the free energy, especially the Umklapp term considered in Eq. (31), could change under pressure.

We have only discussed the commensurate phases related to the transverse ordering of CDW. A commensurate phase is also stabilized under pressure when the chain wavevector component $2k_F = \frac{1}{3}b^*$. For the wavevector $\mathbf{q} = (0, \frac{1}{3}, 0)$, a third-order Umklapp term can be constructed in the free energy

$$\Delta F_{cm}^{(3)} = G_3 \rho^3 \cos 3\theta \delta_{q_b}, \frac{b^*}{3}, \qquad (32)$$

where θ is the phase of the CDW with respect to the lattice. This third-order invariant gives rise to a first-order phase transition (Landau and Lifshitz, 1980), which is observed experimentally (Megtert et al., 1982—see Fig. 38). The shape of the phase diagram in the vicinity of such a commensurate phase has been reviewed on theoretical grounds by Jérome and Schulz (1982a and 1982b). It is interesting to remark that if $2k_F = \frac{1}{3}b^*$, there is Umklapp equivalence between the $2k_F$ and $4k_F$ wavevectors ($4k_F = b^* - 2k_F$). The mixing of these two instabilities has been considered by Bulaevskii et al. (1980).

The phase diagrams of TTF–TCNQ and TSF–TCNQ have been discussed on the basis of dominant Coulomb interactions between CDWs. The near constancy of the upper critical temperature T_H and the anomalous temperature dependence of "T_L" under pressure are in favor of Coulomb coupling between CDWs. Transverse tunneling effects seem to have a marginal role in the phase diagram of TTF–TCNQ, at least in its low pressure range.

b. Phase Transitions of HMTTF–TCNQ and HMTSF–TCNQ

These salts crystallize in orthorhombic (HMTTF–TCNQ) and pseudo-orthorhombic (HMTSF–TCNQ) space groups. Stacks of D and A adopt the chessboard array of Fig. 4a. The general problem of CDW ordering in such a lattice has been considered by Bjelis and Barisic (1978) within the Ginzburg–Landau model as a function of the coupling between neighboring chains. A dominant interaction between (unlike) first neighbor chains along the directions a and c leads to the CDW order $\mathbf{q} = (0, 2k_F, 0)$ already found in TTF–TCNQ under pressure and in TSF–TCNQ under the form of fluctuations. Here, this order is stabilized at ambient pressure in HMTSF–TCNQ. A very interesting situation, relevant for HMTTF–TCNQ, occurs when the

3. STRUCTURAL INSTABILITIES

A and D sublattices are weakly coupled, as suggested by the formation of short-range CDW ordering below 50 K (T_H) (see Fig. 41). Let us consider the extreme situation of a complete decoupling between the two sublattices. In that case, the intrasublattice bilinear term in the free energy is

$$F^{(2)}(\rho, q_a, q_c) = \left\{ a + V_a \cos q_a a + V_c \cos q_c c \right.$$
$$\left. + 2V_d \left[\cos\left(q_a \frac{a}{2}\right) \cos\left(q_c \frac{c}{2}\right) \right] \right\} \rho^2, \quad (33)$$

where V_a, V_c, and V_d are the coupling between the same CDW in the a, c, and $\mathbf{d} = (\mathbf{a} + \mathbf{c})/2$ directions of the HMTTF-TCNQ lattice. The wavevector stabilized by Eq. (33) is given by:

$$\begin{cases} \sin\left(q_a \frac{a}{2}\right) \left[2V_a \cos\left(q_a \frac{a}{2}\right) + V_d \cos\left(q_c \frac{c}{2}\right) \right] = 0 \\ \sin\left(q_c \frac{c}{2}\right) \left[2V_c \cos\left(q_c \frac{c}{2}\right) + V_d \cos\left(q_a \frac{a}{2}\right) \right] = 0. \end{cases} \quad (34)$$

With $c > a$ in HMTTF-TCNQ, it is reasonable to assume that $|V_c| < |V_a|$. If $(V_a, V_c) \gg V_d$, there will be an out-of-phase ordering of CDW along a and c (doubling these lattice periodicities). This situation is not realistic because $|\mathbf{d}| < a, c$. In the other limit where $V_d \gg (V_a, V_c)$, the CDW will be out of phase along the diagonals d, maintaining the a and c periodicities. If $(V_a, V_c) \sim V_d$, an incommensurate order can develop. If $4V_a V_c$ is different from V_d^2, this order occurs in one direction, and a solution of Eq. (34) is, for $q_c = c^*$:

$$\cos\left(q_a \cdot \frac{a}{2}\right) = \frac{V_d}{2V_a}. \quad (35)$$

This order is more stable than the symmetric one, corresponding to an interchange of q_a and q_c, if $V_a > V_c$. It is also more stable than the out-of-phase ordering along the diagonals previously considered if $V_d < 2V_a$. It provides a simple explanation for the (0.42, $2k_F$, 0) CDW order observed below 49 K in HMTTF-TCNQ. In that case, $q_a \sim 0.42a^*$ requires $V_a \sim 2V_d$ on the sublattice that orders first (TCNQ). The coupling (involving the $V_{a/2}$ and $V_{c/2}$ interactions) between the two sublattices leads to solutions more complex than that obtained from Eq. (33). In particular, it is found (Bjelis and Barisic, 1978) that the wavevector q_a must vary in temperature with the growth of the order parameter on the HMTTF sublattice, ρ_D. Within experimental error (0.02a^*) q_a does not vary with temperature (Megtert et al., 1978). However, the temperature variation of q_a can be small if the coupling between the two sublattices remains weak, i.e., if $V_{a/2}^{AD} \rho_D < V_a^{AA} \rho_A$.

Below 49 K (T_H) the coupling within the TCNQ sublattice is certainly weak, because only a short-range CDW order is established in the a and c directions. This order is improved for decreasing T (Fig. 41). On the scale of the experimental resolution, there is long-range order between the CDW only below 42 K (T_L), i.e., when a substantial distortion occurs on the HMTTF sublattice.

Two explanations have been proposed for the sequence of phase transitions observed in HMTTF-TCNQ:

1. At T_H, most of the distortion occurs on the TCNQ sublattice and induces a CDW on the HMTTF chains. The HMTTF sublattice orders at T_L without any change of symmetry. Thus, T_L is not a real phase transition, but a quasi-phase transition, which produces anomalies in physical quantities (Bjelis and Barisic, 1978). In this picture, one can describe the satellite intensity in the form

$$I_s(\mathbf{Q}_c) = |F_s^A(\mathbf{Q}_c)\rho_A + F_s^D(\mathbf{Q}_c)\rho_D|^2, \qquad (36)$$

where $F_s^A(F_s^D)$ and $\rho_A(\rho_D)$ are the acceptor (donor) structure factor of the distorted sublattices and the acceptor (donor) order parameter, respectively. The rapid increase of x-ray intensity below T_L is due to the increase of ρ_D, for which the associated sublattice distortions have the strongest structure factor. In that case, all the satellites intensities must scale with the expression (36) above and below T_L. The coupling between the two sublattices must be weak enough to account for the near constancy of the incommensurate wavevector.

2. The T_H phase transition is associated with the divergence of the response function of one IR, let's say T^+, of the *Pmna* space group for the wavevector (q_a, $2k_F$, 0), having two incommensurate components. At T_L the phase transition is induced by the divergence of the response function (renormalized by the coupling between ρ^+ and ρ^- order parameters [see Eq. (28)]), of the other IR, T^- (Solyom, 1983): ρ^+ and ρ^- are coupled by a fourth-order term in the free energy

$$F_{\text{coupl}}^{(4)} = D\rho^{+2}\rho^{-2}. \qquad (37)$$

This coupling generally does not lead to a change of q_a with temperature. The general shape of the phase diagram that results from such a coupling has been discussed by Imry (1975) as a function of D. If D is not too strongly repulsive, two phase transitions occur, and below the second one the two-order parameters ρ^+ and ρ^- are superimposed. Thus in this picture the satellite intensities will be given by:

$$I_s(\mathbf{Q}) = \begin{cases} |F_s^+(\mathbf{Q}_c)\rho^+|^2, & \text{for } T_L < T < T_H \\ |F_s^-(\mathbf{Q}_c)\rho^- + F_s^+(\mathbf{Q}_c)\rho^+|^2, & \text{for } T < T_L, \end{cases} \qquad (38)$$

where + and − refer to the symmetry stabilized below T_H and T_L, respectively. The phase shift between the T^+ and T^- CDW is given by Solyom (1983). It is possible that, similar to TTF-TCNQ, T^+ corresponds mostly to a distortion of the TCNQ sublattice and T^- corresponds mostly to a distortion of the HMTTF sublattice.

T_L decreases rapidly under pressure. This can be understood very easily by an increase of the biquadratic intersublattice coupling D under pressure. The weakly metallic behavior observed under pressure could be due to the fact that only one sublattice (TCNQ) undergoes a substantial distortion.

T_H also decreases under pressure. This behavior contrasts with that of T_H in TTF-TCNQ or TSF-TCNQ and could indicate that a substantial part of the coupling between CDWs is via the interchain transfer integrals $t_\perp$ (Friend et al., 1978b). In the treatment of Horowitz et al. (1975), the Peierls temperature is depressed when $t_\perp/t_\parallel$ increases, because the increase of the warping of the Fermi surface under pressure reduces the divergence of the CDW response function. In that case, the stabilized wavevector is, instead of that which minimizes the Coulomb interactions V_{ij} in Eq. (33), the one which provides the best nesting of the Fermi surface. It depends crucially on the real shape of the Fermi surface. The Fermi surface of HMTTF-TCNQ is not known, but the decoupling of the TCNQ and TMTTF transitions means that hybridization effects between A and D bands are probably weak at ambient pressure. Under pressure, with the increase of $t_\perp/t_\parallel$, the nesting of the Fermi surface can be only partial, leaving small electron-hole pockets. This could also account for the semimetallic behavior observed above 8 kbar (Friend et al., 1978b).

Whatever the origin of the CDW coupling, it is expected that q_a changes under pressure. A suitable possibility might be that with the increase of the intersublattice coupling, q_a shifts towards the value a^*. In this picture, TMTSF-TCNQ will resemble HMTSF-TCNQ, where the $(0, 2k_F, 0)$ ordering is observed below 24 K at ambient pressure. From its conducting properties, HMTSF-TCNQ also resembles HMTTF-TCNQ around 8 kbar (Friend et al., 1987b).

19. CONCLUSION

Most of the experimental and theoretical studies of the 3D CDW ordering phase transitions have been devoted to TTF-TCNQ, which shows an extremely complex phase diagram. Recently, interesting effects of nonlinear conductivity (Lacoe et al., 1985) have been observed in these modulated phases.

Although the basic features of the phase diagram of TTF-TCNQ are understood at a qualitative and phenomenological level, a quantitative estimate of the various coupling coefficients and of their temperature and pressure dependence is lacking. Very elegant theories, relevant for the general field of incommensurate modulations, have been proposed for the sliding and lock-in phases of TTF-TCNQ and the global hysteresis of the wavevector. The determination of the phase diagram under pressure has required important studies. In spite of this, the pressure range 5–12 kbar and $P > 18$ kbar remains unexplored. The structural refinement of the various modulated phases of TTF-TCNQ has just begun. Such work is, however, necessary in order to determine the phase shift between CDW, which has been predicted on the basis of energetic considerations. The other charge-transfer salts have been less studied. In some of them (see Table IV), the structural phase diagram at ambient pressure has been determined, but its evolution under pressure is not known. More experimental and theoretical work are necessary for a further understanding of the interactions between stacks in organic conductors and their evolution as a function of the molecular species and the crystallographic array.

VI. General Conclusion

This chapter has attempted to cover more than 10 years of structural studies of the $2k_F$ and $4k_F$ instabilities of the 1D electron gas of organic charge-transfer conductors. One of the major issues of these studies was to show the very close relationship between the structural and electronic instabilities, the lattice being the fingerprint of the CDW electronic instability. In this respect, structural studies bring invaluable information on the electronic subsystem by:

- measuring the charge transfer,
- probing the various regimes of CDW fluctuations,
- determining the q dependence of the interchain couplings, and
- establishing the key role of electron–electron interactions in the 1D electron gap.

Part of this information is relevant for the wider field of compounds showing a CDW instability, while the other part concerns the wider class of organic metals.

Acknowledgments

I am very grateful to the photon factory of KEK (Tsukuba, Japan) for its hospitality during a stay in which a part of this chapter was written. The work of many people is summarized in this review. Among them I want to particularly thank R. Comès who initiated me into the structural studies of 1D conductors. During these 10 years of research on organic conductors

I have also benefited from experimental collaborations with P. A. Albouy, A. J. Epstein, L. Forro, S. Kagoshima, S. K. Khanna, S. Megtert, R. Moret, S. M. Shapiro, and G. Shirane. Very useful discussions with S. Barisic, C. Bourbonnais, V. J. Emery, J. Friedel, C. Noguera, H. J. Schulz, and K. Yamaji during the writing of this paper are acknowledged. A. I. Goldman is also thanked for a careful reading of the manuscript.

VII. Appendix

The purpose of this appendix is to review the various physical quantities that can be obtained from x-ray and neutron scattering studies of 1D conductors. A partial introduction to this field can be found in the papers by Megtert *et al.* (1978c), Pouget (1978, 1987), Comès and Shirane (1979), and Moret and Pouget (1986). Here we shall detail some aspects necessary to understand Parts IV and V of this review. Quantities measured above T_P are the subject of Sections 20 (x-ray diffuse scattering) and 21 (neutron scattering) and those measured below T_P are taken up in Section 22. Then, we shall finish, in Section 23, with some considerations of the resolution corrections in x-ray diffuse-scattering experiments.

20. X-Ray Diffuse Scattering

a. General Concepts

In general, any deviation from the periodic crystalline structure that gives rise to Bragg diffraction produces an extra diffuse scattering at general wave-vectors $\mathbf{Q}$ in reciprocal space. Here we shall deal only with displacive deviations from the mean structure that result from a structural instability.

In this case, the x-ray diffuse intensity at the wavevector $\mathbf{Q}$ can be expressed generally in the form (Cochran, 1969; Yamada, 1974; Pouget, 1978; Cowley, 1980)

$$I_d(\mathbf{Q}) = |F_d(\mathbf{Q})|^2 S(\mathbf{q}, t = 0), \tag{A.1}$$

provided that only a single normal mode of the lattice becomes unstable at T_P. In this expression

$$F_d(\mathbf{Q}) = \sum_j f_j(\mathbf{Q}) e^{-w_j(\mathbf{Q})} \frac{\mathbf{Q} \cdot \mathbf{e}_j(q)}{\sqrt{m_j}} e^{i(\mathbf{G} \cdot \mathbf{r}_j)} \tag{A.2}$$

is the diffuse structure factor of the unstable lattice mode (f_j, w_j, m_j, $\mathbf{e}_j$, $\mathbf{r}_j$ are, respectively, the scattering factor, the Debye Waller factor, the mass, the polarization of the displacement, and the position of the atom j in the unit cell). $\mathbf{Q} = \mathbf{G} + \mathbf{q}$, where $\mathbf{G}$ is a reciprocal lattice vector, and

$$S(\mathbf{q}, t = 0) = \langle |A_\mathbf{q}|^2 \rangle \tag{A.3}$$

is the instantaneous correlation of the normal coordinate $A_\mathbf{q}$ of the unstable lattice mode, and $\langle \ \rangle$ means a thermal average. [At this stage, it is useful to recall that with the change in variable usually done to find the normal modes of vibration in a system composed of different masses, the normal coordinate A_q has the dimensions of a displacement multiplied by the square root of a mass, see Eq. (A.38)].

In the classical limit $k_B T > \hbar\omega_c$, where $\hbar\omega_c$ is the characteristic energy of the fluctuations (for example, the quantum of energy of the unstable phonon mode), the theorem of equipartition of the energy yields the relationship

$$S(\mathbf{q}, t = 0) = k_B T \chi(\mathbf{q}). \tag{A.4}$$

$\chi(\mathbf{q})$ is the susceptibility associated with the normal coordinate $A_\mathbf{q}$. It shows a critical behavior at T_P for a special value $\mathbf{q}_c$ of $\mathbf{q}$. This wavevector describes the long-range displacive modulation of the lattice below T_P.

The susceptibility has a maximum for $\mathbf{q} = \mathbf{q}_c$. Thus for $\mathbf{q}$ close to $\mathbf{q}_c$, it has the following expansion in function of $\delta\mathbf{q} = \mathbf{q} - \mathbf{q}_c$:

$$\chi(\mathbf{q}) = \chi(\mathbf{q}_c)[1 - \delta q_1^2 \xi_1^2 - \delta q_2^2 \xi_2^2 - \delta q_3^2 \xi_3^2], \tag{A.5}$$

using as a frame the proper axes (1, 2, 3) of the matrix $\{|\partial^2 \chi(q)/\partial q_i \partial q_j|\}$. In this frame, the squares of the correlation length ξ_i are defined by

$$\xi_i^2 = -\frac{1}{2\chi(\mathbf{q}_c)} \cdot \left.\frac{\partial^2 \chi(\mathbf{q})}{\partial^2 q_i}\right|_{q_c} = -\left.\frac{\partial \log \chi(\mathbf{q})}{\partial q_i^2}\right|_{q_c}, \tag{A.6}$$

where $i = 1, 2, 3$. In the vicinity of $\mathbf{q}_c$, $\chi(\mathbf{q})$ is generally put in the form of a (relatively well-verified) Lorentzian law

$$\chi(\mathbf{q}) = \frac{\chi(\mathbf{q}_c)}{1 + \sum_{i=1,2,3} \xi_i^2 \delta q_i^2}, \tag{A.7}$$

which is equivalent to Eq. (A.5) for small δq.

In practice, it is useful to consider the following simple cases:

1. $\xi_1 = \xi_2 = 0$: one-dimensional fluctuations along the direction 3. This corresponds to x-ray diffuse sheets perpendicular to the direction 3, whose half-width at half maximum (HWHM) along this direction gives ξ_3^{-1}.

2. $\xi_1 = 0$: two-dimensional fluctuations in the plane (2, 3). This corresponds to x-ray diffuse rods in the direction 1. (The HWHM of these rods in the directions 2 and 3 gives, respectively, ξ_2^{-1} and ξ_3^{-1}.)

3. All the ξ_i are different from zero: three-dimensional fluctuations, corresponding to x-ray diffuse spots (ξ_i^{-1} is given by the HWHM of this spot in the direction of the ith component of $\delta\mathbf{q}$.) Isotropic fluctuations correspond to the special case $\xi_1 = \xi_2 = \xi_3$.

In the vicinity of T_P, for a second-order phase transition, $\chi(q_c)$ and ξ diverge as some power of $(T - T_P)$. (See, e.g., Stanley, 1971.)

$$\chi(q_c) \sim (T - T_P)^{-\gamma}$$
$$\xi \sim (T - T_P)^{-\nu}. \qquad (A.8)$$

In the mean-field approximation, the critical exponents γ and ν take, respectively, the values 1 and $\frac{1}{2}$.

To be rigorous, let us mention that the theory of critical phenomena introduces another (small) exponent η correcting the q^2 dependence of $S(q)$ at T_P (see, e.g., Stanley, 1971):

$$S(q) \sim \frac{1}{q^{2-\nu}}. \qquad (A.9)$$

In the mean-field expression (A.7), we have implicitly assumed $\eta = 0$.

b. Application to 1D Conductors

With the classical limit given by Eq. (A.4) for $S(q, t = 0)$, it is useful to define the susceptibility $\chi(q)$, which is measured in an x-ray diffuse scattering experiment performed on 1D conductors. Two order parameters A_q and ρ_q, describing, respectively, the lattice modulation and the modulation of the electron density, are generally considered. These two degrees of freedom are, in fact, linearly coupled by the electron–phonon interaction. To the lowest order, the Landau expansion of the free energy, $F(A_q, \rho_q)$, takes the form

$$F = \frac{1}{\chi_L}|A_q|^2 + \frac{1}{\tilde{\chi}_e}|\rho_q|^2 + \lambda A_q \rho_{-q} + \lambda^* A_{-q} \rho_q. \qquad (A.10)$$

Before discussing the various contributions at the free energy, it is useful to introduce the field conjugate to each order parameter

$$\text{The Peierls gap:} \quad \Delta_q = \frac{\partial F}{\partial \rho_q}$$
$$\text{The ``stress'':} \quad \sigma_q = \frac{\partial F}{\partial A_q}. \qquad (A.11)$$

In the expression (A.10):

1. The first term represents the elastic energy of the lattice not coupled to the electronic degrees of freedom.

$$\frac{1}{\chi_L} = \left.\frac{\partial \sigma_q}{\partial A_{-q}}\right|_{\rho=0} = \left.\frac{\partial^2 F}{\partial A_q \partial A_{-q}}\right|_{\rho=0} \qquad (A.12)$$

amounts to $(\hbar\omega_0)^2$ in the units used to express A_q ($\hbar\omega_0$ is the energy of the bare phonon mode, which becomes unstable by coupling with the 1D electron gas).

2. The second term represents the electronic energy in the absence of coupling with the lattice.

$$\frac{1}{\tilde{\chi}_e} = \left.\frac{\partial \Delta_q}{\partial \rho_{-q}}\right|_{A=0} = \left.\frac{\partial^2 F}{\partial \rho_q \, \partial \rho_{-q}}\right|_{A=0} \tag{A.13}$$

is the inverse electronic susceptibility, which may include the interchain Coulomb coupling between CDW: $g_{1\perp}$. With this interaction, treated in the R.P.A. approximation, χ_e can be put into the form

$$\tilde{\chi}_e = \frac{\chi_e}{1 + g_{1\perp}\chi_e}, \tag{A.14}$$

with χ_e being the electron-hole response function in chain direction. This quantity depends on the strength and on the range of the intrachain Coulomb interactions (see, for example, Solyom, 1979; Emery, 1979; Barisic and Bjelis, 1985). In a 1D electron gas without Coulomb interactions, $\chi_e(q, T)$ has the well-known form

$$\chi_e(q, T) = \sum_k \frac{f(E_k) - f(E_{k+q})}{E_{k+q} - E_k}, \tag{A.15}$$

where $f(E)$ is the Fermi-Dirac function for electrons of energy E. It diverges at $T = 0$ K for $q = 2k_F$. The divergent part of χ_e has the following T and $\delta q (= q - 2k_F)$ dependence:

$$\chi_e^D(q, T) = N(E_F)\left[\ln\left(\frac{E_c}{k_B T}\right) - \frac{7}{16\pi^2}\zeta(3)\left(\frac{\hbar v_F}{k_B T}\right)^2 \delta q^2\right], \tag{A.16}$$

where $N(E_F) = d_{/\!/}/\pi \hbar v_F$ is the density of states at the Fermi level per spin direction, v_F is the Fermi velocity, and E_c is a cut-off energy, of the order of the Fermi energy. To be rigorous, χ_e also contains a nondivergent part, χ_e^N, which also affects the lattice dynamics over a wide range of values of q (i.e., for $q \lesssim 2k_F$).

3. The third term represents the linear electron-phonon coupling energy, with a coupling constant given by:

$$\lambda = \frac{\partial \Delta_{-q}}{\partial A_q} = \frac{\partial \sigma_q}{\partial \rho_{-q}} = \frac{\partial^2 F}{\partial A_q \, \partial \rho_{-q}}, \tag{A.17}$$

3. STRUCTURAL INSTABILITIES

In a coupled system, two response functions are interesting:

- the displacement-displacement response function

$$\frac{1}{\chi_A} = \frac{\partial \sigma_q}{\partial A_{-q}}\bigg|_{\Delta=0}, \qquad (A.18)$$

which gives the susceptibility associated with A_q after elimination of ρ_q in Eq. (A.10) by the condition $\Delta_q = 0$.

- the CDW response function

$$\frac{1}{\chi_\rho} = \frac{\partial \Delta_q}{\partial \rho_{-q}}\bigg|_{\sigma=0}, \qquad (A.19)$$

which gives the susceptibility associated with ρ_q after elimination of A_q in Eq. (A.10) by the condition $\sigma_q = 0$.

Equations (A.18) and (A.19) are linked by the following derivative equality (Landau and Lifshitz, 1980):

$$\frac{\partial \Delta}{\partial \rho}\bigg|_\sigma = \frac{\partial(\Delta, \sigma)}{\partial(\rho, \sigma)} = \frac{\partial(\Delta, \sigma)/\partial(\Delta, A)}{\partial(\rho, \sigma)/\partial(\rho, A)} \cdot \frac{\partial(\Delta, A)}{\partial(\rho, A)} = \frac{\partial \sigma/\partial A|_\Delta}{\partial \sigma/\partial A|_\rho} \cdot \frac{\partial \Delta}{\partial \rho}\bigg|_A.$$

This leads to the thermodynamical relationship

$$\frac{\tilde{\chi}_e}{\chi_\rho} = \frac{\chi_L}{\chi_A}. \qquad (A.20)$$

With the free-energy expansion Eq. (A.10), which treats the electron–phonon coupling in the RPA approximation, the ratio Eq. (A.20) amounts to

$$\frac{\tilde{\chi}_e}{\chi_\rho} = \frac{\chi_L}{\chi_A} = 1 - |\lambda|^2 \chi_L \tilde{\chi}_e. \qquad (A.21)$$

In this approximation, χ_A takes the form

$$\chi_A = \chi_L + \frac{|\lambda|^2 \chi_L^2 \tilde{\chi}_e}{1 - |\lambda|^2 \chi_L \tilde{\chi}_e} = \chi_L(1 + |\lambda|^2 \chi_L \chi_\rho). \qquad (A.22)$$

With Eq. (A.22), the correlation function measured by x-ray diffuse scattering and given by Eq. (A.4) has two contributions:

$$S(q, t = 0) = S^N(q, t = 0) + \delta S^D(q, t = 0). \qquad (A.23)$$

$$S^N(q, t = 0) = \langle |A_q^0|^2 \rangle = k_B T \chi_L (1 + |\lambda|^2 \chi_L \chi_\rho^N) \qquad (A.23')$$

is the contribution of bare phonons of "amplitude" of vibration A_q^0 which can be eventually screened by the nondivergent part, χ_e^N, of the electron-hole response function. It contributes to the normal thermal diffuse scattering, which is generally weakly q dependent.

$$\delta S^D(q, t = 0) = \langle |\delta A_q|^2 \rangle = |\lambda|^2 \chi_L^2 \langle |\rho_q|^2 \rangle = k_B T |\lambda|^2 \chi_L^2 \chi_\rho^D \quad (A.23)''$$

is an extra contribution due to the linear coupling of the lattice degrees of freedom with the divergent part of the electron-hole response function, which drives the CDW electronic instability. The divergence of $\tilde{\chi}_e$ occurring around the wavevector $q \sim 2k_F$ (or $4k_F$), δS^D is well localized in reciprocal space. $\delta S^D(q, t = 0)$ appears in the form of an extra scattering superimposed on the background due to the thermal diffuse scattering [$S^N(q, t = 0)$ plus similar contributions due to other phonon branches] and from additional scattering processes (Compton scattering, etc.) (Fig. 48a). The susceptibility χ_A, which requires knowledge of the contribution $S^N(q, t = 0)$ at the background intensity, cannot be easily extracted from x-ray diffuse scattering experiments. (It can be obtained from inelastic neutron studies—see Section 21.) However, the extra diffuse scattering δS^D can be easily measured. It yields a quantity proportional to the divergent part of CDW response function χ_ρ, whose determination is considered in Part IV of this review.

In the approximation of the free energy done in Eq. (A.10), χ_ρ and χ_A, even for a purely 1D system, diverge at the mean-field temperature of the lattice instability, T_p^{MF}, defined by $1 = |\lambda|^2 \chi_L \chi_e$ in Eq. (A.21). In fact, in the 1D regime, fluctuations suppress this instability. This effect can be accounted for by a fourth-order term in the free energy (see, for example, Dieterich, 1976). In that case, χ_A differs from the prediction of Eq. (A.21), especially in the vicinity of T_p^{MF}. The "Hartree-Fock" correction at χ_A, due to such an anharmonic contribution, has been considered by Jérome and Schulz (1982a) in the case of a noninteracting electron gas.

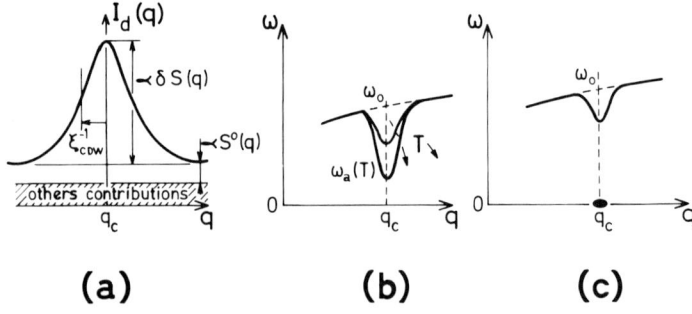

FIG. 48. (a) Schematical representation of the x-ray diffuse scattering intensity, $I_d(q)$, along the q direction, showing an extra contribution [$\alpha \, \delta S(q)$], varying strongly with q around q_c, superimposed on a weakly q-dependent background [$\alpha S^0(q)$]. (b) Softening at q_c of a phonon branch (Kohn anomaly) for T decreasing. (c) Weak Kohn anomaly and quasi-elastic response (central peak) at q_c.

3. STRUCTURAL INSTABILITIES

c. Disorder

In presence of disorder or impurities which induce CDW Friedel oscillations $\langle \delta\rho_q \rangle$ and local (static) displacements $\langle \delta^S A_q \rangle = \lambda \chi_L \langle \delta\rho_q \rangle$, there is an additional contribution at $\delta S^D(q, t = 0)$. It can thus be expressed under the form (Sham, 1978, Girault et al., 1988):

$$\delta S^D(q, t = 0) = \langle |\delta A_q|^2 \rangle + \overline{|\langle \delta^S A_q \rangle|^2}, \qquad (A.24)$$

where the horizontal line on the second term means an average on the sample [omitted in Eq. (A.3) because homogeneous samples are considered]. In Eq. (A.24) the first contribution, due to thermal fluctuations, is assumed to be the same as in the pure sample, where $\langle \delta A_q \rangle = 0$ and:

$$\langle |\delta A_q|^2 \rangle = k_B T |\lambda|^2 \chi_L^2 \chi_\rho^D.$$

The second contribution is due to extra displacements associated with the formation of Friedel oscillations around impurities. In that case one has:

$$\langle \delta^S A_q \rangle = \lambda \chi_L \chi_\rho V_q^i,$$

where V_q^i is the Fourier transform of the impurity potential. In the dilute limit one gets:

$$\overline{|\langle \delta^S A_q \rangle|^2} = |\lambda|^2 \chi_L^2 \chi_\rho^2 c \overline{|V_q|^2}, \qquad (A.25)$$

where c is the impurity concentration and $\overline{V_q^2}$ is the impurity mean square potential averaged on the sample. The contribution Eq. (A.25) is responsible for the enhancement of the x-ray diffuse scattering in disordered samples, with respect to that observed in pure samples.

21. Neutron Scattering

The quantity that characterizes the dynamics of the order parameter fluctuations in the vicinity of a phase transition is the temporal correlation function

$$S(q, t) = \langle |A_q(0) A_{-q}(t)| \rangle.$$

Its Fourier transform $S(q, \omega)$ can be obtained directly in a neutron scattering experiment. The quantity measured, the differential cross section (cross section per unit of solid angle Ω and frequency ω), is given by the Van Hove formula (Cochran, 1969; Cowley, 1980)

$$\frac{d^2\sigma}{d\Omega\, d\omega} = \frac{k_F}{k_I} |F_d(\mathbf{Q})|^2 S(q, \omega),$$

where k_I and k_F are, respectively, the wavevector of the incoming and outcoming neutrons, $F_d(\mathbf{Q})$ is given by Eq. (A.2) where the neutron scattering length b_j replaces the x-ray scattering factor $f_j(\mathbf{Q})$. (Here, also, we still consider only one unstable phonon mode, and we ignore its eventual coupling with other modes of the same symmetry).

Upon regrouping the $\omega > 0$ and $\omega < 0$ terms, the fluctuation dissipation theorem gives

$$S(q, |\omega|) = \hbar \coth(\tfrac{1}{2}\beta\hbar|\omega|) \operatorname{Im} \chi(q, |\omega|), \tag{A.26}$$

where the real and imaginary parts of the generalized susceptibility are related by the Kramers–Kronig relationship

$$\operatorname{Re} \chi(q, \omega) = -\frac{1}{\pi} PP \int_{-\infty}^{+\infty} \frac{\operatorname{Im} \chi(q, \omega')}{\omega - \omega'} d\omega'. \tag{A.27}$$

In the classical limit, $k_B T \equiv 1/\beta \gg \hbar\omega$, $S(q, \omega)$ becomes

$$S(q, |\omega|) = \frac{2k_B T}{\omega} \operatorname{Im} \chi(q, |\omega|). \tag{A.28}$$

In a conventional x-ray experiment, no analysis in energy can be performed because the x-ray photon energy (~ 10 KeV) is much larger than the phonon energy (~ 1 meV). Thus the quantity measured by Eq. (A.3) is an integral over frequency of $S(q, \omega)$

$$S(q, t = 0) = \frac{1}{2\pi} \int_{-\infty}^{+\infty} S(q, \omega) \, d\omega. \tag{A.29}$$

From Eqs. (A.27), (A.28), and (A.29) one derives the expression (A.4)

$$S(q, t = 0) = k_B T \operatorname{Re} \chi(q, \omega = 0), \tag{A.30}$$

where $\operatorname{Re} \chi(q, \omega = 0)$ is the thermodynamic susceptibility $\chi(q)$.

In the classical limit, the energy integration of the neutron cross section of the unstable lattice mode gives the susceptibility χ_A defined in Section 20b.

In the soft mode description, the instability lies in a particular phonon branch, which has, at q_c, a frequency ω_a that drops to zero as T approaches T_P from above (Fig. 48b). This effect can be described by a susceptibility that has, generally, the ω dependence of a damped harmonic oscillator

$$\chi(q_c, \omega) = \frac{1}{\hbar^2} \left[\frac{1}{\omega_a^2 - \omega^2 - 2i\Gamma\omega} \right], \tag{A.31}$$

where the damping constant Γ includes anharmonic contributions. The frequency integration gives, simply

$$\chi_A(q_c) = \frac{1}{(\hbar\omega_a)^2},$$

Using the mean-field expression (A.21) one finds

$$(\hbar\omega_a)^2 = (\hbar\omega_0)^2 \left[1 - \frac{|\lambda|^2}{(\hbar\omega_0)^2}\tilde{\chi}_e\right]. \tag{A.32}$$

The divergence of $\tilde{\chi}_e$ (or χ_e when the chains are not coupled) leads to the formation at $2k_F$ of a Kohn anomaly, generally in the lowest phonon branch that experiences a first-order electron–phonon coupling with the 1D electron gas. Taking, for example, the temperature dependence of χ_e given by Eq. (A.16), the softening of the phonon frequency at q_c can be described in the vicinity of T_P^{MF} by:

$$(\hbar\omega_a)^2 = (\hbar\omega_0)^2 \left[1 - \frac{T_P^{MF}}{T}\right], \tag{A.33}$$

with $k_B T_P^{MF} = E_c \exp[-(\hbar\omega_0)^2/|\lambda|^2 N(E_F)]$ being the mean-field Peierls transition temperature (Rice and Strassler, 1973). Within the soft-mode description the frequencies ω_a and ω_0 can be incorporated into the extra diffuse scattering expression, which thus takes a form sometimes quoted (Khanna et al., 1977)

$$\delta I_d \sim \delta S^D \propto k_B T \left(\frac{1}{\omega_a^2} - \frac{1}{\omega_0^2}\right). \tag{A.34}$$

A more complex description of the structural instability (which might apply to TTF-TCNQ, see Section 11b) is that shown by Fig. 48c, where, in addition to a smooth Kohn anomaly in the phonon spectrum, there is a quasielastic response. When T approaches T_P from above, the phonon branch does not soften completely, but the central response grows critically in intensity at q_c and drives the divergence of χ_A. In that case, the generalized susceptibility $\chi(q, \omega)$ has an ω dependence more complex than that given by the soft-mode expression (A.31).

The central peak can be viewed as an extra scattering coming from domains of the low-temperature phase that are formed above T_P with a finite lifetime. In 1D conductors several mechanisms have been invoked to explain the formation of such domains. Existing theories rely either on extrinsic mechanisms (formation of Friedel oscillations in the vicinity of defects—see, e.g., Sham, 1979) or an intrinsic mechanism [anharmonic motion of the lattice (see, e.g., Dieterich, 1976)]. Let us also mention, in the latter class of mechanisms, that a strong electron–phonon coupling (see, e.g., Le Daeron and Aubry, 1983) or electron–electron interactions (see, e.g., Barisic and Bjelis, 1985) favor, for slow electronic responses ($\omega_{e1} < \omega_0$), the formation of a central peak associated with relaxational dynamics.

22. Satellite Reflections

The Peierls structural phase transition occurring at T_P corresponds to a long-range displacive modulation with the wavevector q_c of the high-temperature atomic positions. The amplitude of the distortion is characterized by the order parameter $|A_{q_c}|$

$$\langle |A_{q_c}| \rangle = 0 \quad T > T_P$$
$$\langle |A_{q_c}| \rangle \neq 0 \quad T < T_P. \tag{A.35}$$

For a second-order phase transition, the order parameter behaves near T_P as (see, e.g., Stanley, 1971)

$$\langle |A_{q_c}| \rangle \sim (T_P - T)^\beta, \tag{A.36}$$

where β is $\frac{1}{2}$ in the mean-field approximation. Satellite reflections measured at the wave vector $\mathbf{Q}_c = \mathbf{G} + \mathbf{q}_c$ have an intensity that can be put in the form

$$I_S(\mathbf{Q}_c) = |F_S(\mathbf{Q}_c)|^2 \langle |A_{q_c}| \rangle^2, \tag{A.37}$$

when the atomic displacements are small (this is usually the case for a Peierls transition). In this case, $I_S(\mathbf{Q}_c)$ is proportional to the square of the order parameter.

$F_S(\mathbf{Q}_c)$ is the satellite structure factor that characterizes the distortion. When the displacement of the jth atom in the nth (high-temperature) unit cell can be described by a single wave

$$\mathbf{u}_j^n = \frac{A_q}{\sqrt{m_j}} \mathbf{e}_j(q) \exp[-i\mathbf{q}(\mathbf{R}_n + \mathbf{r}_j)], \tag{A.38}$$

the intensity diffracted at the wave vector $\mathbf{Q}$ is

$$I_S(\mathbf{Q}) = \left| \sum_{j,n} f_j(\mathbf{Q}) e^{-w_j(\mathbf{Q})} \exp[i\mathbf{Q} \cdot (\mathbf{R}_n + \mathbf{r}_j + \mathbf{u}_j^n)] \right|^2. \tag{A.39}$$

For $\mathbf{u}$ and $\mathbf{Q}$ such that $\mathbf{Q}\mathbf{u}_j \ll 1$, Eq. (A.39) becomes

$$I_S(\mathbf{Q}) = |F_d(\mathbf{Q})|^2 \langle |A_q| \rangle^2 \delta(\mathbf{Q} - \mathbf{q}_c - \mathbf{G}). \tag{A.40}$$

This expression is identical to Eq. (A.37), with $F_S(\mathbf{Q}_c) = F_d(\mathbf{Q}_c)$; $F_d(\mathbf{Q}_c)$ is given by Eq. (A.2). Satellite reflection intensities are measured either by x-ray or by neutron diffraction.

23. Resolution Corrections in X-ray Diffuse Scattering Experiments

One of the most severe limitations for the accurate measurement of the physical quantities discussed in Part IV [$\chi_\rho(q_c)$ and ξ_ρ] is due to the fact that the extra x-ray diffuse scattering represents a weak intensity, typically on the order of 10^{-5}–10^{-6} times the intensity of a main Bragg reflection for the 1D organic conductors (Khanna et al., 1977; Moret and Pouget, 1986). With conventional x-ray sources, such weak intensities can be recorded only with a somewhat relaxed resolution. In practical cases, the HWHM of the resolution ($\Delta_{\text{res.}}$) is not negligible with respect to the HWHM of the experimental signal ($\Delta_{\text{exp.}}$). Thus, the shape and the HWHM of the observed signal, resulting from the convolution of the instrumental resolution with the intrinsic diffuse scattering (generally assumed to be Lorentzian), is influenced by the resolution. If one assumes a Lorentzian resolution, the convoluted signal is also Lorentzian. In that case, the inverse of the intrinsic correlation length, ξ^{-1}, is simply given by

$$\xi^{-1} = \Delta_{\text{exp.}} - \Delta_{\text{res.}}.$$

If one assumes a Gaussian resolution, the observed signal has a Voigt profile, which is intermediate between a Gaussian and a Lorentzian. ξ^{-1} can be obtained from the knowledge of $\Delta_{\text{exp.}}$ and $\Delta_{\text{res.}}$ using the numercial calculations of Langford (1978).

The actual shape of the resolution function to consider in x-ray diffuse scattering experiments, especially those using the "monochromatic" Laue techniques, well adapted for the study of 1D fluctuations (see Part IV), is not straightforward. The experimental determination of the resolution function from "Bragg sheets" gives a shape intermediate between a Lorentzian and a Gaussian, but, however, closer to a Gaussian than a Lorentzian. Thus the Lorentzian and Gaussian resolution corrections considered above give a value of ξ in excess and defect, respectively. A more exact extraction of ξ can be done by a least squares fitting of the convolution product of Eq. (A.7) with the instrumental resolution. However, in most cases, the experimental line shape with poor statistics (see Figs. 17b and 18b, for example) precludes such a procedure.

REFERENCES

Abrahams, E., Solyom, J., and Woynarovich, F. (1977). *Phys. Rev.* B **16**, 5238.
Andersen, J. R., Bechgaard, K., Jacobsen, C. S., Rindorf, G., Soling, H., and Thorup, N. (1978). *Acta Cryst.* B **34**, 1901.
Andrieux, A., Schulz, H. J., Jérome, D., and Bechgaard, K. (1979a). *Phys. Rev. Lett.* **43**, 227 and *J. Physique Lettres* **40**, L 385.
Andrieux, A., Chaikin, P. M., Duroure, C., Jérome, D., Weyl, C., Bechgaard, K., and Andersen, J. R. (1979b). *J. Physique* **40**, 1199.
Aubry, S. (1980). *Ferroelectrics* **24**, 53.
Bak, P. (1976). *Phys. Rev. Lett.* **37**, 1071.
Bak, P., and Emery, V. J. (1976). *Phys. Rev. Lett.* **36**, 978.

Bak, P., and Janssen, T. (1978). *Phys. Rev. B* **17**, 436.
Barisic, S. (1983). *J. Physique* **44**, 1983.
Barisic, S. (1985). *Mol. Cryst. Liq. Cryst.* **119**, 413.
Barisic, S., and Bjelis, A. (1977). *J. Physique Colloque* **38**, C7-254.
Barisic, S., and Bjelis, A. (1983). *J. Physique Lettres* **44**, L 327.
Barisic, S., and Bjelis, A. (1985). In "Theoretical Aspects of Band Structures and Electronic Properties of Pseudo-One-Dimensional Solids" (H. Kanimura, ed.), p. 49. Reidel, Dordrecht.
Barisic, S., and Brazovskii, S. (1981). In "Recent Developments in Condensed Matter Physics" (J. T. Devreese, ed.), Vol. 6, p. 327. Plenum, New York.
Barmatz, M., Testardi, L. R., Garito, A. F., and Heeger, A. J. (1974). *Solid State Commun.* **15**, 1299.
Bates, F. E., Eldridge, J. E., and Bryce, M. R. (1981). *Can. J. Phys.* **59**, 339.
Bechgaard, K., Kistenmacher, T. J., Bloch, A. N., and Cowan, D. O. (1977). *Acta Cryst. B* **33**, 417.
Bechgaard, K., Jacobsen, C. S., and Hessel-Andersen, N. (1978). *Solid State Commun.* **25**, 875.
Beni, G., Pincus, P., and Kanamori, J. (1974). *Phys. Rev. B* **10**, 1896.
Berlinsky, A. J. (1976). *Contemp. Phys.* **17**, 331.
Berlinsky, A. J., Carolan, J. F., and Weiler, L. (1974). *Solid State Commun.* **15**, 795.
Berlinsky, A. J., Carolan, J. F., and Weiler, L. (1976). *Solid State Commun.* **19**, 1165.
Bernasconi, J., Rice, M. J., Schneider, W. R., and Strassler, S. (1975). *Phys. Rev. B* **12**, 1090.
Biljakovic-Franulovic, K., Tomic, S., Prester, M., and Djurek, D. (1979). *J. Physique Lettres* **40**, L 151.
Bjelis, A., and Barisic, S. (1976). *Phys. Rev. Lett.* **37**, 1517.
Bjelis, A., and Barisic, S. (1977). In "Organic Conductors and Semiconductors" (L. Pal, G. Gruner, A. Janossy, and J. Solyom, eds.). *Lecture Notes in Physics* **65**, 291. Springer, Berlin.
Bjelis, A., and Barisic, S. (1978). *J. Physique Lettres* **39**, L 347.
Bjelis, A., and Barisic, S. (1982). *Phys. Rev. Lett.* **48**, 684; *Mol. Cryst. Liq. Cryst.* **85**, 1541.
Blessing, R. H., and Coppens, P. (1974). *Solid State Commun.* **15**, 215.
Bloch, A. N., and Mazumdar, S. (1983). *J. Physique Colloque* **44**, C3-1273.
Bouveret, Y., and Megtert, S. (1988) to be published.
Bozio, R., and Pecile, C. (1980). In "The Physics and Chemistry of Low Dimensional Solids" (L. Alcacer, ed.). NATO ASI, Vol. C. 56, p. 165. Reidel, Dordrecht.
Bozio, R., and Pecile, C. (1981). *Solid State Commun.* **37**, 193; *Chemica Scripta* **17**, 31.
Bozio, R., Pecile, C., and Tosi, P. (1983). *J. Physique Colloque* **44**, C3-1453.
Bray, J. W., Interrante, L. V., Jacobs, I. S., and Bonner, J. C. (1983). In "Extended Linear Chain Compounds" (J. S. Miller, ed.), Vol. 3, p. 353. Plenum, New York.
Bulaevskii, N. L., Buzdin, A. J., and Khomskii, D. J. (1980). *Solid State Commun.* **35**, 101.
Butler, M. A., Wudl, F., and Soos, Z. G. (1975). *Phys. Rev. B* **12**, 4708.
Caron, L. G., and Bourbonnais, C. (1984). *Phys. Rev. B* **29**, 4230.
Chaikin, P. M., Greene, R. L., Etemad, S., and Engler, E. (1976). *Phys. Rev. B* **13**, 1627.
Chaikin, P. M., Kwak, J. F., and Epstein, A. J. (1979). *Phys. Rev. Lett.* **42**, 1178.
Chappel, J. S., Bloch, A. N., Bryden, W. A., Maxfield, M., Poehler, T. O., and Cowan, D. O. (1981). *J.A.C.S.* **103**, 2442.
Chasseau, D., Comberton, G., Gaultier, J., and Hauw, C. (1978). *Acta Cryst. B* **34**, 689.
Chui, S. T., and Bray, J. W.. (1980). *Phys. Rev. B* **21**, 1380.
Clark, W. G., Hammann, J., Sanny, J., and Tippie, L. C. (1979). In "Quasi-One-Dimensional Conductors II" (S. Barisic, A. Bjelis, J. R. Cooper, and B. Leontic, eds.). *Lecture Notes in Physics* **96**, 255. Springer, Berlin.

3. STRUCTURAL INSTABILITIES

Cochran, W. (1969). *Adv. Physics* **18,** 157.
Coleman, L. B., Cohen, M. J., Sandmann, M. J., Yamagishi, D. J., Garito, A. F., and Heeger, A. J. (1973). *Solid State Commun.* **12,** 1125.
Comès, R. (1977). *In* "Chemistry and Physics for One-Dimensional Metals" (H. J. Keller, ed.), p. 315. Plenum, New York.
Comès, R., and Shirane, G. (1979). *In* "Highly Conducting One Dimensional Solids" (J. T. Devreese, R. P. Evrard, and V. E. Van Doren, eds.), p. 17. Plenum, New York.
Comès, R., Lambert, M., Launois, H., and Zeller, H. R. (1973). *Phys. Rev.* B **8,** 571.
Comès, R., Shapiro, S. M., Shirane, G., Garito, A. F., and Heeger, A. J. (1975). *Phys. Rev. Lett.* **35,** 1518.
Comès, R., Shirane, G., Shapiro, S. M., Garito, A. F., and Heeger, A. J. (1976). *Phys. Rev.* B **14,** 2376.
Conwell, E. M. (1978). *Phys. Rev.* B **18,** 1818.
Conwell, E. M. (1979). *In* "Quasi One Dimensional Conductors I" (S. Barisic, J. R. Cooper, and B. Leontic, eds.). *Lecture Notes in Physics* **95,** 270. Springer, Berlin.
Conwell, E. M. (1980a). *Solid State Commun.* **33,** 17.
Conwell, E. M. (1980b). *Solid State Commun.* **36,** 939.
Conwell, E. M., and Howard, I. A. (1986). *Synthetic Metals* **13,** 71.
Cooper, J. R., and Lukatela, J. (1979). *In* "Quasi One Dimensional Conductors I" (S. Barisic, A. Bjelis, J. R. Cooper, and B. Leontic, eds.). *Lectures Notes in Physics* **95,** 174. Springer, Berlin.
Cooper, J. R., Jérome, D., Weger, M., and Etemad, S. (1975). *J. Physique Lettres* **36,** L 219.
Coppens, P. (1975). *Phys. Rev. Lett.* **35,** 98.
Coppens, P., and Gururow, T. N. (1978). *In* "Synthesis and Properties of Low Dimensional Materials" (J. S. Miller and A. J. Epstein, eds.), *NY Acad. Sci.* **313,** 244.
Coppens, P., Petricek, V., Levendis, D., Larsen, F. K., Paturle, A., Yan, G., and LeGrand, N. D. (1987). *Phys. Rev. Lett.* **59,** 1695.
Cowley, R. A. (1980). *Adv. Physics* **29,** 1.
Craven, R. A., Salamon, M. B., De Pasquali, G., Herman, R. M., Stucky, G., and Schultz, A. (1974). *Phys. Rev. Lett.* **32,** 769.
Craven, R. A., Tomkiewicz, Y., Engler, E. M., and Taranko, A. R. (1977). *Solid State Commun.* **23,** 429.
Debray, D., Millet, R., Jérome, D., Barisic, S., Fabre, J. M., and Giral, L. (1977). *J. Physique Lettres* **38,** L 227. [The compressibility data of this paper have been reexamined by Filhol et al. (1981).]
Delhaes, P., Flandrois, S., Amiell, J., Keryer, G., Toreilles, E., Fabre, J. M., Giral, L., Jacobsen, C. S., and Bechgaard, K. (1977). *J. Physique Lettres* **38,** L 233.
Dénoyer, F., Comès, R., Garito, A. F., and Heeger, A. J. (1975). *Phys. Rev. Lett.* **35,** 445.
Devreux, F., Guglielmi, M., and Nechtschein, M. (1978). *J. Physique* **39,** 541.
Dieterich, W. (1976). *Adv. Phys.* **25,** 615.
Dixit, S. N., and Mazumdar, S. (1984). *Phys. Rev.* B **29,** 1824.
Djurek, D., Franvlovic, K., Prester, M., Tomic, S., Giral, L., and Fabre, M. (1977). *Phys. Rev. Lett.* **38,** 715.
Duke, C. B., Lipari, N. O., and Pietronero, L. (1975). *Chem. Phys. Lett.* **30,** 415.
Eldridge, J. E., and Bates, F. E. (1982). *Mol. Cryst. Liq. Cryst.* **85,** 169.
Ellenson, W. D., Comès, R., Shapiro, S. M., Shirane, G., Garito, A. F., and Heeger, A. J. (1976). *Solid State Commun.* **20,** 53.
Ellenson, W. D., Shapiro, S. M., Shirane, G., and Garito, A. F. (1977). *Phys. Rev.* B **16,** 3244.
Emery, V. J. (1976). *Phys. Rev. Lett.* **37,** 107.

Emery, V. J. (1977). *In* "Chemistry and Physics of One Dimensional Metals" (H. J. Keller, ed.), p. 1. Plenum, New York.
Emery, V. J. (1979). *In* "Highly Conducting One Dimensional Solids" (J. T. Devreese, R. P. Evrard, and V. E. Van Doren, eds.), p. 247. Plenum, New York.
Emery, V. J., Bruinsma, R., and Barisic, S. (1982). *Phys. Rev. Lett.* **48**, 1039.
Epstein, A. J., and Miller, J. S. (1978). *Solid State Commun.* **27**, 325.
Epstein, A. J., and Miller, J. S. (1980). *In* "The Physics and Chemistry of Low Dimensional Solids" (L. Alcader, ed.). NATO ASI Vol. C 56, p. 339. Reidel, Dordrecht.
Epstein, A. J., Etemad, S., Garito, A. F., and Heeger, A. J. (1972). *Phys. Rev. B* **5**, 952.
Epstein, A. J., Miller, J. S., and Chaikin, P. M. (1979). *Phys. Rev. Lett.* **43**, 1178.
Epstein, A. J., Miller, J. S., Pouget, J. P., and Comès, R. (1981). *Phys. Rev. Lett.* **47**, 741.
Epstein, A. J., Kaufer, J. W., Rommelmann, H., Howard, I. A., Conwell, E. M., Miller, J. S., Pouget, J. P., and Comès, R. (1982). *Phys. Rev. Lett.* **49**, 1037.
Epstein, A. J., Bigelow, R. W., Miller, J. S., McCall, R. P., and Tanner, D. B. (1985). *Mol. Cryst. Liq. Cryst.* **120**, 43.
Etemad, S. (1976). *Phys. Rev. B* **13**, 2254.
Etemad, S., Penney, T., Engler, E. M., Scott, B. A., and Seiden, P. E. (1975). *Phys. Rev. Lett.* **34**, 741.
Etemad, S., Engler, E. M., Schultz, T. D., Penney, T., and Scott, B. A. (1978). *Phys. Rev. B* **17**, 513.
Farges, J. P. (1985). *J. Physique* **46**, 465.
Ferraris, J. P., Cowan, D. O., Walatka, V. V., and Perlstein, J. M. (1973). *J.A.C.S.* **95**, 948.
Filhol, A., and Thomas, M. (1984). *Acta Cryst. B* **40**, 44.
Filhol, A., Rovira, M., Hawn, C., Gaultier, J., Chasseau, D., and Dupuis, P. (1979). *Acta Cryst. B* **35**, 1652.
Filhol, A., Bravic, G., Gaultier, J., Chasseau, D., and Vettier, C. (1981). *Acta Cryst. B* **37**, 1225 (erratum *B* 37, 2120).
Fincher, C., Shirane, G., Comès, R., and Garito, A. F. (1980). *Phys. Rev. B* **21**, 5424.
Firsov, Yu. A., Prigodin, V. N., and Seidel Chr. (1985). *Physics Report* **126**, 245.
Flandrois, S., and Chasseau, D. (1977). *Acta Cryst. B* **33**, 2744.
Forro, L., Zuppiroli, L., Pouget, J. P., and Bechgaard, K. (1983). *Phys. Rev. B* **27**, 7600.
Forro, L., Bouffard, S., and Pouget, J. P. (1984). *J. Physique Lettres* **45**, L 343.
Friend, R. H., Miljak, M., and Jérome, D. (1978a). *Phys. Rev. Lett.* **40**, 1048.
Friend, R. H., Jérome, D., Fabre, J. M., Giral, L., and Bechgaard, K. (1978b). *J. Phys. C: Solid State Physics* **11**, 263.
Fritchie, C. J. (1966). *Acta Cryst.* **20**, 892.
Gasser, W. (1985). *Solid State Commun.* **56**, 121.
Girault, S., Moudden, A. H., Pouget, J. P., and Godard, J. M. (1988). *Phys. Rev. B* (in press).
Greene, R. L., Mayerle, J. J., Schumaker, R., Castro, G., Chaikin, P., Etemad, S., and La Placa, S. J. (1976). *Solid State Commun.* **20**, 943.
Grobman, W. D., and Silverman, B. D. (1976). *Solid State Commun.* **19**, 319.
Guy, D. R. P., Marseglia, E. A., Parkin, S. S., Friend, R. H., and Bechgaard, K. (1982). *Mol. Cryst. Liq. Cryst.* **79**, 337.
Hardebusch, U., Gerhardt, W., Schilling, J. S., Bechgaard, K., Weger, M., Miljak, M., and Cooper, J. R. (1979). *Solid State Commun.* **32**, 1151.
Hartzstein, C., Zevin, V., and Weger, M. (1980a). *J. Physique* **41**, 677.
Hartzstein, C., Zevin, V., and Weger, M. (1980b). *Solid State Commun.* **36**, 545.
Hawley, M. E., Poehler, T. O., Carruthers, T. F., Bloch, A. N., and Cowan, D. O. (1978). *Bull. Am. Phys. Soc.* **23**, 424.

Henriques, R. H., Alcacer, L., Pouget, J. P., and Jérome, D. (1984). *J. Phys. C: Solid State Phys.* **17**, 5197.
Herman, F. (1977). *Physica Scripta* **16**, 303.
Herman, F., Salahub, D. R., and Messmer, R. P. (1977). *Phys. Rev. B* **16**, 2453.
Herman, R. M., Salamon, M. B., De Pasquali, G., and Stucky, G. (1976). *Solid State Commun.* **19**, 137.
Hirsch, J. E., and Kariotis, R. (1985). *Phys. Rev. B* **32**, 7320.
Hirsch, J. E., and Scalapino, D. J. (1983). *Phys. Rev. B* **27**, 7169.
Hirsch, J. E., and Scalapino, D. J. (1984). *Phys. Rev. B* **29**, 5554.
Hirsch, J. E., and Scalapino, D. J. (1985). *Phys. Rev. B* **32**, 7320.
Horn, P. M., and Rimai, D. (1976). *Phys. Rev. Lett.* **36**, 809.
Horn, P. M., and Guidotti, D. (1977). *Phys. Rev. B* **16**, 491.
Horn, P. M., Herman, R., and Salamon, M. B. (1977). *Phys. Rev. B* **16**, 5012.
Horowitz, B., Gutfreund, H., and Weger, M. (1975). *Phys. Rev. B* **8**, 3174.
Horowitz, B., and Mukamel, D. (1977). *Solid State Commun.* **23**, 285.
Hubbard, J. (1978). *Phys. Rev. B* **17**, 494.
Huizinga, S., Kommandeur, J., Sawatzky, G. A., Kopinga, K., and De Jonga, W. J. (1979). *In* "Quasi One Dimensional Conductor II" (S. Barisic, A. Bjelis, J. R. Cooper, and B. Leontic, eds.). *Lecture Notes in Physics* **96**, 45. Springer, Berlin.
Ikemoto, I., Sugano, T., and Kuroda, H. (1977). *Chem. Phys. Lett.* **49**, 45.
Imry, Y. (1975). *J. Phys. C: Solid State Phys.* **8**, 567.
Ishiguro, T., Sumi, H., Kagoshima, S., Kajimura, K., and Anzai, H. (1980). *J. Phys. Soc. Japan* **48**, 456.
Jacobsen, C. S. (1985). *Mat. Fys. Medd. Dan. Vid. Selsk* **41**, 251.
Jacobsen, C. S. (1986). *J. Phys. C: Solid State Phys.* **19**, 5643.
Jacobsen, C. S. (1987). *In* "Low Dimensional Conductors and Superconductors" (D. Jérome and L. G. Caron, eds.). NATO ASI, Vol. B155, p. 253. Plenum, New York.
Jacobsen, C. S., Mortensen, K., Andersen, J. R., and Bechgaard, K. (1978). *Phys. Rev. B* **18**, 905.
Jafarey, S. (1977). *Phys. Rev. B* **16**, 2584.
Jérome, D. (1978). *In* "Molecular Metals" (W. E. Watfield, ed.). NATO Conf. Series VI: Material Sciences, Vol. 1, p. 105. Plenum, New York.
Jérome, D., and Schulz, H. J. (1982a). *Adv. Phys.* **31**, 299.
Jérome, D., and Schulz, H. J. (1982b). *In* "Extended Linear Chain Compounds" (J. S. Miller, ed.), Vol. 2, p. 159. Plenum, New York.
Jérome, D., Muller, W., and Weger, M. (1974). *J. Physique Lettres* **35**, 277.
Kagoshima, S. (1982). *In* "Extended Linear Chain Compound" (J. S. Miller, ed.), Vol. 2, p. 303. Plenum, New York.
Kagoshima, S., Anzai, H., Kajimura, K., and Ishiguro, T. (1975). *J. Phys. Soc. Japan* **39**, 1143.
Kagoshima, S., Ishiguro, T., and Anzai, H. (1976). *J. Phys. Soc. Japan* **41**, 2061.
Kagoshima, S., Ishiguro, T., Schultz, T. D., and Tomkiewicz, Y. (1978a). *Solid State Commun.* **28**, 485.
Kagoshima, S., Anzai, H., Ishiguro, T., Engler, E. M., Schultz, T. D., and Tomkiewicz, Y. (1978b). *In* "Lattice Dynamics" (M. Balkanski, ed.), p. 591. Flammarion Sciences, Paris.
Kagoshima, S., Ishiguro, T., Engler, E. M., Schultz, T. D., and Tomkiewicz, Y. (1980). *Solid State Commun.* **34**, 151.
Kagoshima, S., Pouget, J. P., and Anzai, H. (1983a). *J. Phys. Soc. Japan* **52**, 1629.
Kagoshima, S., Pouget, J. P., Yasunaga, T., and Torrance, J. B. (1983b). *Solid State Commun.* **46**, 521.

Khanna, S. K., Pouget, J. P., Comès, R., Garito, A. F., and Heeger, A. J. (1977). *Phys. Rev.* B **16**, 1468.
Khidekel, M. L., and Zhilyaeva, E. I. (1981). *Synthetic Metals* **4**, 1.
Kistenmacher, T. J. (1978). In "Synthesis and Properties of Low Dimensional Materials" (J. S. Miller and A. J. Epstein, eds.). *NY Acad. Sci.* **313**, 333.
Kistenmacher, T. J., Phillips, T. E., and Cowan, D. D. (1974). *Acta Cryst.* B **30**, 763.
Klimenko, V. E., Krivnov, V. Ya., Ovchinnikov, A. A., Ukrainskii, I. I., and Shvets, A. F. (1976). *Sov. Phys. JETP* **42**, 123.
Kobayashi, H. (1975). *Bull. Chem. Soc. Jpn.* **48**, 1373.
Kobayashi, H., Ohashi, Y., Marumo, F., and Saito, Y. (1970). *Acta Cryst.* B **26**, 456.
Kobayashi, H., Marumo, F., and Saito, Y. (1971). *Acta Cryst.* B **27**, 373.
Kondo, J., and Yamaji, K. (1977). *J. Phys. Soc. Japan* **43**, 424.
Korin, B., Cooper, J. R., Miljak, M., Hamzic, A., and Bechgaard, K. (1981). *Chemica Scripta* **17**, 45.
Kuzmany, H., and Kundu, B. (1979). In "Quasi One Dimensional Conductors I" (S. Barisic, A. Bjelis, J. R. Cooper, and B. Leontic, eds.). *Lecture Notes in Physics* **95**, 259. Springer, Berlin.
Kuzmany, H., and Elbert, M. (1980). *Solid State Commun.* **35**, 597.
Kwak, J. F., Beni, G., and Chaikin, P. M. (1976). *Phys. Rev.* B **13**, 641.
Lacoe, R. C., Schulz, H. J., Jérome, D., Bechgaard, K., and Johannsen, I. (1985). *Phys. Rev. Lett.* **55**, 2351.
Landau, L. D., and Lifshitz, E. M. (1980). "Statistical Physics," 3rd edition, part 1. Pergamon Press, Oxford.
Langford, J. I. (1978). *J. Appl. Cryst.* **11**, 10.
Le Daeron, P. Y., and Aubry, S. (1983). *J. Physique Colloque* **44**, C3, 1573.
Lederer, P., Jamet, J. P., and Montambaux, G. (1986). *Ferroelectrics* **66**, 25.
Lee, P. A., Rice, T. M., and Anderson, P. W. (1973). *Phys. Rev. Lett.* **31**, 462.
Lee, P. A., Rice, T. M., and Klemm, R. A. (1977). *Phys. Rev.* B **15**, 2984.
Matsuzaki, M., Kuwata, R., and Toyoda, K. (1980). *Solid State Commun.* **33**, 403.
Mayerle, J. J., and Torrance, J. B. (1981). *Acta Cryst.* B **37**, 2030.
Mazumdar, S., and Bloch, A. N. (1983). *Phys. Rev. Lett.* **50**, 207.
Mazumdar, S., and Campbell, D. K. (1985). *Phys. Rev. Lett.* **55**, 2067.
Mazumdar, S., and Soos, Z. (1981). *Phys. Rev.* B **23**, 2810.
Mazumdar, S., Dixit, S. N., and Bloch, A. N. (1984). *Phys. Rev.* B **30**, 4842.
McCall, R. P., Tanner, D. B., Miller, J. S., Epstein, A. J., Howard, I. A., and Conwell, E. M. (1985). *Synthetic Metals* **11**, 231; *Mol. Cryst. Liq. Cryst.* **120**, 59.
Megtert, S. (1984). Thesis, Orsay, France (unpublished).
Megtert, S., Pouget, J. P., Comès, R., Garito, A. F., Bechgaard, K., Fabre, J. M., and Giral, L. (1978a). *J. Physique Lettres* **39**, L118.
Megtert, S., Pouget, J. P., and Comès, R. (1978b). In "Molecular Metals" (W.E. Hatfield, ed). NATO Conference Series VI: Material Sciences, Vol. 1, p. 87. Plenum, New York.
Megtert, S., Pouget, J. P., and Comès, R. (1978c). In "Synthesis and Properties of Low Division Materials" (J. S. Miller and A. J. Epstein, eds.). *NY Acad. Sci.* **313**, 333.
Megtert, S., Comès, R., Vettier, C., Pynn, R., and Garito, A. F. (1979a). *Solid State Commun.* **31**, 977.
Megtert, S., Garito, A. F., Pouget, J. P., and Comès, R. (1979b). In "Quasi One Dimensional Conductors I" (S. Barisic, A. Bjelis, J. R. Cooper, and B. Leontic, eds.). *Lecture Notes In Physics* **95**, 57. Springer, Berlin. [Some of the data of this reference has been obtained with TSF-TCNQ imperfectly purified. Thus, Fig. 2, I/T as a function of T, is incorrect. It must be replaced by Fig. 16 of this review.]

Megtert, S., Comès, R., Pynn, R., and Vettier, C. (1979c). *In* "Physics and Chemistry of Low Dimensional Solids" (L. Alcacer, ed.). NATO ASI, Vol. C56, p. 113. Reidel, Dordrecht.
Megtert, S., Comès, R., Vettier, C., Pynn, R., and Garito, A. F. (1981). *Solid State Commun.* **47**, 875.
Megtert, S., Comès, R., Vettier, C., Pynn, R., and Garito, A. F. (1982). *Mol. Cryst. Liq. Cryst.* **85**, 159.
Megtert, S., Bjelis, A., Przystawa, J., and Barisic, S. (1983). *J. Physique Colloque* **44**, C3-1345.
Megtert, S., Bjelis, A., Przystawa, J., and Barisic, S. (1985). *Phys. Rev. B* **32**, 6692.
Metzger, R. M. (1981). *In* "Crystal Cohesion and Conformational Energies" (R. M. Metzger, ed.). *Topics in Current Physics*, Vol. 26, p. 80. Springer, Berlin.
Metzger, R. M. (1982). *Mol. Cryst. Liq. Cryst.* **85**, 57.
Miljak, M., Andrieux, A., Friend, R. H., Malfait, G., Jérome, D., and Bechgaard, K. (1978). *Solid State Commun.* **26**, 969.
Miller, J. S., and Epstein, A. J. (1978). *J.A.C.S.* **100**, 1639.
Montambaux, G. (1985). Thesis, Orsay, France (unpublished).
Montambaux, G., Heritier, M., and Lederer, P. (1986). *Phys. Rev. B* **33**, 7777.
Monceau, P. (1985). *In* "Electronic Properties of Inorganic Quasi One Dimensional Materials II" (P. Monceau, ed.), p. 139. Reidel, Dordrecht.
Mook, H. A., and Watson, C. R. (1976). *Phys. Rev. Lett.* **36**, 801.
Mook, H. A., Shirane, G., and Shapiro, S. M. (1977). *Phys. Rev. B* **16**, 5233.
Moret, R., and Pouget, J. P. (1986). *In* "Crystal Chemistry and Properties of Materials with Quasi One Dimensional Structures" (J. Rouxel, ed.), p. 87. Reidel, Dordrecht.
Moravitz, H. (1981). *Chemica Scripta* **17**, 75.
Mortensen, K., Jacobsen, C. S., Lindegaard-Andersen, A., and Bechgaard, K. (1983). *J. Physique Colloque* **44**, C3-1349.
Mukamel, D. (1977). *Phys. Rev. B* **16**, 1741.
Noguera, C. (1985). *J. Phys. C: Solid State Phys.* **18**, 1647.
Ovchinnikov, A. A. (1973). *Sov. Phys. JETP* **37**, 176.
Phillips, T. E., Kistenmacher, T. J., Bloch, A. N., and Cowan, D. O. (1976). *J.C.S. Chem. Commun.* 335.
Phillips, T. E., Kistenmacher, T. J., Bloch, A. N., Ferraris, J. P., and Cowan, D. O. (1977). *Acta Cryst. B* **33**, 421.
Pouget, J. P. (1978). "Solid State Transformation on Metal and Alloys," p. 523. Editions de Physique, Orsay. [In this paper, page 547 comes directly after page 550.]
Pouget, J. P. (1981). *Chemica Scripta* **17**, 85.
Pouget, J. P. (1987). *In* "Low Dimensional Conductors and Superconductors" (D. Jérome and L. C. Caron, eds.). NATO ASI, Vol. B155, p. 77. Plenum, New York.
Pouget, J. P., Khanna, S. K., Dénoyer, F., Comès, R., Garito, A. F., and Heeger, A. J. (1976). *Phys. Rev. Lett.* **37**, 437.
Pouget, J. P., Shapiro, S. M., Shirane, G., Garito, A. F., and Heeger, A. J. (1979a). *Phys. Rev. B* **19**, 1792.
Pouget, J. P., Megtert, S., and Comès, R. (1979b). *In* "Quasi One Dimensional Conductor I" (S. Barisic, A. Bjelis, J. R. Cooper, and B. Leontic, eds.). *Lecture Notes in Physics* **95**, 14. Springer, Berlin.
Pouget, J. P., Megtert, S., Comès, R., and Epstein, A. J. (1980a). *Phys. Rev. B* **21**, 486. (In Table I, case V.B., of this reference one should read $2k_F^{TCNQ} = \frac{1}{6}a^*$ instead of $2k_F^{NMP} = \frac{1}{6}a^*$.)
Pouget, J. P., Comès, R., and Bechgaard, K. (1980b). *In* "The Physics and Chemistry of Low Dimensional Solids" (L. Alcacer, ed.). NATO ASI, Vol. C56, p. 113. Reidel, Dordrecht.
Pouget, J. P., Megtert, S., and Comès, R. (1981). *In* "Recent Developments in Condensed Matter Physics" (J. T. Devreese, ed.), p. 295. Plenum, New York.

Pouget, J. P., Comès, R., Epstein, A. J., and Miller, J. S. (1982a). *Mol. Cryst. Liq. Cryst.* **85**, 203.
Pouget, J. P., Moret, R., Comès, R., Bechgaard, K., Fabre, J. M., and Giral, L. (1982b). *Mol. Cryst. Liq. Cryst.* **79**. 129.
Pouget, J. P., Escribe-Filippini, C., Hennion, B., Currat, R., Moudden, A. H., Moret, R., Marcus, J., and Schlenker, C., (1985). *Mol. Cryst. Liq. Cryst.* **121**, 111.
Pouget, J. P., Noguera, C., Moudden, A. H., and Moret, R. (1985b). *J. Physique* **46**, 1731 and erratum **47**, 145 (1986).
Rice, M. J., and Strassler, S. (1973). *Solid State Commun.* **13**, 125.
Rice, M. J., and Mele, E. J. (1982). *Phys. Rev. B* **25**, 1339.
Rybaczewski, E. F., Smith, L. S., Garito, A. F., Heeger, A. J., and Silbernagel, B. (1976). *Phys. Rev. B* **14**, 2746.
Sakai, M., Kagoshima, S., and Engler, E. M. (1983). *J. Physique Colloque* **44**, C3-1313.
Sato, K., Iwabuchi, S., Yamauchi, J., and Nagaoka, Y. (1978). *J. Phys. Soc. Japan* **45**, 515.
Sato, M., Fujushita, H., and Hoshino, S. (1983). *J. Phys. C: Solid State Phys.* **16**, L 877.
Saub, K., Barisic, S., and Friedel, J. (1976). *Phys. Lett. A* **56**, 302.
Sawatzky, G. A., Huizinga, S., and Kommandeur, J. (1979). *In* "Quasi One Dimensional Conductors II" (S. Barisic, A. Bjelis, J. R. Cooper, and B. Leontic, eds.). *Lecture Notes in Physics* **96**, 34. Springer, Berlin.
Scalapino, D. J., and Hirsch, J. E. (1983). *J. Physique Colloque* **44**, C3-1507.
Scalapino, D. J., Sears, M., and Ferrell, R. A. (1972). *Phys, Rev. B* **6**, 3409.
Schafer, D. E., Thomas, G. A., and Wudl, F. (1975). *Phys. Rev. B* **12**, 5532.
Schlenker, C., and Dumas, J. (1986). *In* "Crystal Chemistry and Properties of Materials with Quasi One Dimensional Structures" (J. Rouxel, ed.), p. 135. Reidel, Dordrecht.
Schultz, A. J., Stucky, G. D., Blessing, R. H., and Coppens, P. (1976). *J.A.C.S.* **98**, 3194.
Schultz, T. D. (1977). *Solid State Commun.* **22**, 289.
Schultz, T. D. (1980). *In* "The Physics and Chemistry of Low Dimensional Solids" (L. Alcacer, ed.). NATO ASI, Vol. C 56, p. 1. Reidel, Dordrecht.
Schultz, T. D., and Etemad, S. (1976). *Phys. Rev. B* **13**, 4928.
Schultz, T. D., and Craven, R. A. (1979). *In* "Highly Conducting One Dimensional Solids" (J. T. Devreese, R. P. Evrard, and V. E. Van Doren, eds.), p. 147. Plenum, New York.
Schulz, H. J. (1983). *J. Phys. C: Solid State Physics* **16**, 6769.
Schulz, H. J. (1987). *In* "Low Dimensional Conductors and Superconductors" (D. Jérome and L. C. Caron, eds.). NATO ASI, Vol. B155, p.95. Plenum, New York.
Scott, J. C., Etemad, S., and Engler, E. M. (1978). *Phys. Rev. B* **17**, 2269.
Shaik, S. S., and Whangbo, M. H. (1986). *Inorg. Chem.* **25**, 1201.
Sham, L. J. (1976). *Solid State Commun.* **20**, 623.
Sham, L. J. (1979). *In* "Highly Conducting One Dimensional Solids" (J. T. Devreese, R. P. Evrard, and V. E. Van Doren, eds.), p. 227. Plenum, New York.
Shapiro, S. M., Shirane, G., Garito, A. F., and Heeger, A. J. (1977). *Phys. Rev. B* **15**, 2413.
Shirane, G., Shapiro, S. M., Comès, R., Garito, A. F., and Heeger, A. J. (1976). *Phys. Rev. B* **14**, 2325.
Shitzkovsky, S., Weger, M., and Gurfreund, H. (1978). *J. Physique* **39**, 711.
Silverman, B. D. (1981). *In* "Crystal Cohesion and Conformational Energies" (R. M. Metzger, ed.). *Topics in Current Physics* **26**, 108. Springer, Berlin.
Solyom, J. (1979). *Adv. Physics* **28**, 101.
Solyom, J. (1983). *J. Physique Colloque* **44**, C-1569.
Soos, Z. G. (1974). *Ann. Rev. Phys. Chem.* **25**, 121.

Stanley, M. E. (1971). "Introduction to Phase Transition and Critical Phenomena." Clarendon Press, Cambridge.
Sumi, H. (1977). *Solid State Commun.* **21**, 17.
Swingle II, R. S., Groff, R. B., and Monroe, B. M. (1976). *Phys. Rev. Lett.* **35**, 542.
Sy, H. K., and Mavroyannis, C. (1977). *Solid State Commun.* **23**, 79.
Takahashi, T., Jérome, D., Masin, F., Fabre, J. M., and Giral, L. (1984). *J. Phys. C: Solid State Phys.* **17**, 3777.
Terauchi, H. (1978). *Phys. Rev. B* **17**, 2446.
Thomas, G. A., Moncton, D. E., Davey, S. C., Wudl, F., Kaplan, M. L., and Lee, P. A. (1979). *Bull. Am. Phys. Soc.* **24**, 232.
Thomas, J. F., and Jérome, D. (1980). *Solid State Commun.* **36**, 813.
Thorup, N., Rindorf, G., Soling, H., and Bechgaard, K. (1981). *Acta Cryst. B* **37**, 1236.
Tiedje, T., Haering, R. R., Jericho, M. N., Roger, W. A., and Simpson, A. (1977). *Solid State Commun.* **23**, 713.
Tokumoto, M., Koshizuka, N., Anzai, H., and Ishiguro, T. (1982). *J. Phys. Soc. Japan* **51**, 332.
Tombs, G. A. (1978). *Physics Reports C* **40**, 181.
Tomkiewicz, Y. (1980). *In* "The Physics and Chemistry of Low Dimensional Solids" (L. Alcacer, ed.). NATO ASI, p. 187. Reidel, Dordrecht.
Tomkiewicz, Y., Taranko, A. R., and Torrance, J. B. (1976). *Phys. Rev. Lett.* **36**, 751.
Tomkiewicz, Y., Welber, B., Seiden, P. E., and Schumaker, R. (1977a). *Solid State Commun.* **23**, 471.
Tomkiewicz, Y., Taranko, A. R., and Schumaker, R. (1977b). *Phys. Rev. B* **16**, 1380.
Tomkiewicz, Y., Taranko, A. R., and Torrance, J. B. (1977c). *Phys. Rev. B* **15**, 1017.
Tomkiewicz, Y., Andersen, J. R., and Taranko, A. R. (1978). *Phys. Rev. B* **17**, 1579.
Torrance, J. B. (1978). *Phys. Rev. B* **19**, 3099.
Torrance, J. B. (1979). *Accounts Chem. Res.* **12**, 79.
Torrance, J. B. (1985). *Mol. Cryst. Liq. Cryst.* **126**, 55.
Torance, J. B., and Silverman, B. D. (1977). *Phys. Rev. B* **15**, 788.
Torrance, J. B., Scott, B. A., and Kaufman, F. B. (1975). *Solid State Commun.* **17**, 1369.
Torrance, J. B., Mayerle, J. J., Bechgaard, K., Silverman, B. D., and Tomkiewicz, Y. (1980). *Phys. Rev. B.* **22**, 4960.
Torrance, J. B., Mayerle, J. J., Lee, V. Y., Bozio, R., and Pecile, C. (1981). *Solid State Commun.* **38**, 1165.
Ukei, K., and Shirotani, I. (1977). *Commun. Phys.* **2**, 159.
Van Bodegom, B. (1981). *Acta Cryst. B* **37**, 857.
Van Bodegom, B., and Bosch, A. (1981). *Acta Cryst. B* **37**, 863.
Van Bodegom, B., Larson, B. C., and Mook, H. A. (1981). *Phys. Rev. B* **24**, 1520.
Van Smaalen, S., Kommandeur, J., and Conwell, E. M. (1986). *Phys. Rev. B* **33**, 5378.
Visser, R. J., Oostra, S., Vettier, C., and Voiron, J. (1983). *Phys. Rev. B* **28**, 2074.
Weger, M., and Friedel, J. (1977). *J. Physique* **38**, 241 and 881.
Weger, M., and Gutfreund, H. (1979). *Solid State Commun.* **32**, 1259.
Welber, B., Seiden, P. E., and Grant, P. M. (1978). *Phys. Rev. B* **18**, 2692.
Weyl, C., Engler, E. M., Bechgaard, K., Jehanno, G., and Etemad, S. (1976). *Solid State Commun.* **10**, 925.
Williams, R., Lowe-Ma, C., and Samson, S. (1982). *Applied Physics Comm.* **1**, 223.
Yamada, Y. (1974). *Ferroelectrics* **7**, 37.
Yamaji, K. (1979). *J. Phys. Soc. Japan* **47**, 706.
Yamaji, K. (1982). *J. Phys. Soc. Japan* **51**, 2787.
Yamaji, K., Megtert, S., and Comès, R. (1981). *J. Physique* **42**, 1327.

Yamaji, K., Pouget, J. P., Comès, R., Epstein, A. J., and Miller, J. S. (1982). *Mol. Cryst. Liq. Cryst.* **85,** 1605.

Yamaji, K., Pouget, J. P., Comès, R., and Bechgaard, K. (1983). *J. Physique Colloque* **44,** C3-1321.

Zupanovic, P., Barisic, S., and Bjelis, A. (1985). *J. Physique* **46,** 1751.

Zuppiroli, L., Housseau, N., Forro, L., Guillot, J. P., and Pelissier, J. (1986). *Ultramicroscopy* **19,** 325.

CHAPTER 4

Transport in Quasi-One-Dimensional Conductors

E. M. Conwell

XEROX WEBSTER RESEARCH CENTER
WEBSTER, NEW YORK

I.	THE METALLIC RANGE	215
	1. Nature of Conduction and Scattering Processes	215
	2. Band Structure of a One-Dimensional Metal.	217
	3. Electron-Electron Correlations.	220
	4. Lattice Vibrations	222
	5. Coupling of Electrons to Phonons.	224
	6. Relaxation Times for Phonon Scattering in the Metallic State	229
	7. Derivation of Expressions for One-Dimensional Conductivity and Thermoelectric Power	237
	8. Transport Coefficients for the Cases of Elastic Scattering. .	238
	9. Comparison with Experiment for TTF-TCNQ in the Metallic Range	242
	10. Transverse Conductivity in the Metallic Range.	253
II.	THE PEIERLS TRANSITION: IMPLICATIONS FOR TRANSPORT. . . .	254
	11. Wavefunctions, Energies, and the Gap Equation Below T_P	254
	12. Effective Masses and Densities of States Below T_P. . . .	263
	13. Matrix Element for Acoustic-Mode Scattering Below T_P. .	266
	14. Calculation of Mobility for Acoustic-Mode Scattering Below T_P	268
	15. Ionized Impurity Scattering Below T_P	270
III.	THE SEMICONDUCTING RANGE	274
	16. Comparison of Experiment with Theory Below T_P: Low Fields.	274
	17. Comparison of Experiment with Theory Below T_P: High Fields	280
	REFERENCES	288

I. The Metallic Range

1. NATURE OF CONDUCTION AND SCATTERING PROCESSES

The TCNQ and TTF families of compounds on which this volume is based are metallic at high temperatures and undergo a Peierls transition at a temperature T_P to become semiconductors. For TTF-TCNQ, the prototype of this group of materials, whose structure is shown in Fig. 1 of Chapter 1,

$T_P = 54$ K for the TCNQ chain. The evidence of strong electron–phonon coupling provided by the Peierls transition and attendant phenomena such as the Kohn anomaly, described in Chapter 1, led many people to suggest that transport above T_P in these materials involved collective electron–phonon modes, or fluctuating charge-density waves (Bardeen, 1973; Allender et al., 1974; Heeger, 1977, and 1979). As discussed in Chapter 1, Section 6, there is evidence from conductivity measurements under pressure for a contribution to transport of fluctuating charge-density waves in TTF-TCNQ at ~150 K and below (Andrieux et al., 1979). Above 150 K, however, the fluctuations in TTF-TCNQ are barely visible to x-rays and their coherence length is quite small. In another material of this family, TSeF-TCNQ, the same type of measurements showed relatively little effect of fluctuations on the conductivity (Thomas and Jérome, 1980). The lack of importance of fluctuations in TSeF-TCNQ may be due to the charge-density waves on the TSeF chain, which are positively charged, pinning those on the TCNQ chain, which are negatively charged. TTF-TCNQ may be unique in this respect because, with T_P much larger for the TCNQ chain than for TTF, there is a considerable temperature range in which there would be charge-density waves on TCNQ only. As will be discussed, this also makes it possible to depin the charge-density waves on the TCNQ chain with a moderate electric field below T_P, allowing them to contribute to conductivity. In the metallic range, however, a one-electron picture of transport is certainly valid above 150 K in TTF-TCNQ, and very likely valid at all temperatures but those quite close to T_P in the other members of this family. Even well below 150 K, one-electron processes probably account for most of the conductivity of TTF-TCNQ.

In the metallic range, i.e., above T_P, the variation of resistivity ρ with temperature T for the materials of this family was found to be of the form $\rho \propto T^\lambda$ with $\lambda \simeq 2$. For the case of TTF-TCNQ, shown in Fig. 4 of Chapter 1, $\lambda \simeq 2.3$ (Groff et al., 1974). Early attempts to account for this temperature dependence concentrated on electron–electron scattering (Seiden and Cabib, 1976), two-libron scattering (Gutfreund and Weger, 1977), and scattering by the (relatively) high-energy phonons of molecular (internal mode) vibrations (Conwell, 1977). As regards the first mechanism, it was soon realized that scattering of electrons by electrons on the same chain cannot remove momentum from the electron distribution in TTF-TCNQ and related materials; Umklapp processes are not allowed because the bands are less than half-filled. To give rise to resistivity, the scattering would have to be by electrons or holes on the other chain. A major argument against this mechanism being important is the fact that many compounds in which it is well documented that conduction is by a single chain, either TTF or TCNQ, have resistivity comparable to or less than that of TTF-TCNQ (Somoano et al., 1977).

The two-libron theory proponents asserted that two-phonon scattering predominates. To test this theory it is necessary to compare the contributions to resistivity of one- and two-phonon processes. As will be shown, when this comparison is carried out (Van Smaalen *et al.*, 1986), the conclusion is that two-phonon processes are not important in TTF-TCNQ.

Scattering by internal modes, whose frequencies and coupling constants to electrons are fairly well known for TTF and TCNQ, was found to account well for the temperature dependence of ρ for TTF-TCNQ. However, to obtain the correct magnitude of ρ, the bandwidths on the TCNQ and TTF chains had to be taken as $\frac{1}{4}$ and $\frac{1}{8}$ eV, respectively (Conwell, 1977). These values were quickly objected to as smaller than indicated by various experiments and theoretical calculations. The conclusive argument against this theory, however, was that it could not account for an important experimental fact: ρ decreases rapidly with pressure, 25%/kbar at 300 K, as shown in Fig. 6 of Chapter 1. Thus, although, as will be seen, internal modes can be expected to contribute significantly to ρ in the metallic state of TTF-TCNQ, they do not provide the predominant scattering.

An important step toward understanding the origin of the resistivity came with the realization that above 60 K the thermal expansion of TTF-TCNQ is large, as is true of molecular crystals in general. This fact, coupled with the strong pressure dependence observed for ρ, indicates that a significant portion of the temperature dependence of ρ is actually due to volume dependence of the various quantities determining ρ (Cooper, 1979). It was found, in fact, that ρ at constant volume, ρ_v, is quite close to linear in T, rather than quadratic (Friend *et al.*, 1978). This, of course, is consistent with one-phonon scattering processes being predominant.

Before discussing transport in the metallic state of these materials in more detail, we will present some necessary background. Sections 2–5 are devoted to the band structure of a one-dimensional metal, electron–electron correlations in these materials, the lattice vibrations, and the theory of the coupling of electrons to the various types of phonons that occur. With these results we derive relaxation times for phonon scattering in Section 6 and the formal expressions for conductivity σ and thermoelectric power Q parallel to the chains in Sections 7 and 8. In general, the integrals involved must be evaluated numerically, and the results of this evaluation are compared in Section 9 with experimental results for σ and Q parallel to the chains in the metallic range of TTF-TCNQ. Transverse conductivity in the metallic range will be discussed briefly in Section 10.

2. Band Structure of a One-Dimensional Metal

A useful model of a 1D metal is a uniform stack or chain of like molecules, each of which could accommodate two electrons (with opposite spin) in its lowest unoccupied level (LUMO). In the isolated molecule the wavefunction associated with this level is $\varphi_\alpha(r)$, corresponding to an energy E_α. Due to overlap of the wavefunctions of neighboring molecules in the chain, this level spreads, becoming the conduction band. We derive now the properties of the conduction band neglecting electron–electron correlations, the effect of which will be considered later.

Characteristically the separation b between molecules is large enough in these materials that the overlap of electronic wavefunctions on adjacent molecules is small, suggesting a tight-binding approach to the band structure. A good approximation to the wavefunction for an electron in the conduction band with wave vector k is then the superposition of $\varphi_\alpha(r)$ for each molecule multiplied by a modulating function involving k

$$\psi_k(r) = N^{-1/2} \sum_{n=1}^{N} e^{iknb} \varphi_\alpha(r - nb), \tag{1}$$

where $\varphi_\alpha(r - nb)$ is the orbital centered on the nth molecule and N the number of molecules in the chain. The factor e^{iknb} satisfies the requirements of translational symmetry.

The energy of the electron is then determined by

$$\varepsilon_k = \int \psi_k^* H \psi_k \, dr \tag{2}$$

where dr indicates integration over all space and H is the Hamiltonian operator of the system. Since $\varphi_\alpha(r)$ falls off rapidly with r, the only terms that contribute to the integral in Eq. (2) are those where ψ_k and ψ_k^* involve electrons on the same molecule or on nearest neighbors. The former type of term yields just E_α, the energy of the electron on an isolated molecule. The terms from the nearest neighbors on each side may be combined to give $-2t \cos kb$, where t is the overlap or transfer integral, defined by

$$t = -\int \varphi_\alpha^*(r \pm b) H \varphi_\alpha(r) \, dr. \tag{3}$$

The sum $E_\alpha - 2t \cos kb$ represents, within an arbitrary constant, the energy of a conduction electron. For a half-filled band it is convenient to choose that constant so that the zero of energy is at the middle of the conduction band. There are no metals of this type with half-filled bands, however, for reasons that will become clear subsequently. For arbitrary band filling, it is more convenient to choose the zero of energy at $k = 0$, the bottom of the conduction band. The relation between the conduction-band energy ε_k and k may then be written

$$\varepsilon_k = 2t(1 - \cos kb). \tag{4}$$

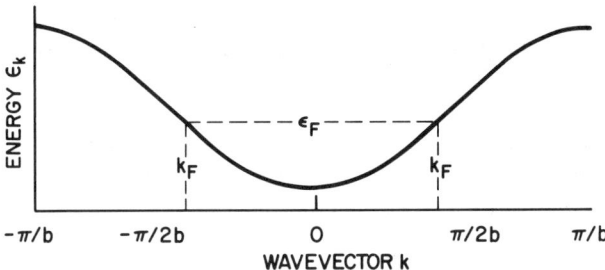

FIG. 1. Electronic energy ε_k vs. wave vector k for a one-dimensional conduction band [from Eq. (4)]. The Fermi energy ε_F and Fermi wavevector k_F are shown for a band $\sim\frac{1}{3}$ full, as is the case for TTF-TCNQ and the compounds in its immediate family.

This relation is plotted in Fig. 1. It is apparent that the bandwidth, or the energy at the top of the band, is $\varepsilon = 4t$. The Fermi energy ε_F and the Fermi wavevector k_F are shown in Fig. 1 about where they would be for the TCNQ chain in TTF-TCNQ, TSeF-TCNQ, HMTTF-TCNQ, and other compounds in this immediate family.

In a crystal with two types of conducting chains, such as TTF-TCNQ, there must be an ε_k vs. k curve for the other chain also. Band structure calculations of Berlinsky et al. (1974) and Herman et al. (1977) suggest that the TTF band has its maximum at $k = 0$, rather than its minimum. This would result in the two bands crossing at k_F. The finite chain interaction, however, should cause the band crossing to be replaced by a small gap at k_F. A gap would make the system semiconducting rather than metallic (Bernstein et al., 1975). The fact that the materials are in actuality metallic can be explained as the result of thermal disorder turning the three-dimensional band structure into a collection of quasi-one-dimensional chains.

The effective mass m^* at the bottom of a band described by Eq. (4) is readily obtained. Defining m^* at the bottom of the band by

$$\varepsilon_k = \frac{\hbar^2 k^2}{2m^*}, \qquad (5)$$

and expanding the right-hand side of Eq. (4) for small kb, we obtain

$$m^* = \frac{\hbar^2}{2tb^2}. \qquad (6)$$

Thus m^* is inversely proportional to the overlap, as expected. A typical value of $4t$ for this family of compounds is $\frac{1}{2}$ eV. For the TCNQ chain in TTF-TCNQ this value arises from theoretical calculations (Berlinsky et al., 1974; Herman et al., 1977) and measurements of the plasma frequency

(Ritsko et al., 1975) and other quantities. With $b \simeq 3.8$ Å, Eq. (6) leads to m^* for the TCNQ chain of the order of $2m_0$, where m_0 is the free electron mass.

Another important quantity that may be deduced from Eq. (4) is the density of states, $\rho(\varepsilon_k)$, the number of states per unit energy range, of a 1D metal. Because the density of states in k space $\rho(k) = Nb/2\pi$ for states of either spin, we find using Eq. (4) that

$$\rho(\varepsilon_k) = 2\rho(k)\frac{dk}{d\varepsilon_k} = \frac{N}{2\pi t \sin kb} = \frac{N}{\pi[\varepsilon(4t - \varepsilon)]^{1/2}}, \quad (7)$$

where the factor 2 takes into account the fact that the states with $+k$ and $-k$ have the same ε_k. Thus $\rho(\varepsilon_k)$ given by Eq. (7) is the number of states in the chain per unit energy range of either spin. We note that $\rho(\varepsilon_k)$ has singularities at the bottom of the band ($\varepsilon_k = 0$) and the top of the band ($\varepsilon_k = 4t$). It has a minimum at the band center.

It is useful to relate the density of states to the velocity v. Because $v = (1/\hbar)\,d\varepsilon_k/dk$, we obtain from Eq. (4)

$$v = \frac{2tb}{\hbar}\sin kb = \frac{Nb}{\pi\hbar\rho(\varepsilon_k)}. \quad (8)$$

3. Electron–Electron Correlations

The simplest model for discussing electron–electron correlations is that due to Hubbard (1963). There are two parameters of importance in this model, the repulsion U for a second electron on a site or molecule and the transfer integral t. Since t is typically 0.1–0.2 eV for the materials we are discussing, while U for one of the isolated molecules (measured as the difference between the first and second ionization levels) is ~ 4 eV, it appears $U \gg t$. This would imply that electron–electron correlations are very important in all the materials of this group. However, as has been pointed out by many authors, the longer range interactions that are omitted in the Hubbard model are not insignificant. Screening, for one thing, is quite important in quasi-one-dimensional metals. In the case of TTF-TCNQ, for example, the calculated Thomas–Fermi screening length is smaller than the dimensions of a molecule. Mazumdar and Bloch (1983) have found that for all these materials screening is essentially complete over distances larger than the nearest-neighbor spacing. Thus it should be necessary to consider, in addition to U, only the nearest neighbor interaction, V. Based on these two interactions, their calculations show that the importance of electron–electron correlations varies greatly with η, the number of conduction electrons per molecule (Mazumdar and Bloch, 1983). They find that short-range correlations, i.e., U and V, are important for $\eta = 1$, the half-filled band, and $\eta = \frac{1}{2}$, the quarter-filled band, while they are relatively unimportant in the range $0.63 \leq \eta \leq 0.8$. These predictions agree well with various types of experiment, as will be discussed below.

Because V is always smaller than U, we will simplify the discussion by neglecting V. A number of effects can be predicted for the limit $U \gg t$. In that limit states that correspond to double occupancy of a site are separated by a large gap from states that correspond to single occupancy. For $\eta = 1$ this gap is at the Fermi energy, which would make materials with one conduction electron/molecule semiconductors ("Mott insulators"). In fact, materials with $\eta = 1$, such as K-TCNQ (Khanna et al., 1947a) HMTSe-TCNQF$_4$ (Hawley et al., 1978, 1979) and HMTTF-TCNQF$_4$ (Torrance et al., 1980) are semiconductors at all temperatures with sizable gaps. For $\eta \neq 1$ the gap due to large U is usually well away from the Fermi energy and the effects of large U are more subtle. One such effect is the enhancement of magnetic susceptibility over what would be predicted for uncorrelated electrons (Torrance et al., 1977). This occurs because U, being large, prevents two electrons (of opposite spin) from occupying the same site. (See Chapter 5 for further discussion.)

Another type of evidence concerning the magnitude of electron–electron correlations comes from the diffuse x-ray scattering due to lattice vibrations. As was discussed in Chapters 1 and 3, the strong interaction of conduction electrons with the phonons that enable transitions from one side of the Fermi surface to the other causes a softening, or decrease in frequency, of these phonons. The softening results in a larger amplitude of vibration for these phonons, as compared with phonons with neighboring wavevectors, giving rise to a peak in the diffuse x-ray scattering. In the absence of electron–electron correlations, this peak would be at the phonon wavevector $2k_F$ and indeed a peak is seen at $2k_F$ for many of these compounds, notably for η in the range 0.63 to 0.8 (Mazumdar and Bloch, 1983). For materials with strong electron–electron correlations, i.e., large U, the restriction to single occupancy of sites, which translates into a restriction to single occupancy of states, results in k_F being twice as large. Thus, in terms of the k_F calculated for the material ignoring electron–electron correlations, the diffuse x-ray spots would appear at $4k_F$ in large U materials. In agreement with the theory of Mazumdar and Bloch (1983), $4k_F$ diffuse x-ray scattering, as well as enhanced magnetic susceptibility, is found for compounds with $\rho = 0.5$, while for those with $0.63 \leq \rho \leq 0.8$, neither of these is found. For compounds with ρ intermediate between 0.55 and 0.63, these effects may or may not appear. In the case of TTF-TCNQ, with $\eta = 0.59$, both $2k_F$ and $4k_F$ scattering are observed. The $4k_F$ scattering has been found to be associated with the TTF chain (Kagoshima, 1980). The fact that $4k_F$ appears for the TTF chain and not for TCNQ is attributed to the smaller bandwidth of the TTF chain. Consistent with this, in TMTSeF-TCNQ, where $\rho = 0.57$, but a larger bandwidth is expected due to the replacement of S by Se, the $4k_F$ scattering is not seen. We conclude that for the TCNQ chain in TTF-TCNQ, and for many of the metallic quasi-one-dimensional compounds of interest, electron–electron correlations are not strong.

4. Lattice Vibrations

Because we are dealing with molecular crystals having more than one molecule per unit cell, the lattice vibration spectrum is quite complicated. The vibrations are conveniently divided into external and internal modes. The latter are based on the molecular vibrations or normal modes, while the former arise from the three translational and three rotational (librational) degrees of freedom of each molecule. For TTF-TCNQ, for example, with four molecules per unit cell, the external lattice vibration spectrum has 24 branches, of which three are acoustic and the remaining 21 optical. A given optical mode might have pure translational or rotational character, but in general it is expected that the character is mixed. Inelastic neutron scattering has provided some data on the dispersion and mixing of the acoustic modes in TTF-TCNQ (Shapiro *et al.*, 1977; see also Chapter 2). Some information on the optical modes, as well as internal modes, is provided by infrared absorption (Bates *et al.*, 1981).

Due to the complexity of the modes and the incomplete information about them, it is usual to discuss the external modes in terms of a simplified model. As shown schematically in Fig. 2, the chains consist of large planar molecules stacked at an angle θ to the chain direction. The simplifications are the

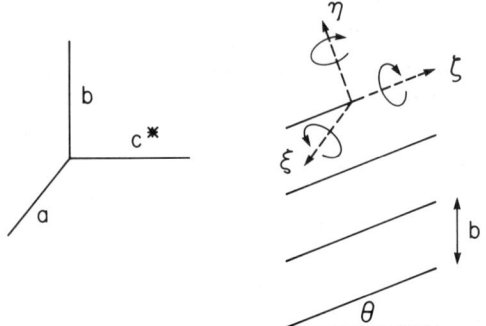

FIG. 2. Coordinate system and model chain of planar molecules showing tilt angle θ, lattice constant b in the stacking direction, and rotation axes for the librations. ζ and ξ ($\|a$) correspond to the long and short axes of the molecule, respectively, and η to the axis perpendicular to the plane of the molecule. [From Conwell (1980b).]

assumption of one molecule per unit cell and the neglect of mode mixing. There are then six external modes or branches of the lattice vibration spectrum. Three of these correspond to rotations, or librations, that are supposed to take place about one of the three inertial axes, indicated on the top molecule in Fig. 2. The other three are acoustic modes. The important acoustic modes for interaction with the electrons are those that propagate along the stacking axis, taken as b in Fig. 2. These consist of longitudinal (LA) modes with displacements along the b axis and two transverse (TA) modes with displacements that we take in the directions indicated c^* and a in Fig. 2, consistent with neutron scattering results (Shapiro et al., 1977). The quanta of the acoustic branches may be called *translons*, in analogy to those of the rotational branches being called *librons*.

In the simple model with one molecule per unit cell and no mode mixing, each possible molecular vibration or normal mode of the molecule gives rise to a single optical branch. For a molecule with m atoms there are $3m-6$ of these optical branches. To a good approximation, particularly for the higher frequency internal modes, these branches will be dispersionless and have frequencies close to those of the normal-mode vibrations of the isolated molecule. For the TCNQ molecule, with 20 atoms, there are 54 normal modes, while for TTF, with 14 atoms, there are 36. Of these, the modes that interact strongly with the electrons, as will be seen, are the symmetric, or a_g, modes, 10 in number for TCNQ, 7 for TTF. Frequencies of these modes are listed in Table I. (See also Chapter 1.)

TABLE I

FREQUENCIES AND COUPLING CONSTANTS FOR INTERNAL PHONONS FOR TTF-TCNQ

TCNQ		TTF	
$\hbar\omega^a$ (K)	g^b	$\hbar\omega^a$ (K)	g^a
213	1.54	353	0.16
485	0.70	684	1.33
882	0.20	1059	0.49
1043	0.24	1574	0.16
1407	0.20	2184	0.62
1721	0.22	2238	0.23
2002	0.20	4436	0.03
2324	0.49		
3174	0.13		
4392	~0		

[a] Lipari et al., 1977
[b] Rice et al., 1977

5. Coupling of Electrons to Phonons

In this section we take a more detailed look at the coupling of the various types of lattice vibrations to the electrons and holes preparatory to discussing conductivity. We start from the usual site Hamiltonian, such as that set up by Hubbard (1963), but limit ourselves to the cases where U may be neglected. We then have, for the electrons on a chain of N identical molecules, plus phonons,

$$H = \sum_{j=1}^{N} \varepsilon_j a_j^+ a_j + \sum_{n=1}^{G} \sum_{q} (b_{q,n}^+ b_{q,n} + \tfrac{1}{2})\hbar\omega_{q,n}$$

$$- \sum_{j=1}^{N} t(a_j^+ a_{j+1} + H.c.), \qquad (9)$$

where ε_j is the energy of the electron on site j, a_j^+ and a_j the creation and annihilation operators, respectively, for an electron on the jth molecule, t the overlap integral defined by Eq. (3) and $b_{q,n}^+$ and $b_{q,n}$ the creation and annihilation operators, respectively, for a phonon with wavevector $\mathbf{q}$ and frequency $\omega_{q,n}$. The subscript n indicates the branch, of which there are G in all.

There are several ways in which the electrons are coupled to the lattice vibrations. First, the transfer integral is a function of the spacing and relative orientation of adjacent molecules. This couples the electrons to the intermolecular or external lattice vibrations. As will be shown, the strength of this coupling to any given mode is proportional to the rate of change of t with respect to the departure from the equilibrium lattice configuration due to that particular mode. Coupling to the internal modes arises from the first term in the Hamiltonian Eq. (9). The energy ε_j of the electron on the jth site depends on the positions of all the atoms in the jth molecule. It is therefore expected, as will be shown in detail, that the coupling to a given internal mode depends on the rate of change of ε_j with the normal coordinate for that mode. The first term of the Hamiltonian also provides another source of coupling to the external modes, because the energy ε_j includes a contribution from the polarization of the neighboring molecules. This depends on the relative positions of those molecules, which is, of course, affected by the external modes. We will calculate the coupling constants arising from the three sources just described in the order given above.

To obtain the coupling constant arising from the modulation of t by the intermolecular vibrations, we make use of the fact that vibration amplitudes are small and expand t in a Taylor series

$$t = t_0 - \left(\frac{\partial t}{\partial u}\right)_0 (u_j - u_{j+1}) + \frac{1}{2}\left(\frac{\partial^2 t}{\partial u^2}\right)_0 (u_j - u_{j+1})^2 + \cdots \qquad (10)$$

Here u_j represents the displacement of the jth molecule from its thermal equilibrium position and $(\partial t/\partial u)$ stands for the rate of change of t with the distance between nearest neighbors. The subscript 0 indicates evaluation at the equilibrium position. For the case of acoustic modes the displacements may be expressed in terms of creation and annihilation operators for these models by (Kittel, 1963)

$$u_j = \sum_{\pm q,n} \hat{e}_{q,n} \left(\frac{\hbar}{2MN\omega_{q,n}}\right)^{1/2} e^{iqR_i}(b_{q,n} + b^+_{-q,n}), \quad (11)$$

where $\hat{e}_{q,n}$ is the (unit) polarization vector of the vibration with wavevector q in the nth acoustic branch, M the mass of a molecule, and R_j the coordinate of the jth molecule. In this and later sections, unless otherwise specified, we limit consideration to acoustic modes in which displacements are parallel to the chains. In order to obtain the matrix element for scattering of an electron from a Bloch state with wavevector k to one with k', it is necessary to use the Fourier expansions for a_j^+ and a_j in terms of the operators a_k^+ and a_k that create and annihilate electrons with wavevector k (Kittel, 1963), i.e.,

$$a_j = N^{-1/2} \sum_{\pm k} e^{ikR_j} a_k \quad (12a)$$

$$a_j^+ = N^{-1/2} \sum_{\pm k} e^{-ikR_j} a_k^+. \quad (12b)$$

With Eqs. (10)–(12) we obtain from the part of the third term in H of Eq. (9) linear in u the perturbing Hamiltonian for the interaction of electrons with external mode phonons of the branch n

$$H'_1 = N^{-1/2} \sum_{q,k} g_{1,n} a^+_{k+q} a_k (b_{q,n} + b^+_{-q,n}), \quad (13)$$

$g_{1,n}$ being the coupling constant for external one-phonon (1p) processes. For acoustic mode or translon scattering this constant may be written (Friedman, 1965).

$$g_{1t,n} = 2i\left(\frac{\partial t}{\partial u_n}\right)_0 \left(\frac{\hbar}{2M\omega_{q,n}}\right)^{1/2} [\sin(k+q)b - \sin kb]. \quad (14)$$

Thus, as suggested earlier, the coupling, and ultimately the scattering, depends directly on the rate of change of t with nearest neighbor distance, which may be considered a deformation potential, and on the amplitude of the displacement, proportional to $(M\omega_{q,n})^{-1/2}$. For degenerate material, i.e., in the metallic state, due to acoustic-mode scattering being essentially elastic, scattering may be considered to take place from $\pm k$ to $\mp k$. We may then write Eq. (14)

$$g_{1t,n} = \pm i\left(\frac{\partial \varepsilon_0}{\partial u_n}\right)_0 \left(\frac{\hbar}{2M\omega_{2k,n}}\right)^{1/2} \sin kb, \tag{15}$$

where we have introduced the bandwidth ε_0, equal to $4t$.

A similar development may be carried out for the linear perturbing term in the Hamiltonian due to the librational modes. This leads again to the expression (13), with the coupling constant

$$g_{1\ell,n} = 2i\left(\frac{\partial t}{\partial \theta_n}\right)_0 \left(\frac{\hbar}{2I_n \omega_{q,n}}\right)^{1/2} [\sin(k+q)b - \sin kb], \tag{16}$$

θ being the angle of libration and I_n the moment of inertia appropriate to the particular libration. For degenerate material this may be approximated in a similar way to Eq. (15), i.e.,

$$g_{1\ell,n} = \pm i\left(\frac{\partial \varepsilon_0}{\partial \theta_n}\right)_0 \left(\frac{\hbar}{2I_n \omega_{2k,n}}\right)^{1/2} \sin kb. \tag{17}$$

It is evident from Eqs. (14) through (17) that an acoustic mode for which $(\partial t/\partial u_n)_0$ vanishes, or a libration for which $(\partial t/\partial \theta_n)_0$ vanishes, will not give rise to $1p$ scattering. It can be seen from Fig. 2 that, by symmetry, the $TA(a)$ mode is an example of the former, while the η and ζ rotations are examples of the latter. These modes can still give rise to two-phonon ($2p$) scattering, however. To study $2p$ scattering we examine the effects of the term in Eq. (10) quadratic in the displacements. Using Eq. (11) and the Fourier expansions given above for a_j^+ and a_j we obtain the perturbing potential for interaction with branch n in the form

$$H_2' = N^{-1} \sum_{k,q,q'} g_{2,n} a_{k+q+q'}^+ a_k (b_{q,n} + b_{-q,n}^+)(b_{q',n} + b_{-q',n}^+), \tag{18}$$

the electron interaction now being with two phonons having wavevectors q and q'. For the acoustic modes the coupling constant is found to be (Conwell, 1980b)

$$g_{2t} = \frac{1}{4}\left(\frac{\partial^2 \varepsilon_0}{\partial u_n^2}\right)_0 \left[\frac{\hbar}{2M(\omega_{q,n}\omega_{q',n})^{1/2}}\right] f_2(k,q,q'), \tag{19}$$

while for the librons it is (Gutfreund and Weger, 1977)

$$g_{2\ell} = \frac{1}{4}\left(\frac{\partial^2 \varepsilon_0}{\partial \theta_n^2}\right)_0 \left[\frac{\hbar}{2I_n(\omega_{q,n}\omega_{q',n})^{1/2}}\right] f_2(k,q,q'), \tag{20}$$

where

$$f_2(k, q, q') = \cos(k + q + q')b - \cos(k + q)b \\ - \cos(k + q')b + \cos kb. \tag{21}$$

In principle, q and q' could come from different libron branches, n and n'. In that case the "deformation potential" $(\partial^2 \varepsilon_0 / \partial u_n^2)$ would be replaced by $(\partial^2 \varepsilon_0 / \partial u_n \, \partial u_{n'})$ in Eq. (19), with the corresponding replacement in Eq. (20).

To obtain the coupling to internal vibrations we expand ε_j in terms of $Q_n(j)$, normal coordinate for the nth internal vibration on the jth molecule (Lipari et al., 1977), i.e.,

$$\varepsilon_j = \varepsilon_{j0} + \left[\frac{\partial \varepsilon_j}{\partial Q_n(j)}\right]_0 Q_n(j) + \frac{1}{2}\left[\frac{\partial^2 \varepsilon_j}{\partial Q_n(j) \, \partial Q_m(j)}\right]_0 Q_n(j) Q_m(j) + \cdots, \tag{22}$$

where ε_{j0} is the electron energy at the jth site (molecular energy plus effect of the surrounding ions) in the absence of internal vibrations. It is convenient to use dimensionless Q_n's, obtained by multiplying the usual normal coordinates by $(\omega_n/\hbar)^{1/2}$ (Lipari et al., 1976). We now make the Fourier expansion

$$Q_n(j) = N^{-1/2} \sum_q e^{iqR_j} Q_n(q), \tag{23}$$

where $Q_n(q)$ is the dimensionless normal coordinate for the nth (internal) phonon branch at wavevector q. Using (Ziman, 1960)

$$Q_n(q) = 2^{-1/2}(b_{q,n} + b^+_{-q,n}), \tag{24}$$

we again obtain from the term in the expansion (22) linear in Q_n a perturbing potential in the form Eq. (13). The coupling constant for this case (subscript i for internal modes) is given by

$$g_{1i,n} = 2^{-1/2}\left[\frac{\partial \varepsilon_j}{\partial Q_n(j)}\right]_0. \tag{25}$$

It is convenient to write this in the form

$$g_{1i,n} = g_n \hbar \omega_n, \tag{26}$$

where g_n is the dimensionless coupling constant defined and calculated for TCNQ by Lipari et al. (1977). As discussed in that reference, the only normal modes coupled to the electrons by terms linear in the Q_n's are those with the symmetry of the molecule (a_g for TCNQ, for example). This limits the modes that can interact with the electrons to 10 for TCNQ and 7 for TTF. Higher order terms need not be considered; two-phonon processes are not likely for the temperature range of interest due to the high energies of the internal-mode branches.

The calculation of the perturbing potential arising from polarization fluctuations due to the thermal vibrations has been carried out by Friedman (1981). For the calculation the polarization energy of a single excess electron on molecule j due to molecule ℓ is taken as

$$P_{j\ell} = -\frac{\alpha_0 e^2}{2R_{\ell j}^4}, \tag{27}$$

when α_0 is the molecular polarizability and $R_{\ell j}$ the intermolecular distance. If molecules j and ℓ are displaced from their equilibrium positions by $\bar{u}_j$ and $\bar{u}_\ell$, respectively, the change in polarization energy for the electron on site j is

$$\Delta P_{j\ell} = \frac{2\alpha_0 e^2 (\bar{R}_\ell - \bar{R}_j) \cdot (\bar{u}_\ell - \bar{u}_j)}{R_{\ell j}^6}. \tag{28}$$

When $\bar{u}_\ell$ and $\bar{u}_j$ are expressed in terms of $b_{q,n}$ and $b_{-q,n}$ by Eq. (11) and the site annihilation and creation operators expressed in terms of a_k and a_k^+, the perturbing Hamiltonian obtained from Eq. (28) is found to be again in the form Eq. (13) with the coupling constant for one-phonon scattering in the degenerate case given by (Friedman, 1981)

$$g_{1P,n} = 2\alpha_0 e^2 \left(\frac{\hbar}{2M\omega_{2k,n}}\right)^{1/2} \sum_\ell \bar{R}_{\ell j} \cdot \hat{e}_{2k,n} \left[\frac{\exp(2i\bar{k} \cdot \bar{R}_{\ell j}) - 1}{R_{\ell j}^6}\right]. \tag{29}$$

As pointed out by Friedman, the sum is independent of the choice of molecule j because of translational invariance. As a result of the strong inverse dependence on distance it is found that only nearest and next-nearest neighbors contribute to the sum in Eq. (29). It is clear that Eq. (29) overestimates the scattering effect of the polarization fluctuations because it neglects screening. As pointed out in Section 3, screening is quite important even for nearest neighbors in TTF-TCNQ and related compounds. Also, Eq. (27) is approximate because it neglects finite size effects, the fact that the molecules are not isotropic, and higher order multipole interactions (Gosar and Choi, 1966).

Second-order, or $2p$, processes for polarization fluctuation scattering have also been considered (Gutfreund et al., 1980), (Kaveh et al., 1982). Carrying out the Taylor series expansion of Eq. (27) to the next term after Eq. (28), we find that the second-order term is $\sim 5\,\Delta R_{\ell j}/R_{\ell j}$ times Eq. (28), where $\Delta R_{\ell j}$ is the change in $R_{\ell j}$. Since $\Delta R_{\ell j}/R_{\ell j} \ll 1$ for the thermal fluctuations and, as will be seen, it is the square of the coupling constant that determines the resistivity ρ, it is not expected that second-order processes of this kind contribute noticeably to ρ; they will not be considered further here.

The quantity required for the calculation of the scattering time due to any of the processes above is the absolute square of the matrix element of H' for

4. TRANSPORT IN QUASI-ONE-DIMENSIONAL CONDUCTORS

that process between the initial and final states of the system. For any of the 1p processes, from Eq. (13), this is given by (Conwell, 1980b)

$$|\langle k'|H'_{1,n}|k\rangle|^2 = N^{-1}|g_{1,n}|^2\delta(k' - k \mp q)\begin{cases} n_q \text{ absorption} \\ (n_q + 1) \text{ emission} \end{cases}, \quad (30)$$

where $g_{1,n}$ is given by expressions (14), (16), (25), or (29) and the upper sign in the δ function is for absorption, the lower for emission. The possibility of Umklapp processes will be neglected here because, as discussed in Section 1 of this chapter, they do not appear important for the materials of concern in this temperature range. For the 2p scattering process we obtain from Eq. (18),

$$|\langle k'|H'_{2,n}|k\rangle|^2 = N^{-1}|g_{2,n}|^2\delta(k' - k \mp q \mp q')\begin{Bmatrix} n_q \\ n_q + 1 \end{Bmatrix}\begin{Bmatrix} n_{q'} \\ n_{q'} + 1 \end{Bmatrix}, \quad (31)$$

where one factor is chosen from each pair of braces, depending on whether the corresponding phonon is absorbed or emitted, and the signs in the argument of the δ function are chosen accordingly to satisfy conservation of crystal momentum. The quantity $g_{2,n}$ is given by Eqs. (19) or (20) and (21).

6. Relaxation Times for Phonon Scattering in the Metallic State

The high conductivity of these materials suggests, as will be verified later, that a Boltzmann equation approach is valid. For dc fields the Boltzmann equation is the statement that in the steady state the rate of change of the distribution f due to applied fields is balanced by the rate of change of f due to collisions. For a small applied electric field E we assume, as usual, a solution of the Boltzmann equation in the form

$$f = f_0(\varepsilon) + f_1(\varepsilon, \bar{E}), \quad f_1 \ll f_0, \quad (32)$$

where $f_0(\varepsilon)$ is the thermal equilibrium distribution

$$f_0 = \left[\exp\left(\frac{\varepsilon - \varepsilon_F}{k_B T}\right) + 1\right]^{-1}. \quad (33)$$

The function f_1 is linear in the electric field and an odd function of the velocity of the carriers.

To calculate the rate of change of f due to collisions we introduce $P(k, k')$, the probability of scattering per unit time from an initial state k to an empty final state. We assume perturbation theory to be valid, leaving its justification for the materials of interest to be discussed later. By the golden rule

$$P(k, k') = \frac{2\pi}{\hbar}|\langle k'|H'|k\rangle|^2 \delta(\varepsilon_f - \varepsilon_i), \tag{34}$$

where ε_i and ε_f are the energies of the initial and final states, respectively. With this, the rate of change of f due to collisions may be written

$$\left[\frac{\partial f(k)}{\partial t}\right]_{\text{coll}} = \int_{-\pi/b}^{+\pi/b} \{f(k')P(k' \to k)[1 - f(k)] - f(k)P(k \to k')[1 - f(k')]\}\rho(k')\, dk' \tag{35}$$

To evaluate Eq. (35) we substitute for f the expression (32). The group of terms obtained by replacing f by f_0, i.e.,

$$\int_{-\pi/b}^{+\pi/b} \{f_0(k')P(k' \to k)[1 - f_0(k)] - f_0(k)P(k \to k')[1 - f_0(k')]\}\rho(k')\, dk', \tag{36}$$

clearly vanishes because it represents the rate of change of f_0 due to collisions. It is easily seen that for the 1-phonon (1p) processes we have considered terms in Eq. (35) with $f_1(k')P(k \to k')$ or $f_1(k')P(k' \to k)$ vanish. This is so because $f_1(k')$ is odd in the velocity v', thus in k', while for all the 1p processes considered the matrix element and therefore $P(k \rightleftarrows k')$ are independent of the final state k'. Thus the only terms left are those involving $f_1(k)$ and the collision term may be written

$$\left[\frac{\partial f(k)}{\partial t}\right]_{\text{coll}} = -\frac{f_1(k)}{\tau} \equiv -f_1(k)\left(\frac{1}{\tau_{\text{abs}}} + \frac{1}{\tau_{\text{em}}}\right), \tag{37}$$

where the relaxation times τ_{em} and τ_{abs} are given by

$$\frac{1}{\tau_{\text{em,abs}}} = \int \{P_{\text{em,abs}}(k \to k')[1 - f_0(k')] + P_{\text{abs,em}}(k' \to k)f_0(k')\}\rho(k')\, dk'. \tag{38}$$

The first term of $1/\tau_{\text{em}}(1/\tau_{\text{abs}})$ actually represents the rate of scattering out of k by emission (absorption), while the second term arises from the change in the transitions into k from k' due to the change $f_1(k)$ in the occupation probability of the state k. τ_{em} and τ_{abs} may be written in different form by making use of the relation derived from the vanishing of (36) or, more specifically, the vanishing of the integrand in (36). This leads to

$$\frac{P_{\text{abs}}(k' \to k)}{P_{\text{em}}(k \to k')} = \frac{f_0(k)}{f_0(k')} \frac{1 - f_0(k')}{1 - f_0(k)}. \tag{39}$$

With Eq. (39), Eq. (38) may be rewritten as (Conwell, 1980b)

$$1/\tau_{em,abs} = \int P_{em,abs}(k \to k') \frac{1 - f_0(k')}{1 - f_0(k)} \rho(k') \, dk'. \tag{40}$$

For elastic scattering $f_0(k') = f_0(k)$ and the factor in braces is unity.

In evaluating the τ's it is convenient, since P_{em} and P_{abs} contain a δ function involving ε_k and $\varepsilon_{k'}$, to replace $\rho(k') \, dk'$ by $\rho(\varepsilon_{k'}) \, d\varepsilon_{k'}$ with the use of Eq. (7). With $P(k, k')$ from Eq. (34), the matrix element Eq. (30) and the density of states in energy Eq. (7) inserted in Eq. (40), we obtain the relaxation time for the state k due to single-phonon-scattering processes

$$\frac{1}{\tau_{em,abs}} = \frac{|g_{1,n}|^2}{\hbar} \begin{Bmatrix} n_{q,n} + 1 \\ n_{q,n} \end{Bmatrix} \int \delta(\varepsilon_{k'} - \varepsilon_k \pm \hbar\omega_{q,n})[\varepsilon_{k'}(\varepsilon_0 - \varepsilon_{k'})]^{-1/2}$$

$$\times \left(\frac{1 - f_0(\varepsilon_{k'})}{1 - f_0(\varepsilon_k)}\right) d\varepsilon_{k'}. \tag{41}$$

Because the relaxation time just derived is the scattering time that appears in the conductivity, as will be shown in the next section, it is worthwhile to examine its detailed properties for the various $1p$ processes we have been considering. For the internal modes, scattering with emission is possible to either $-k'$ or $+k'$, by means of phonons $k' + k$ or $-(k' - k)$, respectively, provided $\varepsilon_{k'} = \varepsilon_k - \hbar\omega_{q,n}$. For scattering by an internal mode n we therefore obtain (Conwell, 1977)

$$\frac{1}{\tau_{em}} = \frac{2}{\hbar}(g_n\hbar\omega_n)^2(n_{q,n} + 1)\frac{H(\varepsilon - \hbar\omega_n)}{\{(\varepsilon - \hbar\omega_n)[\varepsilon_0 - (\varepsilon - \hbar\omega_n)]\}^{1/2}}$$

$$\times \frac{1 - f_0(\varepsilon - \hbar\omega_n)}{1 - f_0(\varepsilon)}, \tag{42}$$

where $g_{1,n}$ has been taken from Eq. (26). The function $H(\varepsilon - \hbar\omega_n) = 1$ provided $\varepsilon > \hbar\omega_n$, 0 otherwise. It is included here (and a similar function later in τ_{abs}) to ensure that the final state lies within the band. By a similar procedure we find (Conwell, 1977)

$$\frac{1}{\tau_{abs}} = \frac{2}{\hbar}(g_n\hbar\omega_n)^2 n_{q,n} \frac{H[(\varepsilon_0 - (\varepsilon + \hbar\omega_n)]}{\{(\varepsilon + \hbar\omega_n)[\varepsilon_0 - (\varepsilon + \hbar\omega_n)]\}^{1/2}} \frac{1 - f_0(\varepsilon + \hbar\omega_n)}{1 - f_0(\varepsilon)}. \tag{43}$$

The variation with $\varepsilon/\varepsilon_0$ of τ_{em} and τ_{abs} is shown in Figs. 3a and 3b for the parameters of the internal modes in TCNQ shown in Table I. It is seen that, in accordance with Eqs. (42) and (43), $\tau_{em} \to 0$ at $\varepsilon = \hbar\omega_n$, while $\tau_{abs} \to 0$ at $\varepsilon = \varepsilon_0 - \hbar\omega_n$. This occurs because of the singularities in the density of final states at $\varepsilon = 0$ and ε_0, respectively. It is not expected to be literally true,

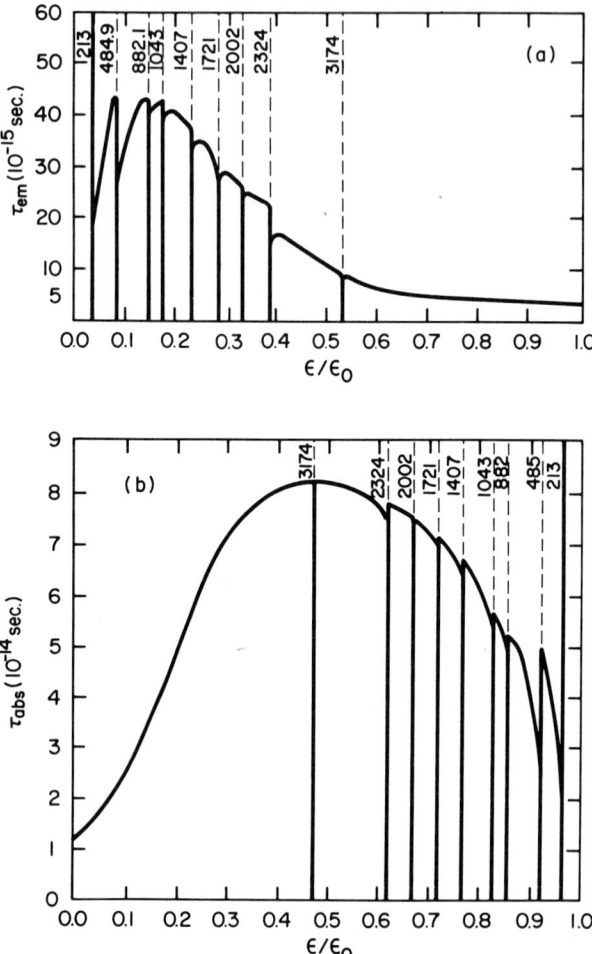

FIG. 3. τ_{em} (a) and τ_{abs} (b) of Eqs. (42) and (43), respectively, plotted for $T = 300$ K, $\varepsilon_0 = 6000$ K, $\varepsilon_F = 12000$ K, and internal-mode frequencies and coupling constants of Table I for TCNQ. [From Conwell (1980b).]

however, because the small lifetime for an electron in levels with $\varepsilon \simeq \hbar\omega_n$ or $\varepsilon \simeq \varepsilon_0 - \hbar\omega_n$ will cause broadening of these levels. Use of Eqs. (42) and (43) would, nevertheless, not be expected to cause a significant error in calculating the conductivity, for example, unless perhaps the Fermi energy happened to coincide with one of the levels concerned. The dips get increasingly narrow as ε increases from 0 for τ_{em}, or decreases from ε_0 for τ_{abs}, because of the increasing number of phonons with which the electron can interact. The interplay between that number of phonons and the density-of-final-states factor is what determines the overall skewed bell shape of these τ's as a function of ε. We note that the τ's of Eqs. (42) and (43) are valid also for nondegenerate material, being somewhat simpler for that case since the factor involving the f_0's may be taken as unity, filling of the states being unimportant in limiting transitions.

In the case of acoustic-mode scattering, which we have noted is essentially elastic, for an electron at $+k_F$ only backward scattering affects the conductivity. From Eqs. (41) and (15), using again the fact that the scattering is elastic, we obtain for the $1t$ case

$$\frac{1}{\tau_{em,1t;abs,1t}} = \left(\frac{\partial \varepsilon_0}{\partial u_n}\right)_0^2 \frac{\sin kb}{M\varepsilon_0 \omega_{2k,n}} \left\{\begin{array}{c} n_q + 1 \\ n_q \end{array}\right\}. \quad (44)$$

Here we have used Eq. (4) to eliminate $[\varepsilon(\varepsilon_0 - \varepsilon)]^{1/2}$. Equation (44) may be put in another form by making use of the dispersion relation for a linear vibrating chain in the harmonic approximation,

$$\omega_{2k} = \omega_0 \sin kb. \quad (45)$$

where ω_0, related to the force constant and M, is the maximum frequency of the branch. With this, combining absorption and emission terms we get

$$\frac{1}{\tau_{1t}} = \left(\frac{\partial \varepsilon_0}{\partial u_n}\right)_0^2 \frac{1}{M\varepsilon_0 \omega_{0,n}} (2n_q + 1). \quad (46)$$

When equipartition is valid,

$$2n_q + 1 \simeq \frac{2k_B T}{\hbar \omega_{2k,n}}. \quad (47)$$

In that case τ_{1t} may be written as

$$\frac{1}{\tau_{1t}} = \left(\frac{\partial \varepsilon_0}{\partial u_n}\right)_0^2 \frac{k_B T}{\hbar M \omega_{0,n}^2} \frac{1}{[\varepsilon(\varepsilon_0 - \varepsilon)]^{1/2}}, \quad (48)$$

a convenient form because it displays explicitly the ε dependence of τ. In particular, according to Eq. (48) $\tau_{1t} \propto v_F$ for the degenerate case. It is useful to trace the origin of this energy dependence. As seen in Eq. (15), the square of the coupling constant is proportional to v_F^2/ω_{2k_F}, ω_{2k_F} coming from the amplitude of oscillation of the molecules. Phonon abundance (i.e., n_q or n_{q+1}) under equipartition contributes another factor of $1/\omega_{2k_F}$ to the matrix element. For the $1t$ case with the usual dispersion, $\omega_{2k_F} \propto v_F$. The matrix element is then energy independent and the sole energy dependence of τ_{1t} comes from the density of final states. Since the translon scattering has been assumed elastic, the relation (8) between $\rho(\varepsilon)$ and v gives $\tau_{1t} \propto v_F$.

The above considerations have been based on acoustic waves travelling either parallel to the chains (LA) or perpendicular to the chains (TA). It is important to note that including in the calculations oblique waves, i.e., waves travelling at an arbitrary angle to the chains, will not have much effect on τ_{1t} calculated above. For all waves only the component of the displacement parallel to the chain causes scattering. Thus $(\partial \varepsilon_0/\partial u_n)_0$ or $(\partial t/\partial u_n)_0$ must be replaced for the oblique waves by $\hat{e}_q \cdot \nabla t$, $\hat{e}_q$ being the unit vector in the direction of polarization of the wave (Friedman, 1965). As a result, when the plane of the molecules is not perpendicular to the chain direction, the maximum of the matrix element occurs not for the waves with displacement parallel to the chains, but rather for those with displacement perpendicular to the plane of the molecules. However, as the displacement moves off the chain direction, $|q|$ required for conservation of energy and momentum increases and ω_q increases, tending to decrease the scattering effect of the wave. In particular, the decrease in n_q that results from the increase in ω_q should be sufficient to make ineffective for scattering waves whose displacement makes a large angle with the chain direction. Although it would be difficult to do an accurate calculation of τ_{1t} taking into account the scattering of oblique waves, the effects just mentioned combine to make the τ_{1t} calculated above a good approximation to τ_{1t} for all waves of a given type, i.e., LA or TA.

For scattering by a single libron the situation is quite similar to that for scattering by a single translon. When the approximations of elastic scattering and equipartition are valid we may write the relaxation time for this case in the form

$$\frac{1}{\tau_{1\ell}} = \frac{2}{\hbar}\left(\frac{\partial \varepsilon_0}{\partial \theta_n}\right)^2 \frac{\sin kb}{I_n \varepsilon_0 \omega_n^2} k_B T, \tag{49}$$

where we have dropped the subscript $2k$ on ω since the libron branches are expected to be fairly flat, i.e., dispersionless, at least for our simple model. For that reason, Eq. (45) does not apply and $\tau_{1\ell}$ cannot be written in a form analogous to Eq. (48) for τ_{1t}. Without dispersion $\tau_{1\ell} \propto v_F^{-1}$ and there would be a constant mean free path for 1ℓ scattering.

For the $2p$ processes the dependence on k' of the collision term cannot be eliminated, and a relaxation time in the sense of Eq. (37) does not exist. To estimate the conductivity for this case it is convenient to calculate a momentum relaxation time, to be denoted by τ_{2p}. This time differs from the mean free time between collisions by the factor $\langle 1 - \cos \chi \rangle^{-1}$, χ being the angle between k and k'. Since only backward collisions are significant in this case, $\langle 1 - \cos \chi \rangle^{-1} = \frac{1}{2}$. We may then take

$$\frac{1}{\tau_{2p}} = 2 \sum_q \sum_{q'} P(k \to k'), \tag{50}$$

where $P(k \to k')$ is given by Eq. (34), with Eqs. (31) and (19) or (20) specifying the matrix element for this case. If we again neglect the phonon energy compared to that of the electron, scattering can only take an electron from $\pm k$ to $\mp k$. To make the argument of the δ function in Eq. (31) vanish, then,

$$q' = -q - 2k. \tag{51}$$

This condition eliminates the summation over q' in Eq. (50). With q' given by Eq. (51) the quantity $f_2(k, q, q')$ defined in Eq. (21) becomes

$$f_2(k, q, q') = 2[\cos kb - \cos(k + q)b]. \tag{52}$$

When the summation over q is changed into an integration we obtain for the case that both phonons are absorbed,

$$\frac{1}{\tau_{2p}} = 2 \frac{2\pi}{\hbar} \frac{Nb}{2\pi} G^2 \rho_k(\varepsilon) \left(\frac{k_B T}{\hbar}\right)^2 \int_{-\pi/b}^{-2k+\pi/b} \frac{f_2^2}{\omega_q^2 \omega_{q'}^2} dq, \tag{53}$$

where $\rho_k(\varepsilon)$ is the density of states of one spin at k or $-k$ [given by Eq. (7)], f_2 is given by Eq. (52) and

$$G = \begin{cases} \dfrac{\hbar}{8NM} \left(\dfrac{\partial^2 \varepsilon_0}{\partial u^2}\right)_0 & \text{translons} \\[2mm] \dfrac{\hbar}{8NI_n} \left(\dfrac{\partial^2 \varepsilon_0}{\partial \theta^2}\right)_0 & \text{librons} \end{cases} \tag{54}$$

The quantities n_q and $n_{q'}$ in the matrix element have been replaced by $kT/\hbar\omega_q$ and $kT/\hbar\omega_{q'}$, respectively. Also, the integration limits have been set in Eq. (53) so that the phonons and the electron stay within the first Brillouin zone. For the libron case we assume ω_q and $\omega_{q'}$ are constants, equal to each other, in fact, if both librons are from the same branch. Integration of Eq. (53) is then straightforward and yields $\tau_{2\ell}$ for the case that q and q' are absorbed. It is readily seen that the other three processes—absorption of q and emission of $-q'$, emission of $-q$ and absorption of q', and emission of q and q'—lead to the same result as Eq. (53). Combining all four processes, we obtain for the momentum relaxation time due to scattering by two librons from the nth branch

$$\frac{1}{\tau_{2\ell}} = \left(\frac{\partial^2 \varepsilon_0}{\partial \theta_n^2}\right)_0^2 \left(\frac{k_B T}{I_n \omega_n^2}\right)^2 \left(\frac{F_\ell(kb)}{\hbar[\varepsilon(\varepsilon_0 - \varepsilon)]^{1/2}}\right), \tag{55}$$

where

$$F_\ell(kb) = \frac{(\pi - kb)(1 + 2\cos^2 kb) - (\tfrac{5}{2}) \sin 2kb}{4\pi}. \tag{56}$$

The results Eqs. (55) and (56) differ from those of Gutfreund and Weger (1977), because in that case integration was carried out over both q and q'. This allowed q and q' to vary independently in violation of conservation of momentum, Eq. (51).

The relaxation time for $1p$ polarization scattering has been evaluated for LA modes only because their contribution is expected to be the largest. For the summation over neighboring molecules in Eq. (29) $\bar{R}_{\ell j}$ may then be taken as $|n|b$, where $n = \pm 1, \pm 2$, etc. The summation yields (Friedman, 1981)

$$\sum_\ell \bar{R}_{\ell j} \cdot \hat{e}_{2k,n} \left[\frac{\exp(2i\bar{k} \cdot \bar{R}_{\ell j}) - 1}{R_{\ell j}^{-6}} \right] = \frac{2}{b^5} S(qb), \qquad (57)$$

where

$$S(qb) = \sum_{n=1}^{\infty} \frac{\cos(qnb) - 1}{n^5}. \qquad (58)$$

It is noted by Friedman that, due to the factor $1/n^5$, most of the sum arises from first and second neighbors. This partially compensates for the neglect of screening in this calculation. With Eq. (57) inserted in Eq. (29) the relaxation time for polarization fluctuation scattering may be obtained from Eq. (38) in the same way as for the other $1p$ processes. The result is (Friedman, 1981)

$$\frac{1}{\tau_{1P,\mathrm{LA}}} = \frac{16\alpha_0^2 e^4}{b^{10}} \frac{k_\mathrm{B} T}{\hbar M \omega_{0,\mathrm{LA}}^2} \frac{S^2(2kb)}{\sin^2 kb} \frac{1}{[\varepsilon(\varepsilon_0 - \varepsilon)]^{1/2}}. \qquad (59)$$

This result is quite similar in form to the expression (48) for $1/\tau_{1t}$. It is of interest to compare the two rates. From Eqs. (59) and (48) we obtain

$$\left(\frac{\tau_{1P}^{-1}}{\tau_{1t}^{-1}} \right)_{k_\mathrm{F}} = \frac{(4\alpha_0 e^2/b^5)^2}{(\partial \varepsilon_0/\partial u_n)_0^2} \frac{S^2(2k_\mathrm{F} b)}{\sin^2 k_\mathrm{F} b}. \qquad (60)$$

Friedman has estimated the ratio $\tau_{1P}^{-1}/\tau_{1t}^{-1}$ for TTF-TCNQ, taking $\alpha_0 = 40$ Å^{-3} (Metzger, 1978), $b = 4$ Å, $(\partial \varepsilon_0/\partial u_\mathrm{LA})_0 = 0.8$ eV/Å (Conwell, 1980b) and $2k = 2k_\mathrm{F} = 0.295(2\pi/b)$ (Shirane et al., 1976). With these values he obtains $S(2kb) \simeq -1.34$ and $(\tau_{1P}^{-1}/\tau_{1t}^{-1})_{k_\mathrm{F}} \simeq 3$. This treatment undoubtedly overestimates the effect of polarization fluctuation scattering because, as noted earlier, it neglects screening which, according to the discussion of Section 3, must be significant in TTF-TCNQ, even for nearest neighbors. Nevertheless, it is reasonable to expect that $1p$ polarization fluctuation scattering makes a significant contribution to the observed resistivity of metallic TTF-TCNQ and similar materials.

7. Derivation of Expressions for One-Dimensional Conductivity and Thermoelectric Power

It is straightforward to solve the one-dimensional Boltzmann equation when a relaxation times exists, as has been shown to be the case in the last section for the $1p$ processes. To terms linear in the electric field intensity, making use of Eq. (32), we have

$$\left(\frac{\partial f}{\partial t}\right)_E = -e\bar{E} \cdot \nabla_P f \simeq -e\bar{E} \cdot \bar{v}\frac{\partial f_0}{\partial \varepsilon}, \qquad (61)$$

where e is the magnitude of the charge on the electron and P the crystal momentum. The velocity $v, = \nabla_P \varepsilon$ where ε is given by Eq. (4), may be written

$$v = \frac{b}{\hbar}[\varepsilon(\varepsilon_0 - \varepsilon)]^{1/2} = (\varepsilon_0 b/2\hbar)\sin kb. \qquad (62)$$

With the use of Eq. (37) for the collision terms, we obtain the solution of the Boltzmann equation for low electric fields

$$f_1 = -e\bar{E} \cdot \bar{v}\tau \frac{\partial f_0}{\partial \varepsilon}. \qquad (63)$$

The current density and energy flux are then given by

$$\bar{j} = N_c e \sum_{\bar{k}} f_1 \bar{v} \qquad (64)$$

and

$$\bar{w} = N_c \sum_{\bar{k}} (\varepsilon - \varepsilon_F) f_1 \bar{v}, \qquad (65)$$

where N_c is the number of chains/cm^2 and the summation over $\bar{k}$ is to be taken over all states (both spins) per unit length of chain. It is convenient to convert the summation into an integration over ε:

$$\sum_{\bar{k}} F(\bar{k}) = \frac{2}{Nb}\int_{-\pi/b}^{+\pi/b} F(k)\rho(k)\,dk = \frac{2}{Nb}\int_0^{\varepsilon_0} F(\varepsilon)\rho(\varepsilon)\,d\varepsilon, \qquad (66)$$

where $2\rho(\varepsilon)$ represents the density of states at $+k$ and $-k$ and with both spin directions, $\rho(\varepsilon)$ being given by Eq. (7). Using Eqs. (63) and (66) we then obtain from Eq. (64)

$$\sigma \equiv \frac{j}{E} = -\int_0^{\varepsilon_0} \sigma(\varepsilon)\frac{\partial f_0}{\partial \varepsilon}\,d\varepsilon, \qquad (67)$$

where

$$\sigma(\varepsilon) = \frac{2N_c e^2 v^2 \tau \rho(\varepsilon)}{Nb}. \qquad (68)$$

To calculate the mobility μ from Eq. (67) it is necessary to divide σ by e times the carrier concentration $N_c n$, where n is the number of carriers per unit chain length. Using Eqs. (66) and (7) we obtain

$$n = \sum_{\vec{k}} f_0 = \frac{2}{\pi b} \int_0^{\varepsilon_0} f_0(\varepsilon)[\varepsilon(\varepsilon_0 - \varepsilon)]^{-1/2} d\varepsilon. \tag{69}$$

Note that, despite the use of Fermi–Dirac statistics, carrier concentration is in general not constant in a two-chain conductor but varies with temperature because charge transfer changes as ε_F and the bandwidths change with temperature. This will be discussed further in Section 9.

The thermopower Q may be evaluated from the Kelvin relation

$$Q = \frac{w}{jT}, \tag{70a}$$

w/j being the Peltier heat. With Eqs. (63)–(66), Eq. (70a) for the thermopower of one type of chain becomes

$$Q = \frac{1}{eT} \frac{\int_0^{\varepsilon_0} (\varepsilon - \varepsilon_F) v^2 \tau \rho(\varepsilon)(\partial f_0/\partial \varepsilon) \, d\varepsilon}{\int_0^{\varepsilon_0} v^2 \tau \rho(\varepsilon)(\partial f_0/\partial \varepsilon) \, d\varepsilon}. \tag{70b}$$

For a material with two types of chain, having individual thermopowers Q_n and Q_p and conductivities σ_n and σ_p, the combined thermopower is the weighted average,

$$Q = \frac{Q_n \sigma_n + Q_p \sigma_p}{\sigma_n + \sigma_p}. \tag{71}$$

When the scattering process is elastic, as is reasonable to assume for translon and libron scattering in degenerate material, integration of the expressions for σ and Q may be simplified by concentrating on properties at the Fermi energy. This will be carried out in the next section. However, when internal-mode scattering is operative, either alone or in combination with elastic processes, it is necessary to integrate Eqs. (67) and (70b) numerically. To do this, it is convenient to replace $(\partial f_0/\partial \varepsilon)$ by $-f_0(1 - f_0)/k_B T$, obtained from Eq. (33), and introduce the dimensionless variable $y = \varepsilon/\varepsilon_0$. The integrals are then conveniently evaluated by Gaussian-type numerical integration.

8. Transport Coefficients for the Cases of Elastic Scattering

Considerable insight may be gained by studying translon or libron scattering alone for the degenerate case. In the preceding section it was found that the

4. TRANSPORT IN QUASI-ONE-DIMENSIONAL CONDUCTORS

integrals that must be evaluated to obtain σ and Q are of the form

$$K_n = -\int_0^{\varepsilon_0} \varepsilon^n \sigma(\varepsilon) \frac{\partial f_0}{\partial \varepsilon} d\varepsilon, \tag{72}$$

$\sigma(\varepsilon)$ being the quantity defined in Eq. (68). To evaluate K_n we take advantage of the fact that $(\partial f_0/\partial \varepsilon)$ is nonvanishing only in the neighborhood of $\varepsilon = \varepsilon_F$ to make a Taylor-series expansion of the coefficient of $(\partial f_0/\partial \varepsilon)$ around ε_F. This leads to the result, to second-order terms in $k_B T/\varepsilon_F$, (Ziman, 1960)

$$K_n = \varepsilon_F^n \sigma(\varepsilon_F) + \frac{\pi^2}{6}(k_B T)^2 \frac{d^2}{d\varepsilon^2} [\varepsilon^n \sigma(\varepsilon)]_{\varepsilon_F}. \tag{73}$$

To first order in T we have, then, from Eqs. (67) and (68),

$$\sigma = [2N_c e^2 v^2 \tau \rho(\varepsilon)]_{\varepsilon_F}. \tag{74}$$

To obtain σ for 1p acoustic scattering we use Eq. (44) for τ and assume equipartition, Eq. (47). Inserting Eq. (62) for v and $\rho(\varepsilon)$ from Eq. (7), we may write for σ due to scattering by the nth acoustic branch (Conwell, 1980b),

$$\sigma_{1t,n} = \frac{N_c e^2 b M \omega_{2k_F,n}^2 \varepsilon_0^2}{h(\partial \varepsilon_0/\partial u_n)^2 k_B T}, \quad \hbar \omega_{2k_F} \ll k_B T. \tag{75}$$

Similarly, we obtain for 1p libron scattering (Conwell, 1980b)

$$\sigma_{1\ell,n} = \frac{N_c e^2 b I_n \omega_{2k_F,n}^2 \varepsilon_0^2}{h(\partial \varepsilon_0/\partial \theta_n)^2 k_B T}. \tag{76}$$

We see that, under equipartition, σ for the 1p processes is proportional to T^{-1} and to the square of the frequency of the phonon with wavevector $2k_F$. Also σ is proportional to the square of the bandwidth.

When equipartition does not hold, the temperature dependence for the 1p processes is steeper than T^{-1}. To derive correctly the temperature dependence for this case requires that more care be exercised in obtaining the relaxation time from Eq. (41). Because the Fermi energy $\varepsilon_F \gg \hbar\omega_q$ for an acoustic phonon or a libron we may still neglect the difference between ε_k and $\varepsilon_{k'}$ in the density-of-states factor $[\varepsilon_{k'}(\varepsilon_0 - \varepsilon_{k'})]^{-1/2}$. We cannot do this, however, in the distribution function $f_0(\varepsilon_{k'})$ because $\varepsilon_{k'} - \varepsilon_F \simeq \hbar\omega_q$. Therefore integration of Eq. (41) for the 1-translon case gives

$$\frac{1}{\tau_{em}} = \frac{2|g_{1,t}|^2}{\hbar} \frac{n_q + 1}{[\varepsilon(\varepsilon_0 - \varepsilon)]^{1/2}} \frac{1 - f_0(\varepsilon - \hbar\omega_q)}{1 - f_0(\varepsilon)},$$

$$\frac{1}{\tau_{abs}} = \frac{2|g_{1,t}|^2}{\hbar} \frac{n_q}{[\varepsilon(\varepsilon_0 - \varepsilon)]^{1/2}} \frac{1 - f_0(\varepsilon + \hbar\omega_q)}{1 - f_0(\varepsilon)},$$

These expressions may be simplified by making use of Eqs. (4) and (45). We may then write the combined τ for translon scattering

$$\tau = \frac{M\omega_0\varepsilon_0}{(\partial\varepsilon_0/\partial u)_0^2 n_{2k}} \left[\frac{n_{2k}+1}{n_{2k}} \frac{1-f_0(\varepsilon-\hbar\omega_{2k})}{1-f_0(\varepsilon)} + \frac{1-f_0(\varepsilon+\hbar\omega_{2k})}{1-f_0(\varepsilon)} \right]^{-1},$$

where $|g_{1,t}|^2$ has been taken from Eq. (15). Inserting for n_{2k} the Bose–Einstein distribution and for f_0 the Fermi–Dirac distribution completes the evaluation of τ. To obtain σ we use Eq. (74). This leads to σ for the case where equipartition does not hold (Conwell and Jacobsen, 1981)

$$\sigma_{1t} = \frac{N_c e^2 bM\varepsilon_0^2 \omega_{2k_F}}{2\pi\hbar^2(\partial\varepsilon_0/\partial u)_0^2} \sinh Z_F, \tag{77}$$

where

$$Z_F = \frac{\hbar\omega_{2k_F}}{k_B T}.$$

In the high-temperature limit $\sinh Z_F$ may be replaced by Z_F and σ of Eq. (77) goes over correctly to that of Eq. (75), derived under the assumption that equipartition is valid. The difference between Eqs. (77) and (75) is small for TTF-TCNQ over the entire metallic range. It is necessary to use Eq. (77), however, at temperatures approaching T_P for materials with low metal-to-semiconductor transition temperatures such as (TMTSF)$_2$PF$_6$ (Conwell and Jacobsen, 1981).

For the $2p$ case the solution (63) of the Boltzmann equation is not valid since a relaxation time does not exist. One should, nevertheless, obtain a reasonable approximation to σ for two-libron scattering by inserting the momentum relaxation time, Eq. (55), for that case in Eq. (74). This leads to (Conwell, 1980b)

$$\sigma_{2\ell} = \frac{N_c e^2 b I_n^2 \omega_n^4 \varepsilon_0^2 \sin^2 k_F b}{4hF_\ell(k_F b)(\partial^2\varepsilon_0/\partial\theta_n^2)^2 (k_B T)^2}, \tag{78}$$

where $F_\ell(k_F b)$ is defined in Eq. (56). As expected, because two phonons are involved, this σ is proportional to ω^4 and T^{-2}, as well as to ε_0^2.

It is informative to write down what the conductivity would be for internal phonon scattering in the limit $\hbar\omega_n \to 0$. Inserting into Eq. (74) τ for the internal modes obtained by letting Eqs. (42) and (43) go to that limit, we obtain

$$\sigma_{\text{int}} \xrightarrow[\hbar\omega\to 0]{} \frac{N_c e^2 b\varepsilon_0^2 \sin^2 k_F b}{2h(g\hbar\omega)^2(2n_q+1)}.$$

Since $k_F b$ depends little on ε_0, σ_{int} in this limit is clearly proportional to ε_0^2. This remains essentially true when $\hbar\omega$ is not assumed small compared to ε_F.

Consider now the thermopower for the case of elastic scattering. When Q, given by Eq. (70), is expressed in terms of K_n, we obtain

$$Q = \frac{1}{eT}\left(\frac{K_1}{K_0} - \varepsilon_F\right). \tag{79}$$

Using Eq. (73) for K_n we obtain a useful general expression for Q, valid to lowest order in $k_B T/\varepsilon_F$,

$$Q = \frac{\pi^2}{3}\frac{k_B^2 T}{e}\left(\frac{d \ln \sigma(\varepsilon)}{d\varepsilon}\right)_{\varepsilon_F}. \tag{80}$$

With Eq. (68) inserted for $\sigma(\varepsilon)$, Q becomes

$$Q = \frac{\pi^2}{3}\frac{k_B^2 T}{e}\left[\frac{1}{v^2\rho}\frac{d}{d\varepsilon}(v^2\rho) + \frac{1}{\tau}\frac{d\tau}{d\varepsilon}\right]_{\varepsilon_F}. \tag{81}$$

When v and ρ are expressed in terms of ε, we obtain for the first term

$$\left[\frac{1}{v^2\rho}\frac{d}{d\varepsilon}(v^2\rho)\right]_{\varepsilon_F} = \frac{1}{2}\frac{\varepsilon_0 - 2\varepsilon_F}{\varepsilon_F(\varepsilon_0 - \varepsilon_F)}. \tag{82}$$

For $1p$ acoustic-mode scattering, with τ given by Eq. (48), where the energy dependence is clearly displayed, we find

$$\left(\frac{1}{\tau_{1t}}\frac{d\tau_{1t}}{d\varepsilon}\right)_{\varepsilon_F} = \left[\frac{1}{v^2\rho}\frac{d}{d\varepsilon}(v^2\rho)\right]_{\varepsilon_F}. \tag{83}$$

Thus the term in $d\tau/d\varepsilon$ makes precisely the same contribution to Q for this case as the first term, the so-called band term, giving (Conwell, 1980b).

$$Q_{1t} = \frac{(\pi^2/3)(k_B/e)(k_B T/\varepsilon_0)(1 - 2\varepsilon_F/\varepsilon_0)}{(\varepsilon_F/\varepsilon_0)(1 - \varepsilon_F/\varepsilon_0)}. \tag{84}$$

Since $\varepsilon_F/\varepsilon_0$ is determined by the charge transfer, which does not vary a great deal with temperature, Q for this case depends essentially on T and ε_0^{-1}. It is easily seen that this expression for Q still holds when there is scattering by more than one type of acoustic phonon. Also, it is valid for the case of internal phonon scattering in the limit $\hbar\omega \to 0$. It will be seen, however, when we discuss the results of numerical integration for Q, that when $\hbar\omega$ is not assumed vanishingly small the internal modes may make quite a different contribution to Q.

The case of one-libron scattering is an interesting one. Here τ is given by Eq. (49), where it is seen that its energy dependence comes from $(\sin k_F b)^{-1}$, which is proportional to $[\varepsilon(\varepsilon_0 - \varepsilon)]^{-1/2}$. The resulting $[(1/\tau)\,d\tau/d\varepsilon]_{\varepsilon_F}$ is the negative of that for the translon case, just canceling the band term and giving $Q_{1\ell} = 0$. In other words, if there were only one-libron scattering present Q would vanish because $\sigma(\varepsilon)$ is independent of ε for this case. Comparing the τ's for the $1t$ and 1ℓ cases, Eqs. (44) and (49), respectively, we see that what makes the $1t$ case different from this one is the dispersion of the acoustic phonons; ω_{2k_F} is a function of ε for the acoustic phonons, but for the librons it is not because we have assumed them dispersionless. Actually, the libron branches can be expected to have some dispersion so $Q_{1\ell}$ may be finite though small.

We can also get an estimate of Q for the two-libron case by inserting Eq. (55) in the expression (81) for Q. This leads to (Conwell, 1980b).

$$Q_{2\ell} = Q_{1\ell} - \frac{\pi^2}{3}\frac{k_B}{e}k_B T\left[\frac{1}{F_\ell(kb)}\frac{d}{d\varepsilon}F_\ell(kb)\right]_{\varepsilon_F}, \qquad (85)$$

where $Q_{1\ell}$ is given by Eq. (84) and

$$\left[\frac{1}{F_\ell(kb)}\frac{d}{d\varepsilon}F_\ell(kb)\right]_{\varepsilon_F} =$$

$$-\frac{1 + 2\cos^2 k_F b + 2(\pi - k_F b)\sin 2k_F b + 5\cos 2k_F b}{2\pi\varepsilon_0 \sin k_F b\, F_\ell(k_F b)}, \qquad (86)$$

$F_\ell(kb)$ being given by Eq. (56). Again, Q increases linearly with T, as anticipated from Eq. (80) but the dependences on ε_F and ε_0 are clearly different from those for the $1p$ cases.

9. Comparison with Experiment for TTF-TCNQ in the Metallic Range

Many parameters must be calculated or obtained from experiment to evaluate the conductivity and thermoelectric power from the expressions just derived. The parameters used in peforming this evaluation for TTF-TCNQ (Conwell, 1980b) are summarized in Tables I through III. A brief description of how they were obtained follows.

TABLE II

Derivatives of the Transfer Integral for the TCNQ-Band of (TTF)(TCNQ), Normalized by the Value of the Transfer Integral[a]

Temp. K	$\frac{1}{t}\frac{\partial t}{\partial u_b}$ Å^{-1}	$\frac{1}{t}\frac{\partial^2 t}{\partial u_b^2}$ Å^{-2}	$\frac{1}{t}\frac{\partial^2 t}{\partial u_a^2}$ Å^{-2}	$\frac{1}{t}\frac{\partial t}{u_{c^*}}$ Å^{-1}	$\frac{1}{t}\frac{\partial^2 t}{\partial u_{c^*}^2}$ Å^{-2}	$\frac{1}{t}\frac{\partial t}{\partial \xi}$ rad^{-1}	$\frac{1}{t}\frac{\partial^2 t}{\partial \xi^2}$ rad^{-2}	$\frac{1}{t}\frac{\partial^2 t}{\partial \eta^2}$ rad^{-2}	$\frac{1}{t}\frac{\partial^2 t}{\partial \eta^2}$ rad^{-2}
60	1.488	1.288	−0.284	1.096	−0.709	2.045	4.554	−0.367	1.043
100	1.509	1.358	−0.286	1.079	−0.730	2.025	4.545	−0.426	1.064
300	1.556	1.529	−0.282	1.048	−0.775	2.001	4.554	−0.415	1.092

[a] From Van Smaalen et al., 1986.

TABLE III

Some Values Used for Calculations on TTF-TCNQ

$(\partial t/\partial u)_{LA}$	-0.20 eV/Å	M	204 amu
$k_F b$	0.295π	b	3.819 ± 10^{-8} cm
$\hbar\omega_{2k_F}(LA)$	85 K	N_c	8.79×10^{13} cm^2
$\hbar\omega_{2k_F}(TA)$	57 K	ε_0(TTF)	3000 (4500) K
$\hbar\omega_0(LA)$	8.6 meV	ε_0(TCNQ)	6000 K
$\hbar\omega_0(TA)$	5.5 meV	I_η	3.8×10^{-37} gm cm^2
$(\partial t/\partial u_{c^*})_{\text{eff}}$	-0.24 eV/Å	I_ζ	7.1×10^{-38} gm cm^2
		I_ξ	3.1×10^{-37} gm cm^2

a. Carrier Concentration and ε_F

Determination of ε_F can be made using Eq. (69) given $n(=p)$ and ε_0 for each of the chains and imposing the condition that the ε_F's for the two chains (conveniently measured from the respective band edges) must be aligned, i.e., must be at the same energy. For the TCNQ chain $\varepsilon_0 \simeq 0.5$ eV according to many experiments and theoretical calculations by Berlinsky *et al.* (1974) and Herman *et al.* (1977). For the TTF chain, although it is generally agreed that ε_0 is smaller than for TCNQ (Schultz and Craven, 1979), there is no agreement as to its value. It must also be taken into account that above 60 K, where the lattice expansion is significant, ε_0, which is determined by overlap, decreases as temperature increases. The values listed in Table III were taken, arbitrarily, to be the values at 60 K; two different values were used for TTF, 3000 K and 4500 K, because of the uncertainty.

The change in bandwidths with temperature results in the charge transfer changing with temperature. The observation that charge transfer increases with application of pressure (Megtert *et al.*, 1976) suggests that charge transfer will decrease as the lattice expands. The amount of change depends mainly on the rate of change of ε_0 with u_b, the distance between adjacent molecules along the b axis. The calculated values of $\partial\varepsilon_0/\partial u_b$ for TCNQ are in the range 0.68 eV Å (Berlinsky *et al.*, 1974) to 0.60 eV Å (Herman *et al.*, 1977). These values are in good agreement with the values deduced from pressure measurements, as will be shown later. Thus either one, or their average, could be used to obtain ε_0 as a function of T for TTF and TCNQ. Given the ε_0's for the two chains at a particular T, and a trial pair of ε_F's that satisfy the condition of alignment, n and p can each be calculated from Eq. (69). This process is iterated until a pair of ε_F's is found for which $n = p$, the condition that must be satisfied by the correct ε_F's. The result of this calculation was that the charge transfer decreases from 0.59 electrons/molecule at 60 K to ~0.48 electrons/molecule at 300 K, more or less independent of the 60 K bandwidth assumed for the TTF chain (Conwell, 1980a).

b. Deformation Potentials

As has been shown in the preceding sections, the σ's for the various $1p$ and $2p$ processes depend on the deformation potentials, quantities of the form $\partial^i \varepsilon_0 / \partial \chi^i$, where $i = 1$ or 2 and χ is a coordinate corresponding to one of the six possible translations or rotations. Because $\varepsilon_0 = 4t$, what must be calculated basically are the derivatives of t, the transfer integral, with respect to the various possible displacements.

The expression for t given in Eq. (3) results from a purely one-electron theory: The Hamiltonian incorporates interactions with the other conduction electrons at most to the extent of the Hartree–Fock scheme. The simple tight-binding model, from which Eq. (3) results, cannot give reliable absolute values for t because of the neglect of correlation. The model usually used when correlation is important, the Hubbard model discussed in Section 3, is also unsatisfactory for determining t because the result depends on the choice, to a large extent arbitrary, of the effective Coulomb repulsion U. A second reason for poor accuracy in the calculation of t is that, even for high-quality wavefunctions, the tail will be poorly determined (Van Smaalen and Kommandeur, 1985); for the distances of importance here it is the tail of the wavefunction that determines the overlap. Nevertheless, calculations for a series of related compounds, the N-substituted morpholinium salts, have shown that, as might be expected, *relative* values of t obtained from Eq. (3) are quite reliable (Van Smaalen and Kommandeur, 1985). With only relative values of t required for calculating σ, because the quantities that enter are ratios $(1/t)\partial^i t/\partial \chi^i$, the use of Eq. (3) should yield sufficiently reliable results to evaluate σ (Van Smaalen et al., 1986).

In calculating t from Eq. (3) the LUMO wavefunction $\varphi_\alpha(r)$ is replaced, still within the tight-binding approximation, by the orthonormal wavefunction

$$\varphi_\alpha^W(r) = \varphi_\alpha(r) - \tfrac{1}{2} S[\varphi_\alpha(r - b) + \varphi_\alpha(r + b)], \tag{87}$$

where S is the overlap integral,

$$S = \int \varphi_\alpha^*(r + b)\varphi_\alpha(r)\, dr. \tag{88}$$

Substitution of Eq. (87) into Eq. (3) and use of the Mulliken approximation (Berlinsky et al., 1974) leads to

$$t = E_\alpha(K - 1)S, \tag{89}$$

where E_α is the orbital energy of the LUMO and K a constant generally taken as 1.75 (Berlinsky et al., 1974). As a result of Eq. (89), $(1/t)(\partial^i t/\partial \chi^i) = (1/S)(\partial^i S/\partial \chi^i)$, and only S and its derivatives need to be calculated. To calculate S, Van Smaalen and Kommandeur (1985) used wavefunctions made up of linear combinations of atomic orbitals (LCAO). The atomic orbitals were self-consistent field Hartree–Fock wavefunctions determined by Jonkman et al. (1974). To obtain the derivatives, S was calculated in the experimentally determined configuration of the molecules and in the configurations with one of the molecules displaced or rotated by a small amount with respect to the required axes; the appropriate differences were then taken. The resulting nonzero derivatives are listed in Table II. The values of the three derivatives expected to vanish by symmetry—$\partial t/\partial u_a$, $\partial t/\partial \eta$, and $\partial t/\partial \zeta$—were about a factor of 100 smaller than those listed. The fact that they are not zero is attributed to the uncertainty in the crystal structure as well as calculational errors. Of the values in the table only $(1/t)(\partial t/\partial u_b)$ had been calculated previously; the value obtained is in excellent agreement with 1.55 Å obtained by Berlinsky et al. (1974) and in reasonable agreement with 1.03 Å obtained by Herman et al. (1977).

c. The Relative Importance of One-Phonon and Two-Phonon Scattering Processes in TTF–TCNQ

To compare the contributions of 2ℓ scattering due to branch n' and $1t$ scattering due to branch n we use the ratio, obtained from Eqs. (75) and (78)

$$\frac{\rho_{1t,n}}{\rho_{2\ell,n'}} = \left\{\frac{[(1/t)(\partial t/\partial u_n)]^2 I_{n'}}{[(1/t)(\partial^2 t/\partial \theta_{n'}^2)]^2 M}\right\} \left\{\frac{I_{n'}\omega_{n'}^2}{k_B T}\right\} \left\{\frac{\omega_{n'}}{\omega_{2k_F,n}}\right\}^2 \left\{\frac{\sin^2 k_F b}{4F_\ell(k_F b)}\right\}, \quad (90)$$

written as a product of four dimensionless factors (in braces). The quantities used for the numerical calculations are shown in Tables II and III. The moments of inertia were calculated from the data of Kistenmacher et al. (1974) and ω_{2k_F}'s taken from Shapiro et al. (1977). With these quantities in Eq. (90) we find the ratio of $\rho_{1t,\text{LA}}$ to $\rho_{2\ell,\eta}$ is 2×10^5 if we assume $\omega_\eta = \omega_{2k_F,\text{LA}}$. If the frequency of phonons of the η branch were half as large, the smallest value that has been considered a possibility (Weger and Gutfreund, 1978; Gutfreund et al., 1979), $\rho_{1t,\text{LA}}/\rho_{2\ell,\eta}$ would be reduced by only an order of magnitude. Thus 2ℓ processes involving the η libron cannot make a noticeable contribution to ρ. The ζ libron, the other for which the first-order scattering vanishes, is almost equally ineffective. The quantity $\rho_{1t,\text{LA}}/\rho_{2\ell,\zeta}$ is 1×10^3 for 300 K if we assume $\omega_\zeta = \omega_{2k_F,\text{LA}}$. The frequency for this libration is probably higher, however, because I_ζ is so small. Some evidence for a higher frequency of this branch is given by Bates et al. (1981), who have seen, in infrared absorption of TTF-TCNQ, modes in the range 85 to 100 cm^{-1} (120 to 144 K) that they identify as mixed ζ libron and optical lattice modes. It is clear that the contribution to ρ of 2ℓ scattering by the η and ζ modes through modulation of the transfer integral is quite negligible. The same conclusion is readily reached for the third 2ℓ process, involving ξ librons. It is also seen that two-translon scattering is negligible compared to one-translon scattering by LA phonons (Van Smaalen et al., 1986).

d. The TTF Chain

Almost all of the discussion so far has dealt with the TCNQ chain, which indeed is much better characterized than the TTF chain. It has been noted for the TTF chain that the bandwidth is smaller and (Section 3) that there is evidence for $U \gtrsim t$, clearly related to the smaller bandwidth. Although the effect of $U \gg t$ on transport has been studied, there are no studies for $U \gtrsim t$, Fortunately, there is good evidence from thermoelectric power and Hall effect measurements that the TTF chains contribute less to transport then TCNQ. When there are two types of conducting particles, here electrons on TCNQ chains and holes in TTF chains, the thermoelectric power Q, as seen in Eq. (71), is an average of Q_n and Q_p, each weighted by the contribution of the corresponding type of carrier to σ. The fact that the thermoelectric power of TTF-TCNQ (Chaikin et al., 1973), shown in Figure 5 of Chapter 1, is negative from 300 K down to close to the temperature where a gap opens up on the TCNQ chain means that σ is much larger for the TCNQ chain in the temperature range of our concern. [Q_n and Q_p are comparable because, as seen for Eq. (70b), they are determined mainly by the average of $(\varepsilon - \varepsilon_F)$.] The Hall coefficient in a situation with two types of carrier may be written in a form similar to Eq. (71), with Q_n replaced by $-\mu_n/(\sigma_n + \sigma_p)$ and Q_p by $\mu_p/(\sigma_n + \sigma_p)$, μ_n being the electron mobility, μ_p the hole mobility. Hall data of Cooper et al. (1977) show that down to ~ 150 K the Hall coefficient is close to what one would expect if only the TCNQ chain were conducting.

e. Calculated σ for TTF-TCNQ

As discussed in Section 4, the lattice vibration spectrum for the external modes of TTF-TCNQ is expected to be considerably more complicated than the simple model of Fig. 2 for which we have been calculating. Nevertheless, the LA branch appears to be well described by the simple model; the dispersion (Shapiro et al., 1977) obeys Eq. (45) and the calculated value of $(\partial t/\partial u_b)$ agrees with that deduced from pressure experiments. Thus it is reasonable to use the value of $\partial t/\partial u_b$ obtained for this branch from the simple model, 0.2 eV Å to one significant figure.

Of the remaining external modes a sizable $1p$ contribution is expected from the transverse acoustic mode travelling in the c^* direction, $TA(c^*)$, because its deformation potential is comparable to that of the LA mode, according to Table II. It is reasonable to expect also a comparable contribution from the ξ libron because its average contribution to the variation of t, the product of $(\partial t/\partial \xi)$ and its rms thermal amplitude (Van Smaalen *et al.*, 1986), is comparable to that for the LA mode. Unfortunately, however, ω_{2k_F} is not known for the ξ libron. Further, although we have a value for $(\partial t/\partial u_{c^*})$ from the simple model of the lattice vibrations, there is evidence for mixing of $TA(c^*)$ with $TA(a)$ (Shapiro *et al.*, 1977), as well as mixing of the ζ librons with translations (Bates *et al.*, 1981). Finally, it can be expected that there will be scattering due to oblique phonons, i.e., phonons travelling in other directions than b and c^*. Although, as discussed in Section 6, the inclusion of oblique phonons should not have a major effect on τ_{1t}, still it could be considered to make some change in the effective values of $(\partial t/\partial u)_0$ and ω_q. In view of these complications, it appears reasonable to lump together all $1p$ scattering other than that of the LA phonons with a deformation potential denoted $(\partial t/\partial u_{c^*})_{\text{eff}}$, to be treated as a parameter in the comparison with experiment.

As indicated earlier, the volume dependence of σ is quite large for TTF-TCNQ. Although the volume dependence of some of the quantities that enter into σ is known, it is unknown for some of the others. The calculations of σ were done, therefore, for constant volume. σ can be evaluated numerically from Eqs. (67) and (68) and Q from Eq. (70). τ is obtained as the reciprocal of the sum of the reciprocals of the individual τ's included. The parameters required are in Tables I through III. In carrying out the calculation (Conwell, 1980b) the number of phonons at any temperature in any given mode was calculated from the Bose–Einstein distribution. Deformation potential values are not available for TTF; they were therefore taken to be the same for the TTF chain as for TCNQ. The results are shown in Fig. 4 for $\varepsilon_0(\text{TTF}) = 3000$ K. To achieve a value of σ at 300 K of 800 to 900 ohm^{-1} cm^{-1}, in agreement with experiment, was found to require $(\partial t/\partial u_{c^*})_{\text{eff}} = -0.24$ eV Å, as shown in Table III. This is a little larger than the value of $(\partial t/\partial u)$ for the LA mode, which seems reasonable in light of the fact that it is supposed to represent all $1p$ scattering other than that due to LA phonons. The fact that this value is reasonable gives support to the thesis that $1p$ scattering due to modulation of the transfer integral, plus internal mode scattering, can account for the entire resistance of TTF-TCNQ above 150 K, where charge-density fluctuations are not expected to have any effect. A reasonable value for $(\partial t/\partial u_{c^*})_{\text{eff}}$ does not prove this thesis, however. In particular, it does not eliminate the possibility that part of the $1p$ scattering is due to polarization fluctuations. As shown in Section 6 τ_{1p} is a constant times τ_{1t}. Thus it is

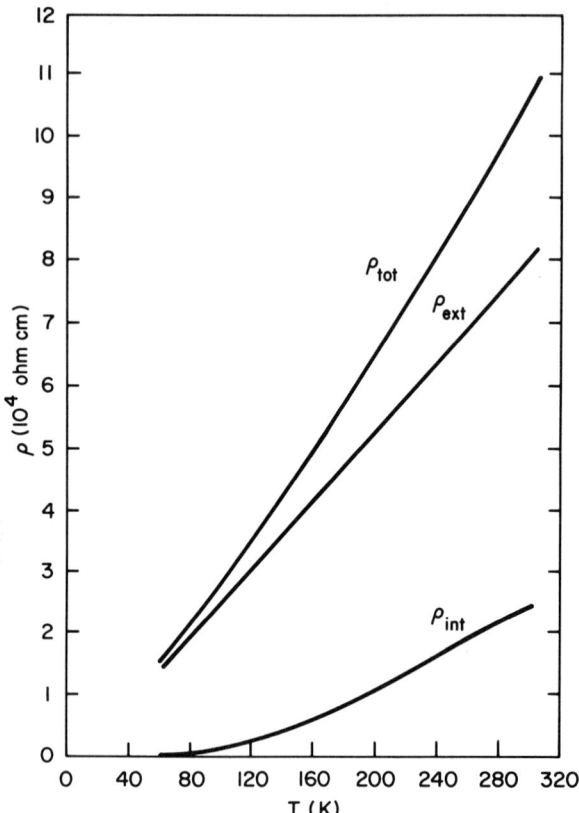

FIG. 4. Calculated resistivity at constant volume ρ_v for $1p$ processes with the parameters of Tables I and III and breakdown into external- and internal-mode scattering. [From Conwell (1980b).]

not possible to separate these two sources of scattering; part of the scattering attributed to $(\partial t/\partial u_{c^*})_{\text{eff}}$ may be due instead to polarization fluctuations.

In Fig. 4 the contributions to the total resistivity of internal and external modes are plotted separately. Because it is ρ at constant volume, ρ_v, that is being plotted, ρ due to the external modes varies linearly with T, except for the lowest temperatures, where departure from equipartition, for the LA phonons primarily, is not negligible. The variation of n_q with T accounts, of course, for the strong temperature dependence of ρ due to the internal modes. The total resistivity ρ_v varies quasilinearly with T, the variation being close to linear at high temperatures and getting steeper at low temperatures, qualitatively in agreement with the deductions of Friend et al. (1978) and Cooper (1979). If ρ_v is characterized as proportional to T^λ, the average λ

4. TRANSPORT IN QUASI-ONE-DIMENSIONAL CONDUCTORS

for the plot in Fig. 4 is 1.23, while that deduced from experiment by Friend *et al.* (1978) is 1.29. This is reasonable agreement, particularly in view of the fact that the deduction of λ from experiment is difficult to do correctly.

It is of interest to examine how the calculated conductivity breaks up into TCNQ and TTF contributions. For the results of Fig. 4 it is found that the ratio of TCNQ to TTF contributions is approximately 4:1 over the entire temperature range. It increases at the higher temperatures because of the strong coupling of TTF to the internal mode at 683 K. This is in good agreement with the deduction from thermopower and Hall measurements, from which a ratio of ~5:1 is deduced for 300 K (Chaikin, 1980). In these calculations the ratio 4:1 is due to the 2:1 ratio assumed for the bandwidths, because as seen in Section 8, $\sigma \propto \varepsilon_0^2$ for all the processes studied. A 4:1 ratio of conductivities could arise also from the deformation potentials for TTF being twice as large as for TCNQ. It is found, in fact, that both σ and Q can be fit equally well with $\varepsilon_0(\text{TTF}) = 4500$, instead of 3000, with correspondingly larger values of the deformation potentials. However, such theoretical estimates as there are have the deformation potential constants smaller for TTF than TCNQ (Berlinsky *et al.*, 1974), lending support to the smaller value of $\varepsilon_0(\text{TTF})$.

From the separate conductivities for TTF and TCNQ it is possible to deduce the scattering times and mean free paths for carriers on the two chains. With the room temperature transfer taken as 0.5, on the basis of the discussion earlier in this section, the calculated mobilities for TCNQ and TTF at 300 K are 4 cm^2/V sec and 1 cm^2/V sec, respectively. A bandwidth of 0.5 eV for TCNQ then leads, with the use of Eq. (74), to $\tau(\varepsilon_F) = 5 \times 10^{-15}$ sec and, with $v_F = 1.2 \times 10^7$ cm/sec, a mean free path ℓ at 300 K of 1.6 lattice constants. It is interesting to note that the effective mass at ε_F is twice the free-electron mass. For TTF at 300 K with $\varepsilon_0 = 0.25$ eV, τ is half as large, v_F half as large, and ℓ a quarter as large, or ~one-half a lattice constant. Certainly the use of simple first-order perturbation theory and the Boltzmann equation to calculate σ is not well justified for TTF at 300 K. Also, there are the complications of level shifts and the resultant band distortions produced by the self energy (Conwell, 1978), which are expected to be significant for TTF at 300 K because of the small bandwidth. However, with decreasing temperature τ and ℓ grow as $T^{2.3}$, the bandwidth increases, decreasing the effect of U, and the self-energy effects decrease, so the justification for using first-order perturbation theory and the Boltzmann equation exists at lower temperatures. In that sense, use of the foregoing theory at room temperature constitutes an extrapolation, but one that can hardly cause much error in the calculated σ since σ of TTF is considerably smaller than σ of TCNQ.

Calculations for the thermopower of TTF–TCNQ are more speculative than those for conductivity for a couple of reasons. First, because Q_n has

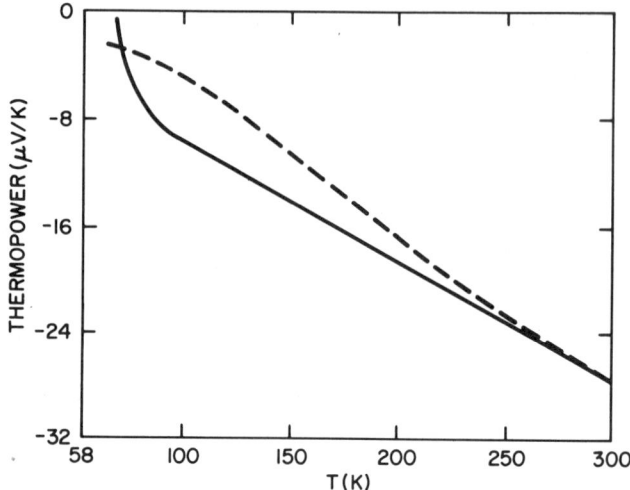

FIG. 5. Measured thermopower (solid line, taken from Chaikin et al., 1973) and calculated thermopower (dashed line) vs. temperature. [From Conwell (1980b).]

the opposite sign to Q_p, Q involves a *difference* of transport coefficients (in this case the product σQ) for the two chains, rather than a sum, as does the conductivity. Accurate knowledge of the properties of the TTF chain is therefore much more important in the calculation of Q than in that of σ. Second, as was shown in Section 8, the dispersion of the libron branches is important in determining their contribution to Q; this dispersion is, of course, unknown. Despite these difficulties, it was possible to obtain a fairly good fit, shown in Fig. 5, to the experimental Q vs. T (Chaikin et al., 1973) with the parameters of Tables I–III and the assumption that all the scattering is due to translons. A significant feature of this fit was that a 4:1 ratio of TCNQ to TTF conductivity, consistent with a 2:1 ratio of bandwidths, appears to be required.

f. Calculations for TSeF–TCNQ

Similar calculations were carried out for TSeF–TCNQ, which has the same type of lattice structure as TTF–TCNQ and somewhat larger conductivity. As expected, because the Se atom is larger than S (Etemad et al., 1978), TSeF–TCNQ has a larger spacing along the chains than TTF–TCNQ. Thus the overlap integral and, consequently, ε_0 must be smaller for the TCNQ chain in TSeF–TCNQ than in TTF–TCNQ. With $\sigma \propto \varepsilon_0^2$ for all the relevant scattering processes, as shown in Section 8, the conductivity of the TCNQ chain must be smaller in TSeF–TCNQ than in TTF–TCNQ. The larger overall conductivity of the former must then be due to larger ε_0 for TSeF than for TTF. Larger overlap is expected because of the larger size of Se; it is also suggested by the smaller magnetic susceptibility of TSeF-TCNQ compared to TTF-TCNQ (Scott et al., 1978; Buravov et al., 1976). The conductivity of TSeF-TCNQ increases with decreasing T about as T^2, somewhat less steeply than that of TTF-TCNQ. So far as transport is concerned, the big difference between TTF- and TSeF-TCNQ is that the thermoelectric power for the latter is small and positive in the metallic range. A small positive Q is maintained until close to the Peierls transition temperature 29 K, where Q drops precipitously. The fact that $Q > 0$ indicates that the donor chain, i.e., TSeF, dominates transport rather than the TCNQ chain, as in TTF-TCNQ. This is corroborated by the data in Fig. 6 for the alloys $(TTF)_{1-x}(TSeF)_x$ TCNQ. Substitution of a few % S for Se in TSeF-TCNQ, which should increase scattering and thus decrease σ on the TSeF chain, is seen to make Q go negative at room temperature (Chaikin et al., 1976). The positive thermoelectric power of unsubstituted TSeF-TCNQ suggests that ε_0 in that compound is larger for TSeF than for TCNQ. This is expected from our previous discussion. For a two-chain conductor $Q > 0$ indicates, according to Eq. (71) that $Q_p\sigma_p > Q_n\sigma_n$. With $\sigma \propto \varepsilon_0^2$ and Q decreasing with ε_0 at most as ε_0^{-1}, according to Section 8, ε_0 must indeed be larger for TSeF than for TCNQ. This is in agreement with band structure calculations (Herman et al., 1977). In light of the uncertainty concerning exact values of bandwidths and coupling constants and even in the value of the room temperature conductivity, quantitative calculations of σ and Q for TSeF-TCNQ are not too meaningful. It is possible to conclude that fits to $\sigma(T)$ and $Q(T)$ in the

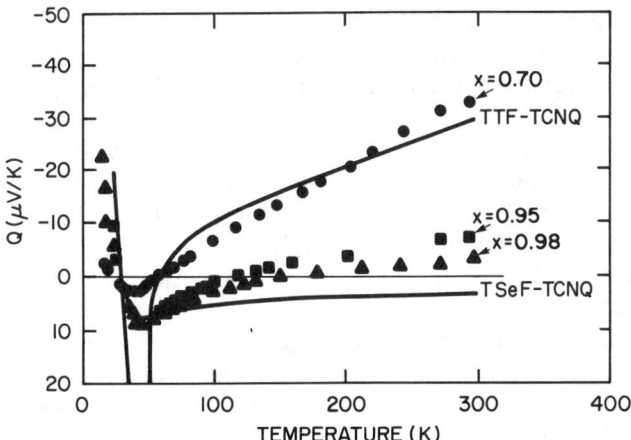

FIG. 6. Thermoelectric power of the alloys $(TTF)_{1-x}(TSeF)_x TCNQ$ for various values of x as a function of temperature. [From Chaikin et al. (1976).]

metallic range can be obtained with reasonable values for the various parameters. The only quantitative conclusion that can be drawn is that if coupling to the acoustic modes is stronger for TSeF than for TTF, as has been suggested by Schultz (1977), the bandwidth for TSeF must be substantially greater than that for TCNQ, probably 20% or more.

g. Pressure Dependence of Conductivity

In addition to fitting the magnitude and T-dependence, any theory of σ must account for the large pressure-dependence, 25–28%/kbar for TTF-TCNQ. (Jérome et al., 1974; Cooper et al., 1975). One quantity with a significant pressure dependence is ε_0. Making use of the measured b-axis compressibility, 4.7×10^{-3}/kbar (Debray et al., 1977) and $b = 3.819$ Å, we find that $\partial \varepsilon_0 \partial u_b = -0.8$ eV/Å translates into a pressure dependence $\partial \varepsilon_0 / \partial P = 0.015$ eV/kbar, or $(1/\varepsilon_0)(\partial \varepsilon_0 / \partial P) = 3\%$/kbar for $\varepsilon_0 = 0.5$ eV. This is in good agreement with the deductions of Cooper (1979) and Jérome et al., (1977). Also in agreement with this, the observed variation of charge transfer with pressure at very low temperatures (Megtert et al., 1981), where the compressibility is known to be smaller by a factor 2 than at 300 K, has been accounted for with $(1/\varepsilon_0)(\partial \varepsilon_0 / \partial P)$ in the range 1% to 1.5% (Conwell, 1980a). Thus, at room temperature, taking into account the fact that σ is proportional to ε_0^2, we find that the pressure dependence of ε_0 leads to σ increasing at 6%/kbar. External phonon frequencies vary considerably with pressure in organic crystals, typically 4% to 6%/kbar in anthracene, naphthalene (Dows et al., 1973), and pyrene (Zallen et al., 1976), with the larger percentage applying to lower frequencies. It has been argued that TTF-TCNQ should not be compared to these materials because of the charge transfer and its metallic nature. However, the elastic and anharmonic properties exhibited by TTF-TCNQ are entirely usual for a molecular crystal. Its compressibility and expansivity, and their anisotropies, are quite comparable to those of anthracene, naphthalene, and pyrene (Zallen and Conwell, 1979).

Further, according to neutron scattering data (Shapiro et al., 1977), for $q \simeq 2k_F$, ω of the TA(a) phonon increases by ~20% as T decreases from 295 to 84 K. With the known compressibility, this temperature decrease is equivalent to ~5 kbars of pressure. Thus ω for TA(a) increases ~4%/kbar, and it is reasonable to expect such behavior at room temperature for other external modes with comparable frequencies. For $1p$ scattering, where $\sigma \propto \omega^2$, this gives an increase of ~8%/kbar. Additional small contributions from the charge transfer, b, and internal-mode frequencies add up to a few %/kbar. Thus one can account quite well for the 18%/kbar observed for TSeF-TCNQ. The origin of the additional 7% to 10% observed for TTF-TCNQ is not clear. The difference could well arise from the smaller bandwidth of TTF than TSeF. Evidence that pressure decreases the ratio $\sigma_{TCNQ}/\sigma_{TTF}$ comes from the rapid decrease of the Hall constant under pressure (Cooper *et al.*, 1977) as well as the large decrease of $|Q|$ under pressure. (Chaikin, 1980). Note that for the same value of $\partial \varepsilon_0/\partial u$ the percentage changes in $\varepsilon_0(TTF)$ and σ_{TTF} are larger because $\varepsilon_0(TTF) < \varepsilon_0(TCNQ)$. One may speculate that the increase in ε_0 of TTF under pressure has an enhanced effect on its contribution to σ due to increasing coherence of transport on this chain, or due to decrease in the effect of U, which, as emphasized earlier, may not be negligible in such a narrow band.

10. Transverse Conductivity in the Metallic Range

Conductivity transverse to the chain direction is, as indicated in Chapter 1, much smaller than that parallel to the chains. In Fig. 7 data for conductivity vs. temperature are shown for the a direction of TTF–TCNQ, the direction where transport is across unlike chains. It is seen that the ratio of conductivity along the b (chain) direction to that along the a direction is $\sim 10^3$ at 300 K and goes up to $\sim 10^4$ near 60 K (Khanna *et al.*, 1974). For the c^* direction the conductivity is intermediate between that for a and b. Strong anisotropy is also found in the thermoelectric power. As shown in Fig. 5 of Chapter 1, the thermoelectric power in the a direction has opposite sign to that in the b direction and quite different temperature-dependence.

It is clear from the small values of σ that the transverse conductivity is due to hopping. Little effort has been expended in studying this hopping process. On the assumption that the ratio of the observed conductivities perpendicular and parallel to the chains equals the ratio of bandwidths perpendicular and

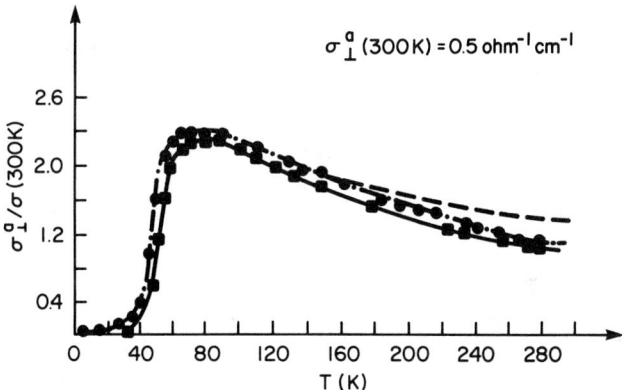

Fig. 7. Temperature dependence of the conductivity ($\sigma_\perp^q$) along the a axis: ●, 10.4 GHz; ■, dc. The dashed curve shows a $T^{-1/2}$ dependence for comparison. [From Khanna *et al.* (1974b).]

parallel to the chains, Khanna *et al.* (1974) obtain a transverse bandwidth for the *a* direction of ~10 K. Soda *et al.* (1976, 1977) estimate, from the interpretation of their NMR data, that the transverse bandwidth is ~60 K.

It is interesting to note that the compound TMTTF–DMTCNQ, which one would expect to have properties similar to TTF–TCNQ, does not have coherent single-electron transport along the chains above ~100 K. Studies of conductivity and thermopower indicate that conduction above ~100 K is by hopping (Jacobsen *et al.*, 1978). The main reason for this departure is that the intermolecular distance along the chains is the largest in this family of compounds, making the bandwidth very small. A second reason suggested is that the nonsymmetric overlap of adjacent molecules in this compound allows stronger first-order libron scattering of the electrons than occurs in the other compounds of this family (Jacobsen *et al.*, 1978).

II. The Peierls Transition: Implications for Transport

11. Wavefunctions, Energies, and the Gap Equation

To study transport below the Peierls transition temperature T_P we need wavefunctions, energy as a function of wavevector, effective masses, densities of states, etc. for the electrons and holes. Wavefunctions and energies in the Peierls-distorted state have been determined for the half-filled tight-banding band, starting from the Hamiltonian Eq. (9) in the site representation, with an additional term for the elastic energy due to nearest-neighbor displacements (Su *et al.*, 1980). As noted earlier, however, there is no case of a half-filled band in the family of materials with which we are concerned; the theory must therefore be generalized to the case of arbitrary band filling. Also, it must be extended to take into account the contributions of other modes–librons and internal vibrations—to the Peierls gap. In this section we carry out the solution to the Hamiltonian, including these other modes, for a non-half-filled band, obtaining wavefunctions and energy as a function of k. We also derive equations for the gap at $T = 0$ and for the T-variation of the gap in terms of t_0, k_F and the parameters of the lattice vibrations. From the energy as a function of k we obtain, in the next section, effective masses and densities of states for the Peierls-distorted band. In the two following sections we will show that the Peierls transition does not affect the matrix element for acoustic scattering and calculate the relaxation time and mobility for acoustic mode scattering below T_P. Another mechanism that could affect the mobility below T_P is ionized impurity scattering. In Section 15 we calculate the relaxation time for this scattering mechanism. Comparison of the theory of these sections with experimental data for the semiconducting range will be carried out in Part III.

As discussed in Chapter 1, the effect of the Peierls transition is to freeze in, with period $2k_F$, distortions due to one or more lattice modes. Denoting a particular mode by the superscript (i), we may take the frozen-in distortions for the acoustic modes and librations, respectively, as

$$u_j^{(i)} = A_{0t}^{(i)} \cos(Qx_j)$$
$$\theta_j^{(i)} = A_{0\ell}^{(i)} \cos(Qx_j), \tag{91}$$

where the A_0's are amplitudes to be determined to give minimum free energy, $Q \equiv 2k_F$ and x_j is the position of the jth molecule in the undistorted lattice. Similarly, for a frozen-in internal mode with normal coordinate Q_n we may write

$$Q_n(j) = A_n \cos(Qx_j), \tag{92}$$

the A_n being amplitudes to be determined also by the condition of minimum free energy. Actually, as has been pointed out by many authors, for an incommensurate case terms with harmonics of Q should be included in Eqs. (91) and (92). It has been shown, however, that for Q either at or well away from commensurability the harmonic components of the lattice displacement are far smaller than the fundamental (Kotani, 1977) and we shall neglect them. Near-commensurate cases require special considerations, such as solitons, which are not relevant to the material of this chapter.

Before using these equations to write the Hamiltonian Eq. (9) in the momentum representation, we note that there is an important difference between the half-filled-band case and all others—there is more than one gap in the other cases. For a tight-binding Hamiltonian symmetric under nearest neighbor interchange, as is Eq. (9), it has been shown that the density of states must have symmetry about the center of the band (Rice and Mele, 1981; Vanderbilt and Mele, 1980). Thus when there is a gap at $\pm k_F$, there is also one at $\pm(\pi/b - k_F)$, where b is the lattice spacing. For the materials of interest here, the gaps are separated by a sizable fraction of the bandwidth, $4t_0$, as a result of the band filling being $\sim \frac{1}{4}$ or $\frac{1}{3}$ and the gaps (2Δ) quite small compared to $4t_0$. Since, as will be seen, the wavefunctions are altered significantly only within an energy range of a few Δ about the gap, it is a good approximation for these materials to deal with only one gap, the one separating filled and empty subbands. With this in mind, we transform the electronic terms of the Hamiltonian Eq. (9) to the momentum representation, using Eqs. (91), (92), (10), (22), and the Fourier transforms Eq. (12), of a_j and a_j^+. The effect of the frozen-in $2k_F$ distortion is to mix the states with k and $k - 2k_F$, giving (Conwell and Banik, 1983)

$$H_{el} = 2 \sum_{0 \leq k \leq k_F} [\varepsilon_k a_k^+ a_k + \varepsilon_{k-Q} a_{k-Q}^+ a_{k-Q} + (\Delta_k' a_k^+ a_{k-Q} + \text{H.c.})] \tag{93}$$

where ε_k, the energy in the undistorted band measured from the Fermi energy (not equal to ε_j in the present case) is given by

$$\varepsilon_k = 2t_0[\cos(k_F b) - \cos(kb)]. \tag{94}$$

The gap parameter Δ'_k consists of two parts

$$\Delta'_k = \Delta' + i\Delta_k, \tag{95}$$

Δ_k incorporating the effects of translational and librational mode distortions and Δ' the effects of internal modes. From Eqs. (10) and (91)

$$\Delta_k = 4\alpha A_0 \sin(k_F b) \cos[(k - k_F)b], \tag{96}$$

where

$$\alpha A_0 = \sum_{i=1}^{3} \left[\left(\frac{\partial t}{\partial u^{(i)}}\right)_0 A_{0\ell}^{(i)} + \left(\frac{\partial t}{\partial \theta^{(i)}}\right)_0 A_{0\ell}^{(i)} \right]. \tag{97}$$

It is seen that for the case of a half-filled band ($k_F = \pi/2b$) and one acoustic mode only Δ_k goes over to the gap parameter obtained by Su et al. (1980) for this case. When there is mode mixing, αA_0 represents the sum over the mixed modes. The quantity Δ' is given by

$$\Delta' = \sum_{n=1}^{G_i} \frac{\beta_n A_n}{2} = \sum_{n=1}^{G_i} \Delta'_n, \tag{98}$$

the sum being over the contributions of the G_i internal modes coupled linearly to the electrons. The fact that the contributions to the Peierls gap of the external and internal modes are $\pi/2$ out of phase, as seen in Eq. (95), was first pointed out by Rice. As a result the gap parameter is the square root of the sum of Δ_k^2 and $(\Delta')^2$.

We have made the approximation in Eq. (93) of replacing the upper limit of the k summation for the upper subband, $\pi/b - k_F$, by $2k_F$. This is reasonable since our treatment is limited to the case where the widths of the subbands are large compared to $k_B T$, and we will be concerned only with Maxwell–Boltzmann distributions of electrons and holes.

The above derivation has been for the case of the Peierls distortion, or charge-density wave (CDW) case. The electronic part of the Hamiltonian for the spin-density wave (SDW) case may be written in a form similar to Eq. (93) with the gap parameter Δ'_k replaced by (Machida, 1981)

$$\Delta_{\text{SDW}} = -\sigma M, \tag{99a}$$

where σ is the spin and M the sublattice magnetization of the SDW state, given by

$$M = \left(\frac{I}{2}\right) \sum_k \sum_\sigma \sigma \langle a_{k\sigma}^+ a_{k-Q,\sigma} \rangle, \tag{99b}$$

I being the exchange integral.

To find the wave functions and energies in the CDW or SDW state we diagonalize H_{el} of Eq. (93) by introducing new operators according to

$$a_k = u_k \alpha_1 + v_k \alpha_2 \tag{100a}$$

$$a_{k-Q} = v_k \alpha_1 - u_k \alpha_2, \tag{100b}$$

where α_1, α_1^+ and α_2, α_2^+ operate on the upper and lower subbands, respectively. The coefficients are required to obey the normalization condition

$$u_k u_k^* + v_k v_k^* = 1. \tag{101}$$

With Eq. (100) in Eq. (93) we find the eigenvalues for both CDW and SDW, measured from the Fermi energy in the metallic state, to be

$$E_{k,i}^\pm = \frac{\varepsilon_k + \varepsilon_{k-Q}}{2} \pm \left\{ \left[\frac{(\varepsilon_k - \varepsilon_{k-Q})^2}{4} \right] + |\Delta_i|^2 \right\}^{1/2}, \tag{102}$$

where + refers to the upper band, − to the lower, and i refers to either CDW or SDW. For $i \to$ CDW

$$\Delta_i = \Delta_k', \tag{103}$$

the gap parameter defined in Eq. (95), while for $i \to$ SDW

$$\Delta_i = -\sigma M. \tag{104}$$

Equation (102) was obtained previously (Boriack and Overhauser, 1977) for a free electron model with $|\Delta_k'|$ replaced by Δ, a constant to be determined empirically or by the condition that the motion of the lattice ions must satisfy Newton's second law. For the half-filled band ($Q = \pi/b$) $\varepsilon_k = -\varepsilon_{k-Q}$ and, for acoustic modes only, $E_k^\pm$ goes over to $\pm(\varepsilon_k^2 + \Delta_k^2)^{1/2}$, the result of Su et al. (1980) and earlier derivations.

The relation between u_k and v_k results from diagonalizing H_{el} subject to the normalization condition (101). For the upper band that relation is

$$\frac{u_1}{v_1} = \frac{\Delta_i}{E_k^+ - \varepsilon_k}, \tag{105}$$

where the subscript k on u and v has been dropped for conciseness.

Combining Eqs. (105) with (102) we obtain

$$u_1 u_1^* = \frac{E_k^+ - \varepsilon_{k-Q}}{2E_k^+ - \varepsilon_k - \varepsilon_{k-Q}} \tag{106a}$$

$$v_1 v_1^* = \frac{E_k^+ - \varepsilon_k}{2E_k^+ - \varepsilon_k - \varepsilon_{k-Q}}. \tag{106b}$$

For the lower band $u_2 u_2^*$ is given by the right-hand side of Eq. (106a) and $v_2 v_2^*$ by that of Eq. (106b), with E_k^+ replaced by E_k^- in both cases. We note that, for $Q = \pi/b$ and no internal modes, u_1 and v_1 go over to the coefficients α_k and β_k of Su et al. (1980).

Having the eigenvalues in the Peierls-distorted state we proceed to find the equilibrium value of the gap at $T = 0$ for the CDW case. For this we need an expression for the total energy at that temperature. The energy of the translational part of the lattice defomation is given by (Su et al., 1980)

$$H_{ac} = \frac{1}{2} \sum_{i=1}^{3} \sum_{j=1}^{N} K^{(i)} (u_{j+1} - u_j)^2, \tag{107}$$

where $K^{(i)}$ is an effective spring constant for the distortion due to the ith acoustic mode. Taking $x_j = jb$ we obtain, using Eq. (91),

$$H_{ac} = 2 \sum_{i=1}^{3} K(A_{0t}^{(i)})^2 \sin^2(k_F b) \sum_{j=1}^{N} \sin^2[(2j+1)k_F b]. \tag{108}$$

With the notation $4k_F b = \eta$ we may write

$$\sum_{j=1}^{N} \sin^2[(2j+1)k_F b]$$

$$= \frac{N}{2} \left\{ 1 - \frac{\cos(\eta/2) \cos[(N+1)\eta/2] \sin(N\eta/2)}{N \sin(\eta/2)} \right.$$

$$\left. + \frac{\sin[(N+1)\eta/2] \sin(N\eta/2)}{N} \right\}. \tag{109}$$

The term in curly brackets equals 2 for $\eta = 2\pi$, the half-filled-band case. For η not an integer times 2π, however, the second and third terms in curly brackets are of order $1/N$ and can be neglected. Thus for the non-half-filled-band case the translational mode contribution to the lattice energy is

$$H_{ac} = \sum_{i=1}^{3} NK(A_{0t}^{(i)})^2 \sin^2(k_F b), \quad k_F b \neq \frac{\pi}{2} \tag{110}$$

less for each mode than half of that for the corresponding mode in the half-filled-band case, $2NK(A_{0t}^{(i)})^2$ (Su et al., 1980). An expression similar to

Eq. (110) may be obtained for the librational modes with $A_{0\ell}^{(i)}$ replaced by $A_{0\ell}^{(i)}$ and $K^{(i)}$ by the elastic constant appropriate to the motion concerned.

For the internal modes the contribution to the lattice energy is

$$H_{\text{int}} = \frac{1}{2} \sum_{n=1}^{G_i} \hbar\omega_n A_n^2 \cos^2(Qx_j), \tag{111}$$

where A_n is the dimensionless amplitude introduced in Eq. (92). By arguments similar to those used for the acoustic energy, we obtain for the band of arbitrary filling

$$H_{\text{int}} = \frac{N}{4} \sum_{n=1}^{G_i} \hbar\omega_n A_n^2, \qquad k_F b \neq \frac{\pi}{2}. \tag{112}$$

For the half-filled band H_{int} is twice that given by Eq. (112). With the use of Eq. (98), Eq. (112) may be written

$$H_{\text{int}} = N \sum_{n=1}^{G_i} \frac{(\Delta_n')^2}{\beta_n^2/\hbar\omega_n}, \qquad k_F b \neq \frac{\pi}{2}, \tag{113}$$

where Δ_n' is the contribution to Δ' of the nth internal mode. This equation may be written in terms of Δ' by using a relation derived by Rice et al., (1975)

$$\frac{\Delta_n'}{\beta_n^2/\hbar\omega_n} = \sum_{n=1}^{G_i} \Delta_n' \bigg/ \sum_{n=1}^{G_i} \frac{\beta_n^2}{\hbar\omega_n} \equiv \frac{\Delta'}{4E_p}, \tag{114}$$

where E_p is the so-called polaron binding energy. With the relation (114) Eq. (113) may be written

$$H_{\text{int}} = \frac{(N/4)(\Delta')^2}{E_p}, \qquad k_F b \neq \frac{\pi}{2}. \tag{115}$$

Thus, for the case of internal modes, since there is no k dependence of the gap parameter, it is possible to express the contribution to the lattice energy in terms of the gap Δ'.

To calculate the gap it is necessary to minimize the total free energy—the sum of the acoustic contribution Eq. (110), a term similar to Eq. (110) for the librons, the internal mode contribution Eq. (115), and the electronic contribution, obtained for $T = 0$ by summing Eq. (102) over the states in the lower band. Because there are two independent parameters, Δ' and Δ_k in the CDW case, the free energy must be minimized with respect to both. The procedure was carried out by Kivelson (1983) with the interesting result that, unless the electron–phonon coupling is strong, the gap is determined either by external mode distortion (bond alternation) or by internal mode distortion (site alternation). The latter situation must hold for the materials we are interested in because the gaps are not large and the coupling to internal modes is larger than that to external modes (Rice et al., 1975). Thus the gap may be determined by minimizing the total free energy with respect to Δ'. The resulting gap equation is

$$\frac{2N\Delta'}{4E_p} - \frac{2N}{\pi}\int_0^{k_Fb} d(kb)\Delta'\left[\frac{(\varepsilon_k - \varepsilon_{k-Q})^2}{4} + (\Delta')^2\right]^{-1/2} = 0, \quad (116)$$

the first term representing the derivative of the internal energy, the second that of the electronic energy, both with respect to Δ'. With the use of Eq. (94), the gap equation may be written (Conwell and Banik, 1983)

$$\frac{\pi t_0 \sin(k_F b)}{2E_p} = a\int_{\pi/2-k_Fb}^{\pi/2} d(kb)[1 - a^2 \sin^2(kb)]^{-1/2}, \quad k_Fb \neq \frac{\pi}{2}, \quad (117)$$

where

$$a^2 = \left[1 + \frac{(\Delta')^2}{4t_0^2 \sin^2(k_Fb)}\right]^{-1}. \quad (118)$$

This can be expressed in terms of complete [$K(m_I)$] and incomplete elliptic integrals of the first kind. For the half-filled band the right-hand side of Eq. (117) becomes $aK(a)$ while the left-hand side must be multiplied by a factor of 2 due to the larger lattice energy of the half-filled band. The result is then in agreement with the gap equation obtained earlier by Rice for the half-filled-band case with internal modes only (Conwell et al., 1978).

$$\frac{\pi t_0}{E_p} = aK(a), \quad k_Fb = \frac{\pi}{2}. \quad (119)$$

Because $\Delta' \ll 2t_0$, a is very close to unity and

$$K(a) \simeq \ln\left[\frac{4}{(1-a^2)^{1/2}}\right] \simeq \log e \frac{8t_0}{\Delta'}, \quad k_Fb = \frac{\pi}{2}. \quad (120)$$

Combining Eqs. (119) and (120) we obtain for the half gap at $T = 0$ in the half-filled band case with internal mode scattering dominant,

$$\Delta'_0 = 8t_0 e^{-1/\lambda_i}, \quad k_Fb = \frac{\pi}{2}, \quad (121a)$$

where

$$\lambda_i = \frac{E_p}{\pi t_0}, \quad (121b)$$

the dimensionless coupling constant to the internal modes.

For the case where acoustic modes dominate, the results of the gap calculation are a little different because the gap parameter, given by Eq. (96), is a function of k. For the half-filled band, with the half-gap Δ taken as the value of Δ_k for $k = k_F = \pi/2b$, minimization of the free energy leads to (Su et al., 1980)

$$\Delta_0 = \frac{8t_0}{e} e^{-1/2\lambda_{ac}}, \quad k_F b = \frac{\pi}{2}, \quad (122a)$$

where the dimensionless coupling constant to the acoustic modes is

$$\lambda_{ac} = \frac{2\alpha^2}{\pi K t_0}. \quad (122b)$$

The equation for the temperature dependence of the gap may also be determined by minizing the free energy. For this case, however, the electronic free energy may no longer be taken as the integral of E_k^- over the states in the lower band; for $T > 0$ thermal energy will allow some electrons to cross the gap. The free energy for the electron gas is then given by (Landau and Lifshitz, 1959)

$$H_e = -k_B T \sum_k \ln(1 + e^{-E_k/k_B T}), \quad (123)$$

where E_k is measured relative to the Fermi energy and the summation over k is taken over both bands. In order to obtain a gap equation that is valid for all types of lattice distortion, we take the lattice free energy as

$$H_{\ell at} = NN(0) \frac{\Delta^2}{2\lambda} \quad (124)$$

where

$$N(0) = \frac{1}{\pi t_0 \sin k_F b}, \quad (125)$$

the density-of-states at the Fermi energy in the metallic state, Δ is the half gap, and λ is the coupling constant defined appropriately for acoustic mode or internal mode distortion. Equating to zero the derivative of the total free energy with respect to Δ we obtain the gap equation for $T \neq 0$

$$\frac{NN(0)}{\lambda} = \sum_k \frac{f(E_k^-)}{E_k^- - (\varepsilon_k + \varepsilon_{k-Q})/2} - \sum_k \frac{f(E_k^+)}{E_k^+ - (\varepsilon_k + \varepsilon_{k-Q})/2}, \quad (126)$$

where the first summation is taken over the lower band and the second over the upper band. Changing the summations to integrations and using Eqs. (102) and (94), we obtain the gap equation in the form

$$\frac{1}{a\lambda} = \int_0^{k_F b} \frac{f(E_k^-) \, dkb}{[1 - a^2 \cos^2(k - k_F)b]^{1/2}} - \int_{k_F b}^{\pi/2 - k_F b} \frac{f(E_k^+) \, dkb}{[1 - a^2 \cos^2(k - k_F)b]^{1/2}} \quad (127)$$

where a is defined in Eq. (118). It will be recognized that the $T = 0$ gap equation Eq. (117) may be recovered from Eq. (127) by setting $f(E_k^-) = 1$, $f(E_k^-) = 0$, changing the variable to $k'b$ where $\sin k'b = \cos(k - k_F)b$ and taking $\lambda = 2E_p/\pi t_0 \sin k_F b$. Equation (127) applies to both charge-density and spin-density wave transitions.

Some interesting deductions can be made from the gap equation. First it should be noted that, because $a \simeq 1$, the major contributions to the integrals in Eq. (127) comes from $k \simeq k_F$, i.e., from levels close to the gap. It can then be seen that Δ will decrease with increasing temperature, slowly as T rises from 0 but more rapidly as T approaches T_P, in agreement with the experimental variation seen in Fig. 10 of Chapter 1, for example. This is the case because, λ in Eq. (127) being fixed, as T goes above 0, with a resulting decrease in $f(E_k^-)$ and increase in $f(E_k^+)$, the equation can only be satisfied if the parameter a increases, requiring a decrease in Δ. When the temperature is high enough so that Δ has decreased perceptibly, this will enhance the rate of decrease of $f(E_k^-)$ and increase of $f(E_k^+)$ with further increase of T, increasing the rate of decrease of Δ and thus making the demise of the gap very rapid as T approaches T_P. Another deduction that can be made from the gap equation is that a nonequilibrium carrier distribution will affect the gap. Injection of excess electrons and holes, by contacts or photogeneration, for example, will decrease $f(E_k^-)$ and increase $f(E_k^+)$ with the result that, as in a superconductor, Δ must decrease to satisfy Eq. (127) (Berggren and Huberman, 1978). Heating the carriers by radiation or by an electric field will have the opposite effect on the distribution functions, tending to move electrons and holes away from their respective band edges, and therefore will tend to increase the gap. However, carrier heating would be expected to have less of an effect than injection precisely because it takes carriers away from the band edges.

The effect of injection on the gap can be quite large. Quantitative results have been obtained by following the method used for superconductors (Owen and Scalapino, 1972). In this method the Fermi energy ε_F in Eq. (127) is replaced by quasi-Fermi energies ε_F^* for electrons and holes. The relation between ε_F^* and excess carrier concentration is readily obtained with the use of the density of states calculated from E_k of Eqs. (102) and (133). To obtain equal electron and hole injection for the non-half-filled band case it would be necessary to use different ε_F^*'s for electrons and holes, but the difference is small and it was neglected. The injection is conveniently described by means of a dimensionless excess carrier concentration η defined by

$$\eta = \frac{n(\varepsilon_F^*) - n(0)}{N(0)\Delta_0}, \qquad (128)$$

where $n(0)$ is the carrier concentration for $\varepsilon_F = 0$, i.e., the thermal equilibrium carrier concentration, $N(0)$ is the density of states at ε_F, given by Eq. (125), and Δ_0 is the thermal equilibrium gap at $T = 0$. An approximate analytic relation between Δ and η at $T = 0$ can be obtained by neglecting the term in $E_k^{\pm}$ [Eq. (102)] outside the radical. This term is small for states near the gap which, as noted earlier, make the maximum contribution to the integrals. With this term neglected, Eq. (127) may be integrated for $T = 0$ with the result (Conwell and Banik, 1982b)

$$\frac{\Delta}{\Delta_0} \simeq \left[\frac{2(\Delta_0^2\eta^2 + \Delta^2)^{1/2} - \Delta_0\eta}{2(\Delta_0^2\eta^2 + \Delta^2)^{1/2} + \Delta_0\eta}\right]^2. \quad (129)$$

In the limit of small injection Eq. (129), combined with Eq. (128) gives

$$\Delta = \Delta_0 - \frac{2}{N(0)}[n(\varepsilon_F^*) - n(0)]. \quad (130)$$

Thus the gap decreases linearly with small injection, as was found also for the case of the superconductor (Owen and Scalapino, 1972). Inserting In Eq. (130) the parameters for TTF-TCNQ and $(TMTSF)_2PF_6$, we find that an injection of $\sim 10^4$ carriers/cm ($10^{18}/cm^3$) is required to have a sizable effect on the gaps of these materials. This is very much higher than the thermal equilibrium concentrations at low temperatures.

For large injection and $T > 0$, it is necessary to integrate Eq. (127) numerically. Some results are shown in Figs. 8 and 9 for $(TMTSF)_2PF_6$. Similar results were obtained for TTF-TCNQ. The parameters used were $\varepsilon_0 = 1$ eV, $\sin k_F b = 0.707$, $b = 3.65 \times 10^{-8}$ cm. The value of λ was obtained by requiring that integration of Eq. (127) give the observed half gap Δ_0 at $T = 0$. Two different values of Δ_0 were used in Fig. 8 because experimental data do not give Δ_0 unambiguously, as will be discussed in Section 16 of this chapter. It is seen from Fig. 8 that Δ/Δ_0 depends somewhat on Δ_0 as well as η. Beyond the values for which the curves are terminated, both in Figs 8 and 9, Δ/Δ_0 is found to be double valued. The solutions for the lower gap are not physically significant, however, as discussed by Berggren and Huberman (1978) and Owens and Scalapino (1972), because they correspond to free energies higher than those in the metallic state. Thus, for $\eta = 0.19$, for example, the semiconducting state is made unstable for T less than $\frac{1}{2}$ the transition temperature. We will return to a discussion of these effects in Section 17.

12. Effective Masses and Densities of States Below T_P

Below the transition temperature the electrons and holes responsible for transport are in the neighborhood of the band edges. For k close to k_F,

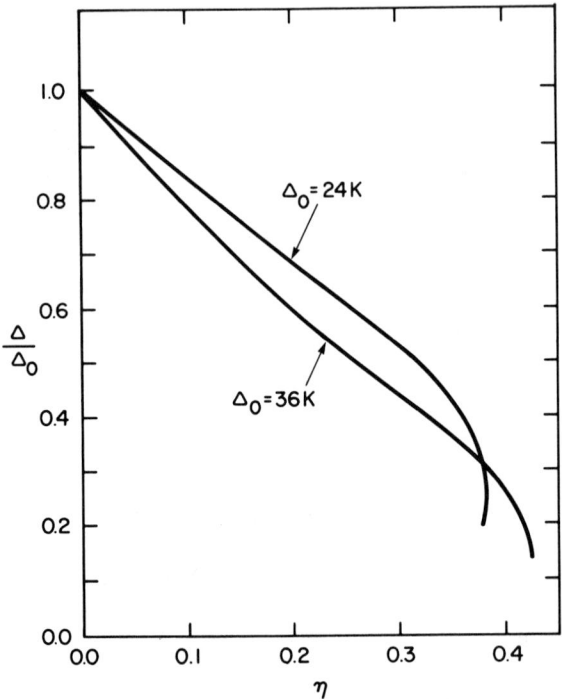

FIG. 8. Variation of Δ/Δ_0 with injection η at $T = 0$ for (TMTSF)$_2$PF$_6$ with two possible values of Δ_0. [From Conwell and Banik (1982b).]

$|\Delta_i|^2 \gg (\varepsilon_k - \varepsilon_{k-Q})^2/4$ and the quantity inside the radical in Eq. (102) may be expanded. To terms quadratic in $(k - k_F)^2$ the expansion gives

$$E_k = \pm \left[\Delta + \frac{\hbar^2(k - k_F)^2}{2m^*} \right], \tag{131}$$

where

$$m^* = \frac{\hbar^2}{2tb^2} \left[\frac{2t}{\Delta} \sin^2(k_F b) \pm \cos(k_F b) \right]^{-1}, \tag{132}$$

the effective mass of an electron or hole. The + sign in Eq. (132) gives m_n^*, the effective mass of an electron at the edge of the upper subband, the −sign m_p^*, the effective mass of a hole at the edge of the lower subband. The subscript i on Δ has been dropped for conciseness of notation. The first factor on the right of Eq. (132) will be recognized as m^* at the bottom of the original undistorted band. Since $\Delta/2t \ll 1$, typically ≤ 0.1, m^* at the edge of either subband is much smaller than the original mass. If the band is less than half filled, as is the case for TTF-TCNQ and its close relatives,

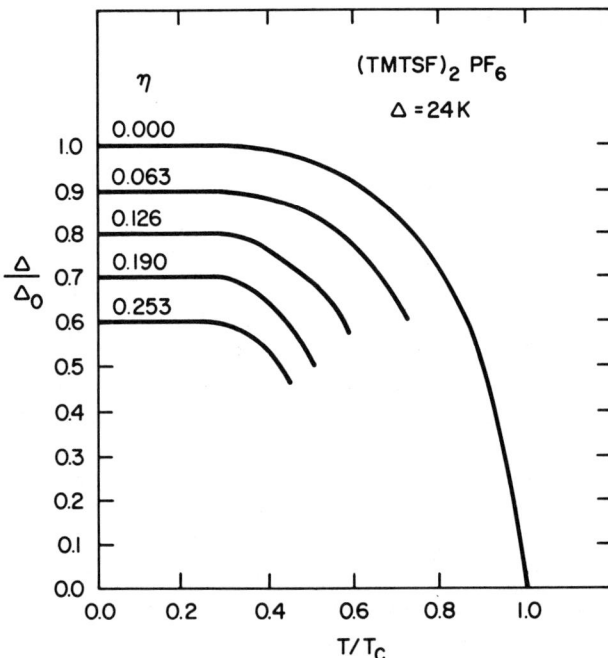

FIG. 9. Variation of Δ/Δ_0 with the ratio of temperature to the transition temperature for (TMTSF)$_2$PF$_6$ with $\Delta = 24$ K. The numbers on the curves indicate the values of the injection η. [From Conwell and Banik (1982b).]

$\cos k_F b > 0$ and $m_n^* < m_p^*$. If the band is more than half filled $m_n^* > m_p^*$. Other things being equal, this would result in a larger mobility for the carriers with the same sign as the dominant carriers above the transition. Thus for a single-chain conductor one would predict the same sign of thermoelectric power below the transition as above, and this is generally found to be the case, with the possible exception of very low temperatures.

For the calculation of transport properties below T_P we need, in addition to m^*, the density of states in energy, $\rho(E_k)$. As shown in Section 2 the density of states for either spin is given by

$$\rho(E_k) = \frac{Nb/\pi}{dE_k/dk}. \tag{133}$$

With E_k given by Eq. (102), $\rho(E_k)$ has a complicated form. For $k_B T \ll \Delta$, where E_k may be written in the form Eq. (131) it simplifies to

$$\rho(E_k) = \frac{m^* Nb}{\pi \hbar^2 |k - k_F|}, \quad \varepsilon_k \ll \Delta, \tag{134}$$

as is appropriate for a one-dimensional parabolic band. In the opposite limit, $\varepsilon_k \gg \Delta$, $\rho(E_k)$ goes over to the one-dimensional density of states $\rho(\varepsilon_k)$ for the undistorted band, which is, of course, much larger.

13. Matrix Element for Acoustic-Mode Scattering Below T_P

The low mobilities at room temperature, a few cm^2/V sec, in this family of quasi-one-dimensional conductors led many people to assume that the low-temperature mobilities would also be small. As will be shown in Section 16, mobilities in TTF-TCNQ and TSeF-TCNQ in the semiconducting range, specifically in the range 10–20 K, are $\sim 10^4$ cm^2/V sec. A Hall mobility of 10^5 cm^2/V sec has been measured at 4 K in the semiconducting range of (TMTSF)$_2$PF$_6$. It should not be totally unexpected that the mobilities are so high at low temperatures in this family of materials. One significant factor is the small effective mass below T_P. More important, the low μ at 300 K is largely due to the many phonon branches characteristic of the large molecules and complex crystal structure, and almost all of these modes freeze out at low temperatures. In fact, because there is a Maxwell–Boltzmann distribution of the carriers at low temperatures, the only phonons with which they can interact are the long-wavelength acoustic modes. In this section we will examine the matrix element for acoustic-mode scattering for $T < T_P$ and show that it can be taken the same as for $T > T_P$. The relaxation time and μ for acoustic-mode scattering will be calculated in the next section.

The perturbing Hamiltonian for 1p acoustic mode scattering derived in Section 5 is

$$H'_{\text{e-ph}} = N^{-1/2} \sum_{q,k} g(k, q) a^+_{k+q} a_k (b_q + b^+_{-q}), \tag{135}$$

where

$$g(k, q) = 2i \left(\frac{\partial t}{\partial u}\right)_0 \left(\frac{\hbar}{2M\omega_q}\right)^{1/2} [\sin(k + q)b - \sin kb]. \tag{136}$$

We will assume that Eq. (136) remains valid below T_P. Actually $(\partial t/\partial u)_0$ is no longer a constant in the Peierls-distorted lattice, but the distortion is quite small, and, particularly for an incommensurate case, its effect should be inconsequential for long-wavelength acoustic modes. We note also that calculations for the incommensurate case show that long-wavelength phonons are little changed below T_P, showing only a slight mixing with low-q phasons, a second-order effect (Walker, 1978). The wavevector of the phonon involved in scattering a thermal electron with wavevector k is found from conservation of momentum and energy, the latter taken from Eq. (131), to be

$$q = \pm 2(k - k_F) \pm \frac{2m^* v_s}{\hbar}, \quad (137)$$

where v_s is the velocity of the phonon branch involved, and the upper signs are for emission, the lower for absorption. Since $v_s \simeq 10^5$ cm/sec, the term in v_s is small compared to the other for thermal electrons at $T \geq 4$ K and will be dropped.

Although we are taking the perturbing Hamiltonian to be the same, a change in the matrix element for scattering could arise from the different wave functions below T_P, i.e., from the mixing of wave functions for k and $k - Q$. As in the case of superconductivity, this results in "coherence factors" that depend on the relative amounts of the different wavefunctions and on the scattering process concerned (Schrieffer, 1964). Since the phonons do not have sufficient energy to cause a transition between lower and upper subbands, the initial and final states are eigenfunctions of the diagonalized H_{el} of Eq. (93), i.e., of the α's rather than the a's. We therefore transform H'_{e-ph} to the α representation by using Eqs. (100). The result can be simplified by making use of the fact that $|q| \ll |k_F|$ to show that $g(k - Q, q) \simeq g(k, q)$. We may then simply add the contributions to the matrix element of a_k and a_{k-Q}, using Eqs. (100), with the result

$$a^\dagger_{k'} a_k + a^\dagger_{k'-Q} a_{k-Q} = \sum_{i=1}^{2} (u_{i,k'} u_{i,k} + v_{i,k'} v_{i,k}) \alpha^\dagger_{i,k'} \alpha_{i,k}, \quad (138)$$

where $k' = k \pm q$. For the transition probability we will need the absolute squared of the coefficient of $\alpha_{k'} \alpha_k$ for both upper and lower bands. Using Eq. (105), we may write for either band

$$|u_{k'} u_k + v_{k'} v_k|^2 = \frac{(E_{k'} + E_k - \varepsilon_{k'-Q} - \varepsilon_k)(E_{k'} + E_k - \varepsilon_{k-Q} - \varepsilon_{k'})}{(2E_{k'} - \varepsilon_{k'} - \varepsilon_{k'-Q})(2E_k - \varepsilon_k - \varepsilon_{k-Q})}. \quad (139)$$

For thermal electrons $E_{k'} \simeq E_k \simeq \Delta$, and the other terms in Eq. (139) are all small compared to Δ. Thus we expect the right-hand side of Eq. (139) to be close to unity. A more detailed examination shows that it differs from unity by quantities of second and high order in $\hbar \omega_q / \varepsilon_k$ and $\hbar \omega_q / E_k$. Since $\hbar \omega_q \leq k_B T/3$ for thermal electrons at 4 K in TTF-TCNQ or (TMTSF)$_2$PF$_6$, the difference from unity of the right-hand side of Eq. (139) is a fraction of a percent, and the coherence factor will be taken as unity. We conclude that the matrix element for acoustic-mode scattering may be taken the same below T_P as above.

14. Calculation of Mobility for Acoustic-Mode Scattering Below T_P

Since the matrix element for acoustic mode scattering below T_P is essentially equal to that above T_P, as shown in the last section, it is independent of the final state k'. It can then be shown, by the methods of Section 6, that a relaxation time exists, given by the rate of scattering out of the state k. The relaxation time $\tau(k)$ can be deduced directly from Eq. (38), for example, by making use of the fact that, for a Maxwell–Boltzmann distribution, $f_0(k') \ll 1$ for all k'. We then find

$$[\tau(k)]^{-1} = \pi(N\hbar)^{-1}|g(k,q)|^2[n_q\rho(E_k + \hbar\omega_q) + (n_q + 1)\rho(E_k - \hbar\omega_q)], \quad (140)$$

where $\rho(E_k \pm \hbar\omega_q)$ represents the density of states after absorption or emission. For thermal electrons, even at temperatures as low as 4 K, the phonon energy, $2\hbar v_s(k - k_F)$ according to Eq. (137), is small enough so that (1) equipartition holds and (2) the scattering may be considered elastic. The terms for emission and absorption in Eq. (140) may then be combined, and with the use of Eqs. (136), (137), (134), and (132) (we neglect the difference between m_n^* and m_p^* here), τ for thermal carriers may be written

$$\tau(k) = \frac{\hbar b M v_s^2 t^2 \tan^2(k_F b)}{\Delta(\partial t/\partial u)_0^2 k_B T} \frac{|k - k_F|^3}{\sin^2[(k - k_F)b]}. \quad (141)$$

To display the k dependence explicitly we have also eliminated ω_q by setting it equal to $2v_s|k - k_F|$. With $|k - k_F|b \ll 1$, $\sin^2(k - k_F)b$ may be approximated by $(k - k_F)^2 b^2$ and $\tau(k)$ is found to increase linearly with $|k - k_F|$, i.e., with the square root of energy ε measured from the band edge. This energy dependence differs from that of τ for acoustic-mode scattering in a 3D semiconductor, which is proportional to $\varepsilon^{-1/2}$, by precisely the factor that gives the difference between the 3D and 1D densities of states. Note that $\tau(k)$ given by Eq. (141) holds for either LA or TA modes, provided the appropriate values of v_s and $(\partial t/\partial u)_0$ are inserted.

In the derivation of Eq. (141) we have assumed propagation of LA and TA modes parallel to the chains. As discussed in Section 6, for TTF-TCNQ and the many members of its family with the planar molecules at an appreciable angle to the chain direction, scattering by oblique waves should not have much effect on $\tau(k)$ of Eq. (141). This is so because of a tendency for cancellation of the effects of increasing matrix element and increasing ω_q as $\bar{q}$ goes off the chain direction. The situation is somewhat different for compounds of the form $(TMTSF)_2X$, X standing for PF_6, NO_3, etc., where the planes of the molecules are essentially perpendicular to the chains. In this case for waves propagating parallel to the chains, i.e., for $\bar{q} = q_1$, only

LA modes can cause perceptible scattering. For an oblique LA phonon, the fact that only the component of the displacement parallel to the chains causes scattering results in $(\partial t/\partial u)_0$ in the matrix element being multiplied by the factor $q_\parallel/(q_\parallel^2 + q_\perp^2)^{1/2}$ when equipartition is valid, a smaller factor when $\hbar\omega_q$ is too large for equipartition. The result is that the square of the matrix element for oblique phonons is smaller than that for phonons propagating parallel to the chain by at least a factor $q_\parallel^4/(q_\parallel^2 + q_\perp^2)^2$. Thus it falls off quite rapidly as $q_\perp$ increases, and scattering by LA phonons with $\bar{q}$ along the chain direction or at small angles to it is greatly favored. For such phonons, because of the large transverse mass, $q_\parallel$ is still essentially equal to $2(k - k_F)$. Thus $q \ll k$ for these phonons, and the final state must have k still quite close to the chain direction. We conclude that the major change in the absolute square of the matrix element to allow for oblique phonons is the multiplying factor $q_\parallel^4/(q_\parallel^2 + q_\perp^2)^2$. For the phonons that scatter significantly this factor is well approximated by

$$F = \frac{16(k - k_F)^4}{[4(k - k_F)^2 + q_\perp^2]^2}. \quad (142)$$

This results in the overall τ with oblique phonons included being given by Eq. (141) times $\langle 1/F \rangle$ for the phonons with not too large $q_\perp$. Thus the average transition rate for all LA phonons as a group decreases by perhaps a factor of 2, because the phonons with $q_\perp \neq 0$ scatter less effectively than those with $q = q_\parallel$, and τ for LA phonons increases by perhaps a factor of 2. This decrease in scattering may be to some extent offset by the contribution of TA waves travelling at an angle to the chains with a component of displacement along the chains.

We conclude that for either the $(TMTSF)_2PF_6$ or the TTF-TCNQ type of structure the actual τ for motion along the chain direction is comparable to the value given by Eq. (141), which was calculated for 1D motion and phonons propagating parallel to the chains.

Given the existence of a relaxation time we may use the formal development of Section 7 to obtain the current

$$\bar{j} = 2ne\mu\bar{E} = N_c e \sum_k f_1 \bar{v}, \quad (143)$$

where f_1 is given by Eq. (63) and we have neglected the small difference between electron and hole mobilities. For thermal electrons or holes with E_k given by Eq. (131) and the mass taken from Eq. (132),

$$v = \frac{4t^2 b \sin^2(k_F b)}{\hbar\Delta}(k - k_F)b. \quad (144)$$

With f_0 taken as the Maxwell-Boltzmann distribution and τ obtained from

Eq. (141) by replacing $\sin[(k - k_F b)]$ with $(k - k_F)b$, we obtain finally for thermal electrons, in the T range where equipartition holds for the phonons (Banik et al., 1981),

$$\mu_{ac} = \frac{8e}{(2\pi)^{1/2}} \frac{t^2 M v_s^2 \sin^3(k_F b)}{\hbar \Delta^{3/2} (\partial t/\partial u)_0^2 \cos^2(k_F b)} (k_B T)^{-1/2}. \tag{145}$$

This is μ due to scattering by phonons moving parallel to the chains. As discussed earlier, however, we expect μ to be much the same if we take into account scattering by oblique phonons.

15. Ionized Impurity Scattering Below T_P

Scattering by impurity ions in a quasi-one-dimensional (1D) conductor is expected to be more important below the Peierls transition temperature T_P, where the smaller number of carriers results in both reduced screening and much smaller, i.e., thermal, velocities. There are some qualitative differences between impurity scattering in 1D and 3D (three-dimensional) cases that should be pointed out before a quantitative calculation is undertaken. In 3D material the scattering effect of repulsive and attractive impurities is the same unless screening is different for the two cases. This is not true in 1D material, however, where the constraint of the carrier to a chain limits the possible deflection. In the 1D case only repulsive impurities can scatter. A carrier going by an attractive impurity ion will be accelerated while in its vicinity, but after having gone by will eventually return ot its original velocity: In the case of a repulsive impurity the carrier will be slowed down as its approaches the ion. If the potential energy due to the ion becomes equal to the original kinetic energy of the carrier, the carrier will be reflected, or scattered backward, ultimately attaining the negative of its original velocity. It is clear that reflection can occur only for ions within some maximum distance R_m from the chain on which the carrier is located. R_m depends on the charge on the ion, the dielectric constant, and the screening. Ions of either sign at a greater distance from the chain than R_m cannot scatter. From this standpoint a repulsive ion on a chain is an impossible barrier for carriers of any energy, equivalent to a chain break. Because continuity of current flow would require that a carrier faced with a break hop to another chain, such breaks have been treated as introducing into the current path resistors with the transverse resistance. This case will not be considered further here. In this section we will calculate the scattering effect on the carriers of impurity ions that are interstitial or on other chains. It should be noted that for compounds like TTF-TCNQ, for example, where the dielectric constant perpendicular to the chains is of the order of unity, R_m is large enough to encompass very many chains for thermal carriers at $T < T_P$.

A difficulty with the calculation for the 1D case is the treatment of screening. It has been pointed out that the constraint on the lateral motion of the carriers decreases their effectiveness in screening (Kuper, 1966; Davis, 1973). In a numerical calculation for a system simulating the 1D conductor KCP, Davis found that in the 1D case the screening length λ was almost an order of magnitude larger than it would be for the same charge density in a 3D case. Nevertheless, he found that the potential energy of an electron at a distance r from the impurity ion is still well represented by the isotropic form

$$V(r) = \pm \frac{Z_i e^2}{\kappa r} e^{-r/\lambda} \tag{146}$$

where $\mp Z_i e$ is the charge on the impurity and κ is the dielectric constant. Equation (146), of course, assumes that κ is isotropic, which is not the case in these materials. To avoid great complications, however, we will assume for the calculation of the scattering time that the perturbing potential is given by Eq. (146) with κ a suitable average. The assumption of isotropic screening should not lead to much error for T well below T_P because the small carrier concentration should make screening relatively unimportant. We first carry out a Fourier analysis of the potential Eq. (146). The scattering is due to the Fourier components that allow conservation of crystal momentum. Obtaining a scattering matrix element incorporating these Fourier components, we use first-order perturbation theory, i.e., the Born approximation, to calculate the relaxation time for impurity ion scattering.

For the calculation the chains are taken to be parallel to the Z axis. We consider the scattering effect of an ion at the origin on an electron or hole located on a chain whose perpendicular distance from the ion is D_i, where $D_i < R_m$. The potential energy Eq. (146) of the carrier may then be written

$$V(z) = \pm \frac{Z_i e^2}{\kappa} \frac{\exp[-(z^2 + D_i^2)^{1/2}/\lambda]}{(z^2 + D_i^2)^{1/2}}. \tag{147}$$

We wish to expand $V(z)$ in the form

$$V(z) = \sum_{q'} V_{q'} e^{-iq'z}. \tag{148}$$

For periodic boundary conditions $V(z + L) = V(z)$ for a chain of length L, the possible values of q' are $2\pi m/L$, where m is an integer. To determine $V_{q'}$ we multiply both sides of Eq. (148) by e^{iqz} and integrate from $-L/2$ to $+L/2$. The resulting integral may be written

$$V_q = \frac{2Z_i e^2}{\kappa L} \int_0^{L/2} \cos qz \frac{\exp[-(z^2 + D_i^2)^{1/2}/\lambda]}{(z^2 + D_i^2)^{1/2}} dz. \tag{149}$$

In the large L limit the integral goes over to the Bessel function

$$K_0\left[D_i\left(q^2 + \frac{1}{\lambda^2}\right)^{1/2}\right].$$

With this, the desired Fourier expansion of $V(z)$ is

$$V(z) = \sum_q \frac{2Z_i e^2}{\kappa L} K_0[D_i(q^2 + 1/\lambda^2)^{1/2}] e^{-iqz}. \tag{150}$$

The effect of the Peierls transition is to mix states with wavevector k with those of $k + 2k_F$, where k_F is the Fermi wavevector. The same mixing will occur for a spin-density wave transition, with the states mixed having opposite spin. The wavefunctions may be taken in the form

$$\Psi_i = u_{i,k}\Psi_k + v_{i,k}\Psi_{k-Q}, \quad i = 1, 2, \tag{151}$$

where 1 refers to states above the gap, 2 to states below,

$$\Psi_k = L^{-1/2} e^{ikz}, \tag{152}$$

and $Q = 2k_F$. As before, we will use the quantities $u_{i,k}$ and $v_{i,k}$ for the non-half-filled band case, i.e., $k_F \neq \pi/2b$ where b is the lattice constant, because the quasi-one-dimensional molecular crystals do not, in general, have half-filled bands. Because impurity scattering cannot cause transitions across the gap, we need these quantities for one band only, which we take to be the upper band. The relations between u_k, v_k, u_k^*, and v_k^* for the upper band are given in Eqs. (105), (106a), (106b), (94), and (102).

The matrix element M_I for scattering is then

$$M_I = \sum_q \int \Psi_{k'}^* V_q e^{-iqz} \Psi_k \, dz, \tag{153}$$

where V_q is the coefficient of e^{-iqz} in Eq. (150). With Eqs. (151) and (152)

$$M_I = \sum_q [(u_{k'}^* u_k + v_{k'}^* v_k)\delta_{q,k-k'} V_q$$

$$+ (u_{k'}^* v_k + v_{k'}^* u_k)\delta_{q,k-k'+Q} V_q]. \tag{154}$$

As noted earlier, scattering results in a reversal of the velocity of the carrier, i.e., $k' = -k$. Since u_k and v_k depend only on energies, as seen in Eqs. (105) and (106), and the energies are the same for $+k$ and $-k$, we may take $u_{k'} = u_k$, $v_{k'} = v_k$. The coefficient $(u_{k'}^* u_k + v_{k'}^* v_k)$ is then unity, as can be deduced from Eqs. (106a) and (106b). The coefficient of $\delta_{q,k-k'-Q} V_q$ may be expressed in terms of $v_k^* v_k$ with the use of Eq. (105). Because we are dealing with thermal electrons, $E_k^+ \simeq \Delta$ and ε_k and ε_{k-Q} may be neglected compared to E_k^+ or Δ. As a result, the coefficient $u_{k'}v_k + v_{k'}u_k$ is well approximated by unity and

$$M_I = V_{2k-2k_F} + V_{2k}, \tag{155}$$

where V_q is the coefficient of e^{-iqz} in Eq. (150). In practice the second term in Eq. (155) wil be negligible because $k \gg (k - k_F)$ and K_0 is a rapidly decreasing function of its argument. As illustration, for an electron with energy $(\frac{1}{2})k_BT$ at 30 K in TTF-TCNQ $|k - k_F| \simeq 10^6 \text{ cm}^{-1}$, whereas $k \simeq k_F = 2.4 \times 10^7 \text{ cm}^{-1}$. With the screening term $1/\lambda^2$ expected to make little contribution, the argument of K_0 is ~ 500 times as large for the second term of Eq. (155). For the half-filled-band case there would be additional terms in Eq. (155) corresponding to q values differing from the two represented in Eq. (155) by reciprocal lattice vectors. Again the argument of K_0 would be much larger for these terms and they can be neglected. The resulting matrix element can therefore be written in general

$$M_I = V_{2|k-k_F|} = \sum_i \frac{2Z_i e^2}{\kappa L} K_0[D_i(4|k - k_F|^2 + 1/\lambda^2)^{1/2}]. \tag{156}$$

It is interesting to note that Eq. (156) is the result that would be obtained if Ψ had been taken as $L^{-1/2}e^{i(k-k_F)z}$.

With M_I independent of the final state for scattering, a relaxation time τ exists. By first-order perturbation theory, Eqs. (34) and (38), τ is given by

$$\frac{1}{\tau} = \frac{2\pi}{\hbar}|M_I|^2 \rho(E_k), \tag{157}$$

where $\rho(E_k)$, the density-of-states in energy for one sign of spin, is

$$\rho(E_k) = \frac{m^*L}{2\pi\hbar^2|k - k_F|}, \tag{158}$$

according to Section (12), m^* being given by Eq. (132). With Eqs. (158) and (156) substituted into Eq. (157), $1/\tau$ becomes

$$\frac{1}{\tau} = \frac{4Z_i^2 e^4 m^*}{\hbar^3 \kappa^2 L} \frac{K_0^2[D_i(4|k - k_F|^2 + 1/\lambda^2)^{1/2}]}{|k - k_F|}. \tag{159}$$

Equation (159) gives $1/\tau$ due to a single ion at a distance D_i from a chain of length L. If the impurities are not too close together we can assume, as usual, that they scatter independently of each other. The total probability of scattering by the impurities at a distance D_i from the chain may then be obtained by multiplying Eq. (159) by the number of impurities at distance D_i. With the density of impurities at D_i per unit chain length given by $n_I(D_i)$, the number at the distance D_i is $n_I(D_i)L$. However, the impurity distribution is expected to be uniform, so $n_I(D_i)$ may be replaced by the number of impurities/cm^3, N_I, divided by the number of chains/cm^2, N_c. Because in principle repulsive impurity ions at all distances $D_i < R_m$ may contribute to the scattering, within the approximation of independent scatterers the relaxation time for ionized impurity scattering in the 1D case is

$$\frac{1}{\tau_I(1D)} = \sum_i \frac{4Z_i^2 e^4 m^* N_I}{\hbar^3 \kappa^2 N_c} \frac{K_0^2[D_i(4|k-k_F|^2 + 1/\lambda^2)^{1/2}]}{|k-k_F|}, \quad D_i < R_m. \tag{160}$$

It is interesting to contrast this with the result for a 3D semiconductor (Ziman, 1960)

$$\frac{1}{\tau_I(3D)} = \sum_i \frac{2\pi Z_i^2 e^4 m^* N_I}{\hbar^3 \kappa^2} \frac{\ln(1 + 4k^2\lambda^2) - (1 + 1/4k^2\lambda^2)^{-1}}{k^3}. \tag{161}$$

It is seen that Eqs. (160) and (161) are quite similar, the only differences being in the dependence of the former on the perpendicular distance between the carrier and the ion, and in the detailed dependence on k. In both cases τ_I is smaller for slower carriers. In the 1D case this is due to the K_0 dependence on $|k - k_F|$, in addition to the explicit proportionality of τ_I to $|k - k_F|$. The monotonic decrease of K_0 with its argument also means that, as expected, the scattering effect of an ion at D_i decreases as D_i increases. It is interesting, however, that the rate of decrease with D_i is quite nonuniform. To illustrate this consider the case of TTF–TCNQ at 30 K. The mass for an electron or hole on the TCNQ chain below T_P is $\sim 0.2\, m_0$, where m_0 is the free electron mass. As noted earlier, for an electron or hole with energy $\tfrac{1}{2}k_B T$ this leads to $|k - k_F| \sim 10^6$ cm^{-1} at 30 K. Neglecting screening and estimating the distance of the closest interstitial to TCNQ chain at 6 Å, we find $2|k - k_F|D_i = 0.12$, corresponding to $K_0 = 2.25$. For an ion at twice this distance $K_0 = 1.58$. Thus at small distances the scattering effect of an ion falls off less than linearly with distance from the chain. At large values of its argument, however, K_0 decreases much more rapidly than linearly. Thus for large D_i ($<R_m$, however) the scattering effect of an ion falls off quite rapidly.

III. The Semiconducting Range

16. COMPARISON OF EXPERIMENT WITH THEORY BELOW T_P: LOW FIELDS

The conductivity of TTF–TCNQ measured in low dc fields, i.e., in the ohmic region, is shown in Fig. 10. It has been plotted as a function of $1/T$ to emphasize the low-temperature region. As noted in Chapter 1, the sharp

4. TRANSPORT IN QUASI-ONE-DIMENSIONAL CONDUCTORS

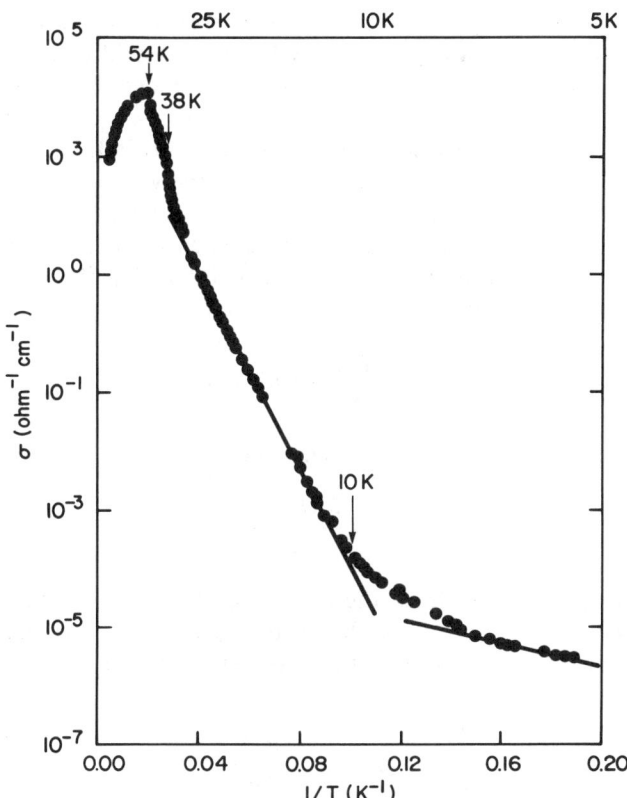

FIG. 10. dc conductivity of TTF–TCNQ between 5 and 300 K. The arrows indicate the anomalies in the conductivity at the structural phase transitions near 54 K and 38 K. [From Cohen and Heeger (1977).]

drops at 54 K and 38 K are due to the opening of Peierls gaps on the TCNQ and TTF chains, respectively. Below these temperatures there are two regions of the plot marked with straight lines, indicating $\log \sigma \propto -C_i(1/T)$, where C_i is a constant in those regions. For the higher temperature region, $25 > T > 10$ K, the data give $C_i \simeq 200$ K. With charge-density wave transport eliminated as a possibility because it would require a large depinning field in this temperature range, as discussed in Section 8 of Chapter 1, the likely source of conduction in this range is electrons and/or holes. These could either be freed from impurities or defects or created by excitation across the gap. In the latter case TTF–TCNQ would be an intrinsic semiconductor and the gap would be $2C_i$ or ~ 400 K. In fact, there is good evidence that the Peierls gap is ~ 400 K. There is strong infrared absorption in this temperature region beginning at an energy of ~400 K (Tanner et al., 1981). In addition, photoconductivity has been found to result from the absorption of radiation of energy ~400 K and above (Eldridge, 1977).

Although the Hall effect is too small to measure at low temperatures in TTF-TCNQ because of the small transverse bandwidth and mobility, the knowledge that σ is due to carriers crossing the gap makes it possible to calculate the carrier concentration. This can be done by using the value we have deduced for the gap, ~400 K, and the density of states calculated in Section 12. From Eq. (66) and the first equality in Eq. (69) the number of conduction electrons per unit volume is given by

$$nN_c = \frac{2N_c}{Nb} \int_0^{4t} f_0(\varepsilon)\rho(\varepsilon)\,d\varepsilon, \tag{162}$$

where N_c is the number of chains/cm^2 and

$$\varepsilon = \frac{\hbar^2(k - k_F)^2}{2m^*} \tag{163}$$

is the energy measured from the band edge. For this calculation we can simplify m^*, given in Eq. (132), by neglecting the small difference between electron and hole masses. We then have m^* equal to the average of the masses:

$$\langle m^* \rangle = \frac{\hbar^2 \Delta}{4t^2 b^2 \sin^2 k_F b}. \tag{164}$$

With this mass used in the density of states Eq. (134) and f_0 taken as the Maxwell–Boltzmann distribution, we find (Banik et al., 1981) the number of electrons or holes per unit volume in the intrinsic range is given by

$$nN_c = pN_c = \frac{N_c}{\pi^{1/2}} \frac{(2\Delta k_B T)^{1/2}}{2tb \sin k_F b} e^{-\Delta/k_B T}. \tag{165}$$

Taking Δ as 215 K (Eldridge, 1977) and the other quantities in Eq. (165) from Table III, with $4t$ the TCNQ bandwidth, we find the concentration of electrons or holes to be 10^9/cm^3 at 10 K. With $\sigma = 10^{-5}$ ohm^{-1} cm^{-1} at 10 K according to Figure 10, this leads to $\mu = 2 \times 10^4$ cm^2/V sec for an electron or hole in TTF-TCNQ at this temperature. For TSeF-TCNQ, where infrared absorption (Bates et al., 1981), $\ell n\sigma$ vs. $1/T$ (Etemad, 1976) and magnetic susceptibility (Scott et al., 1978) all lead to $\Delta \sim 130$ K, a similar calculation from the measured σ (Etemad, 1976) leads to $\mu \simeq 10^4$ cm^2/V sec at 20 K.

We can now compare the value of μ deduced from experimental data for TTF-TCNQ at 10 K with μ_{ac} calculated in Section 14. With $(1/t)(\partial t/\partial u)$

taken as 1.5 eV/Å, the Table II value for LA modes, $v_s = 3.2 \times 10^5$ cm/sec (Schafer et al., 1975), $\Delta = 215$ K and the other values taken from Table III for TCNQ, we find $\mu_{ac} = 2 \times 10^4$ cm/sec for LA modes at 10 K. μ_{ac} for TA modes should be similar. The excellent agreement between the experimental value and the value calculated for acoustic phonon scattering indicates that phonon scattering is predominant even at temperatures as low as 10 K. An average time τ_{ac} for phonon scattering can be deduced from $\mu = (e/m^*) \tau_{ac}$ using the measured μ and $m^* = 0.2 m_0$. The result is $\tau_{ac} = 10^{-12}$ sec. The predominance of phonon scattering means that the average scattering time for any other process, such as impurity ion scattering, must be greater than 10^{-12} sec. This fact, plus the calculated τ_I of Section 15, enables us to obtain an upper limit on the ionized impurity concentration in the TTF-TCNQ sample used to obtain the data of Fig. 10, presumably a typical sample. To obtain an average τ_I we use in Eq. (160) $|k - k_F|$ of a carrier with thermal energy, $\tfrac{1}{2} k_B T$. For $m^* = 0.2 m_0$ $|k - k_F| \simeq 10^6$ cm^{-1}. Taking $Z_i = 1$, neglecting screening (i.e., $\lambda = \infty$) and estimating $\sum K_0^2(4 D_i |k - k_F|)$ as 10^2, we find from Eq. (159) that for τ_I to be greater than 10^{-12} sec,

$$N_I < 27 \times 10^{12} \kappa^2.$$

Unfortunately κ is highly anistropic in TTF-TCNQ and similar crystals. If we take κ as the transverse value, we obtain $N_I < 2 \times 10^{14}$/cm^3. If we take $K = 100$, a kind of average between the transverse value and the value of thousands parallel to the chains, we obtain $N_I < 2 \times 10^{18}$/cm^3. Even at the upper limit these can be considered small values for crystals made from material that was not extensively purified.

Another material for which sufficient measurements have been made below the transition temperature (in this case a spin-density wave transition at 12 K) to permit electron and hole mobilities to be deduced is (TMTSF)$_2$PF$_6$. It should be pointed out that this material is quasi-two-dimensional at low temperatures, actually below ~ 100 K (Jacobsen et al., 1981a) where a plasma frequency has been observed transverse to the chain direction. In fact (TMTSF)$_2$PF$_6$ is sufficiently two-dimensional to allow measurements of magnetoresistance and Hall mobility in the semiconducting range. The Hall mobility μ_H measured at 4 K is 10^5 cm^2/V sec (Chaikin et al., 1981). If only one sign of carrier were present, the number of carriers deduced from the measured Hall constant would require a very large number of impurities or defects, inconsistent with the high mobility. It was concluded, therefore, that the samples are intrinsic, with the consequence that the measured $\mu_H \simeq \mu_p - \mu_n$. Thus the measured μ_H indicates that μ_p and μ_n are greater than 10^5 cm^2/V sec. To determine μ_p and μ_n we compare these results with the theory presented in earlier sections because, despite the observation of a Hall effect, the resistivity of (TMTSF)$_2$PF$_6$ is highly anisotropic, conductivities parallel and perpendicular to the chains differing by a factor of 10^4 or more at low temperatures (Jacobsen et al., 1981b).

The difference between μ_p and μ_n must be due to the difference between m_p and m_n because differences in the wavefunctions and matrix elements for electrons and holes are inconsequential. For carriers near the band edges, since the term in $\cos k_F b$ is small compared to the other, Eq. (132) for the mass may be rewritten

$$m^* = \langle m^* \rangle (1 \mp \alpha), \qquad (166)$$

where $\langle m^* \rangle$ is given by Eq. (164), the upper sign in the parenthesis is for electrons, the lower for holes, and

$$\alpha = \frac{\Delta \cos k_F b}{2t \sin^2 k_F b}. \qquad (167)$$

Note that, whereas $\langle m^* \rangle$ is reduced by a factor $\sim \Delta/2t$ from the effective mass at the bottom of the original, undistorted band, the difference in masses, $\alpha \langle m^* \rangle$, is reduced by another factor $\sim \Delta/2t$. For (TMTSF)$_2$PF$_6$ the slope of $\ell n \sigma$ vs. $1/T$ gives $\Delta = 24$ K (Chaikin et al., 1981), $2t = 0.5$ eV (Bechgaard et al., 1980), $b = 3.65$ Å, and $\sin k_F b = 0.707$, leading to $\langle m^* \rangle = 0.0095 m_0$ and $\alpha = 0.006$. If we assume that $\mu \propto (m^*)^r$, r depending on the scattering mechanism,

$$\mu_H \simeq \mu_p - \mu_n = |2r\alpha \langle \mu \rangle|, \qquad (168)$$

where $\langle \mu \rangle$ is the average of electron and hole mobilities (Conwell and Banik, 1982a). Whatever the scattering mechanism, the value of r should lie in the range 0.5 to ~ 3. For acoustic-mode scattering we can find r by substituting for Δ in terms of m^* [given with sufficient accuracy by $\langle m^* \rangle$, Eq. (164)] in Eq. (145). This gives $r = -\frac{3}{2}$. With this value for r, $\mu_H = 10^5$ cm^2/V sec and $\alpha = 0.006$, Eq. (168) yields $\langle \mu \rangle = 6 \times 10^6$ cm^2/V sec. This large value of $\langle \mu \rangle$ suggests that phonon scattering is predominant down to 4 K! That is not entirely unexpected because it is known that crystals of the TMTSF salts are particularly pure (Bechgaard, 1982). It has been suggested that the electrochemical procedure by which the crystals are made is self-purifying (Engler et al., 1981). An approximate value for μ_{ac} of (TMTSF)$_2$PF$_6$ may be calculated from Eq. (145) by taking M as 408 amu, intermediate between the masses of TMTSF and PF$_6$, and $\partial t/\partial u$, not known for this material, as 0.2 eV Å, the value for TTF–TCNQ. These values in Eq. (145) lead to $\mu_{ac} = 3 \times 10^6$ cm^2/V sec, in reasonable agreement with the value of 6×10^6 cm^2/V sec deduced from μ_H.

It can be concluded from the above discussion that conduction in the intrinsic range of this family of quasi-one-dimensional materials—exemplified by TTF–TCNQ down to ~ 10 K, (TMTSF)$_2$PF$_6$ down to ~ 4 K—is well understood, at least qualitatively. The same cannot be said for conduction at lower temperatures. It is seen from Fig. 10 that from $0.09 \leq (1/T) \leq 0.14$ $\ell n \sigma$ vs. $1/T$ has a continuously decreasing slope. Below that a straight line has been drawn corresponding to a slope of 14 K. It was suggested by Cohen and Heeger (1977) that the 14 K slope represents the formation energy of a naturally occurring defect in a charge-density wave, called a Φ-particle. It had been predicted theoretically (Rice *et al.*, 1976) that Φ-particles, which are phase kinks of $\pm 2\pi$, could be excited thermally in a charge-density wave. Corresponding to compressions or rarefactions of the charge-density wave, these phase kinks were predicted to have charge $\pm 2e$ and to be mobile, thus capable of carrying current. It was Cohen and Heeger's suggestion that the Φ-particles are the source of conduction in TTF–TCNQ below ~ 10 K. One difficulty with this suggestion is that the formation energy estimated for this defect by Rice *et al.* (1976) was ~ 100 K, much greater than 14 K. A more serious difficulty is that the 14 K activation energy has not been seen in the samples of other investigators. Rather, they see below the intrinsic range a continuously decreasing slope of $\ell n \sigma$ vs. $1/T$ down to whatever temperature the measurements are carried out. In fact, the sample of Fig. 10 might also have a continuously varying slope; data to lower temperatures would be required to settle this point. A continuously decreasing slope for $\ell n \sigma$ vs. $1/T$ has also been seen for $(TMTSF)_2PF_6$ below 4 K (Mortensen *et al.*, 1981). In fact, such a continuously varying slope in the low-temperature limit is characteristic of materials in this family (Gunning *et al.*, 1977).

A source of the continuously varying slope of σ vs. $1/T$ is spatial variation of the gap (Conwell and Banik, 1981a, 1982a). For the materials under discussion the gaps are much more likely to be spatially varying than in the usual semiconductor for two reasons. First, because they are molecular crystals, their energy levels are greatly affected by strain, as is evident in the strong pressure-dependence of gaps and other properties. Second, the charge-density waves or spin-density waves are much affected by defects.

As regards strain effects, the rate of gap change with pressure for TTF–TCNQ has been measured as -20 ± 10 K/kbar or $\sim 10\%$/kbar (Cooper *et al.*, 1975). For $(TMTSF)_2PF_6$ with a gap ~ 48 K at ambient pressure that vanishes under ~ 9 kbar (Jerome *et al.*, 1980) the average rate of change is also $\sim 10\%$/kbar. Strains equivalent to a kbar of pressure are found at ~ 20 lattice constants from an edge dislocation in one of these materials (Conwell and Banik, 1982a). The resulting shift in band gap would give rise to a barrier between a region characterized by such strain and an unstrained region of $\sim 0.05 \Delta$ or 10 K for TTF–TCNQ, 1 to 2 K for $(TMTSF)_2PF_6$. As will be seen, this appears to be a reasonable barrier height to explain a number of aspects of the low-temperature behavior of these materials. Closer to the dislocation axis than 20 lattice constants the strain must be greater than that stated above. Of course the strains need not be due to single-edge dislocations but could be due to groups of dislocations such as subgrain boundaries. In any case the strong tendency of the crystals in this family of materials to crack with changes in temperature suggests large internal strains. Indeed, large internal strains and dislocations have been seen in typical TTF–TCNQ crystals by x-ray transmission (Begg *et al.*, 1980). As discussed above, such strains must lead to substantial gap variation and barriers between regions.

The effect of random defects or impurities on transitions of the type that occur in these materials has been studied by many people. It is well known that the transition in a system with random impurities or defects is not sharp but occurs over a range of temperatures. The system behaves like an ensemble of systems with a distribution of transition temperatures. Harris (1974) has studied the dependence of the critical temperature T_c on the fraction of defects for the Ising model with random bond defects. He finds that the width of the distribution in T_c is proportional to the "width" (mean-square fluctuation within the coherence length) of the defect concentration. Since the gap is proportional to T_c, this implies a range of gaps dependent on the "width" of the defect concentration. The case of random bond defects is similar to the weak pinning case for charge-density waves, as pointed out by Fukuyama and Lee (1978). In that case domains of coherent charge-density waves are formed since the system can gain energy by taking advantage of the fluctuations in impurity concentration; it follows that the gap in each domain will depend on its defect concentration. In the strong-pinning case there are strong local distortions of the CDW. These may be considered as local strains, affecting the gap as discussed earlier. Perhaps the most direct demonstration of the effect of point defects on the gap is the smearing out of the transition by weak to moderate irradiation, indicating the creation of regions of smaller and larger gaps than that observed in the unirradiated sample (Chiang *et al.*, 1977; Mutka *et al.*, 1981). In fact, it has been shown that weak irradiation of TMTSF–DMTCNQ, which at ambient pressure has a sharp metal-to-semiconductor transition at 42 K (Jacobsen *et al.*, 1978), stabilizes the metallic state, as evidenced by the Hall coefficient remaining small and constant from 300 K down to 2 K (Zuppiroli *et al.*, 1982).

From the above discussion we conclude that spatially varying gaps, due to strains and other defects, are most likely in these materials. They can explain the continuously decreasing slope of $\ln\sigma$ vs. $1/T$ seen below ~ 10 K in TTF–TCNQ, ~ 4 K in $(TMTSF)_2PF_6$ (Mortensen *et al.*, 1981). Spatially varying gaps could explain also the observation that the microwave conductivity of TTF–TCNQ is greater than the dc conductivity below 15 or 20 K by an amount that increases with decreasing temperature to ~ 5 orders of magnitude at 5 K (Cohen *et al.*, 1978). In particular, barriers of <20 K between regions of different gap could account for this observation; barriers are ineffective in hindering conductivity at frequencies high enough so that the carriers do not reach the barriers during the time of a half cycle. Such barriers could also account for part of the very high dielectric constants observed around 4 K in the TTF-TCNQ family of materials. For $(TMTSF)_2PF_6$ also, conductivity has been observed to increase with frequency, starting at 1 GHz (Hardiman et al., 1981); this also could be attributed to barriers. Spatially varying gaps could also account for the continuous background of photoconductivity and infrared absorption observed for frequencies below "the gap" in TTF-TCNQ and TSeF-TCNQ (Bates et al., 1981). An additional, more direct, piece of evidence for spatially varying gaps in $(TMTSF)_2PF_6$ is the observation by Chaikin et al. (1976) of an increase in σ with E at 20 K, well above the transition temperature of 12 K, of comparable magnitude and at comparable fields to what is observed below the transition temperature. There is no way such an increase could be due to metallic $(TMTSF)_2PF_6$. The increases below the transition temperature, it will be shown in the next section, can be attributed to hot electron effects due to the high mobility in the semiconducting phase. In the metallic phase, however, the mobility, deduced from the measured σ and known carrier concentration, is less than 1000 cm^2/V sec. Thus the observed increase must be due to semiconducting regions existing in series with metallic regions at these temperatures, well above the transition temperature. In the next section we will see that a spatially varying gap is needed to account for many features of the behavior of these materials in high electric fields, i.e., in the nonlinear conductivity region.

17. Comparison of Experiment with Theory Below T_P: High Fields

Nonlinear conduction below T_P has been seen in TTF-TCNQ and in $(TMTSF)_2PF_6$. The former material will be discussed first. In TTF-TCNQ two types of nonlinear conduction, in different temperature ranges, have been seen below T_P for the TCNQ chain. One type is seen from the neighborhood of the transition temperature, 54 K, down to ~34 K and is characterized by a threshold voltage (Lacoe et al., 1985). The threshold ranges from a minimum of ~0.25 V/cm just below 54 K to 6.4 V/cm at 34 K, below which it increases rapidly. Above the threshold, conductivity increases roughly linearly with field. As was noted earlier, with T_P = 38 K for the TTF chain, in the temperature range over which this nonlinearity is seen there is a well-formed charge-density wave on the TCNQ chains only. It is expected that in this T range the charge-density wave would be pinned mainly by impurities. Indeed, the small threshold field just below T_P on the TCNQ chains is not much different from that required in $NbSe_3$, for example (Monceau *et al.*, 1976), where the charge-density waves are pinned by impurities. The rapid increase in threshold field as T decreases toward T_P for the TTF chain would be expected for conduction arising from charge-density waves on TCNQ. As charge-density waves form on the TTF chain, with opposite charge from those in TCNQ, their electrostatic attraction would cause a very strong pinning. It was concluded, therefore, by Lacoe *et al.* (1985) that the nonlinearity seen in the range $54 K > T > 34 K$ is due to depinned charge-density waves on TCNQ chains.

A large nonlinearity with quite different properties is seen in TTF–TCNQ at much lower temperatures, from above 15 K (Banik *et al.*, 1981) on down. This nonlinearity was first seen by Kahlert (1975), who found a very large increase in the 4 K conductivity of TTF–TCNQ with electric field. He suggested it could be due to impact ionization of impurities by hot carriers. This suggestion was seconded by Bloch *et al.* (1976), although the calculations they provided in support are not applicable to TTF–TCNQ. They assumed a low-field mobility of 40 cm^2/V sec, due to impurity scattering. This low mobility permitted carrier heating only because the rate of energy loss was made artificially low by excluding the loss to acoustic phonons. Extensive experiments of Cohen and Heeger (1977) provided a more complete description of this nonlinearity. There is no threshold field, and in their samples the increase in σ became measurable at a few V/cm at 4 K, somewhat higher fields at lower temperatures. As shown in Fig. 11, the departure from Ohm's law begins slowly but grows more rapid with increasing field. At the highest field for which measurements were taken, ~ 400 V/cm (limited by the occurrence of sample heating at higher fields), σ has increased by a factor greater than 10^4. The nonohmic effects are limited to the high conductivity (b) direction; no increase in σ is seen in the a direction for fields up to a few hundred V/cm.

Cohen and Heeger suggested that the initial, slower part of the rise in σ is associated with Φ particles, or phase kinks of $\pm 2\pi$ in the charge-density waves, to which they ascribed an activation energy or formation energy of 14 K. In particular, they suggested that the Φ particles are responsible for the low-field conduction below ~ 10 K in TTF–TCNQ. The increase in σ with increasing E was then attributed to decrease of the Φ-particle formation energy with field. The difficulty with these suggestions, as discussed in Section 15, is that the 14 K activation energy has not been seen in the samples of other investigators, who nevertheless see the same increase in σ by a factor of 10^4 or more. In addition, measurements by Jacobsen (Banik *et al.*, 1981) at temperatures above 4 K show that, although somewhat higher fields are required for a measurable increase in σ, the increase is still found at 15 K. If the source of the nonlinearity were Φ particles with a low field activation energy of 14 K, no effect on σ of field should be seen at 15 K.

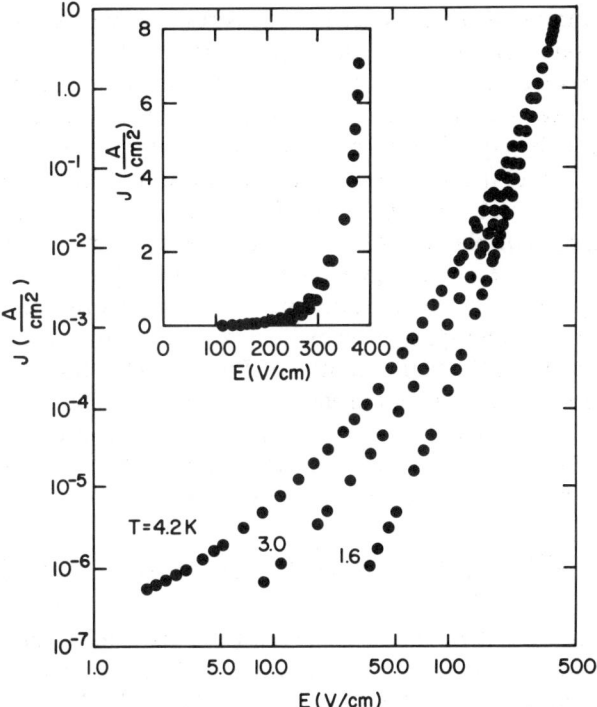

FIG. 11. Current density vs. field for TTF-TCNQ at the temperatures indicated. dc techniques were used for $j < 5 \times 10^{-2}$ A/cm^2; pulsed measurements covered the range above 10^{-2} A/cm^2. [From Cohen and Heeger (1977).]

At the time of Cohen and Heeger's work, it was not generally recognized that the mobility of carriers in TTF-TCNQ has such high values at low temperatures. As noted in Section 13, the high value, $\sim 10^4$ cm^2/V sec, is due to the low effective mass of the carriers below T_P and the freezing out of the many optical phonon branches. A value of μ two orders of magnitude larger still is found for (TMTSF)$_2$PF$_6$ at 4 K, according to Section 16. These high values of μ suggest the possibility of conductivity changes due to heating of the carriers by the field. It is readily seen that such heating would increase the relaxation time τ of thermal carriers and therefore σ. As shown in Eq. (141) the dependence on carrier speed of τ is given by the factor $|k - k_F|^3/\sin^2[(k - k_F)b]$. For thermal carriers $(k - k_F)b \ll 1$ and $\sin[(k - k_F)b]$ may be approximated by $|k - k_F|b$. As a result $\tau \propto |k - k_F|$ and increases as the speed of the carriers increases. Thus, as indicated above, heating of the carriers should increase τ, μ, and therefore σ, at any rate so long as Eq. (141) holds.

To determine whether carrier heating is responsible for the increase in σ it is necessary, of course, to do a quantitative calculation. A simple and reasonably accurate way of accomplishing this is to assume that the carriers have a Maxwell–Boltzmann distribution at an effective temperature T_e and determine T_e by equating the average rate of energy gain from the field to the average rate of energy loss to the phonons. The rate of energy loss to phonons may be obtained by first calculating $\partial n_q/\partial t$, the rate of change in the net number of phonons of wave vector q emitted by all the carriers, and then summing the net energy emitted at q over all q. With $P(k, k')$, the probability per unit time of scattering from k to k' provided the carrier is certainly in state k initially, given by Eq. (34), the rate of emission of phonons of wavevector q is the sum over all k of $P(k + q, k)$ times the probability of the carrier being initially in the state $k + q$, $f(k + q)$. Similarly, the rate of absorption of phonons of wavevector q is the sum over all k of the rate of transition from $k - q$ to k, multiplied by the probability of the carrier being initially in the state $k, f(k)$. Thus (Conwell, 1967)

$$\frac{\partial n_q}{\partial t} = \sum_k [P(k + q, k)f(k + q) - P(k, k - q)f(k)]. \tag{169}$$

To evaluate Eq. (169), $P(k, k')$ is taken from the golden rule Eq. (34), the matrix element for acoustic phonon scattering being given by Eqs. (135) and (136) and E_k by Eq. (102). $f(k)$ is, of course, the Maxwell–Boltzmann distribution. Once $(\partial n_q/\partial t)$ is obtained, the average rate of energy loss of a carrier to acoustic modes is given by

$$\left\langle \frac{d\varepsilon}{dt} \right\rangle_{ac} = \frac{1}{nN_cV} \sum_q \hbar v_s q \frac{\partial n_q}{\partial t}, \tag{170}$$

where nN_cV is the total number of carriers in a sample of volume V. With E_k given by the general expression (102) involving ε_k and ε_{k-q}, the integrals in Eqs. (169) and (170) must be evaluated numerically. Before discussing the results of a numerical approach it is worthwhile to consider the case of low enough fields, i.e., the warm electron region, so that the parabolic approximation Eq. (131) for E_k and the resulting simple density of states Eq. (134) may be used. It is then possible to obtain an analytic expression for the rate of energy loss,

$$\left\langle \frac{d\varepsilon}{dt} \right\rangle_{ac} = \frac{8}{\pi} \frac{ev_s^2}{\mu_0} \left(\frac{T_e}{T} \right)^{1/2} \left(1 - \frac{T}{T_e} \right), \tag{171}$$

where μ_0 is the low-field acoustic mode mobility, given by Eq. (145), and T is the lattice temperature. To determine T_e, Eq. (171) must be equated to the rate of energy gain from the field, $(eE)[\mu(T_e)E]$. For not too high fields, where $|k - k_F|b \ll 1$, the proportionality of τ to $|k - k_F|$ means $\mu \propto T_e^{1/2}$. (The $T^{-1/2}$ dependence of μ for thermal carriers given in Eq. (145) is the result of a factor $T^{1/2}$ for the carrier temperature and T^{-1} from the phonon abundance when equipartition holds.) We may then write

$$\mu = \mu_0 \left(\frac{T_e}{T}\right)^{1/2}, \tag{172}$$

where again μ_0 is the low-field mobility, given by Eq. (145). With this and Eq. (171) the energy balance equation may be written

$$\frac{8}{\pi} \frac{e v_s^2}{\mu_0} \left(\frac{T_e}{T}\right)^{1/2} \left(1 - \frac{T}{T_e}\right) = e\mu_0 \left(\frac{T_e}{T}\right)^{1/2} E^2. \tag{173}$$

This leads to

$$\frac{T_e}{T} = \left[1 - \frac{\pi}{8}\left(\frac{\mu_0 E}{v_s}\right)^2\right]^{-1}. \tag{174}$$

Because we are dealing with the warm electron range, where T_e is not much greater than T, Eq. (174) may be written

$$\frac{T_e}{T} \simeq 1 + \frac{\pi}{8}\left(\frac{\mu_0 E}{v_s}\right)^2. \tag{175}$$

Making use of Eq. (175) in Eq. (172) we obtain for the fractional increase in mobility with field in the warm electron region

$$\frac{\mu(E) - \mu_0}{\mu_0} \simeq \frac{\pi}{16}\left(\frac{\mu_0 E}{v_s}\right)^2. \tag{176}$$

It should be noted that although we have referred only to LA modes so far, Eq. (176) would still be valid if μ_0 included TA modes also, i.e., if μ_0 represented the actual low-field mobility.

Equation (176) is convenient for comparison of theory with experimental data. For TTF-TCNQ $v_s = 3 \times 10^5$ cm sec. As shown in Section 16, the low-field mobility at 10 K in Cohen and Heeger's samples is 2×10^4 cm^2/V sec. Allowing for the $T^{-1/2}$ variation of μ_0 for acoustic phonon scattering gives $\mu_0 = 3 \times 10^4$ cm^2/V sec at 4 K. With these numbers in Eq. (176) we predict for a field of 10 V/cm a fractional change in μ of $\sim 20\%$, in good agreement with the data of Fig. 11. It should be noted that the field at which nonlinearity becomes observable is somewhat sample-dependent, being greater by a factor of ~ 3 for the samples of Jacobsen (Banik et al., 1981). The differences could be due to smaller μ_0, or perhaps nonuniform fields due to defects that make the gap nonuniform, as discussed in Section 15.

The first report of nonlinear conductivity in $(TMTSF)_2PF_6$ was made by Walsh et al. (1980), who found that the dc conductivity starts to increase with field at a few mV cm at 4 K. This feature, plus the reappearance of the ESR line at 4 K under a small microwave field ("spin resurrection") led Walsh et al. to conjecture that the nonlinear increase in σ is due to spin-density waves depinned by the electric field, reminiscent of charge-density waves so depinned in $NbSe_3$. Detailed measurements of the frequency, electric, and magnetic field dependences by Chaikin et al. (1976, 1981) showed, however, that the nonlinear behavior is quite different from that of $NbSe_3$, there being no threshold field, for example. The "spin resurrection" was later attributed to heating by the microwave field (Janossy et al., 1983). Chaikin et al. (1976) suggested that the nonlinearity could be due to single carrier, rather than collective, effects, specifically heating of the carriers by the field and, resulting from that, impact ionization. In fact, the $(TMTSF)_2PF_6$ data at relatively low electric fields, i.e., the warm electron region, are in reasonably good agreement with Eq. (176). Data for 4.2 K are shown in Fig. 12. On the reasonable assumption that the entire change of resistance with field at the lowest fields shown is due to change in mobility, we find from Fig. 12 that the dc field for which $\Delta\mu/\mu_0 = 0.05$ is 20 mV/cm at 4 K. With $v_s = 3 \times 10^5$ cm sec, Eq. (176) leads to $\mu_0 = 7.5 \times 10^6$ cm^2/V sec. This value is quite close to the low-field mobility of 6×10^6 cm^2/V sec deduced in Section 16 from the measured Hall mobility. It can also be seen from part b of Fig. 12 that $\Delta\mu/\mu_0$ varies with a power of E between 1.0 and 2.0, in fair agreement with Eq. (176). We believe the discrepancy is due to nonuniform resistivity, and therefore electric field intensity, in the sample, due to a spatially varying gap. Nevertheless, there is reasonable agreement between experiment and theory for the warm electron range in $(TMTSF)_2PF_6$ at 4 K, as well as for TTF-TCNQ at ~10 K.

For fields beyond the warm electron range the parabolic approximation Eq. (131) that has been made for E_k as a function of $(k - k_F)$ and for the density of states, as well as equipartition for the phonons, are no longer valid. Calculations of T_e from the energy balance equation, and of μ from T_e, must then be done numerically. It is found that μ increases with increasing T_e by about a factor 2 for TTF-TCNQ, only ~15% for $(TMTSF)_2PF_6$, and then decreases with further increase in T_e. This is not unexpected. Beyond the parabolic region the density of final states for scattering increases and, with increasing T_e, the size of the phonons with which the carriers interact also increases. Both these effects make phonon scattering more effective as T_e increases, causing μ to decrease. How then is it possible to account for the larger increases seen, 10^4 times as large in TTF-TCNQ? Cohen and Heeger's suggestion for the large increase was the depinning of charge-density waves at high fields. The magnitude of electric field required for depinning at

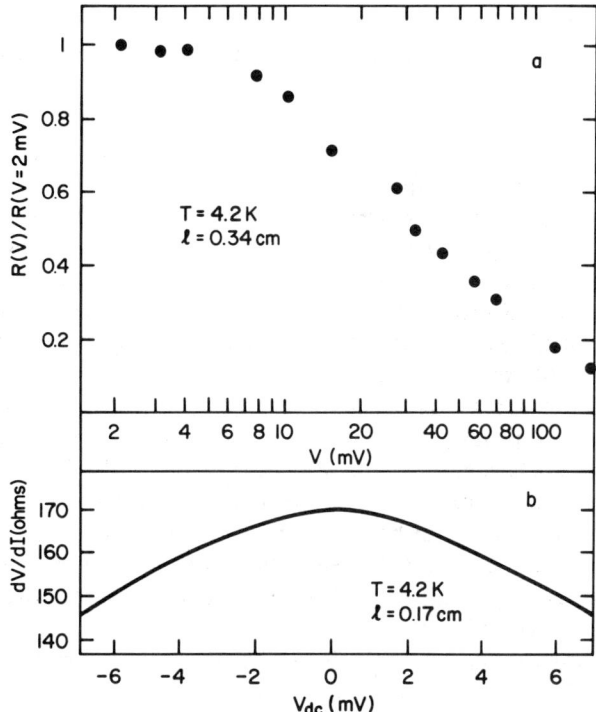

FIG. 12. (a) Electric-field dependence of the resistance of $(TMTSF)_2PF_6$. (b) Derivative of voltage with respect to current vs. electric field in the very-low-field regime. The sample length ℓ is indicated in both cases. [From Chaikin et al. (1980).]

temperatures where there are well-formed charge-density waves on TTF chains as well as TCNQ has been calculated by Lacoe et al. (1985) as 5×10^4 V/cm, a value consistent with the optically observed phase mode frequency (Tanner et al., 1981). The field required for σ to increase by a factor $>10^4$ is, however, two orders of magnitude smaller (Cohen and Heeger, 1977). In addition, Cohen and Heeger showed that the nonlinearity was not much affected by strong irradiation of the samples. Such irradiation, however, is known to cause strong pinning of charge-density waves (see Chapter 7). Impact ionization of defects, although the possibility of a small amount of this occurring cannot be eliminated, is an equally unlikely candidate for increases in σ by a factor $>10^4$. For example, if one were to assume that the lower temperature region of Fig. 10 actually had a constant slope corresponding to an activation energy of 14 K, due to some type of defect level, release of all the carriers in these levels would increase σ at 4 K by less than three orders of magnitude.

A prerequisite for understanding high-field conductivity in a given temperature range is the understanding of low-field conductivity in that range. As discussed in Section 16 we believe that the properties in the low-temperature limit are dominated by spatial nonuniformity of the gap, the result of strains and defects interacting with the charge-density waves and spin-density waves. The spatial nonuniformity we envision is on a scale greater than many mean free paths of the carriers (consistent with the high mobilities) so that it can be considered that there are sizable regions of constant gap separated by barriers. Barrier heights are expected to range from less than $k_B T$ to many times $k_B T$. This model gives rise to a number of sources for nonlinear conductivity. Heating of electrons and holes would permit more of them to surmount the barriers. This would increase σ, augmenting the increase due to increasing μ over the range of fields for which that occurs, and probably continuing to high fields. If there are carriers in shallow defect levels, as might possibly be the case for the TTF-TCNQ, their ionization could also make a contribution at not too high electron temperatures and fields. It should be noted that this mechanism for conductivity increase does not occur in $(TMTSF)_2 PF_6$ at 4 K because the constant slope (corresponding to Δ) of $\ln\sigma$ vs. $1/T$ to 4 K means that there is no contribution of impurities to σ above that temperature.

None of the mechanisms mentioned above could account for the really large increases seen in σ. A possible mechanism for a large increase is creation of electron-hole pairs by the hot electrons, first across smaller gaps and then across progressively larger gaps. To satisfy conservation of energy and momentum in this process a carrier would have to have energy $>4\Delta$. Estimates of the electron temperature at high fields, based on various assumptions about the low-frequency optical phonons with which the hot carriers can interact, in addition to the acoustic phonons, indicate that it is very unlikely that high enough temperatures could be achieved for electron-hole pair creation to occur. For $(TMTSF)_2 PF_6$ even the most favorable assumption, that the energy loss is to acoustic modes only, did not result in high enough temperatures for pair creation in the field range of the observed increases.

Another mechanism that has been suggested for the large increases in σ is based on gap decrease due to injection across the internal barriers (Conwell and Banik, 1982b). A favorable situation for injection would be that of a large-gap region in between two regions with smaller gaps. When a large voltage is applied, the number of carriers injected into the large gap region from one of the smaller gap regions could be considerably greater than the thermal equilibrium number in the large gap region, particularly if the carriers were hot. As shown in section 11, the increase in carrier concentration would cause a decrease in the gap of the large-gap region and thus an increase in σ. Although this effect in itself might not be very large, it could trigger a much larger effect. Decrease in the gap means a decrease in the height of the barrier between the small- and large-gap regions, which allows more injection, etc. The gap might decrease abruptly, but should not be destroyed; a stable situation would be achieved when it is close in magnitude to those of its injecting neighbors, for then injection would drop off greatly. Once this process is initiated, i.e., once the applied voltage is high enough to cause a sizable decrease in one or more barriers, the voltage will be redistributed so that there are larger drops across the remaining large barriers. With further increase in applied voltage, injection over these will increase more rapidly, pushing other barriers toward instability. Thus as this process proceeds, the rise of current with voltage will grow progressively steeper. We suggest that it is this process that gives rise to the very steep rise in σ with E observed in TTF–TCNQ above 150 V/cm and in $(TMTSF)_2PF_6$ above 0.6 V cm at 2 K.

It is clear that a great deal of work must yet be done to understand the properties of TTF–TCNQ, $(TMTSF)_2PF_6$ and related materials in the semiconducting range, and indeed to understand the properties of carriers and of charge- and spin-density waves in real samples with defects.

References

Allender, D., Bray, J. W., and Bardeen, J. (1974). *Phys. Rev. B* **9**, 119.
Andrieux, A., Schultz, H. J., Jerome, D., and Bechgaard, K. (1979). *Phys. Rev. Lett.* **43**, 227.
Banik, N. C., Conwell, E. M., and Jacobsen, C. S. (1981). *Solid State Commun.* **38**, 267.
Bardeen, J. (1973). *Solid State Commun.* **13**, 359.
Bates, F. E., Eldridge, J. E., and Bryce, M. R. (1981). *Can. J. Phys.* **59**, 339.
Bechgaard, K., Jacobsen, C. S., Mortensen, K., Pedersen, H. J., and Thorup, N. (1980). *Solid State Commun.* **33**, 1119.
Bechgaard, K. (1982). *Mol. Cryst. Liq. Cryst.* **79**, 1.
Begg, I. D., Narang, R. S., Roberts, K. J., and Sherwood, J. N. (1980). *J. of Crystal Growth* **49**, 735.
Berggren, K. F., and Huberman, B. A. (1978). *Phys. Rev. B* **18**, 3369.
Berlinsky, A. J., Carolan, J. F., and Weiler, L. (1974). *Solid State Commun.* **15**, 795.
Bernstein, U., Chaikin, P. M., and Pincus, P. (1975). *Phys. Rev. Lett.* **34**, 271.
Bloch, A. N., Carruthers, T. F., Poehler, T. O., and Cowan, D. O. (1976). *In* "Chemistry and Physics of One-Dimensional Metals" (H. J. Keller, ed.), pp. 47–86. Plenum, New York.
Boriack, M. L., and Overhauser, A. W. (1977). *Phys. Rev. B* **15**, 2847.
Buravov, L. I., Lyubovskaya, R. N., Lyubovsky, R. B., and Khidekel, M. L. (1976). *Zh. Eksp. Teor. Fiz.* **70**, 1982 (*Sov. Phys.-JETP* **43**, 1033).
Chaikin, P. M., Kwak, J. F., Jones, T. E., Garito, A. F., and Heeger, A. J. (1973). *Phys. Rev. Lett.* **31**, 601.
Chaikin, P. M., Kwak, J. F., Greene, R. L., Etemad, S., and Engler, E. M. (1976). *Solid State Commun.* **19**, 1201.

Chaikin, P. M. (1980). *In* "Proceedings of the NATO Advanced Study Institute on the Physics and Chemistry of Low-Dimensional Solids, Tomar, Portugal" (L. Alcacer, ed.), p. 53. Reidel, Holland.
Chaikin, P. M., Grüner, G. Engler, E. M., and Greene, R. L. (1980). *Phys. Rev. Lett*. **45**, 1874.
Chaikin, P. M., Haen, P., Engler, E. M., and Greene, R. L. (1981). *Phys. Rev. B* **24**, 7155.
Chiang, C. K., Cohen, M. J., Newman, P. R., and Heeger, A. J. (1977). *Phys. Rev. B* **16**, 5163.
Cohen, M. J., and Heeger, A. J. (1977). *Phys. Rev. B* **16**, 688.
Cohen, M. J., Gunning, W. J., and Heeger, A. J. (1978). *In* "Quasi-One-Dimensional Conductors I" (Proceedings Dubrovnik, 1978) (S. Barisic, A. Bjelis, J. R., Cooper, and B. Leontic, eds.), *Lecture Notes in Physics* **95**, 279. Springer, Berlin.
Conwell, E. M. (1967). "High Field Transport in Semiconductors," p. 120. Academic Press, New York.
Conwell, E. M. (1977). *Phys. Rev. Lett.* **39**, 777.
Conwell, E. M. (1978). *Solid State Commun.* **27**, 817.
Conwell, E. M. (1980a). *Solid State Commun.* **33**, 17.
Conwell, E. M. (1980b). *Phys. Rev. B* **22**, 1761.
Conwell, E. M., Epstein, A. J., and Rice, M. J. (1978). *In* "Quasi One-Dimensional Conductors I" (Proceedings Dubrovnik, 1978) (S. Barisic, A. Bjelis, J. R. Cooper, and B. Leontic, eds.), *Lecture Notes in Physics* **95**, 204. Springer, Berlin.
Conwell, E. M., and Banik, N. C. (1981a). *Phys. Rev. B* **24**, 4883.
Conwell, E. M., and Banik, N. C. (1981b). *Solid State Commun.* **39**, 411.
Conwell, E. M., and Banik, N. C. (1982a). *Mol. Cryst. Liq. Cryst.* **79**, 95.
Conwell, E. M., and Banik, N. C. (1982b). *Phys. Rev. B* **26**, 530.
Conwell, E. M., and Banik, N. C. (1983). *Phys. Rev. B* **27**, 7420.
Conwell, E. M., and Jacobsen, C. S. (1981). *Solid State Commun.* **40**, 203.
Cooper, J. R., Jerome, D., Weger, M., and Etemad, S. (1975). *J. Phys. (Paris)* **36**, 219.
Cooper, J. R., Miljak, M., Delplanque, G., Jerome, D., Weger, M., Fabre, J. M., and Giral, L. (1977). *J. Phys. (Paris)* **38**, 1097.
Cooper, J. (1979). *Phys. Rev. B* **19**, 2404.
Davis, D. (1973). *Phys. Rev. B* **7**, 129.
Debray, D., Millet, R., Jerome, D., Barisic, S., Giral, L., and Fabre, J. M. (1977). *J. Phys. Lett. (Paris)* **38**, L-277.
Dows, D. A., Hsu, L., Mitra, S. S., Brafman, O., Hayek, M., Daniels, W. B., and Crawford, R. K. (1973). *Chem. Phys. Lett.* **22**, 595.
Eldridge, J. E. (1977). *Bull. Am. Phys. Soc.* **22**, 288.
Engler, E. M., Tomkiewicz, Y., Mortensen, K., and Greene, R. L. (1981). *Bull. Am. Phys. Soc.* **26**, 214.
Etemad, S. (1976). *Phys. Rev. B* **13**, 2254.
Etemad, S., Engler, E.M., Schultz, T. D., Penney, T., and Scott, B. A. (1978). *Phys. Rev. B* **17**, 513.
Friedman, L. (1965). *Phys. Rev. A* **140**, 1649.
Friedman, L. (1981). *Solid State Commun.* **40**, 41.
Friend, R. H., Miljak, M., Jerome, D., Decker, D. L., and Debray, D. (1978). *J. Phys. Lett. (Paris)* **39**, L-134.
Fukuyama, H., and Lee, P. A. (1978). *Phys. Rev. B* **17**, 535.
Gosar, P., and Choi, S.-I. (1966). *Phys. Rev.* **150**, 529.
Groff, R. P., Suna, A., and Merrifield, R. E. (1974). *Phys. Rev. Lett.* **33**, 418.
Gunning, W. J., Khanna, S. K., Garito, A. F., and Heeger, A. J. (1977). *Solid State Commun.* **21**, 765.
Gutfreund, H., and Weger, M. (1977). *Phys. Rev. B* **16**, 1753.

4. TRANSPORT IN QUASI-ONE-DIMENSIONAL CONDUCTORS

Gutfreund, H., Kaveh, M., and Weger, M. (1979). *In* "Quasi-One-Dimensional Conductors" (Dubrovnik, 1978) (S. Barisic, A. Bjelis, J. R. Cooper, and B. Leontic, eds.), *Lecture Notes in Physics* **95**, 105. Springer, Berlin.
Gutfreund, H., Hartzstein, C., and Weger, M. (1980). *Solid State Commun.* **36**, 647.
Hardiman, M., Grüner, G., and Greene, R. L. (1981). *Bull. Am. Phys. Soc.* **26**, 213.
Harris, A. B. (1974). *J. Phys. C* **7**, 1671.
Hawley, M. E., Poehler, T. C., Carruthers, T. F., Bloch, A. N., Cowan, D. O., and Kistenmacher, T. J. (1978). *Bull. Am. Phys. Soc.* **23**, 424.
Hawley, M. E., Bryden, W. A., Bloch, A. N., Cowan, D. O., Poehler, T. O., and Stokes, J. P. (1979). *Bull. Am. Phys. Soc.* **24**, 232.
Heeger, A. J. (1977). *In* "Chemistry and Physics of One-Dimensional Metals" (H. J. Keller, ed.), p. 87. Plenum Press, New York.
Heeger, A. J. (1979). *In* "Highly Conducting One-Dimensional Solids" (DeVreese *et al.*, eds.), p. 69. Plenum, New York.
Herman, F., Salahub, D. R., and Messmer, R. P. (1977). *Phys. Rev. B* **16**, 2453.
Hubbard, J. (1963). *Proc. Roy. Soc. A* **276**, 238.
Jacobsen, C. S., Mortensen, K., Andersen, J. R., and Bechgaard, K. (1978). *Phys. Rev. B* **18**, 905.
Jacobsen, C. S., Tanner, D. B., and Bechgaard, K. (1981a). *Phys. Rev. Lett.* **46**, 1142.
Jacobsen, C. S., Mortensen, K., Weger, M., and Bechgaard, K. (1981b). *Solid State Commun.* **38**, 423.
Janossy, A., Hardiman, M., and Grüner, G. (1983). *Solid State Commun.* **46**, 21.
Jerome, D., Muller, W., and Weger, M. (1974). *J. Phys. Lett. (Paris)* **35**, L-277.
Jerome, D., Soda, G., Cooper, J. R., Fabre, J. M., and Giral, L. (1977). *Solid State Commun.* **22**, 319.
Jonkman, H. T., VanderVelde, G. A., and Nieupoort, W. C. (1974). *Chem. Phys. Lett.* **25**, 62.
Kagoshima, S. (1980). *J. Phys. Soc. Japan* **49**, Supp. A, 857.
Kahlert, H. (1975). *Solid State Commun.* **17**, 1161.
Kaveh, H., Weger, M., and Friedman, L. (1982). *Phys. Rev. B* **26**, 3456.
Khanna, S. K., Bright, A. A., Garito, A. F., and Heeger, A. J. (1974a). *Phys. Rev. B* **10**, 2139.
Khanna, S. K., Ehrenfreund, E., Garito, A. F., and Heeger, A. J. (1974b). *Phys. Rev. B* **10**, 2205.
Kistenmacher, T. J., Phillips, T. E., and Cowan, D. O. (1974). *Acta Cryst. B* **30**, 763.
Kittel, C. (1963). "Quantum Theory of Solids." Wiley, New York.
Kivelson, S. (1983). *Phys. Rev. B* **28**, 2653.
Kotani, A. (1977). *J. Phys. Soc. Japan* **42**, 4081.
Kuper, C. G. (1966). *Phys. Rev.* **150**, 189.
Lacoe, R. C., Schulz, H. J., Jerome, D., Bechgaard, K., and Johannsen, I. (1985). *Phys. Rev. Lett.* **55**, 2351.
Landau, L. D., and Lifshitz, E. M. (1959). "Theory of Elasticity." Pergamon, Oxford.
Lipari, N. O., Duke, C. B., and Pietronero, L. (1976). *J. Chem. Phys.* **65**, 1165.
Lipari, N. O., Rice, M. J., Duke, C. B., Bozio, R., Girlando, A., and Pecile, C. (1977). *Int. J. Quantum Chem.* **11**, 583.
Long, J. P., and Slichter, C. P. (1980). *Phys. Rev. B* **21**, 4521.
Machida, K. (1981). *J. Phys. Soc. Japan* **50**, 2195.
Mazumdar, S., and Bloch, A. N. (1983). *Phys. Rev. Lett.* **50**, 207.
Megtert, S., Comès, R., Vettier, C., Pynn, R., and Garito, A. F. (1976). *Solid State Commun.* **19**, 925.
Megtert, S., Comès, R., Vettier, C., Pynn, R., and Garito, A. F. (1981). *Solid State Commun.* **37**, 875.

Metzger, R. M. (1978). In *Synthesis and Properties of Low-Dimensional Materials. NY Acad. Sci.* **313**, 145.
Monceau, P., Ong, N. P., Portis, A. M., Meerschaut, A., and Rouxel, J. (1976). *Phys. Rev. Lett.* **37**, 602.
Mortensen, K., Tomkiewicz, Y., Schultz, T. D., Engler, E. M., Patel, V. V., and Taranko, A. R. (1981). *Solid State Commun.* **40**, 915.
Mutka, H., Zuppiroli, L., Molinie, P., and Bourgoin, J. C. (1981). *Phys. Rev. B* **23**, 5030.
Owen, C. S., and Scalapino, D. J. (1972). *Phys. Rev. Lett.* **28**, 1559.
Rice, M. J., Duke, C. B., and Lipari, N. O. (1975). *Solid State Commun.* **17**, 1089.
Rice, M. J., Bishop, A. R., Krumhansl, J. A., and Trullinger, S. E. (1976). *Phys. Rev. Lett.* **36**, 432.
Rice, M. J., and Mele, E. J. (1981). *Chem. Scr.* **17**, 121.
Ritsko, J. J., Sandman, D. J., Epstein, A. J., Gibbons, P. C., Schnatterly, S. E., and Fields, J. (1975). *Phys. Rev. Lett.* **34**, 1330.
Schafer, D. E., Thomas, G. A., and Wudl, F. (1975). *Phys. Rev. B* **12**, 5532.
Schrieffer, J. R. (1964). "Theory of Superconductivity." Benjamin, Reading, Massachusetts.
Schultz, T. D. (1977). *Solid State Commun.* **22**, 289.
Schultz, T. D., and Craven, R. A. (1979). In "Highly Conducting One-Dimensional Solids" (J. T. DeVreese, R. P. Evrard, and V. E. Van Doren, eds.), Chap. 4. Plenum Press, New York.
Scott, J. C., Etemad, S., and Engler, E. M. (1978). *Phys. Rev. B* **17**, 2269.
Seiden, P. E., and Cabib, D. (1976). *Phys. Rev. B* **13**, 1846.
Shapiro, S. M., Shirane, G., Garito, A. F., and Heeger, A. J. (1977). *Phys. Rev. B* **15**, 2413.
Shirane, G., Shapiro, S. M., Comés, R., Garito, A. F., and Heeger, A. J. (1976). *Phys. Rev. B* **14**, 2325.
Soda, G., Jerome, D., Weger, M., Fabre, J. M., and Giral, L. (1976). *Solid State Commun.* **18**, 1417.
Soda, G., Jerome, D., Weger, M., Alizon, J., Gallice, J., Robert, H., Fabre, J. M., and Giral, L. (1977). *J. Phys. (Paris)* **38**, 931.
Somoano, R. B., Gupta, A., Hadek, V., Novotny, M., Jones, M., Datta, T., Deck, R., and Hermann, A. M. (1977). *Phys. Rev. B* **15**, 595.
Su, W. P., Schrieffer, J. R., and Heeger, A. J. (1980). *Phys. Rev. B* **22**, 2099.
Tanner, D. B., Cummings, K. D., and Jacobsen, C. S. (1981). *Phys. Rev. Lett.* **47**, 597.
Thomas, J. F., and Jerome, D. (1980). *Solid State Commun.* **36**, 813.
Torrance, J. B., Tomkiewicz, Y., and Silverman, B. D. (1977). *Phys. Rev. B* **15**, 4738.
Torrance, J. B., Mayerle, J. J., Bechgaard, K., Silverman, B. D., and Tomkiewicz, Y. (1980). *Phys. Rev. B* **22**, 4960.
Vanderbilt, D., and Mele, E. J. (1980). *Phys. Rev. B* **22**, 3939.
Van Smaalen, S., and Kommandeur, J. (1985). *Phys. Rev. B* **31**, 8056.
Van Smaalen, S., Kommandeur, J., and Conwell, E. (1986). *Phys. Rev. B* **33**, 5378.
Walker, M. B. (1978). *Can. J. Phys.* **56**, 127.
Walsh, Jr., M. W., Wudl, F., Thomas, G. A., Nalewajek, D., Hauser, J. J., Lee, P. A., and Poehler, T. (1980). *Phys. Rev. Lett.* **45**, 829.
Weger, M., and Gutfreund, H. (1978). *Comments Solid State Physics* **8**, 135.
Zallen, R., and Conwell, E. M. (1979). *Solid State Commun.* **31**, 557.
Zallen, R., Griffiths, C. H., Slade, M. L., Hayek, M., and Brafman, O. (1976). *Chem. Phys. Lett.* **39**, 85.
Ziman, J. M. (1960). "Electrons and Phonons." Oxford University Press, London.
Zuppiroli, L., and Bouffard, S. (1981). *Chem. Scri.* **17**. 199.
Zuppiroli, L., Mutka, H., and Bouffard, S. (1982). *Mol. Cryst. Liq. Cryst.* **85**, 1.

CHAPTER 5

Optical Properties

Claus S. Jacobsen

TECHNICAL UNIVERSITY OF DENMARK
DK-2800 LYNGBY, DENMARK

I.	INTRODUCTION	293
	1. Coverage	293
	2. Optical Properties of Solids	295
	3. The Problem of Time Scale in Spectroscopy	303
II.	METHODS	304
	4. General Remarks	304
	5. Specific Techniques	304
III.	GENERAL OPTICAL PROPERTIES	309
	6. Dimer Optics	309
	7. Segments and Chains: Electron–Electron Correlations	313
	8. Intramolecular and Intermolecular Transitions: Experimental	315
	9. Plasma Edge and Band Structure	330
	10. Plasmon Excitations	343
IV.	INFRARED PROPERTIES	345
	11. Plasma Edge Relaxation	346
	12. Role of Molecular Vibrations	348
	13. Lattice Vibrations	360
	14. Infrared Excitation Spectrum	360
V.	CONCLUSION	376
	LIST OF SYMBOLS AND CONVERSION FACTORS	377
	REFERENCES	377

I. Introduction

1. COVERAGE

The primary aim of the present review is to describe the optical properties of highly conducting, organic charge-transfer salts, and to consider what kind of information is obtained. The words *highly conducting* mean that we will place special emphasis on materials with room-temperature conductivities in excess of 100 Ω^{-1} cm^{-1}. Where illuminating we will also discuss properties of less conducting materials, as well as the effects of phase transitions and the properties of the usually insulating low-temperature phases of the conducting materials.

A recent review on optical properties of one-dimensional systems in general has been given by Tanner (1982). Our approach will be to consider what happens to the optical absorption spectra in passing from the isolated molecule to molecular dimers and on to stacks of molecules.

We will then concentrate on optical transitions related to the transfer of electrons between molecules and electron coupling to phonons and intramolecular vibrations. Thus the frequency range of interest goes from basically zero frequency through the visible.

Many experimental techniques have been employed. We shall include results obtained by transmission and reflectance spectroscopy, photoconductivity, bolometry, and Raman scattering. We will also briefly discuss electron energy-loss experiments, since these bear directly on optical properties. Not included are photoemission and related techniques.

As an introduction, Fig. 1 shows the complexity of the overall optical properties of TTF-TCNQ at 300 K. The considerable anisotropy is striking. Most of the transitions in the frequency range above 10,000 cm^{-1} are intramolecular in nature. The absorption band centered around 3000 cm^{-1} and only observed along the stacking axis (b) is the band of main interest, associated with the charge-transfer processes along the stacks and thus related to the static conductivity.

To be able to characterize this band we will, in the next section, briefly discuss simple, conventional theory for optical properties of solids and present the qualitative features of semiconductors, metals, and superconductors. Theory more directly aimed at the molecular metals will be discussed and referenced in the main test where appropriate.

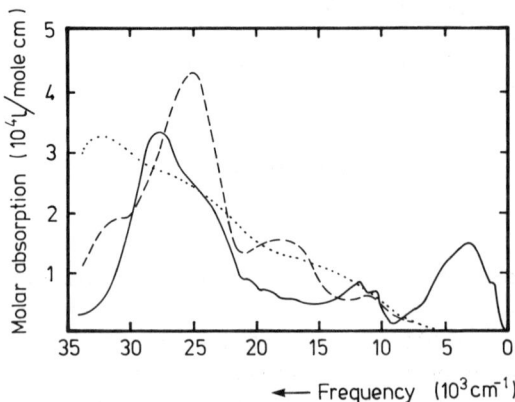

FIG. 1. Molar absorption coefficients for TTF-TCNQ as obtained by analyzing polarized reflectance data. Note that low frequencies are to the right. [From Tanaka *et al.* (1978).]

2. Optical Properties of Solids

In this section we give a simple introduction to the optical properties of isotropic solids. For a more thorough treatment the reader is referred to, for example, Wooten (1972). A plane electromagnetic wave in a medium propagates with a complex wave vector $\tilde{\mathbf{q}}$ and a cyclic frequency ω. If, for example, $\tilde{\mathbf{q}} \parallel \hat{x}$, the electric field can be described by the function

$$\mathbf{E}(x, t) = \hat{y} E_0 \exp[i(\tilde{q}x - \omega t)]. \tag{1}$$

E_0 is the amplitude at $x = 0$, and $\hat{y}$ indicates one possible polarization of the field (perpendicular to $\mathbf{q}$). $\tilde{q}$ and ω are connected through a dispersion relation

$$\tilde{q}^2 = \frac{\omega^2}{c^2} \tilde{\varepsilon}(\omega), \tag{2}$$

where c is the speed of light in vacuum and $\tilde{\varepsilon}(\omega)$ is a complex response function, the transverse dielectric function, which characterizes the optical properties of the medium. Equation (2) is obtained from the macroscopic Maxwell equations.

To see how $\tilde{\varepsilon}(\omega)$ affects the field propagation in the medium, it is useful to introduce the complex index of refraction by

$$\tilde{N} = \sqrt{\tilde{\varepsilon}} = n + i\kappa, \tag{3}$$

where n is the refractive index and κ is the extinction coefficient. Equations (1) and (3) give

$$\mathbf{E}(x, t) = \hat{y} E_0 e^{-(\omega/c)\kappa x} e^{i(\omega/c)nx} e^{-i\omega t}. \tag{4}$$

Hence the light propagates with a wavelength $\lambda = 2\pi c/n\omega$ and an attenuation per unit length $\omega\kappa/c$, corresponding to an energy absorption coefficient

$$\alpha = \frac{2\omega\kappa}{c}. \tag{5}$$

An optical experiment will involve one or more boundaries between solid and vacuum (air). The appropriate reflection and transmission coefficients can be determined from the boundary conditions on the fields and the geometry of the problem. For example, the normal incidence energy reflection coefficient, normally called *reflectance*, is given by

$$R = \left| \frac{\tilde{N} - 1}{\tilde{N} + 1} \right|^2 = \left| \frac{\sqrt{\tilde{\varepsilon}} - 1}{\sqrt{\tilde{\varepsilon}} + 1} \right|^2. \tag{6}$$

If R is known over a wide frequency range, $\tilde{\varepsilon}(\omega)$ can be determined by use of dispersion relations (see Section 5).

The $\tilde{\varepsilon}(\omega)$ used above contains contributions from current in phase with the electric field. It is often useful, especially for metals, to describe the imaginary part of $\tilde{\varepsilon}$ by a frequency-dependent conductivity, $\sigma(\omega)$:

$$\tilde{\varepsilon}(\omega) = \varepsilon_1(\omega) + i\varepsilon_2(\omega) = \varepsilon(\omega) + i\frac{\sigma(\omega)}{\varepsilon_0 \omega}, \quad (7)$$

where now ε and σ are real functions. We note that σ defined in this way describes all lossy phenomena.

The dielectric function for an electronic solid-state system, which is described in terms of Bloch functions, $u_{\mathbf{k},\ell}$, can most easily be obtained through the self-consistent field (SCF) or random phase approximation (RPA) (Ehrenreich and Cohen, 1959; see also Wooten, 1972). $\tilde{\varepsilon}(\omega)$ obtained in this way is actually describing the longitudinal response to a test charge, but in the optical, long-wavelength limit the transverse and longitudinal response must be the same for the same direction of the electric field. This also holds for principal axes in anisotropic crystals.

Neglecting local field effects (as may be appropriate for a metal) we may write the result, in general,

$$\tilde{\varepsilon}(\omega) = 1 + \Delta\varepsilon_{\text{intraband}} + \Delta\varepsilon_{\text{interband}}, \quad (8)$$

where

$$\Delta\varepsilon_{\text{intraband}} \cong -\frac{e^2}{\varepsilon_0 \omega^2}\left(\sum_{\mathbf{k}} \frac{1}{\hbar^2}\frac{\partial^2 \varepsilon_{\mathbf{k}}}{\partial k_e^2} f(\varepsilon_{\mathbf{k}})\right) \quad (9)$$

and

$$\Delta\varepsilon_{\text{interband}} \cong \frac{2e^2}{\varepsilon_0 m^2}\left(\sum_{\mathbf{k},\ell,\ell'} \frac{|P^{\mathbf{k}}_{\ell'\ell}|^2/\hbar\omega_{\ell'\ell}(\mathbf{k})}{\omega_{\ell'\ell}^2(\mathbf{k}) - \omega^2} f(\varepsilon_{\mathbf{k},\ell})\right). \quad (10)$$

Equation (9) is only included if one band, ℓ, is partially filled (metal). $\varepsilon_{\mathbf{k}}$ is the one-electron energy in this band, and $\mathbf{k}$ runs over the first Brillouin zone. $f(\varepsilon_{\mathbf{k}})$ is the Fermi–Dirac occupation number. $\partial^2 \varepsilon_{\mathbf{k}}/\partial k_e^2$ is to be calculated along the direction of the electric field. Equation (9) may be written in the simple form

$$\Delta\varepsilon_{\text{intraband}} = -\frac{\omega_p^2}{\omega^2}, \quad \omega_p^2 = \frac{ne^2}{\varepsilon_0}\left(\frac{1}{m^*}\right)_{\text{opt}}, \quad (11)$$

where ω_p is the plasma frequency, n is the density of carriers in the band, and $(m^*)^{-1}_{\text{opt}}$ is the effective mass in the field direction, averaged over occupied states according to Eq. (9). In Eq. (10) m is the free electron mass, ℓ and ℓ' are band indices, $\omega_{\ell'\ell}(\mathbf{k}) = [\varepsilon_\ell(\mathbf{k}) - \varepsilon_{\ell'}(\mathbf{k})]/\hbar$, and $P^{\mathbf{k}}_{\ell'\ell}$ is a matrix element of the momentum **p**

$$P^{\mathbf{k}}_{\ell'\ell} = \frac{1}{\Omega} \int_{\text{unit cell}} u^*_{\mathbf{k},\ell} (\mathbf{e} \cdot \mathbf{p}) u_{\mathbf{k},\ell'} \, d^3r, \tag{12}$$

where **e** is the field polarization vector and Ω is the unit cell volume. Note that momentum conservation requires a virtually unchanged **k** vector in the optical transition (direct optical transition).

If we have a metal, where all $\omega_{\ell\ell'} \gg \omega_p$, Eq. (10) yields a nearly frequency-independent contribution to ε for $\omega \lesssim \omega_p$. Then

$$\tilde{\varepsilon}(\omega) \cong \varepsilon_\infty - \frac{\omega_p^2}{\omega(\omega + i\Gamma)}. \tag{13}$$

Here a relaxation rate, Γ, is introduced to describe damping. This is the often used Drude expression for the optical properties of a metal. The reflectance, ε and σ are shown schematically in Fig. 2. Note the characteristic plasma edge in reflectance at $\omega_p/\sqrt{\varepsilon_\infty}$, where ε crosses zero.

For a semiconductor Eq. (9) does not contribute and Eq. (10) gives a continuum of Lorentzian oscillators. Important are the joint density of states describing the density of possible wave-vector-conserving transitions, as well as the matrix elements. Since these often vary slowly over the Brillouin zone, structure in ε tends to be dominated by critical points (including the semiconducting gap, if direct) in the joint density of states. The optical properties can typically look like those shown in Fig. 3.

The above picture is a crude description of the optical properties. More accurate models must consider the damping mechanism and include phonon absorption, indirect phonon assisted transitions, local field effects, electron-electron interactions, etc.

As an example, Eqs. (8)-(10) can not describe the properties of metal in the superconducting state. It is of some interest for our later discussion to see how ε and σ are modified when a transition from a normal to a superconducting state occurs. BCS theory allows numerical calculations to be done (Mattis and Bardeen, 1958), and the result is shown schematically in Fig. 4 for $T = 0$. The conductivity now contains a δ-function contribution at zero frequency, but it should be noted that the oscillator strength in this mode is rather small. This strength has been taken from the gap region $0 < \omega < 2\Delta/\hbar$. For conventional superconductors the gap is situated in the very far infrared (typically below 50 cm^{-1}).

However, even a metal in the normal state shows important deviations from Drude behavior at low temperature. First, consider an ideal metal with a rigid lattice. Here $\Gamma = 0$, and the frequency-dependent conductivity, $\sigma(\omega)$, simply consists of a δ-function at zero frequency holding the full oscillator

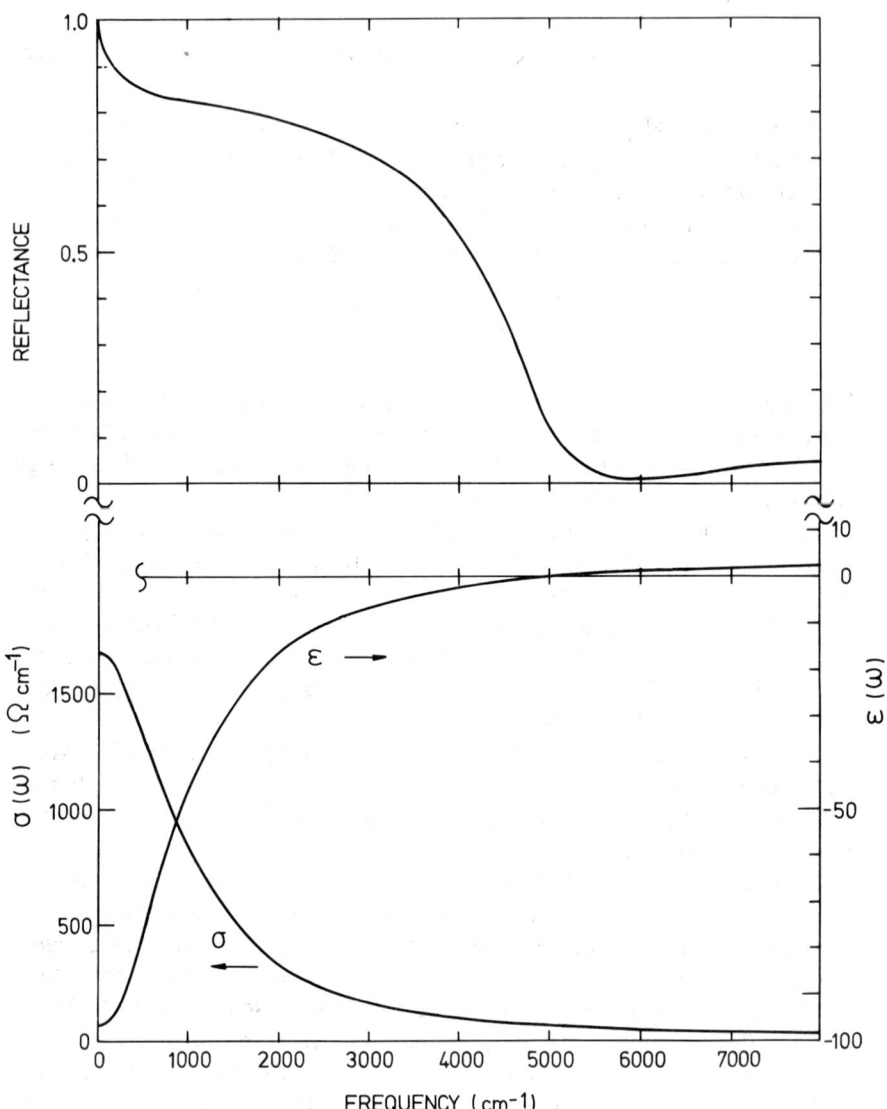

Fig. 2. Optical properties of a metal based on the Drude formula. Parameters used: $\omega_p = 10{,}000\ \text{cm}^{-1}$, $\Gamma = 1000\ \text{cm}^{-1}$, $\varepsilon_\infty = 4.0$.

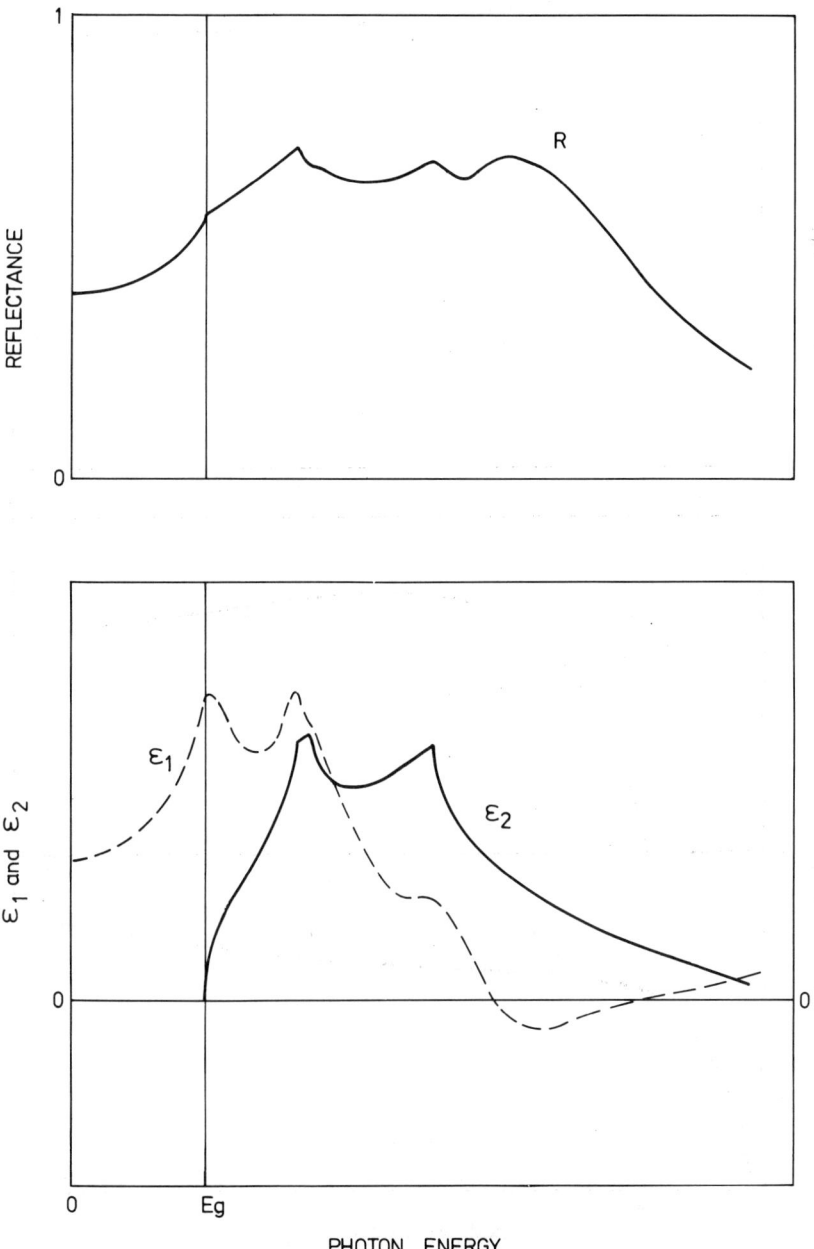

FIG. 3. Optical properties of a three-dimensional semiconductor with a direct energy gap E_g.

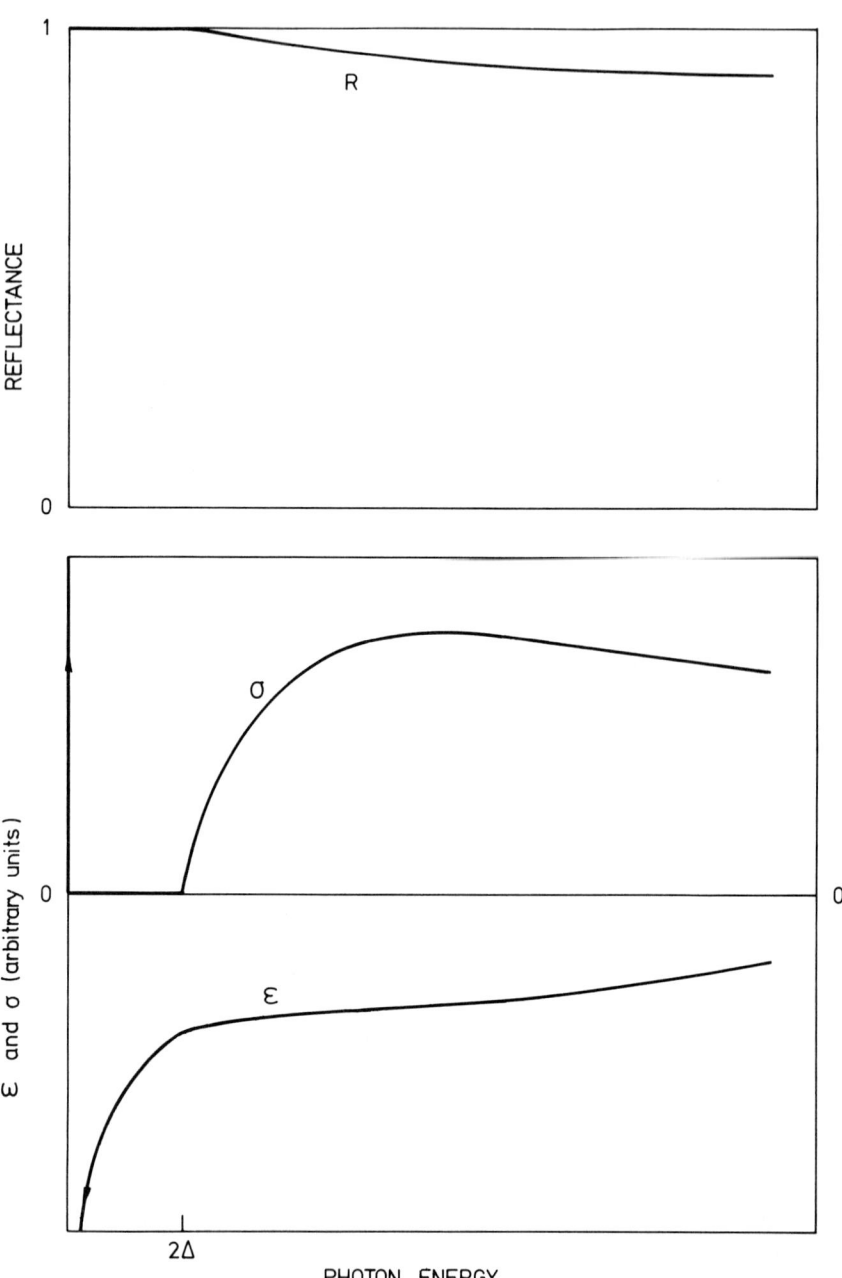

FIG. 4. Optical properties of a superconductor at $T = 0$. The solid bar in the σ plot indicates a true δ function. 2Δ is the superconductor gap. The energy range is typically in the far infrared.

strength ω_p^2. Dumke (1961) introduced an *ad hoc* elastic scattering mechanism in a quantum treatment and found Drude behavior at all frequencies. He also found that higher bands usually do not influence the intraband absorption significantly.

A more realistic description takes into account the electron–phonon interaction, which gives rise to inelastic scattering processes. Holstein (1954, 1964) did the pioneering work on this problem, and Allen (1971) has given a good qualitative discussion. The main deviations from Drude behavior arise in the pure metal at low temperature, where the electron–phonon coupling does not contribute to the dc resistivity simply because no current-degrading phonon states are occupied. Then the dc relaxation rate is determined by residual impurities and defects, and a near δ-function contribution to $\sigma(\omega)$ is obtained (typical width $\ll 10 \, \text{cm}^{-1}$). As ω is increased and reaches typical phonon frequencies, the Holstein emission process becomes possible. In this process a photon is absorbed and a phonon is emitted, thus transferring a corresponding momentum to the electron. It is schematically shown in Fig. 5a.

The Holstein processes produce infrared absorption thresholds with onset at the various phonon frequencies. Since the total oscillator strength is conserved, the new strength is removed from the narrow mode at dc, not by a change in the width, but rather by renormalization of the optical mass in Eq. (11).

At low frequencies the electrons drag along a polarization cloud of phonons. At frequencies well above all important phonon frequencies, Drude behavior is again found, but now with a relaxation rate determined by the combined effect of all phonon emission processes. The rate may be

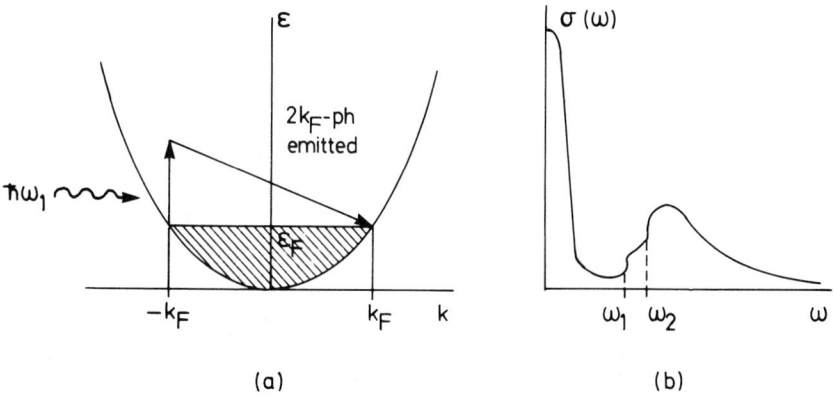

FIG. 5. Holstein absorption ($T \ll$ Debye temperature). (a) Schematic absorption process. (b) $\sigma(\omega)$ for rather strongly coupled metal.

characterized by an appropriately defined dimensionless electron–phonon coupling constant, λ_{tr} (Allen, 1971). Then at high frequencies, $\Gamma \to \pi\lambda_{tr}\langle\omega_{ph}\rangle$, where $\langle\omega_{ph}\rangle$ is an average phonon frequency. The dc effective mass is roughly renormalized to $\tilde{m}^* = m^*(1 + \lambda_{tr})$. Thus a fraction $\lambda_{tr}/(1 + \lambda_{tr})$ of the full intraband oscillator strength appears as infrared absorption. A typical spectrum reminiscent of that of lead is shown in Fig. 5b.

As the temperature is increased and $k_B T$ becomes larger than the phonon energies, phonon scattering contributes increasingly to the dc scattering rate. Consequently the low-frequency contribution broadens and gradually merges with the Holstein contribution. For $k_B T \gg \hbar\omega_{ph}$ a single Drude absorption is reestablished (corresponding to the elastic scattering limit). Then the relaxation rate is given by

$$\Gamma = \frac{2\pi\lambda}{\hbar}\left(k_B T + \frac{\hbar\langle\omega_{ph}\rangle}{2}\right). \tag{14}$$

Thus λ_{tr} may be deduced from the high-temperature properties (Hopfield, 1970).

We will conclude this section with a short description of dispersion relations and sum rules. $\tilde{\varepsilon}(\omega)$ is a linear response function and causality implies the existence of the Kramers–Kronig relations connecting real and imaginary parts (see, e.g., Wooten, 1972)

$$\varepsilon_1(\omega) - 1 = \frac{2}{\pi} P \int_0^\infty \frac{\omega' \varepsilon_2(\omega')}{\omega'^2 - \omega^2} d\omega' \tag{15a}$$

$$\varepsilon_2(\omega) = -\frac{2\omega}{\pi} P \int_0^\infty \frac{\varepsilon_1(\omega') - 1}{\omega'^2 - \omega^2} d\omega'. \tag{15b}$$

P indicates principal value. At very high frequencies, $\omega \gg \omega_{\ell\ell'}$ for all ℓ, ℓ', the electrons can be considered free. From Eq. (11)

$$\varepsilon_1(\omega) \cong 1 - \frac{\Omega_p^2}{\omega^2}, \quad \Omega_p^2 = \frac{n_{total} e^2}{\varepsilon_0 m}, \tag{16}$$

where m is the free electron mass. Comparing Eqs. (15a) and (16) we deduce the sum rule

$$\int_0^\infty \omega\varepsilon_2(\omega)\, d\omega = \frac{1}{\varepsilon_0}\int_0^\infty \sigma(\omega)\, d\omega = \frac{\pi}{2}\Omega_p^2, \tag{17}$$

proportional to the total number of electrons. If we have intraband transitions well separated from interband transitions, it is meaningful to introduce a partial sum rule based on Eqs. (11) and (17)

$$\int_0^\omega \sigma(\omega')\,d\omega' = \frac{\pi}{2}\frac{n_{\text{band}}^{\text{eff}}(\omega)e^2}{m_{\text{opt}}^*}. \tag{18}$$

Here $n_{\text{band}}^{\text{eff}}(\omega)$ can be interpreted as the effective number of electrons participating in transitions up to frequency ω. We can see that the area below the $\sigma(\omega)$ curve in a natural way corresponds to the optical oscillator strength.

3. The Problem of Time Scale in Spectroscopy

Consider a molecular metal with one carrier per two molecules. If the mean free path deduced from conductivity is several lattice constants, as is often the case, it is natural to talk about delocalized carriers. At optical frequencies, however, the distinction between localized and delocalized carriers is not unambiguous, but depends on the character of the excitation observed.

As a simple example we can consider a molecular vibration, which in an uncharged molecule has a frequency ω_1 and in a charged molecule ω_2. Typically the ionization shift $\Delta\omega = |\omega_2 - \omega_1| \ll \omega_1$, and ω_1, ω_2 are lower than the hopping rate of charge carriers, Ω, which is governed by the transfer integral, t, by $\Omega = t/\hbar$. Under these assumptions, clearly only one spectral line will be observed, i.e., vibrational spectroscopy suggests the carriers are delocalized. Now we can estimate the linewidth assuming that the original vibrations have infinite lifetimes.

In the random hopping of the electrons to and from the molecule in question, the phase in the vibration oscillation will become dephased with a mean square dephasing angle $\langle \phi^2 \rangle$, which by elementary statistics is given by

$$\langle \phi^2 \rangle = \Omega \left(\frac{1}{\Omega}\frac{\Delta\omega}{2}\right)^2 = \frac{\Delta\omega^2}{4\Omega} \tag{19}$$

per unit of time. The line width is the reciprocal of the time, which gives a dephasing of 1 radian, so the observed width should be

$$\Delta\omega_{\text{obs}} = \frac{\Delta\omega^2}{4\Omega} = \Delta\omega\,\frac{\hbar\Delta\omega}{4t} \tag{20}$$

This result implies that a single line will be observed as long as $\hbar\Delta\omega \ll 4t$, i.e., the splitting between the original lines must be much smaller than $4t$, which is just the electronic bandwidth.

When the initial splitting is comparable to the bandwidth, instead, two lines will be seen. This is typically what is observed in the charge-transfer salts for the electronic transitions in the visible, where large ionization shifts are encountered. Thus in this range, band theory is inappropriate, and the spectra should be understood as a superposition of localized excitations.

The problem is somewhat similar to that of motional narrowing in magnetic resonance spectroscopy. It is important to emphasize that it is not the actual frequency in the measurement that matters, but instead the relation between splitting and bandwidth.

II. Methods

4. General Remarks

The linear chain compounds we are concerned with are, of course, anisotropic in the extreme (cf., Fig. 1). In a correct treatment of the optical properties, we are consequently dealing with tensorial response functions, $\tilde{\bar{\varepsilon}}(\omega)$ and $\tilde{\bar{\sigma}}(\omega)$. Thus polarized measurements on oriented single crystals are called for. Some information still can be obtained from unpolarized measurements. This is especially the case in the infrared, where both reflectance and absorbance is almost completely dominated by the charge-transfer excitations along the stacks. Thus the resulting spectra will resemble polarized chain axis spectra.

In the more accurate polarized experiments it is most convenient to use the principal axes as polarization directions. Again, because of the extreme anisotropy in the infrared, the stacking axis is expected to be a principal axis for $\tilde{\bar{\varepsilon}}$ as well as for $\tilde{\bar{\sigma}}$.

In general the principal axes for $\tilde{\bar{\varepsilon}}$ and $\tilde{\bar{\sigma}}$ will coincide with the crystallographic axes only in crystals with orthorhombic or higher symmetries (Landau and Lifshitz, 1960). All charge-transfer salts of interest here have lower symmetry. In monoclinic crystals, one axis is fixed (e.g., the stacking axis in TTF-TCNQ), but the two others can have any orientation in the transverse plane. In the triclinic symmetry class, which includes most compounds, nothing can be said *a priori*. The principal axes of $\tilde{\bar{\varepsilon}}$ and $\tilde{\bar{\sigma}}$ may not coincide, and they may rotate with frequency. Such rotation of the principal axes is common on going from the infrared range to the visible. In the infrared, the charge-transfer excitations will lock one axis to the stacking axis. In the visible the axes are more likely to be determined by the orientation of the molecules (Helberg, 1976, 1985). Thus a complete determination of the optical properties may be a formidable task and has indeed not been accomplished for any material.

5. Specific Techniques

A wide range of sample preparation forms and of instrumentation is used in the studies. We cannot go into complete detail, but below we shortly discuss some of the techniques, which are either commonly used or of

The principal importance. There are a number of others, which have not been employed often, but nevertheless may be highly useful, like photoacoustic spectroscopy (Rosencwaig, 1975), thermoreflectance (Perov and Fisher, 1974), piezoreflectance (Jordan, 1981), and other modulation spectroscopic methods.

Of external parameters usually only temperature is varied, but in a few cases the effects of hydrostatic pressure and magnetic field have been studied.

a. Reflectance and Transmittance of Samples Based on Powders and Films

These techniques are quite versatile, since they allow studies on materials, which are not available as single crystals. Samples are usually prepared following standard methods and are therefore suitable for measurements in commercial spectrophotometers.

The KBr pellet method is by far the most common method. KBr is used as host material for a powder of the compound in question. Typically ~ 1 mg

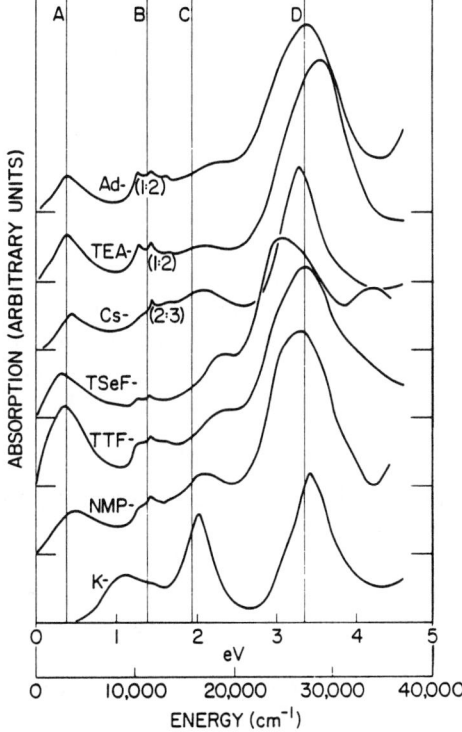

FIG. 6. Powder absorption spectra of a number of TCNQ compounds, as measured with the KBr disc method and normalized per mole TCNQ. [From Torrance *et al.* (1975a).]

is mixed with KBr powder and then pressed to a $\frac{1}{2}$-inch disc in a die. KBr is easy to use in this way and transmits radiation from 300 cm^{-1} and into the ultraviolet. The refractive index is low, about 1.5. The main practical disadvantage is the tendency of KBr to adsorb water. Water has strong absorption bands in the infrared, which may disturb the measurement.

A transmittance spectrum of a KBr pellet will contain absorption characteristics of the compound. Lineshapes may be somewhat distorted, due to the inhomogeneous nature of the sample. An example of the KBr technique is shown in Fig. 6. Here it is used to compare a range of materials. The A band is attributed to intraband transitions. Indeed, high absorption in the infrared of a KBr disc indicates a material with reasonably high conductivity. This empirical fact is used in the search for new, promising materials.

There are other sample preparation methods. Powder can be compressed to a pellet of the material. Such pellets (compactions) are commonly used for reflectance studies. The measured spectrum is somewhat complicated to interpret, since it is unpolarized and again is based on an inhomogeneous sample.

Some materials can be sublimed to form films. A film of suitable thickness (a fraction of a micron) on a KBr disc constitutes another type of sample suitable for transmittance and reflectance measurements. At long wavelengths a polycrystalline film appears reasonably homogeneous. At shorter wavelengths diffuse scattering is serious, but a quantitative analysis can, in some cases, be carried rather far (Tanner et al., 1976).

b. Reflectance and Transmittance of Single Crystals

Polarized single-crystal studies are clearly superior when quantitative measurements are needed. In the near infrared and visible, measurements can be done on needle-shaped crystals less than 0.1 mm wide. For mid-infrared and, especially, far-infrared investigations, a larger surface area is needed. It is possible to align optically several single crystals to form a mosaic with acceptable area.

The samples when used as grown single crystals are usually too thick to allow transmission measurements, except for selected polarizations in limited spectral ranges. Thus most studies are done as reflectance.

If the power reflectance, R, is measured along a principal axis and over a wide spectral range, it is possible to obtain fairly accurate values for the complex dielectric function, or for ε and σ. A dispersion (or Kramers–Kronig) analysis can be used on the logarithm of the complex reflection coefficient,

$$\log(re^{i\theta}) = \ln r + i\theta = \tfrac{1}{2} \ln R + i\theta. \tag{21}$$

Here θ is the phase shift on reflection and $r = \sqrt{R}$. θ may now be calculated from a wide-range R spectrum

$$\theta(\omega) = \frac{\omega}{\pi} P \int_0^\infty \frac{\ln R(\omega')}{\omega^2 - \omega'^2} d\omega'. \tag{22}$$

Extrapolations outside the measured range are employed (see, e.g., Wooten, 1972). The low-frequency extrapolation is not so critical, provided the data cover nearly the entire range of interest. However, it is adjusted to make the result agree with independent information on low-frequency properties (e.g., dc conductivity and microwave dielectric constant). A frequently-adopted low-frequency extrapolation for conductors is the Hagen–Rubens form $R = 1 - A\sqrt{\omega}$, which follows the Drude spectrum for $\omega \to 0$.

At high frequencies the measured data should cover a range well beyond the frequencies of main interest. Even so, the result of, especially, sum rule calculations will, to some extent, depend on the choice of high-frequency extrapolation. Therefore, the latter is chosen so that Eq. (22) agrees with independently-measured values of $\theta(\omega)$ at a few frequencies near the plasma frequency. $\theta(\omega)$ especially for quasi-one-dimensional systems, may be measured with simple ellipsometric methods. A good method has been described by Young and Walker (1977). Now $\varepsilon(\omega)$ and $\sigma(\omega)$ are calculated from

$$\varepsilon(\omega) + i\frac{\sigma(\omega)}{\varepsilon_0 \omega} = \left(\frac{1 + \sqrt{R(\omega)}e^{i\theta(\omega)}}{1 - \sqrt{R(\omega)}e^{i\theta(\omega)}}\right)^2. \tag{23}$$

In the calculation of the phase, Eq. (22), we note that $\theta(\omega)$ is independent of a constant multiplicative factor on $R(\omega')$. It is the relative variation in $R(\omega')$ that contributes to $\theta(\omega)$. This means that ranges with low reflectance levels should be measured with the same relative accuracy as ranges with high reflectance. When $R \lesssim 1$ and θ is small, the denominator in Eq. (23) is small. Thus small errors under such circumstances can lead to large errors in ε and σ. This is the case for $\omega \to 0$ in metals, where R always approaches unity. Special care should therefore be taken to obtain correct absolute values of R in the infrared to far infrared.

If $R(\omega)$ is only determined over a limited spectral range, it is usually better to analyze the data by a fitting procedure than to attempt the Kramers–König scheme. Standard models include the Drude expression for metals, Eq. (13), and Lorentzian oscillators for interband transitions.

c. Measurements Based on Photoconductivity and Bolometry

These are specialised techniques used to study the low-temperature, semiconducting phase. The sample consists of a single crystal with leads attached for the resistance to be measured.

Photons may excite carriers across the semiconducting gap, leading to a decrease in resistance (photoconductivity). On the other hand, the sample may be heated when the photons are absorbed. Due to the activated conductivity, the heating, likewise, will give rise to a diminished resistance (bolometric effect). The two effects can be separated by their frequency and temperature dependence and give complementary information.

Photoconductivity constitutes, presumably, the most accurate way of determining the size of direct electronic gaps. Bolometry is a sensitive method to study, for example, vibrational modes, when transmission experiments are not possible.

When absorbance is high and reflectance is close to unity, the bolometric technique basically measures $1 - R$ (Eldridge and Bates, 1983). Thus, in that case the method may be used to determine R with high accuracy. Eldridge (1985) and Eldridge and Bates (1985) have further developed a composite bolometer technique, where a small germanium bolometer is attached to the sample. This extends the applicability to conducting crystals (although the method is still only useful at very low temperatures).

d. Raman Spectroscopy

The method is not aimed at a determination of $\tilde{\varepsilon}$ but is a light-scattering technique. An incident photon absorbs or emits a phonon and is reflected or transmitted with a shift in frequency corresponding to the phonon energy.

The technique is widely used for studying molecular vibrations, as a complement to infrared absorption spectroscopy. A specific mode may be either Raman active, infrared active, or both.

Of special interest here is the fact that amplitude modes of charge-density waves are Raman active.

e. Electron Energy Loss Spectroscopy

This technique does not use light as a probe. Still it has direct bearings on the optical properties. Fast electrons probe the elementary excitations of the solid by their momentum shift and loss of energy, and it is possible to determine $-\text{Im}(1/\tilde{\varepsilon}_L)$, where $\tilde{\varepsilon}_L$ is the longitudinal dielectric function. From a dispersion relation the full frequency and wavevector-dependent $\tilde{\varepsilon}_L(\omega, \mathbf{q})$ can in principle be obtained.

For $q \to 0$ it is of interest to compare the results with those obtained by light spectroscopy. This is a direct check on the equality of the transverse and longitudinal dielectric function at small momentum transfers.

For general $\mathbf{q}$ supplementary information is obtained (e.g., on plasmon dispersion). The method has limited resolution (typically, 0.1 eV or 1000 cm^{-1}).

The method requires elaborate equipment.

III. General Optical Properties

In this part we shall focus on the overall optical properties. We are, in particular, interested in the charge-transfer or intraband excitations and how they develop from transitions in a simple building block like the molecular dimer.

The key parameters in a dimer are the energy-transfer integral, t, and the on-site Coulomb repulsion energy, U, which both influence the optical properties. The transition to an organic conductor with uniform stacks and high dc conductivity is not trivial and in no way completely understood.

Some questions one would like to answer include: (1) To what extent can the intraband transitions be described by the one-electron band theory, on which the SCF dielectric function of Eq. (13) is based? If the plasma frequency is correctly given, then the one-electron bandwidth can be derived from near infrared spectra. (2) How does the short-range Coulomb repulsion influence the spectra?

As we shall see, it is presently not possible to reach a decisive conclusion on these points. However, it does seem that the oscillator strength associated with charge-transfer bands involving U in all studied, highly conducting materials is quite small.

In Sections 6–10 we take the following approach: The theoretically expected dielectric function for the isolated dimer with one and two carriers is introduced. Then the theoretical approach to larger assemblies and to the uniform chain with strong Coulomb interactions is shortly reviewed.

Next we describe how the experimental spectra compares with the expectations, and what changes are actually found on going to highly conducting substances with uniform chaims. The main objective is to investigate whether the plasma edge in reflectance can be properly analyzed in terms of a Drude model based on the SCF dielectric function.

Finally we address the question of the corresponding plasmons. These are observed in electron energy-loss spectroscopy and are, under certain circumstances, excitable optically.

6. Dimer Optics

The dimer is the simplest entity in which intermolecular interactions can be studied.

For localized electrons the appropriate dielectric function can be written in the form (see, e.g., Atkins, 1970):

$$\tilde{\varepsilon}(\omega) = \varepsilon_\infty + \frac{2N}{\hbar\varepsilon_0} \sum_n \frac{\omega_{n0}|d_{0n}|^2}{\omega_{n0}^2 - \omega^2 - i\omega\gamma_n}, \qquad (24)$$

where the summation runs over final states, n. N is the density of dimers, and d_{0n} is a matrix element characterizing the dimer transitions for a field polarized along **e**

$$d_{0n} = \langle 0| \sum_i e\mathbf{r}_i |n\rangle \cdot \mathbf{e}. \tag{25}$$

As usual $\omega_{n0} = (E_n - E_0)/\hbar$.

a. The Dimer With One Electron

We consider a single nondegenerate orbital on each of the molecules, a and b. The interaction Hamiltonian is

$$\tilde{H} = -t(\tilde{a}^+\tilde{b} + \tilde{b}^+\tilde{a}), \tag{26}$$

which defines the transfer integral, t. The symmetrized wave functions are

$$\psi_1 = \frac{1}{\sqrt{2}}(\tilde{a}^+ + \tilde{b}^+)|0\rangle$$

$$\psi_2 = \frac{1}{\sqrt{2}}(\tilde{a}^+ - \tilde{b}^+)|0\rangle \tag{27}$$

in which $\tilde{H}$ is diagonal,

$$E_1 = -t \quad \text{and} \quad E_2 = +t. \tag{28}$$

Introducing Eqs. (27)–(28) in Eq. (24) yields

$$\tilde{\varepsilon}(\omega) = \varepsilon_\infty + \frac{N_{\text{dimer}}}{\hbar^2 \varepsilon_0} \frac{ta^2 e^2}{4t^2 - \omega^2 - i\omega\gamma}. \tag{29}$$

a is the distance between equivalent points on the molecules. In this model the dimers are assumed noninteracting. We can now think of the corresponding uniform chain being obtained by moving the dimers together until the interdimer overlap equals the intradimer overlap. We expect about twice the number of transitions, since transitions between the previous dimers are also possible.

The proper solid-state description is that of a quarter-filled tight-binding band. The SCF dielectric function of Eqs. (8)–(11) is in that case

$$\tilde{\varepsilon}(\omega) = \varepsilon_\infty - \left(\frac{2\sqrt{2}}{\pi}\right) \frac{2n_{\text{electr}}}{\hbar^2 \varepsilon_0} \frac{ta^2 e^2}{\omega(\omega + i\gamma)}. \tag{30}$$

Since $n_{\text{electr}} = N_{\text{dimer}}$, the expectations are clearly fulfilled: $\tilde{\varepsilon}(\omega)$ of Eq. (30) shows an 80% increase in oscillator strength over that of Eq. (29). In addition, the oscillator frequency has gone to zero, corresponding to the levels broadening into bands and merging, when the uniform chain is formed. It is important to note that the only difference in oscillator strength is through the numerical prefactor. The dependence on n, t, and a is unchanged.

b. *The Dimer with Two Electrons*

In the discussion above the Coulomb interaction between electrons was irrelevant. When we are dealing with a dimer with two electrons, the Coulomb repulsion energy will be higher by an amount U, when the two electrons reside on the same molecule than when there is one on each. Following Harris and Lange (1967) (see also Rice, 1979), the proper Hamiltonian is

$$\tilde{H} = -t \sum_\sigma (\tilde{a}_\sigma^+ \tilde{b}_\sigma + \tilde{b}_\sigma^+ \tilde{a}_\sigma) + U(\tilde{n}_{a\uparrow}\tilde{n}_{a\downarrow} + \tilde{n}_{b\uparrow}\tilde{n}_{b\downarrow}). \tag{31}$$

σ is the spin variable and $\tilde{n}$ are number operators. Equation (31) is the dimer version of the Hubbard model for narrow-band systems (Hubbard, 1963). A properly symmetrized basis set of wave functions is constituted by

$$\psi_1 = \frac{1}{\sqrt{2}}(\tilde{a}_\uparrow^+ \tilde{a}_\downarrow^+ + \tilde{b}_\uparrow^+ \tilde{b}_\downarrow^+)|0\rangle$$

$$\psi_2 = \frac{1}{\sqrt{2}}(\tilde{a}_\uparrow^+ \tilde{a}_\downarrow^+ - \tilde{b}_\uparrow^+ \tilde{b}_\downarrow^+)|0\rangle$$

$$\psi_3 = \frac{1}{\sqrt{2}}(\tilde{a}_\uparrow^+ \tilde{b}_\downarrow^+ + \tilde{a}_\downarrow^+ \tilde{b}_\uparrow^+)|0\rangle \tag{32}$$

$$\psi_4 = \frac{1}{\sqrt{2}}(\tilde{a}_\uparrow^+ \tilde{b}_\downarrow^+ - \tilde{a}_\downarrow^+ \tilde{b}_\uparrow^+)|0\rangle$$

$$\psi_5 = \tilde{a}_\uparrow^+ \tilde{b}_\uparrow^+ |0\rangle$$

$$\psi_6 = \tilde{a}_\downarrow^+ \tilde{b}_\downarrow^+ |0\rangle.$$

ψ_4, ψ_5, and ψ_6 are triplet states. All matrix elements involving these are zero. For ψ_1, ψ_2, and ψ_3, the matrix elements of $\tilde{H}$ are

$$\begin{bmatrix} U & 0 & 2t \\ 0 & U & 0 \\ 2t & 0 & 0 \end{bmatrix} \tag{33}$$

It is now straightforward to calculate eigenvalues and eigenfunctions and determine $\tilde{\varepsilon}(\omega)$ from Eq. (24). The level diagram is shown in Fig. 7. The only allowed optical transition from the ground state is to ψ_2 with energy U.

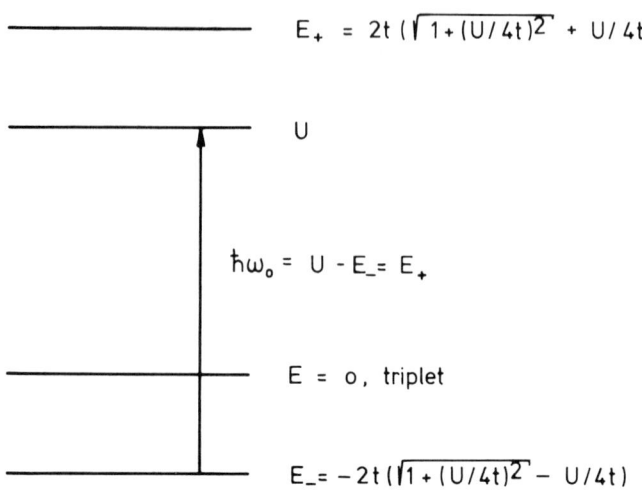

FIG. 7. Energy-level diagram for the dimer with two electrons. The allowed optical transition from the ground state is shown.

The dielectric function is

$$\tilde{\varepsilon}(\omega) = \varepsilon_\infty + \frac{2N_{\text{dimer}}}{\hbar^2 \varepsilon_0} \frac{1}{\sqrt{1 + (U/4t)^2}} \frac{ta^2 e^2}{\omega_0^2 - \omega^2 - i\omega\gamma}, \quad (34)$$

where

$$\omega_0 = \left(2t\sqrt{1 + \left(\frac{U}{4t}\right)^2} + \frac{U}{2}\right)\Big/\hbar. \quad (35)$$

For $U \to 0$ Eq. (34) and Eq. (29) give the same result, except that the oscillator strength of Eq. (34) is twice that of Eq. (29) (we have twice the number of electrons). In this limit the SCF dielectric function for the corresponding uniform chain (half-filled band) is

$$\tilde{\varepsilon}(\omega) = \varepsilon_\infty - \frac{n_{\text{electr}}}{\hbar^2 \varepsilon_0} \left(\frac{4}{\pi}\right) \frac{ta^2 e^2}{\omega(\omega + i\gamma)}. \quad (36)$$

The oscillator strength is now only 30% over that of Eq. (34), due to banding effects.

In the other limit, for $U/4t \gg 1$, $\hbar\omega_0 \cong U$ and the oscillator strength vanishes with $1/U$. Hence the optical transition (position and intensity) gives information on both U and t. We shall see that this is a general feature of the more realistic systems.

7. Segments and Chains: Electron–Electron Correlations

Most approaches start out from the extended Hubbard Hamiltonian (see, e.g., Hubbard, 1978):

$$\tilde{H} = -t \sum_{i,\sigma} (\tilde{c}^+_{i+1,\sigma} \tilde{c}_{i,\sigma} + \tilde{c}^+_{i,\sigma} \tilde{c}_{i+1,\sigma}) + \tfrac{1}{2} \sum_{i \neq j} V_{i,j} \tilde{n}_i \tilde{n}_j + \tfrac{1}{2} \sum_{i,\sigma} U \tilde{n}_{i,\sigma} \tilde{n}_{i,-\sigma}. \quad (37)$$

This Hamiltonian is analogous to that of Eq. (31), but of course much more complex. i indicates the sites, σ is the spin, and t is again a transfer integral. U is the properly screened Coulomb energy for two carriers on the same site, while $V_{i,j}$ is the corresponding energy for two carriers on site i and j, respectively.

The exact solution of the problem given by Eq. (37) is not known. For $V_{i,j} = 0$ the ground state was found by Lieb and Wu (1968). For a half-filled band the chain is insulating with an excitation gap $\Delta E \cong U - 4t$, valid for $U \gg 4t$ (Ovchinnikov, 1969). Thus the optical absorption should set on near $\omega = (U - 4t)/\hbar$. For a band filling away from $\tfrac{1}{2}$, the chain is conducting, and additional optical absorption should appear at low frequencies.

For a thorough treatment of various approximative schemes used, see Lyo and Gallinar (1977), Maldague (1977), and Gasser and Höfling (1979) and references therein.

Lyo and Gallinar (1977) consider the half-filled band in the $U \gg 4t$ limit. They find optical absorption extending from $U - 4t$ to $U + 4t$, i.e., over twice the one-electron bandwidth. In the atomic limit ($t \to 0$) the total oscillator strength is proportional to t^2/U.

More general band fillings are treated by Maldague (1977) and by Gasser and Höfling (1979). In all cases absorption appears at or near U.

Again in the half-filled band case, Lyo (1978) included a finite nearest neighbor interaction and found pronounced asymmetry in the absorption: For reasonable values of V there is a strong onset of absorption at $U - 4t$ and a tail extending to $U + 4t$. Physically this results from formation of excitons. The lower peak corresponds to a charge-transfer excitation where the extra electron and the hole are neighbors, while the tail corresponds to a further separation (therefore being energetically more costly).

Hubbard (1978) argued that $V_{i,j}$ in Eq. (37) can in no way be neglected. Numerical estimates indicate that for a TCNQ chain both U and $V_{1,2}$ are larger than $4t$, while $V_{1,3}$ may be comparable. Then the approach is first to neglect banding effects ($t \to 0$).

For $U + V_{1,3} > 2V_{1,2}$ the ground state is found to contain no double occupancy. Then the problem is classical and concerns how to distribute the carriers to minimize Eq. (37). The result is what Hubbard named *generalized Wigner lattices*. For a $\tfrac{1}{3}$-filled band, for example, there will be two occupied sites, on empty, two occupied, and so on. Now the optical spectrum is obtained by superimposing with proper weight, contributions from all possible transitions. Intramolecular transitions depend on whether the molecule in question is charged or not, and also on the surroundings. The various possible intermolecular transitions are given by combinations of U and $V_{i,j}$.

A similar localized approach was previously used in a less rigorous form by Tanaka et al. (1976) to analyze spectra of a range of TCNQ salts.

Models neglecting t have also included further degrees of freedom. Král (1976a,b) has considered strong coupling polaronic effects involving interactions with excitons and intramolecular vibrations (see Section 12) and applied the results to TCNQ compounds.

To make any quantitative calculation of the distribution of oscillator strength, models neglecting t are hardly useful, since the charge-transfer absorption vanishes with t.

Some quantitative information on the total amount of oscillator strength may be obtained from a sum rule, valid if only a single, nondegenerate level need to be considered (see, e.g., Mazumdar and Soos, 1981)

$$\int_{CT} \sigma(\omega)\, d\omega = \frac{\pi}{2} \frac{a^2 e^2}{\hbar^2} \langle -H_t \rangle, \tag{38}$$

where H_t is the transfer part of the Hamiltonian, Eq. (37), only. In the absence of electron–electron interactions, the variational principle states that Eq. (38) has its maximum value ($H = H_t$). Localization of the wavefunctions in the presence of correlations increases the kinetic energy (which here is a negative number) and thus induces a reduction in oscillator strength. It has been suggested (Jacobsen, 1986) that this reduction can be used as a measure of the importance of Coulomb interactions. Experimentally, the sum rule, Eq. (38), is calculated over the infrared range with some reasonable extrapolation, while the corresponding oscillator strength for the system with no electron–electron interaction may be estimated from an analysis in the plasma-edge region (see Section 9). Under certain assumptions estimates of Hubbard parameters may then be given (Jacobsen, 1986). It is useful to note that rather small systems, for example, segments with just four molecules, give good hints to the effect of Coulomb interactions on oscillator strength. It is also possible to calculate the expectation value of H_t, Eq. (38), in various infinite chain models (Baeriswyl et al., 1986).

Calculations on small systems, segments, and rings have been reported by a number of authors, including Maldague (1977), Mazumdar and Soos (1981), Yartsev (1982, 1984), and Malek et al. (1984). Yartsev also considers the electron–molecular vibration coupling (see below), while Malek et al. include excited molecular states in calculations aimed at TCNQ-salts.

All these studies and the other theoretical work previously quoted suggest that the spectrum of an organic conductor with general ($p < 1$) band filling and important Coulomb interactions can be quite complicated.

Limiting the discussion to on-site and near-neighbor interactions, it is found that, in addition to a possible low-lying Drude-like absorption, a number of correlation bands, which basically measure U, V, $U - V$, $U - 2V$, etc., appear. Although all studies tend to agree on the total (integrated) oscillator strength, the relative distribution depends on the model details. Especially for the small systems, the boundary conditions play an important role. What remains clear is that while in the half-filled band case virtually all intensity appears in the $U - V$ and U bands, the relative intensity in these bands rapidly drops when lowering the band filling. Near the quarter-filled band ($p = 0.5$), where most good organic metals can be found, the $U - V$ and U bands are expected to contain at most a few percent of the total oscillator strength. It is less clear what interpretation should be given to the experimentally found, broad infrared absorption band, extending to dc, which could be either normal metallic (Drude-like) with phonon interference or basically charge-transfer/correlation-like (with position at V). We will return to this in the final section after having presented a number of experimental cases.

8. Intramolecular and Intermolecular Transitions: Experimental

a. Single-Stack Conductors

The possibly clearest and least unambiguous experimental example of how the spectrum of a highly conducting organic salt develops from the various excitations and interactions discussed above is that of the TTF-halides, dealt with extensively by Torrance et al. (1979). The reason is that the TTF molecule and ion have relatively simple spectra. In addition, materials exist that contain quasi-isolated dimers of TTF^+ ions. Finally, the crystal structure of the highly conducting, mixed valence compounds is such that the molecular plane is perpendicular to the stacking axis. This means that the low-lying intramolecular transitions, which tend to be polarized in the molecular plane, do not couple with charge-transfer excitations along the stacks. This is different in the TCNQ salts, where the molecules usually are tilted with respect to the stacking axis.

In Fig. 8 are shown solution spectra of TTF^0, TTF^+, and $(TTF^+)_2$. The neutral molecule (long dashed line) absorbs only at relatively high frequencies, beginning at photon energies around 3 eV. Note the rather drastic change going to the spectrum of the monomer ion (short dashed line). Now transitions are found at much lower energies. This is a general phenomenon for the molecules of interest to this paper. On cooling the TTF^+ containing

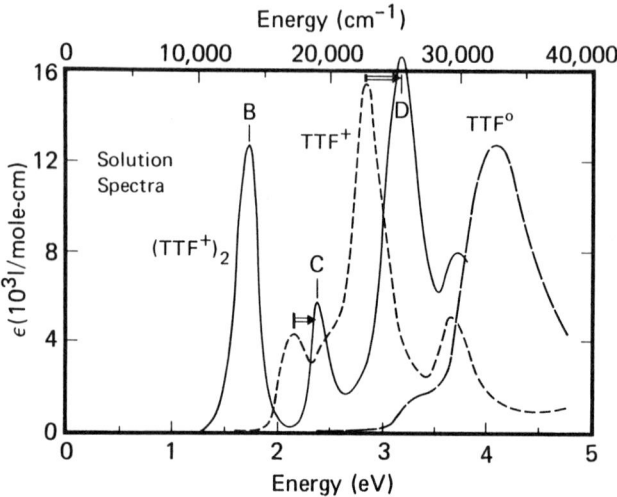

FIG. 8. Absorption spectra of TTF⁰, TTF⁺, and (TTF⁺)₂ in solution. The arrows indicate Davydov shifts upon dimerization. [From Torrance et al. (1979).]

solution, dimerization occurs, and the spectrum changes again (solid line). Spectral structure as that of TTF⁺ can clearly be identified (peaks C and D), but is shifted to a slightly higher energy. This is the well-known Davydov shift (Davydov, 1971) due to interactions between the two molecules. The direction of the shift is consistent with a relative orientation of the two molecules, where the overlap is maximized.

In addition the dimer spectrum shows a prominent absorption peak (B) not found in the monomer spectrum and lying below the monomer absorptions. Peak B can only be assumed to be a new intermolecular transition. Equation (39) tells us to expect such a transition with an energy

$$E_{CT} = 2t\sqrt{1 + \left(\frac{U}{4t}\right)^2} + \frac{U}{2}. \tag{39}$$

In principle, if the oscillator strength and the interplanar distance in a dimer are known well enough, Eq. (39) would allow a determination of t and U. In the present case we may expect t to be rather small. Then $E_{CT} \cong U \cong$ 1.7 eV from Fig. 8.

We can now proceed to the solid state. In Fig. 9 is shown the same solution dimer spectrum as in Fig. 8, together with a powder spectrum of TTF–Cl. This material contains dimers of TTF⁺ (Scott et al., 1977). As judged from the crystal structure, the interdimer contact is rather weak. Indeed the absorption spectrum (dashed line) is close to that of dimers in solution

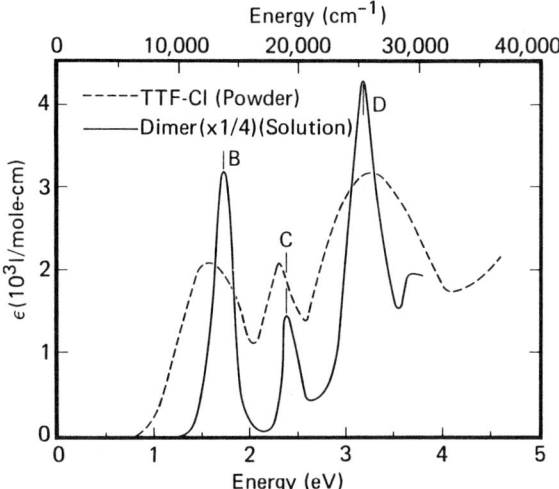

FIG. 9. Powder absorption spectrum of TTF-Cl compared to absorption spectrum of $(TTF^+)_2$ in solution. [From Torrance et al. (1979).]

(solid line), the main changes being slight shifts arising from the additional solid-state interactions as well as a certain broadening, which may be due to the finite interdimer contact. The considerable downshift of the charge-transfer band, B, may be caused by a lowering of U in the solid state, induced by additional screening.

Still following Torrance et al. (1979), we go on to the highly conducting materials. In Fig. 10 is shown the powder absorption spectrum of the mixed valence salt TTF-Br$_{0.79}$ (long dashed line) together with data from Figs. 8 and 9. TTF-Br$_{0.79}$ has uniform stacks of the TTF molecules with eclipsed overlap and a room-temperature conductivity of 400 Ω^{-1} cm^{-1} (Scott et al., 1977). Thus we are now dealing with a ~40% filled band, acting like a true organic metal. There is little doubt that the high-energy transitions C and D are of the same origin as in the dimer spectra, i.e., they are basically transitions inside the individual molecule. As expected, a new strong low-energy absorption peak has appeared (A). This peak must be related to the high dc conductivity and represent the Drude absorption of a metal (possibly with correlation bands superimposed). For noninteracting carriers, the Drude absorption exhausts the sum rule for intraband transitions. Here, however, a weak absorption is still seen near B. Since it has the same position as the strong absorption in TTF-Cl, it may be assumed to have the same origin, basically representing the on-site Coulomb interaction. It is somewhat surprising that next-nearest neighbor interactions and metallic screening hardly change the position. In any case, the strength of the B transition is

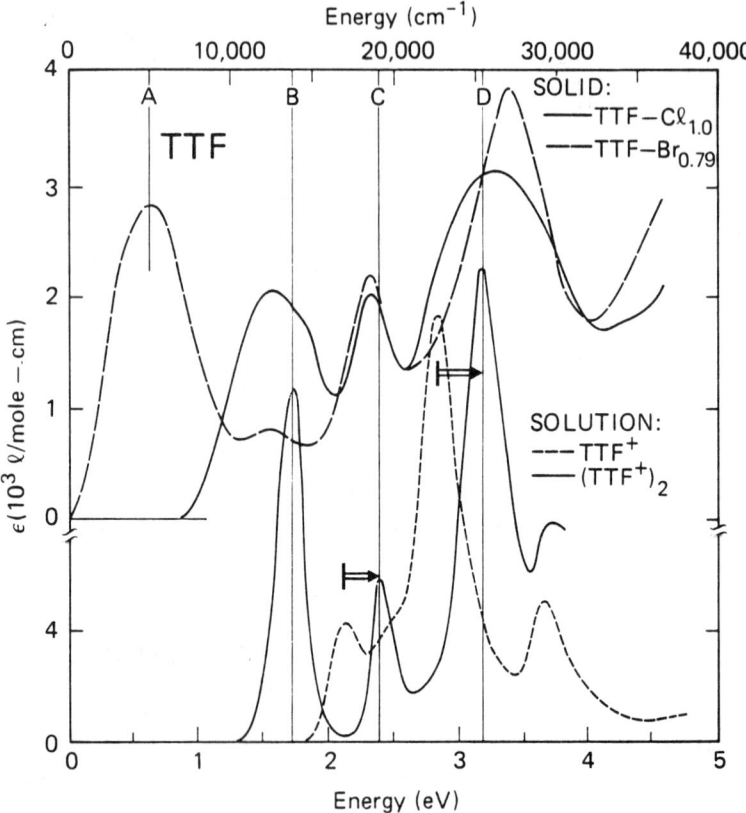

FIG. 10. Powder absorption spectrum of TTF-Br$_{0.79}$ compared to spectra from Figs. 8 and 9. [From Torrance et al. (1979).]

rather small: Of the total oscillator strength available for intraband transitions, most has shifted to low frequencies.

More information can, of course, be obtained from polarized measurements. Torrance et al. (1979) have also measured the polarized reflectance of TTF-Br$_{0.79}$ parallel and perpendicular to the stacking axis. The results are reproduced in Fig. 11. The perpendicular component ($R_\perp$) shows the C and D transitions very clearly. They are completely absent in the parallel component ($R_\parallel$), thus confirming that the transition dipoles are in the molecular plane. The A and B transitions are not seen in $R_\perp$, while $R_\parallel$ displays a typical plasma edge with the B transition appearing as a weak shoulder. The plasma edge is associated with the strong A transition in Fig. 10, i.e., with the metallic behavior. The polarization dependence thus fully supports the assignments by Torrance et al.

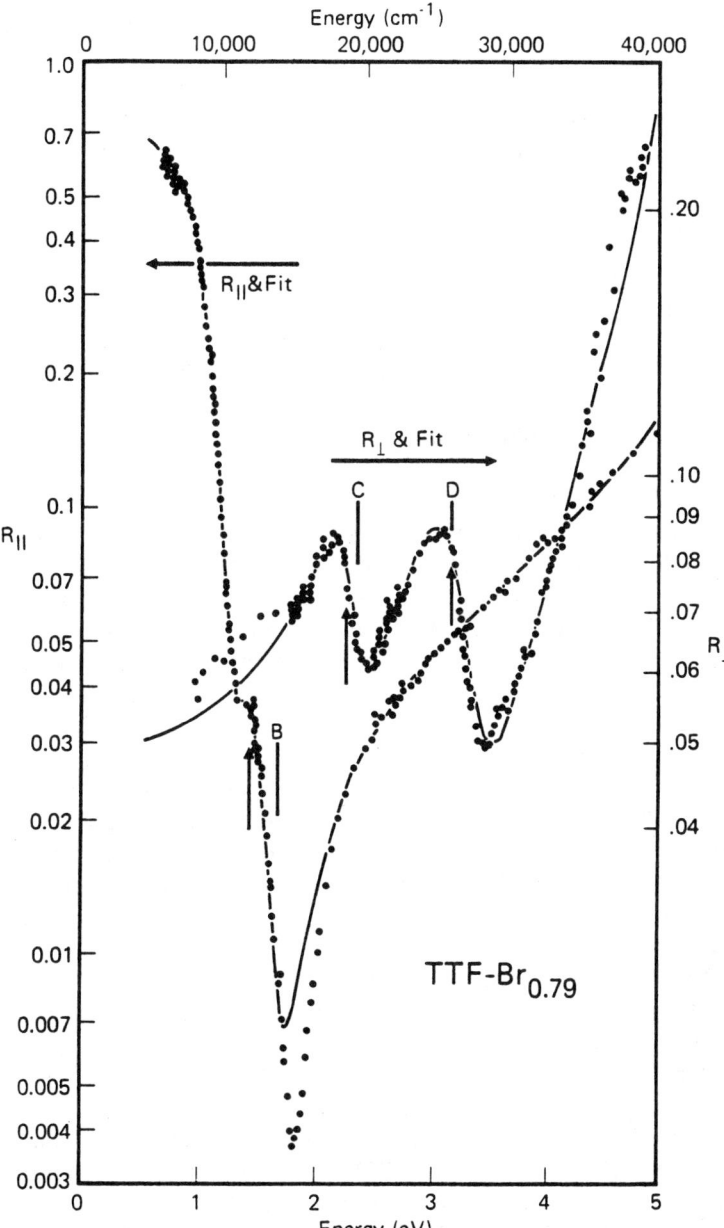

FIG. 11. Polarized single crystal reflectance of TTF-Br$_{0.79}$ at 300 K. Solid lines indicate fits to Lorentz oscillator models. [From Torrance *et al.* (1979).]

As indicated in Fig. 11, they also performed a least squares fit to the reflectance data, using a three-oscillator model. Here we just want to quote the ratio of the strengths in the A and B transitions in $R_\parallel$: $(\omega_p^A/\omega_p^B)^2 = 17$. Hence in this case the error of omitting the strength of the B transition in a sum rule calculation would amount to no more than 6%.

In a first approximation it may therefore be reasonable to use Eq. (11) (see also Section 9) to obtain the transfer integral, t, and then Eq. (39) to find U. Doing this, Torrance et al. (1979) find $t = 0.28$ eV and $U = 1.25$ eV. These numbers indeed suggest that the organic metals may be in the intermediate coupling regime ($U \simeq 4t$).

There have been other studies of TTF single-stack conductors. Warmack and Callcott (1976) were the first to report the reflectance spectrum of TTF-$I_{0.7}$, which in many ways is analogous to TTF-$Br_{0.79}$. Their data shows no identifiable B transition, but the overall behavior resembles that of the bromide. The same holds for the studies of TTF$_{12}$-SeCN$_7$ by Somoano et al. (1977).

Kuroda and coworkers have also made rather extensive optical studies of TTF salts with emphasis on the overall behavior. Sugano et al. (1978) measured polarized absorption spectra of TTF-$Br_{0.71}$, TTF-$I_{0.71}$, and TTF-$SCN_{0.57}$. They identify transitions in the spectra typical for TTF^0 as well as TTF$^+$ and use the relative spectral weight to calculate the degree of charge transfer, ρ, or equivalently the fraction of TTF-molecules ionized. Their results for ρ are in accordance with that chemically expected for TTF-$Br_{0.71}$ and TTF-$SCN_{0.57}$, but not so for TTF-$I_{0.71}$. Here they find $\rho \simeq 0.58$. XPS data independently have indicated $\rho \simeq 0.52$ (Ikemoto et al., 1977), thus the charge transfer in TTF-$I_{0.71}$ seems to be incomplete.

Sugano et al. (1978) find a correlation between the strength of peak B in the spectra and ρ. The intensity is higher when ρ is high, just as expected theoretically (see above). For example, the B transition is of medium strength in TTF-$Br_{0.71}$, but quite weak in TTF-$I_{0.71}$ and TTF-$SCN_{0.57}$. This is easy to understand in the localized picture. As argued by Sugano et al. the B transition is basically the charge-transfer transition in a dimer with two carriers. Thus the intensity will be proportional to the density of TTF$^+$ pairs at any particular moment. If we think in terms of generalized Wigner lattices (Hubbard, 1978) this density is 0 at $\rho = 0.5$, while for $\rho = 0.67$ the chain will contain alternating dimers and neutral molecules.

Cao et al. (1980a) have measured the temperature dependence of the polarized reflectance spectrum of TTF-$I_{0.71}$. Surprisingly they find a strong temperature dependence of the strength in the B band, shown in Fig. 12. From being very weak above 200 K, it is growing in strength as the temperature is lowered. Other measurements suggest that ρ may change at a transition near 160 K (Sugano and Kuroda, 1977), being 0.71 below. Assuming

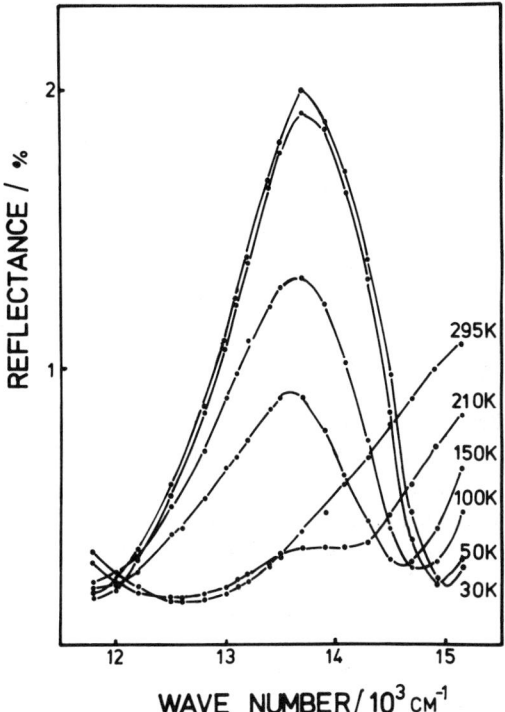

FIG. 12. Temperature dependence of the chain axis reflectance spectrum of TTF-$I_{0.71}$ around 14,000 cm^{-1}. [From Cao et al. (1980a).]

this to be the case, immediately accounts for the intensity variation in terms of the ideas quoted above.

In a recent paper, Kuroda et al. (1982) report the temperature dependence of reflectance spectra of TTF-$Br_{0.76}$ and TTF-$SCN_{0.58}$. In the latter compound they fail to observe the B transition. In the first it is observed with an almost temperature-independent intensity. This is expected from a fixed charge transfer, as assumed in the bromide. The position is found to vary from 12,200 cm^{-1} to 13,100 cm^{-1} on cooling from 300 K to 30 K. The shift could arise from an increase in U on cooling, cf., Eq. (39), associated with the metal-insulator transition.

We can sum up the overall optical properties of TTF single-stack conductors as follows: Optical transitions are either intramolecular or charge-transfer excitations. The intramolecular transitions are situated above ~16,000 cm^{-1}, and the results indicate that, viewed spectroscopically, the chains consist of neutral and ionized TTF molecules. No real banding effects are seen. The charge-transfer excitations are found below 16,000 cm^{-1}.

A strong dominating band in the infrared connects to the high dc conductivity. In some materials a second, rather weak band is found, always around 12,000–14,000 cm^{-1}. There is considerable evidence that this band corresponds to the B transition: $2TTF^+ \rightarrow TTF^0 + TTF^{2+}$. For $\rho < 0.6$ the band is weak, and for $\rho \leq 0.5$ it should be completely absent. If we accept this interpretation, it seems that U for TTF at optical frequencies does not vary much from material to material.

There are other highly conducting single-stack materials. The TMTSF$_2$X family is famous for its superconducting members. Since the charge transfer is 0.5 by stoichiometry, there should be no B transition. In Fig. 13 the reflectance spectrum of TMTSF$_2$AsF$_6$ is shown at three temperatures. The data clearly confirms the expectation.

Other materials investigated are TTT$_2$-I$_3$ (Kamarás et al., 1977; Somoano et al., 1978; Kamarás and Grüner, 1979), TSeT-Br$_{0.5}$ (Kaplunov et al., 1977), and (perylene)$_2$X$_{1.1}$ (Wilckens et al., 1982). The latter are interesting in that they seem to possess a distinct transition with B characteristics (position, polarization, and intensity), in spite of estimated ρ of 0.55. However, the molecules are not perpendicular to the stacking axis in this case (Keller et al., 1981), so the interpretation may not be simple.

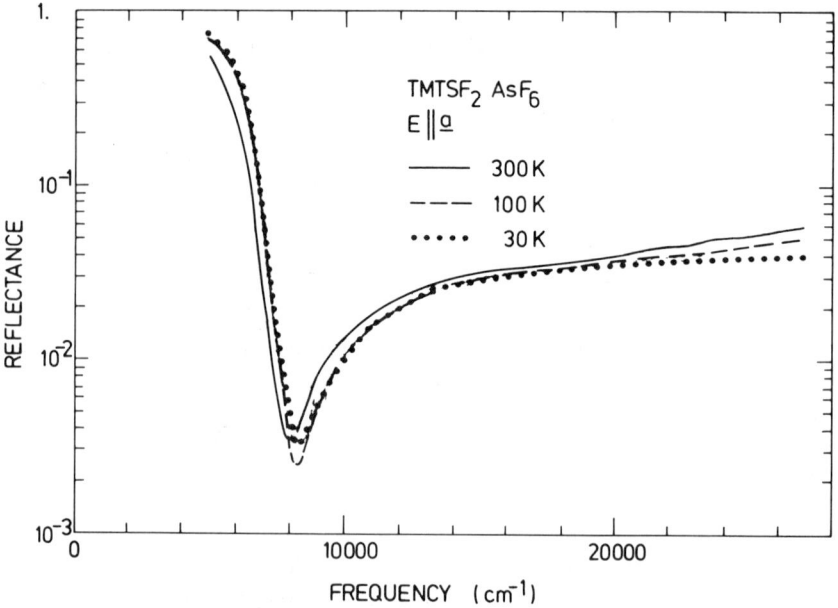

FIG. 13. Polarized chain axis reflectance of TMTSF$_2$AsF$_6$ at 300 K and 30 K. [From Jacobsen et al. (1983).]

That the spectra become complicated when low-lying intramolecular and charge-transfer transitions couple is also observed in materials of the (BEDT-TTF)$_2$X family. In the two crystalline modifications, α- and β-(BEDT-TTF)$_2$I$_3$, appreciable mixing can occur only in the latter, and indeed the B-type absorption band appears here in spite of the low $\rho = 0.5$ (Koch et al., 1985; Sugano et al., 1985). The point will be further discussed in the following section on TCNQ-compounds.

b. TCNQ-based Conductors

For many years most organic conductors of fair quality were based on the TCNQ-molecule, and virtually all excellent organic metals were of the TTF-TCNQ family, i.e., with two conducting stacks (double-stack conductors). These materials continue to play an important role, but, as we shall see, their near infrared to visible spectra are not well understood.

Complicating features include: (1) Two absorption bands of different physical origin have the same position. (2) In most materials the TCNQ-molecules stack in a slipped fashion so that charge-transfer processes can mix with the intramolecular excitations (cf., above). (3) In the double-stack conductors there are low-lying excitations on both kinds of stack.

We shall first place emphasis on the TCNQ molecule and some less conducting materials (including single-stack compounds), which are illuminating. The absorption spectrum of TCNQ$^-$ and of (TCNQ$^-$)$_2$ has been determined by Boyd and Phillips (1965). In Fig. 14 their results are compared with some powder absorption spectra of TCNQ-salts (Torrance et al., 1979). The absorption bands of TCNQ^0 (not shown) start at 23,000 cm^{-1} (Hiroma et al., 1970). The dashed line in the bottom part of the figure shows the characteristic monomer spectrum. The low-energy band from 11,000–16,000 cm^{-1} has a pronounced vibronic splitting with three resolved sub-bands, the strongest being at 11,000 cm^{-1}. The next-lowest band (D) is found around 24,000 cm^{-1}. In the dimer spectrum this transition is shifted up to 27,000 cm^{-1} (Davydov shifted as in the TTF-case). The lowest band is shifted by about 4000 cm^{-1}, but now the vibronic structure has vanished. The new charge-transfer band, which is a rough measure of U, is centered at 11,000 cm^{-1}, exactly where the monomer spectrum is most intense. That these two bands happen to be in the same position is the first of the complicating factors mentioned. The large shift of the monomer band and the vanishing of the vibronic splitting in the dimer may be tentatively associated with strong mixing of the intramolecular transition with the charge-transfer excitation.

It is reasonable to assume that the TCNQ ions in solution dimerize with an overlap similar to what is almost universally seen in the solids (see, e.g.,

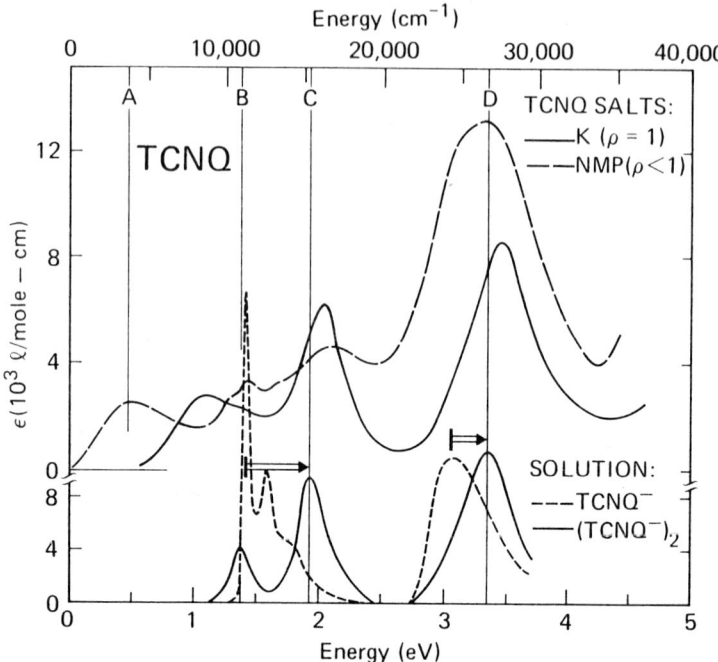

FIG. 14. Solution spectra of TCNQ⁻ and (TCNQ⁻)₂ (Boyd and Phillips, 1965). Also shown are powder absorption spectra of the semiconducting K-TCNQ and the conducting NMP-TCNQ. [From Torrance et al. (1979).]

Fritchie, 1966). Then the angle between the long axis of the molecule (the polarization direction of the low-energy monomer transition) and a line connecting identical points on the two ions is about 50°–60°, accounting for the mixing. The importance of this mixing in understanding the spectra of TCNQ-salts has been stressed by Tanaka et al. (1976). In consequence the dimer bands may not be cleanly polarized. For example, the band at 11,000 cm^{-1} in Fig. 14 is predominantly a charge-transfer excitation. However, it has some character of the monomer transition and will be polarized slightly off the "stacking" direction.

In the top part of Fig. 14 the solid line is a powder absorption spectrum of K–TCNQ. This material is semiconducting and contains dimerized chains of TCNQ ions (Hoekstra et al., 1972). The individual dimers are not isolated, but still the absorption spectrum should resemble that of the dimer in solution. This is indeed what is found: Peaks C and D are shifted slightly up in frequency, while the charge-transfer band is centered at 9000 cm^{-1}. The downshift from 11,000 cm^{-1} could be due to additional screening in the solid state. Actually a second weak band in KTCNQ appears near 11,000 cm^{-1}, as noted by Tanner et al. (1977). It can barely be seen in Figure 14. Tanner et al. attribute this band to the monomer transition. However, it seems more likely that the charge-transfer band is split: There can be transfers either inside a "dimer" or between "dimers." The latter process is expected to have the smallest strength (the overlap integral is smaller) and the highest energy (taking nearest-neighbor interactions into account).

An additional or alternative interpretation for the shape of the B band in KTCNQ (Yakushi et al., 1979) is in terms of the theory by Lyo (1978), where a significant near-neighbor interaction splits the band into an exciton-absorption and weak interband transition (see above).

The dashed line in Fig. 14 (top) finally represents a metallic compound, NMP-TCNQ ($\rho < 1$). At this point we just want to note the appearance of the A band in the infrared, similar to what is found in the TTF-materials.

Before further discussing the conducting materials it is useful to consider the properties of another semiconductor: $Cs_2(TCNQ)_3$. The electronic structure is interesting in the context of optical properties. Detailed crystal structure studies (Fritchie and Arthur, 1966) show that the TCNQ chains consist of $TCNQ^-$ dimers alternating with $TCNQ^0$. Thus the carriers are truly localized, as also evidenced by the low dc conductivity, $\sigma(300\,K) \simeq 10^{-3}\,\Omega^{-1}\,cm^{-1}$ (Siemons et al., 1963). The dimers are unusual in the sense that the direction of charge transfer is almost perpendicular to the long axis of the molecules. Thus the mixing described above should not occur.

The infrared and optical properties of $Cs_2(TCNQ)_3$ have recently been studied extensively by Cummings et al. (1981) and by Yakushi et al. (1981). We reproduce from Cummings et al. the frequency-dependent conductivity in Fig. 15. Concentrating on the chain axis component (solid line), the dominant absorption features are two strong bands centered at $4000\,cm^{-1}$ and $10,000\,cm^{-1}$. With reference to the crystal structure the interpretation is straightforward: The lower band corresponds to a charge-transfer process, where an electron hops from $TCNQ^-$ to $TCNQ^0$. The upper band arises from the charge-transfer process within the dimer. Thus the two bands closely correspond to the usual A and B transitions. Using the dimer theory (Section 6), Cummings et al. deduce the dimer parameters $t = 0.17\,eV$ and $U = 1.14\,eV$ from the position and strength of the B band. In another system, which contains isolated dimers. DMeFc-TCNQ, a similar analysis leads to $t = 0.27\,eV$ and $U = 1.0\,eV$ (Tanner et al., 1980). Such a value of U is also consistent with the K-TCNQ data in Fig. 14.

We now return to $Cs_2(TCNQ)_3$. A detailed polarized reflectance spectrum has been obtained by Yakushi et al. (1981), and part of it is shown in Fig. 16. Data are given for a direction that is approximately along the long axis of TCNQ (1) and for the chain axis (4), both at two temperatures, 27 K and 298 K. The bands A and A' clearly are the same as discussed above. The

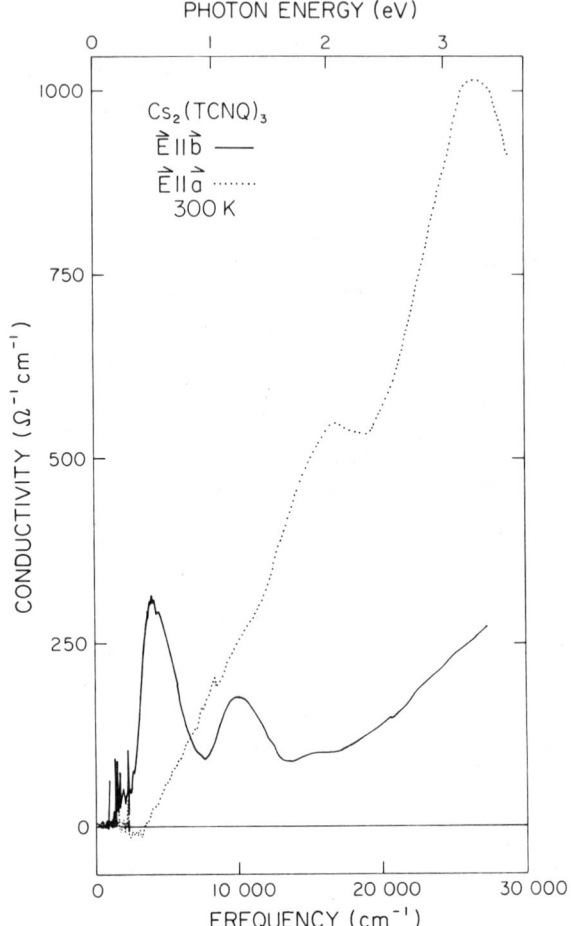

FIG. 15. Frequency-dependent conductivity of $Cs_2(TCNQ)_3$ as obtained from Kramers–Kronig analysis of polarized reflectance data. The chain axis is b (solid line). [From Cummings et al. (1981).]

B band can only be assigned to the lowest monomer transition. The vibronic splitting is visible at low temperature. This is consistent with the absence of interaction between the monomer transition and the dimer charge-transfer band. The transition can still mix with the low-energy charge transfer band, A, which may explain why the vibronic structure is not resolved at room temperature and why B is seen in the chain axis polarization at low temperature. We note that in this case the dimer charge-transfer band and the monomer band are not in exactly the same position.

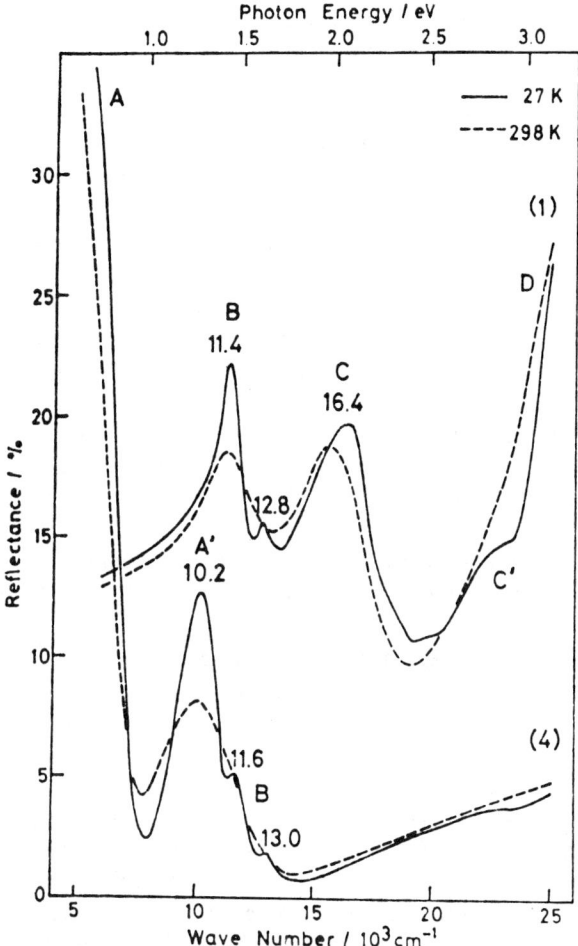

FIG. 16. Reflectance spectra of Cs_2TCNQ_3 single crystal polarized along the $[\overline{2}01]$ (1) and [010] (4) directions. [From Yakushi *et al.* (1981).]

The band C polarized along the long axis of the molecule can not be accounted for by any previously discussed transition. Yakushi *et al.* suggest a new interpretation for the sequence C–C′–D, where C′ is only observed at low temperature. The C and D transitions are attributed to the interacting pair of TCNQ⁻ and TCNQ⁰. The physical mechanism should be a charge transfer from the next highest occupied orbital in TCNQ⁻ to the lowest unoccupied orbital in TCNQ⁰. There are two such transitions: one where the excitation corresponds to TCNQ⁰ in the lowest singlet excited state, which should be D, and one which corresponds to TCNQ⁰ in the lowest triplet excited state, which should be C. The configurations are expected to mix to some extent. The C' band is assigned to the usual next-lowest monomer transition.

The interpretation is supported by the identification of the triplet excited state of $TCNQ^0$ at 2.0 eV in electron energy-loss spectroscopy by Ritsko et al. (1977). Even if the assignment by Yakushi et al. does not hold, the experimental data of Fig. 16 unambiguously calls for a hitherto unknown optical absorption process, polarized basically perpendicular to the chain direction.

Yakushi et al. (1981) have also investigated various complex 1:2 salts ($\rho = 0.5$). These too show the monomer band and the C transition simultaneously. The same is evident from the powder absorption spectra of a series of 1:2 compounds by Kamarás et al. (1978). From the polarization dependence, where Yakushi et al. find a considerable stacking axis component, it is clear that there must be some mixing with charge-transfer transitions, but the shift and broadening observed in the 1:1 materials are not present.

The classical highly conducting double-stack material is TTF–TCNQ. The room-temperature conductivity is of the order $600\,\Omega^{-1}\,cm^{-1}$ (Cohen et al., 1974), and $\rho = 0.55$ at 300 K according to diffuse x-ray measurements (Khanna et al., 1977). Recent detailed polarized reflectance measurements have been reported by Cao et al. (1980b). Figure 17 presents these results (see also the data by Tanaka et al. (1978) in Fig. 1). The chain axis is b, while the long axis of the molecules has its largest component along c, but also a significant component along b.

The transition seen near $10{,}000\,cm^{-1}$ clearly bears the signature of the monomer transition of $TCNQ^-$. The vibronic structure, as well as the polarization dependence, are strong arguments for this interpretation, although the considerable oscillator strength along the stacks suggests some mixing with charge-transfer processes. A C band is polarized basically along the long axis and may have the origin discussed above. However, the lowest intramolecular excitation in TTF^+ may also contribute here.

The monomer band as identified by the vibronic splitting is observed in several TTF–TCNQ derivatives like TMTTF– and TMTSF–DMTCNQ (Jacobsen et al., 1978). These compounds have in common with TTF–TCNQ a rather low ρ. Of more interest is the material HMTSF–TCNQ, where $\rho = 0.74$ (Weyl et al., 1976). We might expect a more intense "U" band interacting with the monomer transition. The reflectance spectrum has been determined by Jacobsen et al. (1977) and is shown in Fig. 18. The monomer transition cannot be safely identified in the chain axis spectrum, in spite of the fact that the TCNQ molecules in this material possess the usual tilt (Phillips et al., 1976). In the other TTF–TCNQ-type materials, the monomer

5. OPTICAL PROPERTIES

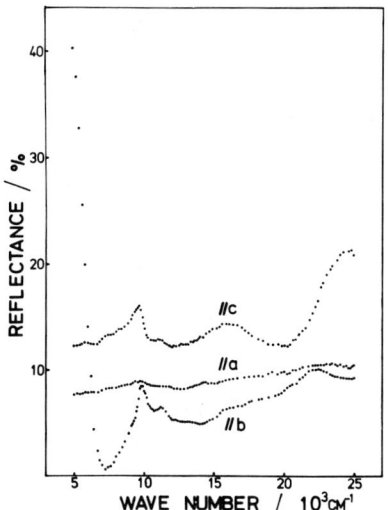

FIG. 17. Polarized reflectance spectra along the a, b, and c axes of TTF-TCNQ at 300 K. [From Cao et al. (1980b).]

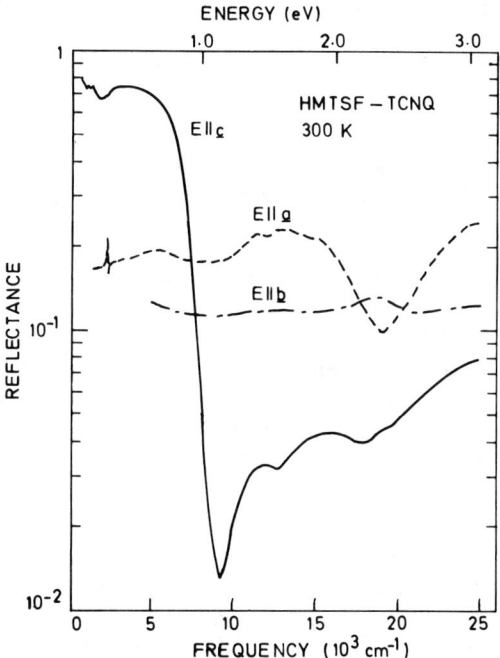

FIG. 18. Polarized reflectance spectra along the a and c axes of HMTSF-TCNQ at 300 K. [From Jacobsen et al., 1977).]

transition has its strongest band at 10,000 cm^{-1}. Here, instead, a rather broad absorption is found near 12,000 cm^{-1}. Thus it is possible that this band is a mixture of "U" band and monomer transition, the latter having lost the vibronic features. We note that in this case it is only weakly observed in the perpendicular spectrum. The strong band at 17,000 cm^{-1} seen in both a and c polarizations is presumably a HMTSF$^+$ excitation. The ionized parent molecule, TSF$^+$, is known to have a strong transition at this position (Torrance et al., 1979).

In conclusion of this lengthy discussion, it can be stated that a full assignment of a spectrum of a double-stack conductor may be quite complicated, and more work, theoretical and experimental, is needed, especially with regards to TCNQ stacks. However, one point to stress is that there is good evidence that virtually all the intraband oscillator strength in organic metals (i.e., associated with transfers among the conduction electron orbitals alone) is contained in the A band, which therefore can be used to characterize transfer integral and carrier density as in conventional solids.

9. Plasma Edge and Band Structure

The present section is concerned with the most spectacular feature in the polarized reflectance data of quasi-one-dimensional conductors: the sharp plasma edge appearing along the chain direction. We shall here deal with its position only or with the frequency of long-wavelength oscillations in the carrier density, i.e., the plasmons.

As will be discussed in Part IV, the optical properties in the infrared deviates strongly from the simple Drude behavior described by the dielectric function of Eq. (13)

$$\tilde{\varepsilon}(\omega) = \varepsilon_\infty - \frac{\omega_p^2}{\omega(\omega + i\Gamma)}. \tag{40}$$

However, in the vicinity of $\omega_p/\sqrt{\varepsilon_\infty}$, where the reflectance edge is found, Eq. (40) describes the chain-axis optical properties rather well. It is the main purpose of this section to discuss to what extent band structure parameters, as obtained from analysis of the reflectance edge, vary consistently with material, temperature, pressure, etc.

Thus fitting Eq. (40) to the data should just be considered a convenient mathematical tool in estimating the zero-crossing in $\varepsilon(\omega)$ (at $\sim \omega_p/\sqrt{\varepsilon_\infty}$) and the influence of background screening (ε_∞). The zero-crossing basically identifies the plasmon frequency, which we shall then assume is related to band structure and charge density as given by the simple relations of Section 2. Since the plasmons are long-wavelength oscillations, their frequency is indeed expected to be insensitive to, e.g., short-range correlations and electron–phonon coupling, which strongly affects size as well as distribution of oscillator strength at low frequencies.

For a tight-binding band of the form

$$\varepsilon(k) = 2t(1 - \cos ka), \qquad -\frac{\pi}{a} \leq k \leq \frac{\pi}{a}, \qquad (41)$$

and a degree of charge transfer ρ, Eq. (9) yields

$$\omega_p^2 = \frac{4ta^2 e^2 \sin(\pi\rho/2)}{\pi \varepsilon_0 \hbar^2 V_m}. \qquad (42)$$

This formula applies to a single-stack conductor with a uniform chain and a chain axis lattice constant a. V_m is here the total volume per molecule in the stack. In case of a double-stack conductor, t is the average transfer integral for the two types of stack, and V_m is the average total volume per molecule. The same remark applies to slightly dimerized systems. However, here a should be interpreted as the average distance between molecules along the stack.

a. Chain-Axis Transfer Integral

In Table I are collected Drude parameters and transfer integrals, as obtained by fitting Eq. (40) to experimental data and using Eq. (42) to determine t. An example showing data and fit is given in Fig. 19. Note the low, almost dispersionless perpendicular reflectance component.

The values of ω_p in Table I varies from about 5000 cm^{-1} to about 15,000 cm^{-1}, while transfer integrals vary from 0.04 eV to 0.37 eV. The first important issue to discuss is the scatter in Drude parameters found by different authors on the same material. For TTF-TCNQ at room temperature, as an example, ω_p values range from 9540 cm^{-1} to 11,400 cm^{-1}, leading to a spread in $\langle t \rangle$ of order 30%. The problem arises from the high correlation between ε_∞ and ω_p found in fitting Eq. (40) to reflectance data. The position of the reflectance edge is basically given by a zero-crossing in $\varepsilon(\omega)$, which for $\Gamma = 0$ occurs at $\omega_p/\sqrt{\varepsilon_\infty}$. Thus the position depends on the polarizability of the background. Indeed the quantity $\omega_p/\sqrt{\varepsilon_\infty}$ show less that 2% scatter for TTF-TCNQ as taken from different analyses.

An accurate estimate of ε_∞ depends both on the edge shape and, in particular, on good data on the high-frequency side of the minimum, where reflectance is quite low. Hence the experiment here should be carried out carefully, and the fitting procedure should be performed in terms of $\log(R)$ rather than R to emphasize ranges with low reflectance.

An accurate value of ω_p will also depend on the validity of Eq. (40). In many cases other optical transitions are located near or even on the

TABLE I

Drude Parameters and Transfer Integrals of Highly Conducting Organic Solids (Stacking Axis)

Material	Temp. K	σ_{dc} Ω^{-1} cm^{-1}	ε_∞	ω_p cm^{-1}	Γ cm^{-1}	ρ	$\langle t \rangle$ eV	Reference
TTF-I$_{0.7}$	300	100–450^a	1.32	10,530	2060	0.52^b	0.18^c	Warmack and Callcott (1976)
TTF-I$_{0.71}$	295	100–450^a	1.87	12,900	2590	0.52^b	0.28	Cao et al. (1980a)
—	230	10^{-1a}	1.70	12,700	2520	0.52^b	0.27	—
—	210	10^{-2a}	1.65	12,500	2370	0.52^b	0.26	—
TTF-SeCN$_{0.58}$	300	750	2.63	14,400	1850	0.58	0.31^c	Somoano et al. (1977)
TTF-Br$_{0.79}$	300	400	1^d	15,100	2340	0.79	0.28	Torrance et al. (1979)
TMTSF$_2$ClO$_4$	290	670^f	2.44	9700	1350	0.50	0.23	Kikuchi et al. (1982)
—	145	2600^f	2.46	10,100	940	0.50	0.25	—
—	30	3 × 10^{4f}	2.46	10,200	680	0.50	0.26	—
TMTSF$_2$AsF$_6$	300	500^e	2.56	9940	1230	0.5	0.25	Jacobsen et al. (1983)
—	100	3 × 10^{3e}	2.52	10,270	1030	0.5	0.27	—
—	30	10^{4e}	2.55	10,470	1160	0.5	0.28	—
TMTSF$_2$PF$_6$	300	500^e	2.42	9700	1520	0.5	0.25	—
TMTTF$_2$PF$_6$	300	20^f	2.50	8860	1380	0.5	0.20	—
TTI$_2$-I$_3$	300	1000	2.61	10,700	610	0.50	0.37^f	Somoano et al. (1978)
TSeT-Br$_{0.5}$	300	1000	1.0^g	5500^g	3300^g	0.50	0.04^c	Kaplunov et al. (1977)
TSeT-Br$_{0.5}$	300	1000	2.0^g	8300^g	3500^g	0.50	0.10^c	Delhaes et al. (1980)
(perylene)$_2$-(PF$_6$)$_{1.1}$	300	880^h	2.55	11,100	1450	0.55	0.23^c	Wilckens et al. (1982)
Ni(pc)I	300	500	2.04	5042	780	0.33	0.21	Martinsen et al. (1984)
TTF-TCNQ	300	600^k	2.43	9540	1500	0.55^i	0.11^c	Bright et al. (1973)
—	85	10^{4k}	2.43	9540	900	0.59^i	0.10^c	—
TTF-TCNQ	300	600^k	3.3	11,100	2300	0.55^i	0.16^c	Grant et al. (1973)
—	84	10^{4k}	2.8	11,400	2100	0.59^i	0.16^c	—
—	20	10^{-1m}	3.2	12,300	1800	0.59^i	0.18^c	—
TTF-TCNQ	295	600^k	3.0^n	10,400	1480	0.55^i	0.14^c	Cao et al. (1980b)
—	200	2 × 10^{3k}	3.0^n	11,100	1320	0.58^i	0.15^c	—

5. OPTICAL PROPERTIES

Compound	T (K)							Reference
—	100	—	—	—	—	—	—	
—	30	—	—	—	—	—	—	
TTF-TCNQ	300	$6 \times 10^{3\,k}$	3.0^n	11,300	1000	0.59^j	0.15^c	
TSF-TCNQ	300	10^m	3.0^n	11,500	860	0.59^j	0.16^c	Tomkiewicz et al. (1977)
HMTTF-TCNQ	300	600^k	—	9700	—	0.55^i	0.11^c	
TTF-TCNQ	300	800°	—	11,300	—	0.63^p	0.15^c	
HMTTF-TCNQ	300	500°	—	10,000	—	0.72^s	0.13^c	
TTF-TCNQ	300	600^k	3.27	11,400	1430	0.55^i	0.15	Jacobsen (1985)
TMTSF-TCNQ	300	1000^u	—	12,000	—	0.57^g	0.22	
TMTSF-DMTCNQ	300	500^u	2.91	11,200	1180	0.50^q	0.22	
TMTTF-DMTCNQ	300	120^u	—	10,700	—	0.50^v	0.19	
HMTTF-TCNQ	300	500°	3.15	12,400	1260	0.72^s	0.20	
HMTSF-TCNQ	300	2000°	3.30	14,200	1030	0.74^f	0.26	
DBTTF-TCNQCl$_2$	300	40^w	2.51	7500	2000	0.56^x	0.10	
HMTSF-TNAP	300	2400^y	2.95	11,300	1100	0.56^v	0.21	
DIPSΦ$_4$I$_{2.28}$	300	250	2.8^g	5800^g	1700^g	—	—	Strzelecka et al. (1979)
ΦPh(TCNQ)$_2$	300	75	2.05^g	6900^g	2900^g	—	—	Strzelecka and Rivory (1981)
BIPA(TCNQ)$_2$	300	170	2.0^g	9700^g	3000^g	—	—	

[a] Warmack et al. (1975)
[b] From XPS studies, Ikemoto et al. (1977)
[c] Recalculated value, based on Eq. (41)
[d] Model with several oscillators
[e] Bechgaard et al. (1980)
[f] Bechgaard et al. (1981)
[g] Data from compaction studies
[h] Keller et al. (1981)
[i] Khanna et al. (1977)
[k] Cohen et al. (1974)
[m] Cohen and Heeger (1977)
[n] ε_∞ was fixed in the analysis
[o] Bloch (1977)
[p] Weyl et al. (1976)
[q] Pouget (1981)
[r] Assumed value
[s] Megtert et al. (1978)
[t] Delhaes et al. (1979)
[u] Jacobsen et al. (1978)
[v] J.-P. Pouget, private communication
[w] Jacobsen et al. (1980)
[x] Mortensen et al. (1983)
[y] Bechgaard et al. (1978)

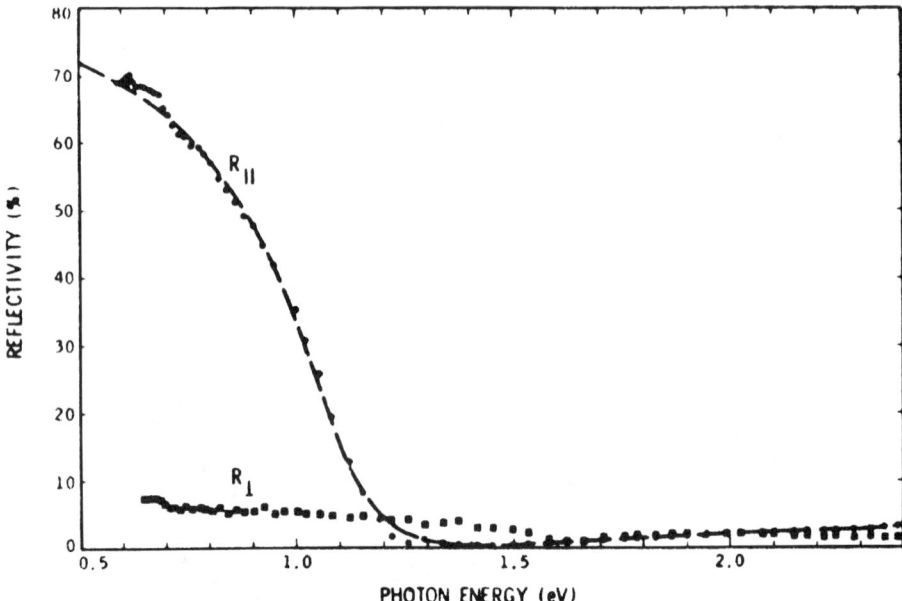

FIG. 19. Reflectivity of (TTF)$_{12}$(SeCN)$_7$ at 300 K measured for polarizations along the stacking axis ($R_\parallel$) and perpendicular to the stacking axis ($R_\perp$). The dashed line is a Drude fit. [From Somoano *et al.* (1977).]

reflectance edge. Then ε_∞ cannot be considered frequency independent, and a proper analysis must include one or more Lorentzian oscillators instead of ε_∞. Such an analysis was, for example, done by Torrance *et al.* (1979) for TTF-Br$_{0.79}$ (see Fig. 11).

Now assuming that reliable parameters are obtained, we can investigate whether the numbers in Table I are qualitatively consistent. In a number of cases, pairs of analogous materials like TTF-TCNQ and TSF-TCNQ exist. While TTF contains sulphur atoms, TSF contains the bigger selenium atoms. Thus $\langle t \rangle$ for the selenium analogue is always expected to be the higher, provided the overlap pattern between molecules is the same. This is indeed found for the pair TTF/TSF-TCNQ, when values obtained by a particular group (Tomkiewicz *et al.* 1977) are used. The same tendency is seen for the pairs HMTTF/HMTSF-TCNQ and TMTTF/TMTSF-DMTCNQ.

Another qualitative confirmation of the relation between ω_p and t comes from the TTF single-stack conductors. Here t is about 0.3 eV, a much higher value than $\langle t \rangle \cong 0.1$–$0.2$ eV in TTF-TCNQ, where actually the TCNQ-stack is believed to possess the biggest t. This difference can be directly related to the mode of overlap in the TTF stacks. The mode of overlap is slipped in TTF-TCNQ, but eclipsed in the TTF single-stack materials.

Molecular orbital calculations have estimated $t \simeq 0.05$ eV and $t \simeq 0.25$ eV for the two cases (Salahub et al., 1976), in reasonable agreement with the values estimated from optics.

It is more difficult to decide whether real quantitative agreement is found. Theoretical estimates of transfer integrals show a wide scatter (see, e.g., Herman et al., 1977). Standard measurements, which probe the density of states at the Fermi level, include magnetic spin susceptibility and thermoelectric power. The magnetic susceptibility is usually enhanced and shows strong temperature dependence (cf., the chapter on magnetic properties). The thermoelectric power is more promising. In some materials it is almost linear in temperature, as appropriate for a metal, and in single-stack compounds a transfer integral can be estimated if an assumption on the dominating scattering mechanism is made. For $TMTSF_2PF_6$, assuming acoustic phonon scattering, $t \cong 0.35$ eV from the thermoelectric power (Mortensen, 1982), in fair agreement with the results in Table I. Unfortunately, all of the TTF single-stack conductors have phase transitions at rather high temperatures, so no true metallic behavior is observed in the transport properties, However, in TTT_2-I_3 a broad metallic range is observed, and the value of the thermopower (Chaikin et al., 1979) leads, under the same assumptions as above, to $t \cong 0.34$ eV, again in good agreement with the number given in Table I.

Under certain assumptions data for the single-stack conductors may even be exploited to decompose average bandwidths in double-stack conductors into separate stack contributions (Jacobsen, 1985).

We now proceed to discuss the temperature and pressure dependence of the plasma edge. Due to thermal contraction, t is in general expected to increase with decreasing temperature, thus blue-shifting the plasma edge. As Table I shows, this is indeed found for $TMTSF_2X$ and TTF-TCNQ, while, in fact, a slight red-shift is observed in TTF-$I_{0.71}$. Inspecting Eq. (42), it is evident that factors other than t are in principle temperature dependent, namely a, V_m, and ρ. Hence a detailed analysis can only be made if the temperature dependences of these variables are known. The same remarks apply for the pressure dependence.

TTF-TCNQ is the most thoroughly investigated organic conductor. Data is available on temperature dependence of ω_p, a, V_m, and ρ, as well as the pressure dependence of ω_p, and the compressibilities.

The most detailed study of the temperature dependence of the plasma edge has been reported by Cao et al. (1980b). Figure 20 presents their data. It is clear that the edge is blue-shifted on cooling, and it sharpens considerably. Cao et al. have performed a Drude analysis of their data, fixing ε_∞ to 3.0 (see also Table I). They find a total increase in ω_p of 10.1% or 21% in ω_p^2.

This increase should be understood on the basis of Eq. (42). From studies

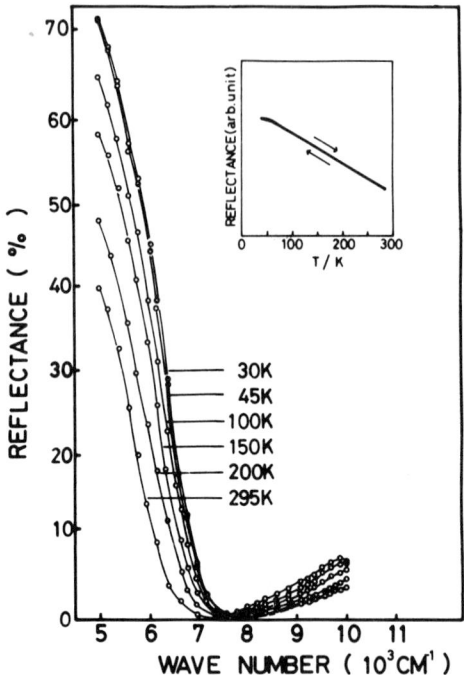

FIG. 20. Temperature dependence of the plasma edge of TTF-TCNQ. The inset shows the relative temperature dependence of $R_\parallel$ at 6000 cm^{-1}. [From Cao et al. (1980b).]

of the crystal structure (Schultz et al., 1976) it is found that the relative change in a^2/V_m is -0.7%. ρ is, from diffuse x-ray measurements, known to increase from 0.55 to 0.59 on cooling (Khanna et al., 1977). The corresponding relative change in $\sin(\rho\pi/2)$ is $+4.9\%$. Thus the data implies $\Delta t/t = +17\%$. The corresponding change in the chain axis lattice constant is -2.3%. A theoretical estimate for $\Delta t/t$ with this change is $+10\%$ (Herman, 1977).

A similar analysis can be done on the basis of the pressure dependence of the plasma edge, which was reported by Welber et al. (1978). Their results are shown in Fig. 21. The pressure range covered is considerable. The highest pressure, 62 kbar, corresponds to a relative change in the chain axis lattice constant of -12%. To compare with the analysis above we will focus on a pressure 6 kbar, which has the same effect on the lattice constant as cooling to low temperature (Debray et al., 1977; Welber et al., 1978). The same pressure gives a relative change in a^2/V_m of $+0.6\%$. The charge transfer increases $+5\%$ at low temperature for 6 kbar (Megtert et al., 1979). Assuming the same to be the case at 300 K, $\sin(\rho\pi/2)$ will change $+3.9\%$.

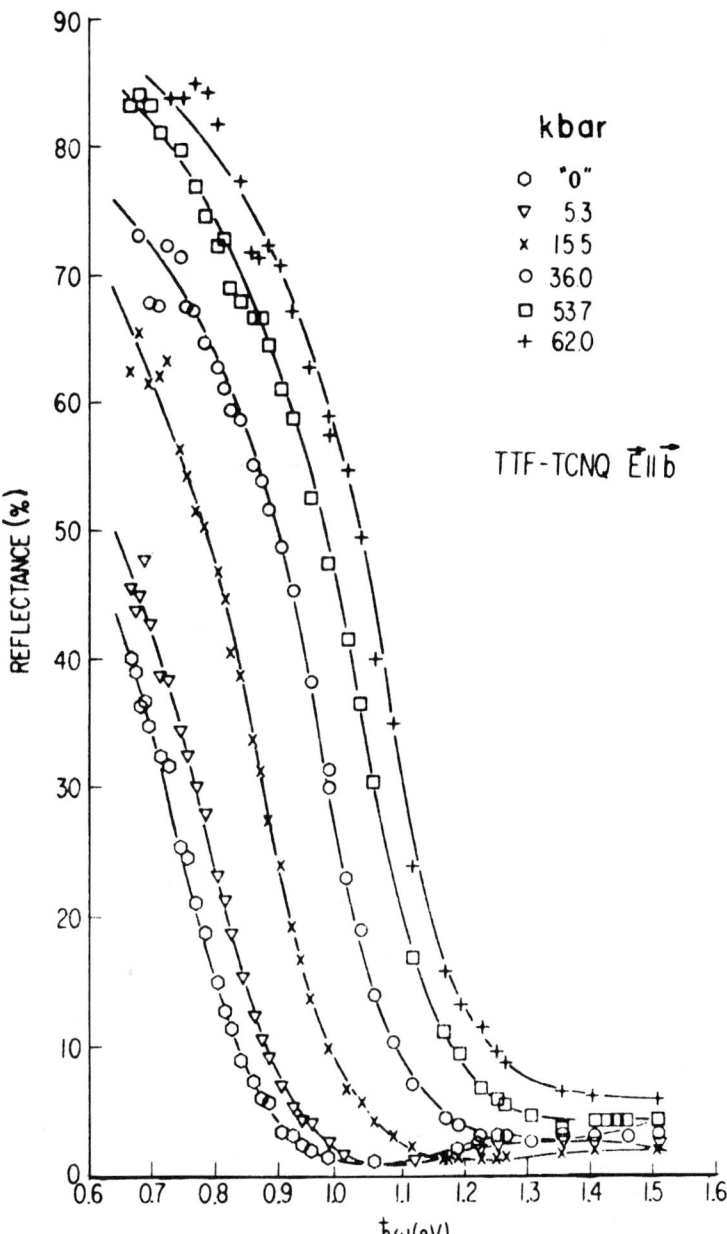

FIG. 21. Pressure dependence of the plasma edge of TTF-TCNQ. [From Welber *et al.* (1978).]

The observed shift in the plasma edge leads to $\Delta(\omega_p^2)/\omega_p^2 \simeq 10.5\%$. Thus from this experiment $\Delta t/t \simeq +6\%$.

The considerable discrepancy between the two results can have several causes. The thermal contractions and the compressibility show somewhat different anisotropies; thus it is conceivable that the mode of overlap between molecules changes differently with temperature and pressure. Another, and perhaps more likely, cause arises from the possible errors encountered in the Drude analysis. For example, Cao et al. (1980b) use a fixed ε_∞ but also report a blue shift of the near-infrared B transition on cooling. Such a shift should result in a decreasing contribution to ε_∞ with decreasing temperature. This effect could lead to an overestimate of the shift in ω_p. Similar factors may influence the analysis of the pressure data.

The above discussion merely serves to point out that while the plasma-edge analysis apparently gives reliable results for the bandwidths, great care must be exercised if differences of a few percent are to be taken seriously.

b. Dependence on Charge Density

The small changes in charge transfer encountered in the previous discussion do not prove the relation between ω_p and ρ predicted by Eq. (42). There is one interesting example, in which it has been possible to vary the charge density considerably. In the organic alloy-series $(NMP)_x(Phen)_{1-x}(TCNQ)$, x can be varied from 0.5 to 1 (Miller and Epstein, 1978; Epstein and Miller, 1978). Phen=phenazine is a neutral, closed-shell molecule. Based on diffuse x-ray scattering experiments (Pouget et al., 1982), it is found that for $x = 1$, there is $\frac{2}{3}$ charge on the TCNQ molecule and $\frac{1}{3}$ on the NMP molecule. Below about $x = 0.67$ the charge transfer suddenly approaches unity, hence the NMP–Phen stack becomes inactive and the charge on TCNQ equals x.

Weinstein et al. (1981) have studied the behavior of the plasma edge in these systems. In Fig. 22 their results are shown for $x = 0.51$ and $x = 1.0$. The shift observed is reasonably consistent with the overall change in total charge density (proportional to x).

c. Band Structure Anisotropy

In principle the Drude picture should apply to the transverse polarization as well as to the chain axis polarization. However, the transverse transfer integral, $t_\perp$, is usually several orders of magnitude below $t_\parallel$ so that $\omega_{p\perp} \ll \omega_{p\parallel}$. With typical relaxation rates this leads to overdamped plasmons, and experimentally all that is seen is a slight increase in the far-infrared reflectance with increasing conductivity.

As one exception, in the $TMTSF_2X$ series of materials a plasma-edge-like feature is observed at low temperature along the b directions. This direction

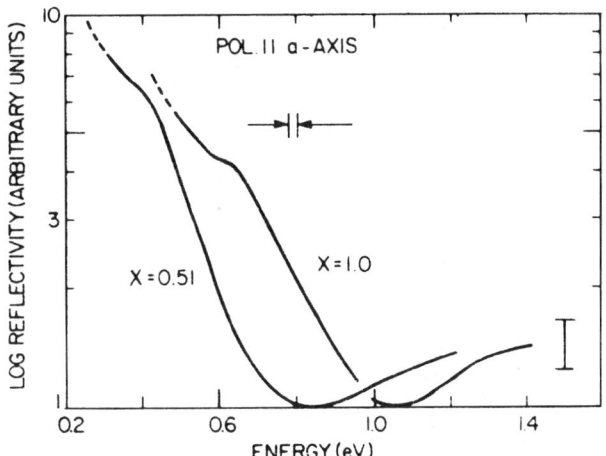

FIG. 22. Plasma reflectance edge for $(NMP)_x(Phen)_{1-x}TCNQ$, $x = 1.0$ and $x = 0.51$. Note the logarithmic reflectance scale. Dashed lines indicate rise due partly to decreased scattered light. [From Weinstein et al. (1981).]

corresponds to a transfer from a TMTSF stack to the neighbor TMTSF stack. Figure 23 presents the results of Jacobsen et al. (1981) on $TMTSF_2PF_6$. They interpret the spectral feature around 1000 cm^{-1} as the true transverse plasma edge. The sharp line near 800 cm^{-1} is a vibrational mode in the PF_6^- ion. Later measurements (Jacobsen et al. 1983) have revealed that no edge is observed in the third direction, c, consistent with the crystal structure and the conductivity anisotropy (Bechgaard et al., 1980).

Jacobsen et al. (1981) performed a Drude fit to the transverse edge and found a fairly temperature-independent $\omega_{p,b} \simeq 2000 \text{ cm}^{-1}$. They applied the simplest possible effective mass model and got $t_b \simeq 3$ meV.

However, as pointed out by Kwak (1982), this model does not represent the actual band structure. If we use a simple two-dimensional, rectangular-lattice, tight-binding model

$$\varepsilon(k_a, k_b) = 2t_a \cos k_a a + 2t_b \cos k_b b \qquad (43)$$

and assume $t_b \ll t_a$, Eq. (9) leads to

$$\omega_{p,b}^2 = \frac{2}{\pi \sin(\rho\pi/2)} \frac{e^2 b^2}{\varepsilon_0 \hbar^2 V_m} \frac{t_b^2}{t_a}. \qquad (44)$$

The basic difference from Eq. (42) is that ω_p^2 is reduced with the bandwidth anisotropy. The physical reason is that only the vicinity of the Fermi surface contributes to the optical effective mass, in contrast to what is the case for the chain direction.

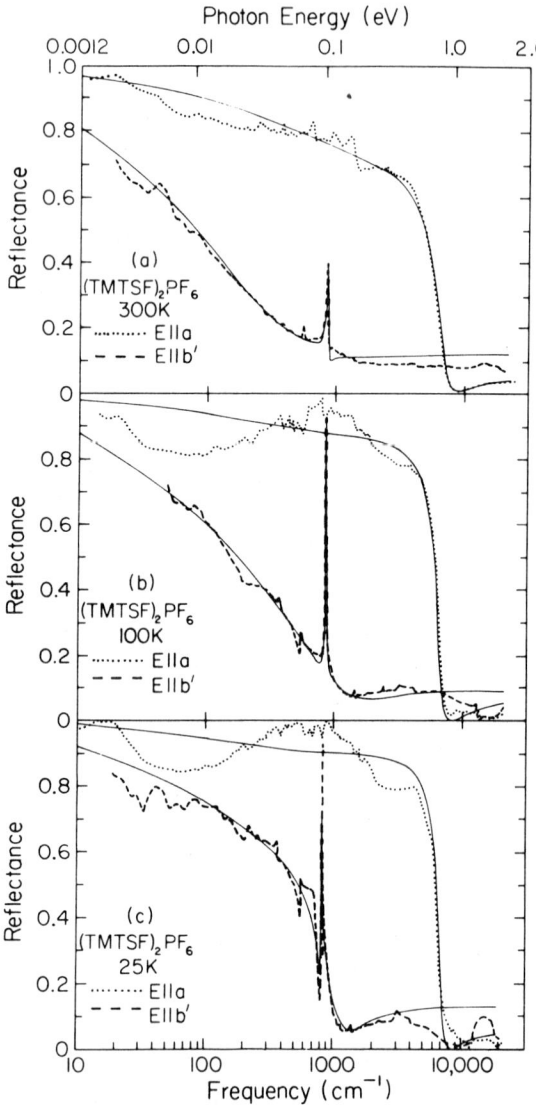

FIG. 23. Polarized reflectance of TMTSF$_2$PF$_6$ for $E \parallel a$ (chain axis) and $E \parallel b'$ (perpendicular to a in the sheets of TMTSF stacks). Data are shown for three temperatures: (a) 300 K, (b) 100 K, and (c) 25 K. Solid lines indicate Drude fits. Note the logarithmic frequency scale. [From Jacobsen et al. (1981).]

With $\rho = 0.5$, $t_a = 0.28$ eV (Table I), and $\omega_{p,b} = 1800\,\text{cm}^{-1}$ (Jacobsen et al., 1983), Eq. (44) yields $t_b \cong 22$ meV, an order of magnitude higher than the original estimate and in better agreement with calculated values (Grant, 1982).

The value is based on the model of Eq. (43). However, since only the vicinity of the Fermi surface contributes to the optical mass, the result is presumably quite model dependent. A more accurate band-structure model is needed for a reliable estimate.

A somewhat similar behavior is observed in the BEDT-TTF family of materials, which is of great current interest due to its superconducting members.

In these materials one finds loosely packed BEDT-TTF stacks with rather close interstack contacts due to a large number of peripheral sulphur atoms (see, e.g., Shibaeva et al., 1985). The infrared and optical properties have been studied by a number of groups (see Tajima et al., 1984, 1985, 1986; Sugano et al., 1985; Kaplunov et al., 1985; Koch et al., 1985; Vlasova et al., 1985; Jacobsen et al., 1985; Menghetti et al., 1986).

In the α-form of $(\text{BEDT-TTF})_2\text{I}_3$, a well-developed plasma edge is found for a polarization *perpendicular* to the stacks, while in the superconducting β-phase, it is along the stacks. In the β-phase plasma edges are observed for two polarizations at low temperatures, see Fig. 24. Here the anisotropy in

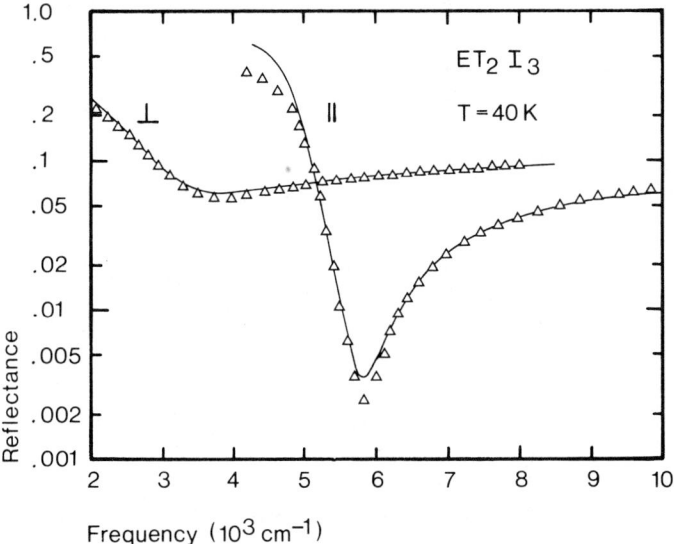

FIG. 24. Drude fits to the plasma edges in $\beta\text{-(BEDT-TTF)}_2\text{I}_3$. Solid lines are models, and triangles are experimental points at $T = 40$ K. [From Jacobsen et al. (1985).]

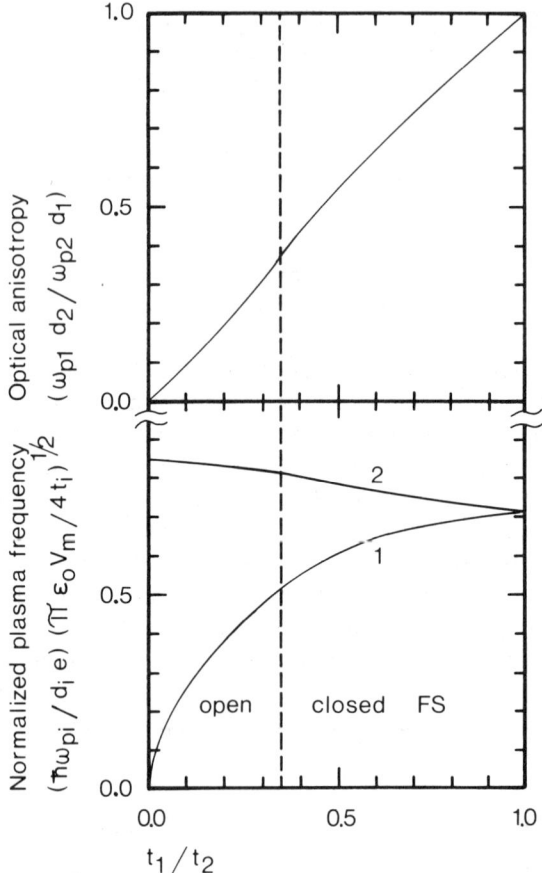

FIG. 25. Anisotropic plasma behavior in two-dimensional orthorhombic, quarter-filled tight-binding model. The bottom graphs show normalized plasma frequencies. [From Jacobsen et al. (1985).]

plasma-edge position is only about 2, thus Eq. (44) cannot be used. Equation (43) leads, for $\rho = 0.5$, to the general results shown in Fig. 25. It is interesting to note that ω_{p1}/ω_{p2} is roughly proportional to t_1/t_2 so that one finds $t_\parallel \cong 0.19$ eV and $t_\perp \cong 0.08$ eV, suggesting a closed Fermi surface in this case.

Again such analysis is oversimplified, since the real band structure is much more complicated than that of Eq. (43). For example, the crystal structure shows fairly dimerized stacks so that the conduction band splits into two. Then the interband contribution has to be considered separately (Tajima et al., 1985).

10. Plasmon Excitations

The plasma edge in reflectance signifies a zero crossing in $\varepsilon_1(\omega)$, while simultaneously $\varepsilon_2(\omega) \ll 1$. At this frequency longitudinal oscillations in the charge density can occur. As mentioned above, these excitations are the plasmons, which are directly observable in electron energy-loss spectroscopy. A few observations of plasmons in highly conducting organics have been reported, namely, by Ritsko et al. (1975) on TTF-TCNQ and by Nücker et al. (1986) on α- and β-(BEDT-TTF)$_2$I$_3$.

Some of the results for the chain axis in TTF-TCNQ are shown in Fig. 26 for two momentum transfers. The optical experiment corresponds to a

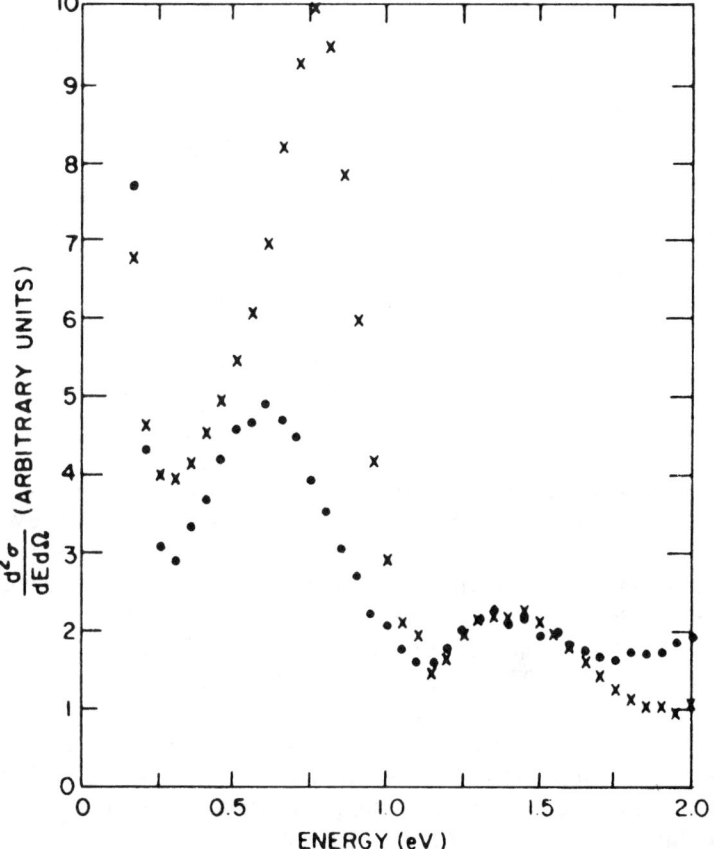

FIG. 26. Inelastic electron scattering cross section for TTF-TCNQ. Momentum transfer along b: Crosses ~ 0.1 Å^{-1}, circles ~ 0.7 Å^{-1}. The cross section is proportional to the loss function $-\text{Im}(1/\bar{\varepsilon})$. [From Ritsko et al. (1975).]

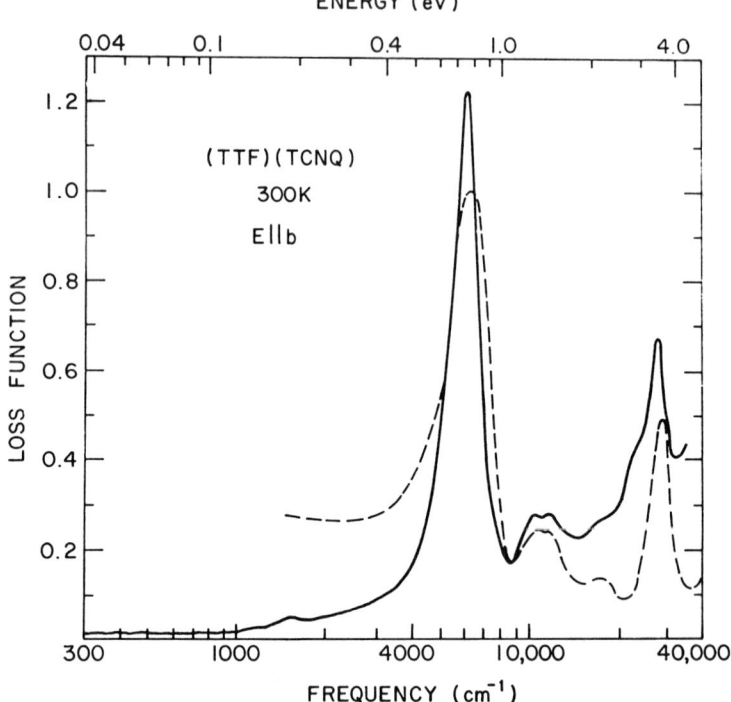

FIG. 27. — $\mathrm{Im}(1/\bar{\varepsilon}(\omega))$ for the b-axis of TTF-TCNQ as obtained from optical spectroscopy (solid line, data from Tanner et al., 1976) and from electron energy loss spectroscopy (dashed line). (Data from Ritsko et al., 1975.) The latter is drawn on an arbitrary scale.

vanishing momentum transfer. Here Ritsko et al. find a plasmon energy $\hbar\Omega_p \simeq 0.75$ eV. This should be compared to $\hbar\omega_p/\sqrt{\varepsilon_\infty} \simeq 0.76$ eV from Table I.

A more detailed comparison can be made by drawing the energy-loss function by Ritsko et al. together with $-\mathrm{Im}(1/\bar{\varepsilon})$ as obtained from a Kramers–Kronig transformation of polarized single crystal reflectance (Tanner et al., 1976). These functions should be proportional. The result is shown in Fig. 27. Allowing for the limited resolution of the energy-loss experiment and a background rising toward zero energy, the agreement is excellent. The comparison confirms the equality of the longitudinal and the transverse dielectric function in the long wavelength limit and also supports the validity of the Kramers–Kronig procedure.

For higher momentum transfers (Fig. 26), the plasmon energy is lower: A negative plasmon dispersion is observed in contrast to normal three-dimensional metals, where the dispersion is always positive.

The energy-loss experiment also give additional information on other excitations. The band at 1.2–1.6 eV (Fig. 26) was thoroughly discussed in Section 8, and was attributed to the lowest monomer transition in TCNQ$^-$. The lack of dispersion of this excitation evident from Fig. 26 supports the view that it is intra- rather than intermolecular in nature.

The negative plasmon dispersion is well understood. The frequency and wavevector-dependent dielectric function of quasi-one-dimensional conductors has been calculated in various approximations with special emphasis on the plasmons (see, e.g., Williams and Bloch, 1974; Ovchinnikov and Ukrainskii, 1974; Williams and Bloch, 1976; Campos *et al.*, 1977; Liebmann *et al.*, 1977; Kahn *et al.*, 1978; Nobile and Tosatti, 1980; Brosens *et al.*, 1982). The negative dispersion, as well as some of the angular dependence observed, can be accounted for.

Another interesting aspect of the plasmon spectrum of TTF–TCNQ is the apparent absence of Landau damping: The plasmon excitations never seem to merge with the single-particle excitations. However, it is possible that the plasmon damping throughout the momentum range is dominated by processes assisted by high-energy phonons (see below).

Williams and Bloch (1974) and Williams *et al.* (1974) have suggested that plasmons may be excited optically in the usual normal incidence experiment and thus contribute to the optical absorption. The excitation is made possible via static or thermal disorder (Hopfield, 1965). Presently no plasmon absorption has been safely identified in the optical spectra of organic metals, but the theory suggests that the absorption contribution can be broad and without structure. Thus is may be difficult to observe.

Plasmons can also be excited in special optical geometries (Bulaevskii and Kukharenko, 1972). This effect has been unambiguously observed in platinum chain complexes (Brüesch, 1973) and has presumably been seen in Cs$_2$TCNQ$_3$ (Vlasova *et al.*, 1975).

IV. Infrared Properties

The role of the key parameters U and t has been discussed extensively above. Here we shall concentrate on the low-energy excitations in the range of band A: the intraband transitions, the vibrational modes, and effects of their coupling. First we consider possible relations between the optical conductivity and the dc conductivity then the role of molecular vibrations, and subsequently the effects of the various phase transitions occurring in organic metals. Some general features in the frequency-dependent conductivity are commented on in relation to theoretical work.

11. Plasma-Edge Relaxation

If we for a moment return to the Drude expression

$$\varepsilon(\omega) = \varepsilon_\infty - \frac{\omega_p^2}{\omega(\omega + i\Gamma)}, \qquad (45)$$

the relaxation rate Γ was introduced as a phenomenological, frequency-independent parameter. Assuming Eq. (45) to hold, the corresponding dc conductivity can be calculated from Eq. (7)

$$\sigma_{\text{opt}}(0) = \frac{\varepsilon_0 \omega_p^2}{\Gamma}. \qquad (46)$$

For TTF-TCNQ we have from the plasma-edge analysis, $\Gamma(300\,\text{K}) \cong 1500\,\text{cm}^{-1}$, $\omega_p \cong 10{,}000\,\text{cm}^{-1}$, which leads to $\sigma_{\text{opt}}(0) \simeq 1100\,\Omega^{-1}\,\text{cm}^{-1}$. Actually $\sigma_{\text{dc}} \simeq 600\,\Omega^{-1}\,\text{cm}^{-1}$ (Cohen et al., 1974). For many other materials $\sigma_{\text{opt}}(0)$ is typically about twice σ_{dc}. Thus the question arises whether the comparison is meaningful.

The temperature dependence of Γ for TTF-TCNQ has been determined in some detail by Bright et al. (1974) and by Cao et al. (1980b). Some of the data by Bright et al. are shown in Fig. 28. It is immediately clear that the temperature dependence of $\Gamma = 1/\tau$ is much weaker than that of σ_{dc}. Similar discrepancies are found in other materials, like $TMTSF_2PF_6$ (cf., Table I).

Intraband absorption in metals is usually believed to be due to second-order processes involving phonons. By including intermediate states in higher bands, the appropriate matrix element can be written (Dumke, 1961)

$$M = \sum_{n'} \frac{\langle n, k+q|H_{\text{em}}|n', k+q\rangle \langle n', k+q|H_{\text{ep}}|n, k\rangle}{\varepsilon_{n'}(k+q) \pm \hbar\Omega_q - \varepsilon_n(k)} \\
+ \sum_{n'} \frac{\langle n, k+q|H_{\text{ep}}|n', k\rangle \langle n', k|H_{\text{em}}|nk\rangle}{\varepsilon_{n'}(k) - \varepsilon_n(k) - \hbar\omega} \qquad (47)$$

for initial and final states in band n. H_{ep} is the electron–phonon coupling term, and H_{em} the electromagnetic term in the Hamiltonian. $\hbar\Omega_q$ is the phonon energy and $\pm$ in Eq. (46) refers to emission and absorption, respectively. Energy conservation further requires

$$\hbar\omega = \varepsilon_n(k+q) - \varepsilon_n(k) \pm \hbar\Omega_q. \qquad (48)$$

Bright et al. (1974) argue that the electron–phonon coupling matrix element in Eq. (47) is basically the same as that appearing in a resistivity calculation with one-phonon scattering, at least if higher bands can be neglected. Bright et al. then conclude that the dc conductivity is controlled by another scattering mechanism.

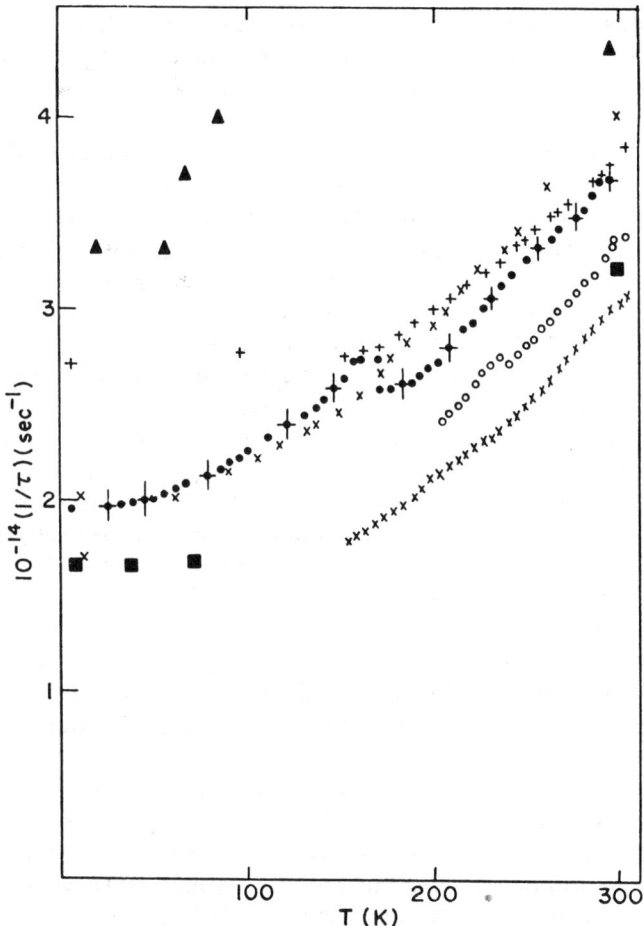

Fig. 28. Optical relaxation rate as function of temperature for several samples of TTF-TCNQ. The breaks in some of the curves are caused by strain-induced defects in the samples. [From Bright et al., 1974).]

If Eq. (46) holds and higher bands can be ignored, a simple relation can be derived for the optical relaxation rate:

$$\Gamma = \left(\frac{2\pi}{\hbar}\right)\lambda\left(k_B T + \frac{k_B \theta_D}{2}\right), \quad (49)$$

where λ is the dimensionless electron–phonon coupling constant, and θ_D is the Debye temperature. This relation was derived by Bright et al. (1974) for the one-dimensional case, based on the calculation by Holstein (1954) in three dimensions [cf., Eq. (14)]. The last term in Eq. (48) represents phonon emission, which also contributes at low temperatures. Bright et al. derive a value of $\lambda = 1.3$ from the data in Fig. 28 and Eq. (48), while Cao et al. (1980b) get $\lambda = 0.68$.

Seiden and Cabib (1976) and Welber et al. (1978) have suggested a different interpretation. They decompose the optical scattering rate into a large temperature-independent contribution, tentatively proposed to be due to emission of high-energy phonons, and a strongly temperature-dependent part, which is interpreted as arising from electron–electron scattering. The latter part should correspond closely to the dc conductivity.

There seems to be a serious problem with these interpretations. For the elastic or quasi-elastic scattering considered, absorption beyond the band-width is not possible, cf., Eq. (47). However, with bandwidth estimates based on the results of Table I, the plasmon frequency frequently exceeds or is in the vicinity of the bandwidth. This may be connected to a commonly observed decrease of relaxation rate with frequency throughout the near infrared (compare, e.g., Drude fit and experimental data for $\bar{E} \parallel$ stacks in Fig. 24). As the frequency is increased, more and more phonon-assisted processes become impossible.

That the scattering mechanisms at dc and at optical frequencies are different in origin is also supported by irradiation experiments reported by Gunning and Heeger (1979). They find that a defect level, which reduces the dc conductivity with a factor 2, does not influence Γ within experimental accuracy.

Studies at high doses on TMTSF-materials also support this conclusion (Zuppiroli et al., 1985).

Instead other approaches must be considered to understand the near infrared absorption. Williams and Bloch (1974 and 1976) have suggested that emission of high-energy intramolecular optical phonons, or excitation of plasmons via the electron-lattice coupling, may play a role. Ritsko et al. (1975) point to the possibility of simultaneous excitation of two electron-hole pairs, again based on the electron–electron interactions.

Whatever the exact mechanism is, the similarity of the plasma edges in a wide range of materials points to a common source of absorption.

12. Role of Molecular Vibrations

The constituent molecules of the conducting charge-transfer salts have a substantial number of internal vibrational modes. Their frequencies are typically lying in the 100–3000 cm^{-1} range. In the isolated molecule some modes are infrared active, while others are Raman active or both. In the crystals, coupling between the conduction electrons and certain modes gives rise to remarkable optical features. These effects allow the determination of electron–molecular vibration coupling constants, and also give information on the nature of phase transitions.

The basis for a proper interpretation is the complete assignment of modes for the individual molecule (neutral or ionized). Such assignments have, for example, been accomplished for TCNQ and TCNQ$^-$ (Girlando and Pecile, 1973; Bozio and Pecile, 1980a; Bozio et al., 1978), for TTF and TTF$^+$ (Bozio et al., 1979), and for TMTXF/TMTXF$^+$ (X = S,Se) (Meneghetti et al., 1984), while a preliminary analysis has appeared for BEDT-TTF (Meneghetti et al., 1986).

a. Determination of the Degree of Charge Transfer from Vibrational Spectroscopy

In many highly conducting organics, like TTF–TCNQ, the charge transfer is incomplete and cannot be determined from stoichiometry. The degree of charge transfer, ρ, is an important quantity, which usually is found by interpreting diffuse x-ray scattering experiments. It is of considerable value to have alternative methods for obtaining ρ, and vibrational spectroscopy constitutes one.

It is found from calculation as well as experiment that the exact frequency of a certain mode depends on whether the molecule is charged or not. Thus, in principle, a simple frequency determination provides a method to determine ρ.

Rather than observing two lines corresponding to neutral and ionized molecules, and with ρ equal to the relative weight of intensities, a single sharp line in between is observed, and its relative position simply gives ρ. That only a single line is found is due to the electron hopping rate being much higher than the ionization shift (cf., Section 3). In a few cases two lines have been reported in highly conducting materials (Kapachina et al., 1978). However, since the measurements are done on powders, sample preparation problems cannot be excluded.

In order to actually use the method for a specific molecule with any accuracy, it is necessary to select a mode that has a large ionization shift and a low sensitivity to variations in the surrounding environment (crystal field). For these reasons totally symmetric Raman modes have usually been preferred.

A carefully selected infrared active mode may also be employed. Chappell et al. (1981) used a C=N stretching mode in TCNQ to determine ρ for a series of TCNQ-salts. Their results are reproduced in Fig. 29. The study includes a number of materials, where ρ is known independently (black dots), as well as some where it is not (open circles). As evident from the scatter and size of the error bars, the method is not quite as accurate as an analysis of diffuse

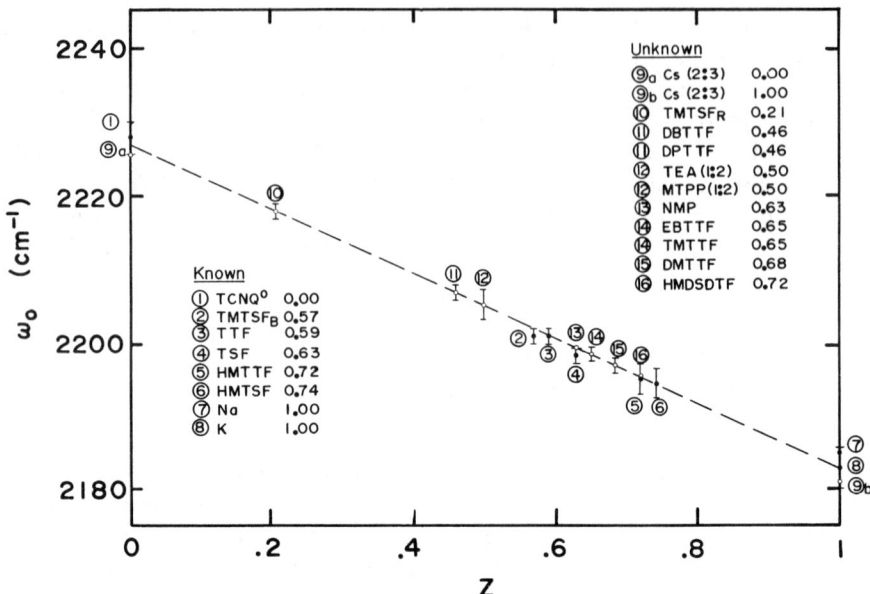

FIG. 29. Frequency of specific C=N stretching mode for a range of TCNQ compounds as determined by infrared spectroscopy. The dashed line indicates the established connection between frequency and degree of charge transfer. (Reprinted with permission from Chappell et al., 1981. Copyright 1981 American Chemical Society.)

x-ray scattering. However, the latter constitutes a difficult experiment and does not always lead to a result.

A comment is in order for Cs_2TCNQ_3. Here lines are observed for both $TCNQ^0$ and $TCNQ^-$. This is consistent with Cs_2TCNQ_3 being nonconducting and with the results of a bond length analysis [Fritchie and Arthur (1966), see also Section 8].

In Table 2 most available ρ data found by vibrational spectroscopy are collected. For comparison ρ values from other sources are included, mostly based on diffuse x-ray scattering. From the numbers given, it is clear that an accuracy of the order 0.05 in ρ may be obtained.

b. Electron–Intramolecular Vibration (emv) Coupling

The conduction electron orbital energy is, in general, a function of the exact atomic configuration in the molecule. Hence the intramolecular vibrations will modulate the one-electron energies, giving rise to the emv coupling. It can be shown by symmetry arguments (Duke et al., 1975) that for a non-degenerate level only the totally symmetric, A_g modes couple linearly to the

TABLE II

Degree of Charge Transfer in Highly Conducting Charge Transfer Salts by Vibrational Spectroscopy

Material	ρ	Mode (R = Raman, I = Infrared)	ρ from other source
TTF-TCNQ	0.5[a]	R, several averaged	0.55[k]
—	0.59[b]	R, TCNQ(C=C)	—
—	0.63[c]	R, TTF(C=C)	—
—	0.6[c]	R, TCNQ(C=C)	—
—	0.59[d]	I, TCNQ(C=N)	—
TMTSF-TCNQ	0.59[d]	I, TCNQ(C=N)	0.57[k]
TSF-TCNQ	0.65[d]	I, TCNQ(C=N)	0.63[k]
HMTTF-TCNQ	0.72[d]	I, TCNQ(C=N)	0.72[k]
HMTSF-TCNQ	0.74[d]	I, TCNQ(C=N)	0.74[k]
HMDSDTF-TCNQ	0.72[d]	I, TCNQ(C=N)	
TMTTF-TCNQ	0.65[d]	I, TCNQ(C=N)	
—	0.55[e]	R, TMTTF(C=C)	
—	0.4[e]	R, TCNQ(?)	
—	0.50[f]	R, TCNQ(C=C)	
NMP-TCNQ	0.63[g]	R, TCNQ(C=C)	0.67[m]
—	0.62[b]	R, TCNQ(C=C)	—
TTF-I$_{0.71}$	0.67[h]	R, TTF(C=C)	0.71[n], 0.52[p]
TTF-Br$_{0.76}$	0.71[h]	R, TTF(C=C)	0.76[n]
TTF-Cl$_{0.80}$	0.79[h]	R, TTF(C=C)	0.80[n]

[a] Kuzmany and Stolz (1977)
[b] Matsuzaki et al. (1980a); see also Matsuzaki et al. (1985)
[c] Bozio and Pecile (1980a)
[d] Chappell et al. (1981)
[e] Torrance et al. (1981)
[f] Tokumoto et al. (1982)
[g] Kuzmany and Elbert (1980)
[h] Bozio and Pecile (1980a); see also Matsuzaki et al. (1980b)
[k] From diffuse x-ray scattering; see Table I for references
[m] Pouget et al. (1980)
[n] From stoichiometry
[p] Ikemoto et al. (1977)

electrons. From a solid state point of view, these optically inactive (but Raman active) modes constitute a set of high-energy, dispersionless optical phonons. The emv coupling adds a term to the Hamiltonian (see, e.g., Duke, 1978)

$$H_{\text{emv}} = \sum_{\ell} \tilde{c}_\ell^+ \tilde{c}_\ell \sum_n g_n(\tilde{b}_{n\ell}^+ + \tilde{b}_{n\ell}). \tag{50}$$

$\tilde{c}_\ell^+$ and $\tilde{b}_{n\ell}^+$ are the electron and phonon creation operators, respectively. ℓ is the site index, while n specifies the A_g modes, which couple to the electrons through the coupling constants, g_n.

The emv coupling can only be observed optically when charge can move to or from the molecule, as in the solid state or in molecular complexes in solution. The effect has been known for a long time (Ferguson and Matsen, 1958) and plays a particularly important role for the organic conductors. For a recent thorough discussion of the theoretical framework, see Painelli and Girlando (1986).

The effect may also be observed by other than optical means, for example, by electron tunneling experiments (Simonsen and Coleman, 1973; Cooper et al., 1982).

That the emv coupling may influence the instabilities in the molecular metals was first suggested by Gutfreund et al. (1974a). The same authors also proposed how a specific mode with a large coupling constant might perturb the optical properties of quasi-one-dimensional metals (Gutfreund et al., 1974b). Torrance et al. (1975b) pointed out that the emv coupling could give rise to antiresonances in the optical conductivity. Since then, more detailed and quantitative work has been carried out, especially for TCNQ-salts, and optical experiments play an important part in this work.

(1) *Determination of emv Coupling Constants.* Much of the effort followed the suggestion by Rice et al. (1975a) that intramolecular distortions may stabilize the charge-density-wave state in TTF–TCNQ. The charge-density wave is then considered to correspond to a complicated superposition of distortions, intermolecular as well as intramolecular. The important point for the optical properties is that each phonon band that couples to the electrons will become infrared active along the molecular chain: The molecular vibrations are excited by phase oscillation of the charge-density wave (Rice, 1976).

The observed positions and absorption intensities will, of course, reflect the coupling strengths, and the optical conductivity can be written (Rice, 1976; Rice et al., 1977a):

$$\sigma(\omega) = \frac{\varepsilon_0 \omega_p^2}{i\omega} \left[f\left(\frac{\omega}{2\Delta}\right) - f(0) - \left(\frac{\omega}{2\Delta}\right)^2 f^2\left(\frac{\omega}{2\Delta}\right) \lambda D_\phi(\omega) \right], \quad (51)$$

where ω_p denotes the plasma frequency of the noninteracting conduction electrons. 2Δ is the energy gap of the stabilized semiconducting state. The functions

$$f(x) = \left(\pi i + \ln\left[\frac{1-S}{1+S}\right]\right) \frac{1}{2Sx^2}, \quad S = (1 - x^{-2})^{1/2}, \quad (52)$$

$$D_\phi(\omega)^{-1} = D_0(\omega)^{-1} + 1 - \frac{V}{\Delta} + \left(\frac{\lambda \omega^2}{4\Delta^2}\right) f\left(\frac{\omega}{2\Delta}\right), \quad (53)$$

and

$$D_0(\omega) = -\sum_n \frac{(\lambda_n/\lambda)\omega_n^2(q_0)}{\omega_n^2(q_0) - \omega^2 - i\omega\Gamma_n}. \tag{54}$$

The summation runs over all involved phonon bands. $2V$ is the energy gap in the absence of emv coupling, $\lambda_n = N(0)g_n^2/\hbar\omega_n(q_0)$ are the dimensionless emv coupling constants and $\lambda = \Sigma\lambda_n$. $N(0)$ is the density of states at the Fermi level for the original metallic system. $\omega_n(q_0)$ is the frequency of the nth phonon at the wavevector of the charge-density wave, q_0, and Γ_n is the natural width.

In order to apply Eqs. (51)–(54) the quantity V/Δ must be determined. V/Δ is obtained from the low-temperature, static dielectric constant

$$\varepsilon_s = 1 + \left(\frac{\omega_p}{2\Delta}\right)^2\left(\frac{2}{3} + \frac{\lambda\Delta}{V}\right), \tag{55}$$

in terms of ω_p and Δ.

A spectacular experimental example of this effect is the complex salt TEA(TCNQ)$_2$, which has been studied experimentally by Kaplunov et al. (1972) and by Brau et al. (1974). Very strong bands are observed for a polarization along the TCNQ stacks, while the polarizations perpendicular to the stacks, where vibrational lines are normally expected to be strongest, show only weak bands at slightly different frequencies. In effect, the 10 otherwise forbidden A_g vibrations borrow oscillator strength from the conduction electron system.

Rice et al. (1977a) have analyzed the chain axis optical conductivity by Brau et al. (1974) in terms of the phase phonon theory cited above. A reasonably good fit, which is shown in Fig. 30, was obtained.

Kamarás et al. (1980) later studied the effects of irradiation on the vibrational structure in TEA(TCNQ)$_2$ and Qn(TCNQ)$_2$. Rather high damage levels are required to affect the spectra. Thus it is possible that a localized, rather than a charge-density wave, model may be appropriate.

Rice and coworkers (Rice et al., 1977b; Rice, 1979; Rice et al., 1980) have published models for dimerized systems with one and two carriers per dimer. Correspondingly, experimental data on various semiconducting TCNQ compounds have been analyzed. For the dimerized systems with two carriers, important temperature dependence in the spectra has been reported by Graja et al. (1981). One factor to take into account is the finite population of the low-energy triplet levels (cf., Section 6).

Yartsev (1982, 1984) has done model calculations for trimerized and tetramerized systems with emv coupling (two electrons on segments of three and four molecules, respectively).

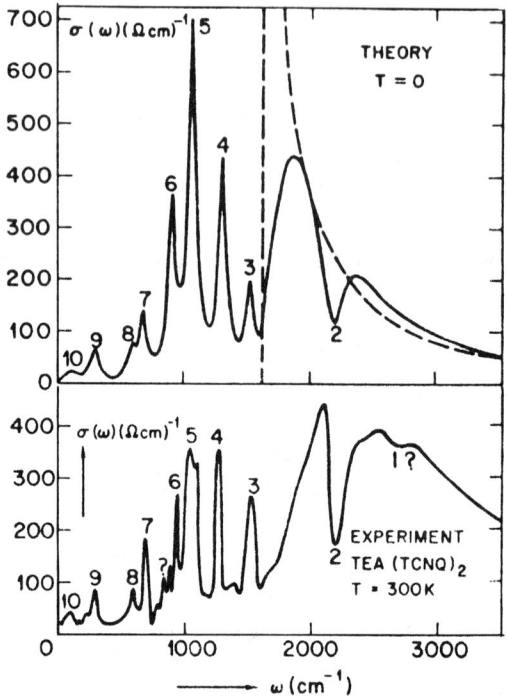

FIG. 30. Observed and theoretically calculated frequency dependent conductivity of TEA(TCNQ)$_2$. [From Rice et al. (1977a).]

For the metallic materials, some evidence for emv coupling in TTF–TCNQ was presented by Aharom-Shalom et al. (1977), based on an interpretation of isotope shifts in spectra, which were mainly polarized perpendicular to the chain direction.

Etemad (1981) interpreted low-temperature powder data on TTF–TCNQ within the framework of the phase-phonon theory, partly in the regime where $\omega > 2\Delta$.

Estimates of emv coupling constant for TCNQ and TTF are given in Tables III and IV, respectively. In Table III we give values obtained by quantum chemical calculations by model fitting to frequency-dependent conductivity and by a new method proposed by Painelli et al. (1984). The latter can be used on powdered samples of dimerized materials and seems to yield quite reproducible results. The bare coupling constants are basically material independent, specific for a particular molecule. Thus the differences between the columns of Table III reflect the error margins involved. The two experimental sets agree reasonably well. Taking the last column, which is

TABLE III

EMV COUPLING CONSTANTS FOR TCNQ

Number character[a]	Frequency in TCNQ^0 (cm^{-1})	g_n (cm^{-1})		
		Calculated[b]	MEM(TCNQ)$_2$[c]	Revised exp.[d]
1 (C—H)	3048	31	44	40
2 (C≡N)	2229	423	350	380
3 (C=C, C—H)	1602	1057	540	470
4 (C=C)	1454	392	500	430
5 (C=C, C—H)	1207	229	300	290
6 (C—C)	948	237	85	130
7 (C—C)	711	263	190	210
8 (C—CN$_2$, C—C)	602	18	50	120
9 (ring deform.)	334	194	180	240
10 (C—C≡N)	144	78	75	110

[a] Girlando and Pecile (1973)
[b] Lipari et al. (1976)
[c] Rice et al. (1980)
[d] Painelli et al. (1984). The values are averages over several materials.

presumably the most reliable, we can calculate the contributions to the dimensionless electron–phonon coupling constant in the metallic phase of TTF-TCNQ. With an estimated TCNQ bandwidth of 0.8 eV (Jacobsen, 1985), we get $\Delta\lambda_{\text{int}}(\text{TCNQ}) \cong 0.2$, with the biggest contributions ($\Delta\lambda_n \cong 0.04$) from modes 3, 4, and 9.

TABLE IV

EMV COUPLING CONSTANTS FOR TTF

A_g mode[a] n, character	Frequency[b] TTF^0 (cm^{-1})	g_n (cm^{-1})	
		Calculated[a]	Experimental[c]
1 (C—H)	3083	90	—
2 (C=C)	1555	360	230
3 (C=C)	1518	940	1000
4 (S—C—H)	1094	170	80
5 (S—C)	735	360	350
6 (S—C, C—S—S)	474	630	560
7 (S—C—S, C—S—C)	244	40	80

[a] Lipari et al., 1977
[b] See also Bozio et al., 1979
[c] Painelli et al. (1984); average values for TTF-Br and TTF-Chloranil

There is much less data available on TTF, but calculated and experimental values are in fair agreement (Table IV). With a TTF-bandwidth of only 0.4 eV (Jacobsen, 1985), $\Delta\lambda_{int}$(TTF) = 0.8, of which modes 3 and 6 each contribute about 0.34. Equivalent modes will appear in all TTF-based molecules, including the TMTSF and BEDT-TTF types. No reliable estimates for coupling constants are available for these molecules, but many studies are in progress. A review paper on the infrared properties of associated materials is planned for a later volume of this series.

(2) Fano Interference Effects. Most of the work discussed above has been concerned with the semiconducting state at frequencies below the gap ($\omega < 2\Delta$). At frequencies above the gap the emv coupling will result in Fano (1961) type interferences, as first discussed by Torrance *et al.* (1975b). This also follows from the phase-phonon theory (Rice *et al.*, 1977a). An A_g mode above the gap will usually appear as an indentation in the frequency-dependent conductivity. One clear example is the TCNQ(v_2) mode in TEA(TCNQ)$_2$, see Fig. 30.

The infrared and Raman properties in this range have been analyzed theoretically by Horovitz *et al.* (1978), who find very strong effects of the emv coupling, especially when several vibrational modes are considered simultaneously. Such strong effects have so far not been observed experimentally.

Another question that must be addressed is whether or not the A_g modes are infrared active in the metallic state. Normally a charge-density wave or other symmetry-breaking effect is required to obtain the anomalous infrared activity. An incipient Peierls transition may give such an effect, as reported by Gutfreund *et al.* (1974b). From a different viewpoint, Brazovskii and Finkel'stein (1981) have discussed the activation of intramolecular modes by charge-density wave fluctuations.

Gor'kov and Rashba (1978) have discussed a purely metallic state. They consider a process where an energetic conduction electron excites an A_g mode assisted by emission or absorption of an acoustic phonon. Again, typical Fano antiresonances appear. Gor'kov and Rashba suggest that the effect is seen in TTF-TCNQ at 300 K, the strongest feature being the dip in $R(\omega)$ centered around 1400 cm^{-1} (see Fig. 36).

c. Nature of the Peierls Transition

The intramolecular vibrations can be considered microscopic probes for structural changes occurring at the Peierls transition. Thus vibrational spectroscopy provides information on the precise nature of this transition as it occurs in TTF-TCNQ and other materials.

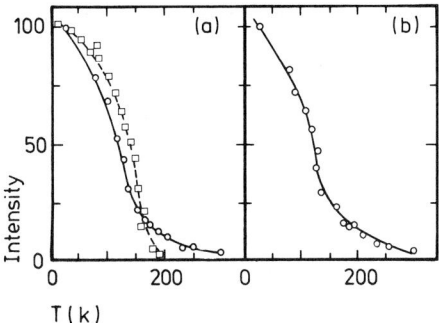

FIG. 31. Normalized peak intensity of the 470 cm^{-1} (ν_6) absorption band of TTF against temperature for (a) TTF-SCN$_{0.58}$ and (b) TTF-SeCN$_{0.58}$. Broken curve: x-ray superlattice intensity (Thomas et al., 1978). [From Bozio and Pecile (1980b); copyright The Institute of Physics.]

The method is based on the fact that the intensity of the phase phonon absorption, or the activation of an otherwise forbidden mode, is related to the amplitude of the charge-density wave, or other symmetry-breaking effect. Thus it should be possible to observe the growth of charge-density waves and changes in crystal symmetry with decreasing temperature by just monitoring the absorption intensity of one of the A_g modes. This was confirmed in a study by Bozio and Pecile (1980b) on TTF-halides and pseudohalides. These have rather high transition temperatures, so the effect is particularly clear. In Fig. 31 the results for TTF-SCN$_{0.58}$ and TTF-SeCN$_{0.58}$ are shown. The transition temperatures are 170 K for both materials (Somoano et al., 1977), close to where the intensity increases sharply. The data for TTF-SCN$_{0.58}$ are compared with a direct measure for the amplitude of the charge-density wave namely the intensity of the x-ray superlattice reflections.

When the vibrational modes are properly identified, additional information can be obtained. In double-stack systems it is possible, as demonstrated by Bozio and Pecile (1981), for the case of TTF-TCNQ to gain knowledge of the role of the separate stacks. Their experimental data is shown in Fig. 32. The intensity of the TCNQ band starts growing just above 50 K. The intensity of the TTF band increases suddenly below 50 K and has a sharp rise at 38 K. This behavior is consistent with the present understanding of the nature of the transitions at 53, 49, and 38 K, and how they involve the two-stack systems (see, e.g., Schultz and Craven, 1979).

Etemad (1981) reports a growth of the intensities of the phase-phonon modes in TTF-TCNQ from 150 K and down, indicating a wide regime with charge-density wave fluctuations. Etemad also finds evidence for coupling to some normally infrared active modes. His study of the C=N stretching

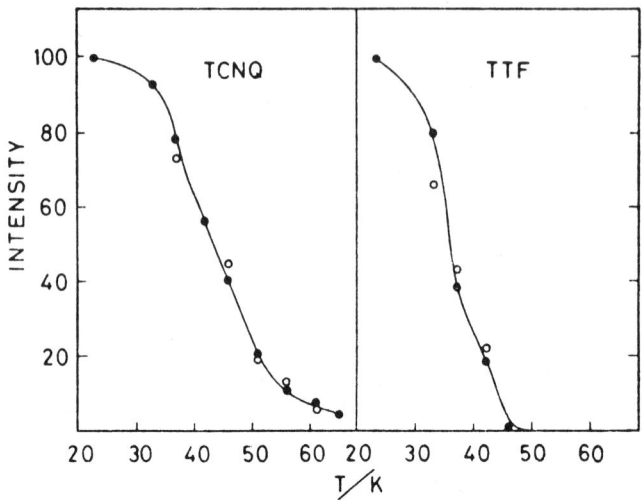

FIG. 32. Temperature dependence of the absorption intensities at 317 cm^{-1} (v_9, TCNQ) and 253 cm^{-1} (v_7, TTF) in TTF-TCNQ. Full lines are visual aids. [From Bozio and Pecile (1981).]

modes is reproduced in Fig. 33. As the temperature is lowered from 300 K the 2216 cm^{-1} absorption is broadened and splits up exactly at the phase transition. Etemad argues from supplementary measurements that the anomalously splitoff mode is polarized along the long axis of the TCNQ molecule and that this is a signature for molecular displacements in the same direction. This finding is consistent with analysis of x-ray diffraction experiments by Khanna *et al.* (1977). A similar effect is observed in TSF-TCNQ.

Bates *et al.* (1981) have recently undertaken an extensive investigation of the far-infrared vibrational spectrum of TTF-TCNQ and TSF-TCNQ. The methods include bolometric and transmission experiments on single crystals. The modes are identified by comparing the spectra of the two compounds and by using deuterated TTF and TCNQ selectively in the crystals.

Interestingly it is found that anomalous infrared activity occurs not only along the chain direction but also in the *a* direction. The lattice constant in this direction does indeed increase at the phase transitions, i.e., the symmetry is lowered. The activity in the *a* direction is presumably mediated by electron transfer between TTF and TCNQ. The usual growth in intensities with decreasing temperature is observed. In addition, some non-totally symmetric modes are found to show anomalous activity (see, e.g., Eldridge and Bates, 1982).

In both TTF-TCNQ and TSF-TCNQ, it is found that two specific modes involving the donor molecules show strong activity. They are assigned to (1) a libration around the long axis of the molecule and (2) a torsion of the two

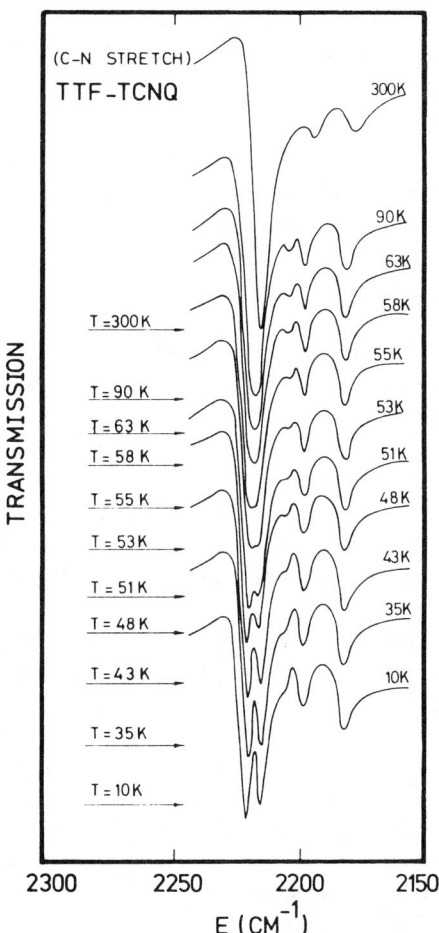

FIG. 33. Temperature dependence of the C=N stretching mode absorption in TTF-TCNQ. [From Etemad (1981).]

rings around the central bond. They should show no or little normal infrafred activity. Therefore it is concluded that such distortions strongly participate in the Peierls state. A final point in the paper by Bates *et al.* is the absence of any anomalous far-infrared activity on the TCNQ stacks in TSF-TCNQ. Therefore the TCNQ stacks are assumed to play little role in the Peierls transition in this material. This is somewhat in conflict with the interpretation of Etemad (1981) of the C=N vibrational behavior mentioned above.

More recently, anion ordering transitions have been studied by this technique in TMTSF$_2$X (Bozio *et al.*, 1982a) and TMTTF$_2$X (Bozio *et al.*, 1982b);

Garrigou-Lagrange et al., 1984; Bozio et al., 1985), while Meneghetti et al. (1986) characterized a metal-insulator transition in α-(BEDT-TTF)$_2$I$_3$.

13. Lattice Vibrations

The lattice vibrations in a complicated system like TTF-TCNQ have not been fully analyzed (Morawitz, 1981). The modes can be translational (acoustic and optic) or rotational (librations) or mixtures. The frequency range of interest is about 0–100 cm^{-1}, and experimentally the lattice modes can be detected by neutron diffraction, by far-infrared spectroscopy, and by Raman scattering.

Kuzmany and Stolz (1977) have reported the Raman scattering of TTF-TCNQ and find five out of six Raman active lattice modes from 39 cm^{-1} to 96 cm^{-1}. There have been a few other studies on TTF-TCNQ (Kuzmany et al., 1978; Temkin and Fitchen, 1978), but the experiments are difficult due to sample deterioration (Kuzmany, 1978).

Raman scattering experiments on the (TMTSF)$_2$X group have shown these to be very weak scatterers (Iwahana et al., 1982), but recent work has identified about six external modes (in the range 36–64 cm^{-1}) (Krauzman et al., 1986).

The far-infrared study by Bates et al. (1981) on TTF-TCNQ and TSF-TCNQ does give some tentative assignments of external modes. A large number of low-energy modes are seen, and isotope shifts indicate that most of them are of mixed character.

The same group (Eldridge et al., 1985) has studied TMTSF$_2$X compounds and observe extra structure below various anion ordering temperatures. This is attributed to activation of zone-boundary phonons by zone folding.

14. Infrared Excitation Spectrum

In this section we shall first deal with the determination of small energy gaps like the Peierls gap by optical methods. Then we will present data on the few highly conducting organic solids, where the full infrared excitation spectrum is known, and establish connection to theoretical models.

a. Identification of Energy Gaps

Most of the conducting organic materials exhibit metal-insulator transitions. To properly understand the mechanism, knowledge of the single-particle gap size is crucial. Under certain assumptions the gap may be obtained indirectly from activation energies of transport or magnetic properties. However, the direct observation of the gap in an optical experiment is much more satisfactory. However, as we shall see, the proper identification of a gap is not without problems.

One of the first attempts to obtain by optical means a value for the Peierls gap in TTF–TCNQ was a photoconductivity experiment by Poehler et al. (1974). They found a rapid increase in the 4 K photoconductive signal at $\simeq 180\text{ cm}^{-1}$, corresponding to $E_g \simeq 22$ meV.

Later Tanner et al. (1976) obtained data on the frequency-dependent conductivity, $\sigma(\omega)$. The data did not show an onset of absorption at 180 cm^{-1}, but rather an absorption edge at 300 cm^{-1}. However, Tanner et al. argued that $\sigma(\omega)$ would be dominated by the square-root singularity arising from the density of states in a one-dimensional material (see, e.g., Denner et al., 1974). Then the proper gap identification would be at the maximum in $\sigma(\omega)$, which was found near 1100 cm^{-1}, or $E_g = 140$ meV.

Subsequently Eldridge in a series of papers (Eldridge, 1976, 1977, 1978) reported the results of still more refined bolometry/photoconductivity experiments. The bolometric part of the signal can be eliminated by working at very low temperatures and fairly high frequencies. Some of the data are shown in Fig. 34 for both the chain axis (b) and the a axis polarization.

The vibrational structure is only seen in the bolometric response (as expected), and the purely photoconductive response clearly shows the onset of single-particle excitations starting at ~ 300 cm^{-1} ($E_g \simeq 37$ meV). This is very close to twice the activation energy in the conductivity (Etemad, 1976). A small photoconductive signal is noted for $E \parallel b$ below 300 cm^{-1}. This may be impurity photoconductivity or may be related to the charge-density wave, pinned at low frequencies (see Subsection b.).

Some care must be exercised in the interpretation of these experiments. When the absorption coefficient is very high ($\alpha d \gg 1$, where d is the sample thickness), the signal saturates to $1 - R$, where R is the reflectance. Most

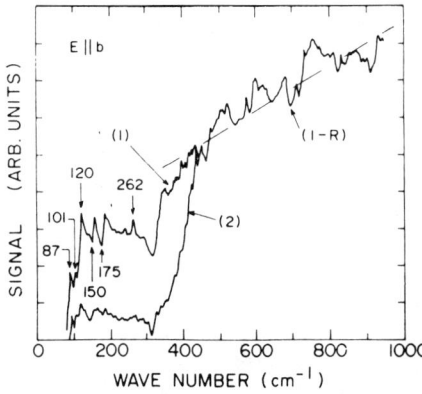
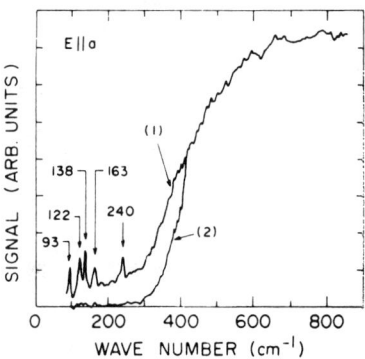

FIG. 34. Photoconductive and bolometric signals of TTF–TCNQ at low temperature for $E \parallel a$ and $E \parallel b$. Curves (1) are the combined signals, while curves (2) are the photoconductive signals alone. The resolution is 7 cm^{-1}. [From Eldrige (1978).]

of the structure above 430 cm^{-1} for $E \parallel b$ is thought to be due to saturation, so in this range the signal is not simply related to the absorption.

Still photoconductivity is presumably the best method for gap determinations. Bates et al. (1981) obtained by the same method a value for the Peierls gap in TSF-TCNQ, $E_g \simeq 22$ meV.

A particularly clear example of the opening of an energy gap was recently reported for TMTSF$_2$ReO$_4$, which has a sharp metal-insulator transition at 182 K (Jacobsen et al., 1982). The frequency-dependent conductivity in the metallic as well as the insulating state was found by analysis of reflectance data and is shown in Fig. 35. Here the edge in the low-temperature conductivity is quite sharp. The gap value deduced from the onset of the edge is 210 meV, while the maximum in $\sigma(\omega)$ is at 235 meV. The difference of 25 meV is relatively small in this case.

The apparent absence of a real square-root singularity in the excitation spectrum can have several causes. Banik (1981) considered the influence of local field effects and found a rounding off of the divergence. Instead an edge is found at the gap. Three-dimensional coupling, disorder, and phonon-assisted absorption processes may also broaden the onset of absorption.

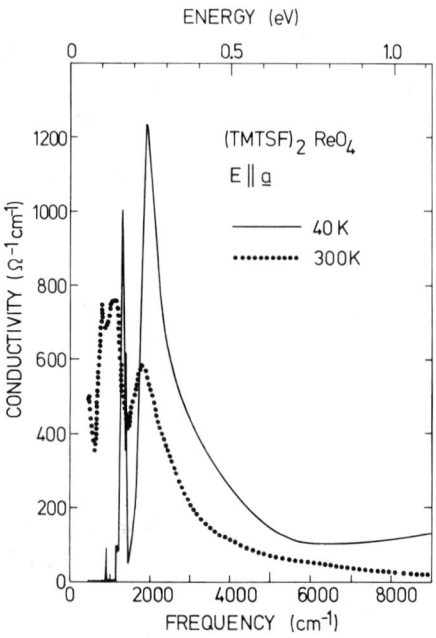

Fig. 35. Frequency-dependent conductivity of (TMTSF)$_2$ReO$_4$ in the chain direction. Dotted curve, 300 K and full curve, 40 K. [From Jacobsen et al. (1982).]

b. TTF-TCNQ

The material in which the most complete knowledge of the infrared, complex dielectric function has been obtained is TTF-TCNQ. We shall here review the experimental data and then discuss theoretical models.

(1) *Experimental.* Early information on the infrared absorption of TTF-TCNQ was based on investigations of thin films and powders (Tanner *et al.*, 1974; Torrance and Nicoli, 1974; Chaudhari *et al.*, 1974). The main result of these studies was the discovery of a prominent maximum in the infrared conductivity, situated at 1000 cm^{-1} or somewhat below. Apart from a sharpening on cooling, the conductivity peak was found at all temperatures. In the metallic regime the far-infrared conductivity is lower than the dc conductivity, as well as the mid-infrared conductivity, a clear indication of non-Drude behavior. In the insulating state at low temperature, Tanner *et al.* (1974) found a broad far-infrared absorption band centered around 80 cm^{-1}. This band was interpreted as arising from a pinned charge-density wave (see below).

Later polarized reflectance measurements at 300 K and further thin film measurements were performed and found to confirm the picture described (Jacobsen *et al.*, 1974; Tanner *et al.*, 1976). Other measurements on films and powders did not add to the basic features (Wozniak *et al.*, 1975; Benoit *et al.*, 1976).

Subsequently the pinned charge-density wave has been the subject for two far-infrared investigations. Coleman *et al.* (1976) measured the far-infrared reflectance at low temperature, but concluded that the pinned mode was situated outside the experimental range (12–100 cm^{-1}), presumably at 2 cm^{-1}. Eldridge and Bates (1979) found with the bolometric technique a peak at 7 cm^{-1}. This was interpreted as being due to a pinned mode at 3.4 cm^{-1}.

Later Jacobsen (1979), Tanner *et al.* (1981), and Tanner and Jacobsen (1982) have made a fairly complete study of the polarized reflectance at various temperatures. The availability of large single crystals of good quality and refined experimental techniques make these measurements superior to earlier reflectance studies. We shall use these results in the following discussion.

However, more recently Eldridge and Bates (1983) and Eldridge (1985) have used the simple bolometric as well as the composite bolometer techniques on thick crystals in attempts to measure $1 - R$ directly. Their latest results resemble those presented here, but deviate in giving better resolution of vibrational structure and also show a different sharpness of some of the features. Likewise, new submillimeter transmission measurements

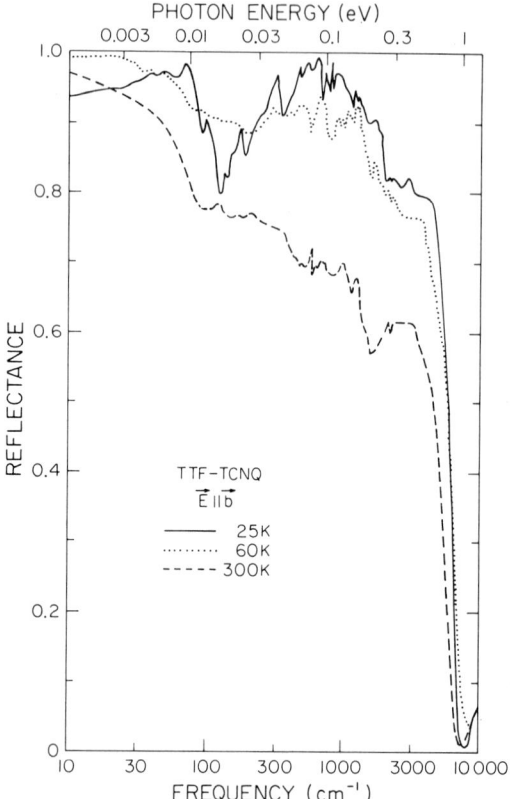

FIG. 36. Reflectance of TTF-TCNQ for $E \parallel b$. Data are shown on a logarithmic frequency scale from 10 cm^{-1} to 10,000 cm^{-1} for three temperatures: 25 K, 60 K, and 300 K. [From Tanner and Jacobsen (1982).]

(Gorshunov et al., 1986) are not entirely consistent with the reflectance measurements, thus we should be aware that revision of the experimental data will presumably continue. This serves to illustrate the difficulties encountered in far-infrared, low-temperature work.

The reflectance data of Tanner and Jacobsen (1982) are shown in Fig. 37. The data are for a polarization along the chain axis and for three temperatures in the frequency range 10–20,000 cm^{-1}. The most noteworthy feature is the qualitative change in the far-infrared spectrum between 25 K and 60 K: This is undoubtedly the signature of the Peierls metal-insulator transitions, which occur at 38–53 K. Other interesting features include the high mid-infrared reflectance at low temperatures and the indications of vibronic structure. Note, for example, the dip at 1400 cm^{-1}, previously mentioned in the context of Fano antiresonances (see Section 12).

The data of Fig. 36 has been analyzed by the Kramers–Kronig transformation to give $\sigma(\omega)$ and $\varepsilon(\omega)$, shown in Fig. 37 and Fig. 38, respectively.

The room-temperature conductivity (Fig. 37, dashed line) does not show strong structure. A broad, low maximum is found around 800 cm^{-1}. The conductivity at 10 cm^{-1} is 900 Ω^{-1} cm^{-1}, reasonably consistent with the dc value. In the far-infrared range the conductivity has a minimum of order 300 Ω^{-1} cm^{-1}, centered around 100 cm^{-1}. The dielectric function (Fig. 38, dashed line) is negative below ~70 cm^{-1} but is positive in a range in the far to mid infrared; it is again negative approaching the near-infrared range and the plasma edge.

At 60 K, where the maximum in the dc conductivity is found, the excitation spectrum shows more pronounced features (dotted lines). There is a peak in $\sigma(\omega)$, now centered at 300 cm^{-1}, a dip in the far infrared, but only to about 1500 Ω^{-1} cm^{-1}, and a rapid rise to ~6000 Ω^{-1} cm^{-1} in the very far infrared. Typical dc values are 10^4 Ω^{-1} cm^{-1} (Cohen *et al.*, 1974). The dielectric function is qualitatively similar to that at 300 K, but with much sharper behavior.

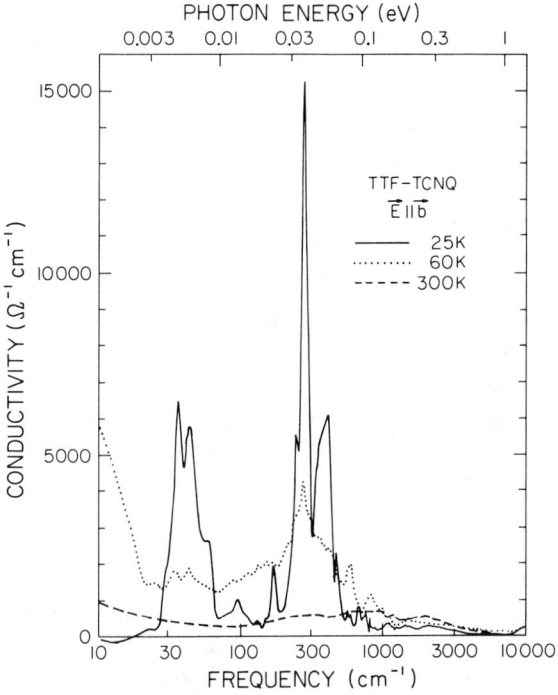

FIG. 37. Frequency dependent *b*-axis conductivity of TTF-TCNQ. Data are shown on a logarithmic frequency scale from 10 cm^{-1} to 10,000 cm^{-1} for three temperatures: 25 K, 60 K, and 300 K. [From Tanner and Jacobsen (1982).]

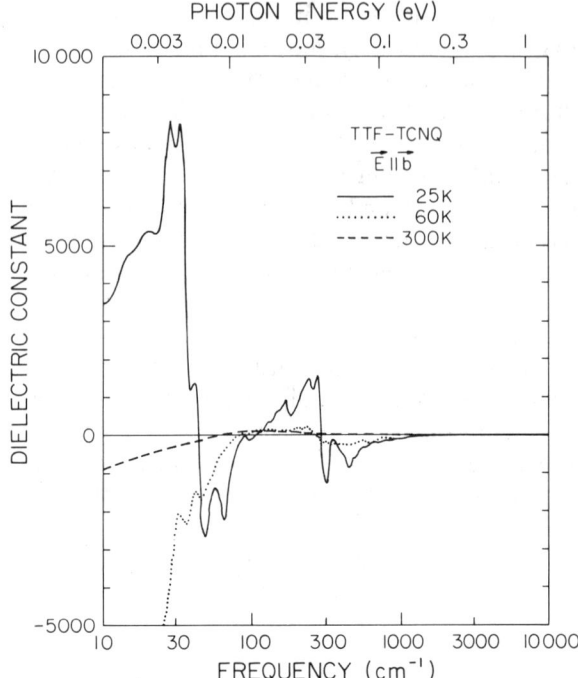

FIG. 38. Frequency dependent b axis dielectric function of TTF–TCNQ. Data are shown on a logarithmic frequency scale from 10 cm^{-1} to 10,000 cm^{-1} for three temperatures: 25 K, 60 K, and 300 K. [From Tanner and Jacobsen (1982).]

In the semiconducting range at 25 K (solid lines) the conductivity shows a very strong peak still centered at 300 cm^{-1}. The far-infrared conductivity is low apart from a second strong peak, which has appeared near 40 cm^{-1}. The dielectric function has correspondingly a strong resonance in the far infrared, pushing the low frequency value as high as 3500. This is consistent with reported microwave values of order 3000–4000 (Gunning *et al.*, 1977).

(2) *Interpretation.* It is the general consensus that TTF–TCNQ at low temperature is in a Peierls–Fröhlich charge-density wave state. This state has been described by for example, Fröhlich (1954), Lee *et al.* (1974), Allender *et al.* (1974), and Rice *et al.* (1975b). The optical properties have been studied theoretically by Schuster (1975), Fukuyama (1976), Kurihara (1976), Buzdin and Bulaevskii (1977), Rice (1978), and others.

The basic optical excitation spectrum consists of (1) phase oscillations of a pinned charge-density wave (phase mode), and (2) single-particle transitions across a Peierls gap. The latter gap is due to the $2k_F$ periodicity of the charge-density wave. Rice (1978) pointed out that there may be many components in the charge-density wave corresponding to various intermolecular and intramolecular modes with coupling to the electrons. Each will have an optically active phase mode. However, most of the oscillator strength should be contained in the low-frequency pinned mode, which is associated with external lattice modes.

It is straightforward to interpret the 25 K conductivity in Fig. 37 in this model. The previously discussed photoconductivity experiments by Eldridge (1978) suggest the existence of a single-particle gap at 300 cm^{-1}. Thus the strong peak in $\sigma(\omega)$ at 300 cm^{-1} is simply due to electronic transitions across the Peierls gap. The other pronounced feature, the peak at 40 cm^{-1}, is then assigned to be the pinned charge-density wave. The differing results of earlier experiments have been suggested being due to experimental difficulties (Tanner et al., 1981).

The theoretical models predict the following pinned mode contribution to the dielectric function (Lee et al., 1974):

$$\Delta\tilde{\varepsilon}(\omega) = \frac{m^*}{M^*}\omega_p^2 \frac{1}{\omega_T^2 - \omega^2 - i\omega\gamma}. \tag{56}$$

Here ω_T is the pinning frequency and γ is the relaxation rate of the pinned-mode excitations. The effective mass of the charge-density wave is related to the band mass by (Rice, 1978)

$$\frac{M^*}{m^*} = 1 + \frac{(2\Delta)^2}{\lambda(\hbar\Omega_0)^2}. \tag{57}$$

2Δ is the Peierls gap and Ω_0 is a weighted average of bare vibration frequencies: $\Omega_0^{-2} = \sum (\lambda_n/\lambda)\omega_n^{-2}(2k_F)$.

The 25 K data yields $\Omega_p = \omega_p(m^*/M^*)^{1/2} = 2200$ cm^{-1} and $M^* = 60m_0 = 20m^*$ (Tanner et al., 1981). Equation (57) then gives $\Omega_0\sqrt{\lambda} = 70$ cm^{-1}, which seems to be consistent with typical vibration frequencies and coupling strengths (cf., Section 12). A further separation into separate stack and mode contributions is presently not possible.

In the Peierls–Fröhlich model the static dielectric constant is given by (Lee et al., 1974)

$$\varepsilon(0) = \varepsilon_\infty + \frac{2}{3}\left(\frac{\hbar\omega_p}{2\Delta}\right)^2 + \frac{m^*}{M^*}\frac{\omega_p^2}{\omega_T^2}, \tag{58}$$

which represents contributions from the background polarizability, transitions across the Peierls gap, and the pinned mode, respectively. A quantitative estimate yields

$$\varepsilon(0) = 2 + 700 + 2900 \simeq 3600, \tag{59}$$

using $\omega_p = 9700 \text{ cm}^{-1}$, $\omega_T = 40 \text{ cm}^{-1}$, and $2\Delta = 300 \text{ cm}^{-1}$. Such a value agrees fully with the microwave measurements (Gunning *et al.*, 1977) and with the experimental low-frequency value in Fig. 38.

The amplitude mode of the charge-density wave has not been safely identified. Kuzmany and Stolz (1977) tentatively assign a Raman line at 56 cm^{-1} to this mode.

Turning now to the conducting phase, $T > T_c = 53$ K, the interpretation of the data is less unambiguous. A widely accepted model is the sliding charge-density wave picture (Heeger, 1979), which connects directly to the Peierls–Fröhlich model. In this picture the charge-density waves are thermally depinned above T_c but still possess a coherence length of many lattice constants. The charge-density waves contribute to the dc conductivity, as described phenomenologically by Eq. (56) with $\omega_T = 0$. The temperature range in which the fluctuating charge-density waves are important may extend to the mean-field scale temperature, which is related to the low-temperature Peierls gap by the BCS relation

$$2\Delta = 3.5 k_B T_p^{\text{MF}}. \tag{60}$$

$2\Delta = 300 \text{ cm}^{-1}$ corresponds to $T_p^{\text{MF}} = 125$ K. Diffuse x-ray scattering experiments are consistent with appreciable fluctuations between 53 K and 125 K (Khanna *et al.*, 1977).

Now $\sigma(\omega)$ for 60 K (Fig. 37) can be interpreted in the following way (Tanner *et al.*, 1981). The far infrared minimum level of 1500 Ω^{-1} cm^{-1} is the single particle contribution to σ and can be extrapolated to dc. Since the temperature is well below T_p^{MF}, the oscillator strength previously in the pinned mode may be expected to be shifted to zero frequency with a relatively small reduction. Assuming $\sigma_{\text{dc}} = 10^4 \,\Omega^{-1}$ cm^{-1} and Drude behavior, the predicted relaxation rate $\gamma = 8 \text{ cm}^{-1}$, which from Fig. 38 seems to be a reasonable value. In this case $\sigma_{\text{collective}}/\sigma_{\text{single particle}} \cong 6$, close to the value estimated by Jérome (1980) from the pressure dependence of the conductivity.

The peak at 300 cm^{-1} still represents transitions across a now fluctuating Peierls gap. Qualitatively, similar features can be seen at 160 K, although much weaker. The sliding charge-density wave picture cannot account for the structure still seen at 300 K, especially not for the peak in $\sigma(\omega)$ moving to higher frequencies.

The dominant peak in $\sigma(\omega)$ in the metallic phase has been the subject for a number of theoretical papers. They do not account in detail for the low-frequency behavior, and we postpone a review until Subsection d., since they may be relevant for several materials.

The sliding charge-density wave picture for the optical properties of TTF–TCNQ just above 53 K has been questioned by Marianer *et al.* (1982).

They account for the narrow low-frequency contribution to $\sigma(\omega)$ by a one-dimensional phonon-drag theory, which is, in principle, a single-particle approach. The 300 cm^{-1} peak in $\sigma(\omega)$ is explained by postulating the existence of a damaged surface layer of thickness $\sim 1\mu$m in the sample. Then the infrared conductivity maximum basically identifies the crossover frequency for broken chains (Bernasconi et al., 1974). At very low frequencies the penetration depth exceeds the thickness of the surface layer, hence the intrinsic properties are probed. So far no independent observation of a damaged surface layer has been reported.

c. $TMTSF_2X$ and $(BEDT-TTF)_2X$

Single-stack conductors based on TMTSF and BEDT-TTF have been studied extensively. There are numerous superconductors and materials that are metallic to very low temperatures. Observed ground states also comprise spin-density waves with transition temperatures of order 12–15 K. Theoretically, the spin-density-wave state is expected to have an infrared spectrum somewhat similar to that of the Peierls (charge-density-wave) state, but the electron-phonon interference lines and gap structure depend sensitively on the amplitude of charge-density-wave harmonics of the spin-density wave (Fenton and Psaltakis, 1983; Fenton and Aers, 1985).

In spite of the fascinating physical properties and considerable efforts by spectroscopists, it has so far not been possible to obtain generally agreed upon far-infrared, low-temperature spectra of $TMTSF_2X$ compounds. Such spectroscopic results would be highly useful in the characterization of both superconductivity and spin-density wave state but may have to await the availability of larger single crystals, such as have been used in the studies of TTF-TCNQ.

The first results were on $TMTSF_2PF_6$ at 25 K (Jacobsen et al., 1983) and were based on the reflectance data of Fig. 23. The derived conductivity was fairly Drude-like for the b polarization (perpendicular to the stacks), but showed a sharp gap structure at 180 cm^{-1} along a (stacking direction). Since the metal-insulator (spin-density-wave) transition in this material occurs at 12 K, the results implied a large amplitude spin-density-wave state, not pinned to the lattice. The high dc conductivity at 25 K was assumed associated with sliding-density waves, thus resulting in a collective mode feature in the conductivity spectrum (analogous to the situation in TTF-TCNQ at 60 K). The width of the collective mode, if modelled by a Drude term, would be of order 0.3 cm^{-1}.

Microwave measurements by Javadi et al. (1985) have not confirmed the existence of such a mode, but could be consistent with a mode of width 1 cm^{-1}.

Ng et al. (1983, 1984, 1985a,b) have measured far-infrared, low-temperature reflectance spectra of the ClO_4^--compound (which has a superconducting ground state) and of the SbF_6^-- and AsF_6^--compounds, which have metal-insulator transitions at 12–15 K. Common features include implications for collective modes of small widths, and a low far-infrared conductivity, which is rising gradually on going to higher frequencies. The latter behavior is attributed to the Holstein mechanism (see Section 2). They further observe unusual and sharp, temperature-dependent phonon structure in the ClO_4^- material and, likewise, in $TMTSF_2SbF_6$ below the metal-insulator transition sharp features, which could be due to the "phase-phonon" mechanism in a spin-density-wave state (Fenton and Psaltakis, 1983). Such features are not observed in $TMTSF_2AsF_6$, although the two compounds are expected to have closely similar electronic structures. The spin-density-wave gap was first taken to be 180 cm^{-1} in $TMTSF_2SbF_6$, but the data was later reinterpreted to correspond to a gap of order 50 cm^{-1}, more in accord with mean field behavior and a transition temperature of 12–15 K. However, the gap is in none of the compounds well developed. This could be due to gap anisotropy, impurities, etc.

Eldridge and Bates (1985, 1986) have studied several $TMTSF_2X$ materials using the composite bolometer technique to obtain $1 - R$ at 6 K. Their $TMTSF_2ClO_4$ data are similar to those of Ng et al., but in $TMTSF_2PF_6$ they do observe a rather well-developed gap at 33 cm^{-1}. No strong features are observed at high frequencies. Eldridge et al. (1985) have also used powder transmission measurements to study the effect of phase transitions on various phonon lines in a range of materials.

As is evident, the experimental situation is not very clear, but, interestingly, all studies report much closer agreement and simpler behavior for $\mathbf{E} \perp$ stacks (b) than for $\mathbf{E} \parallel$ stacks. Since the crystal dimension is much smaller along b than a, the measurements along b would actually be expected to be the more difficult (due to diffraction effects, etc.). The problems associated with measurement techniques, collective modes, and low far-infrared conductivity in these materials have recently been reviewed by Timusk (1987).

The BEDT–TTF materials have so far not been studied in great detail in the far infrared. α-$(BEDT-TTF)_2I_3$, which has a metal-insulator transition at 135 K, shows a marked decrease in absorption at low temperature and all the vibrational features sharpen considerably (Meneghetti et al., 1986).

The superconducting β-$(BEDT-TTF)_2AuI_2$ ($T_c = 5$ K) has a stacking axis, frequency-dependent conductivity, as shown in Fig. 39 for $T = 300$ K and 30 K (Jacobsen et al., 1987). Most of the BEDT–TTF materials have quite low room-temperature conductivities ($\simeq 30 \, \Omega^{-1} \, cm^{-1}$), and it is seen that the far-infrared conductivity matches this value fairly well. However, instead of overdamped Drude behavior, $\sigma(\omega)$ has a broad maximum just above

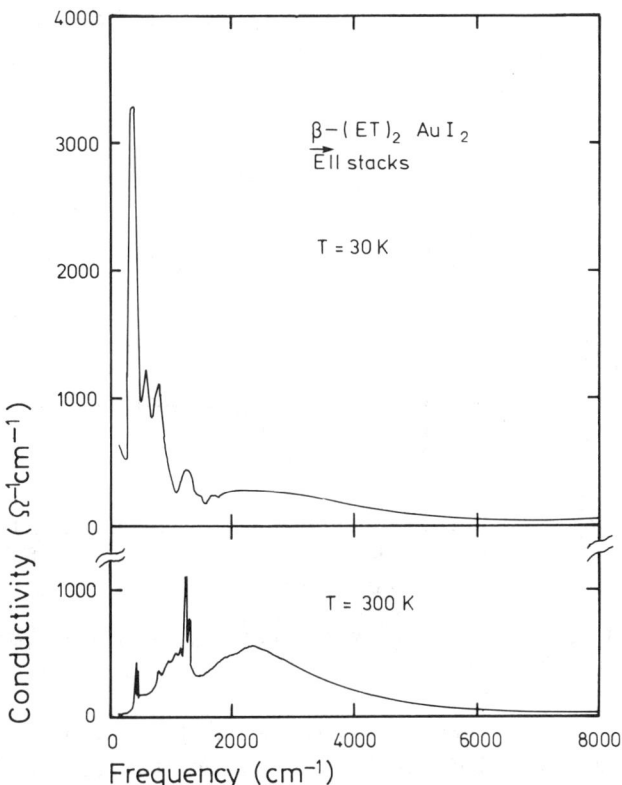

FIG. 39. Frequency-dependent conductivity of β-(BEDT-TTF)$_2$AuI$_2$ along the stacks, at $T = 30$ K and 300 K. [From Jacobsen et al. (1987).]

2000 cm^{-1} and several vibrational features, which are most distinct around 400 cm^{-1} and 1300 cm^{-1}. The broad maximum, which has been observed in many BEDT-TTF-salts, was attributed to an interband transition by Tajima et al. (1985). In a localized model, such an interband transition is just the intradimer excitation at $2t$ (Section 6). However, it should be of equal strength at low temperature, where $\sigma(\omega)$ peaks around 300 cm^{-1} (Fig. 39). Interestingly, the conductivity at low frequencies again extrapolates towards the correct, independently measured dc/microwave value (Tanner et al., 1987), i.e., no collective or narrow mode is implied. The change with temperature has no obvious explanation but agrees with a general observation for organic conductors: The higher the dc conductivity, the lower the frequency of maximum conductivity. This hints that some sort of dynamic localization comes into play, indeed consistent with carrier mean free paths of order or smaller than a lattice constant. We shall discuss this further below.

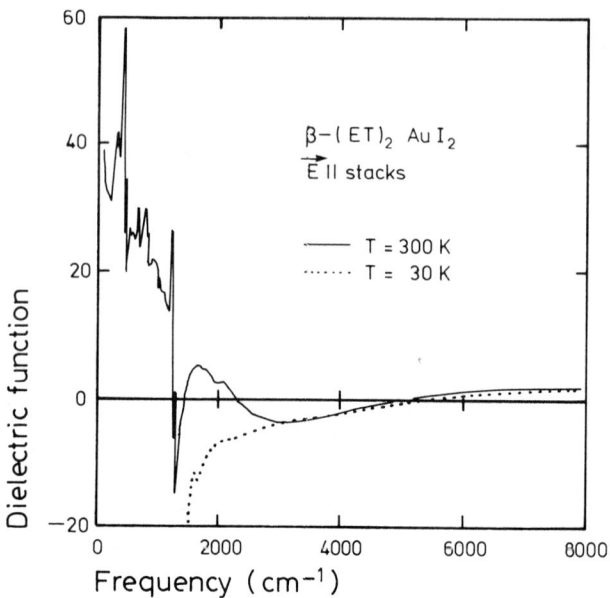

FIG. 40. Dielectric function of β-(BEDT-TTF)$_2$AuI$_2$ along the stacks, at $T = 30$ K and 300 K. [From Jacobsen et al. (1987).]

In Fig. 40 we finally present the dielectric function of β-(BEDT-TTF)$_2$AuI$_2$ corresponding to the conductivity curves of Fig. 39. Again it is clear that the spectrum becomes much more "metallic" at low temperature. Another important observation is that the high-frequency zero-crossing in $\varepsilon(\omega)$ (i.e., the plasmon frequency) only shifts slightly with temperature, another indication that the non-Drude features at low frequencies do not affect the plasmon position (cf., Section 9).

d. Infrared Absorption Maximum

No investigated organic metal has so far displayed simple Drude behavior with a frequency-independent relaxation time. As we have seen, a broad maximum usually appears in the frequency-dependent conductivity. At low temperatures the structure is sharp and may be associated with characterized phase transitions. However, at higher temperature it is more difficult to account for.

As another example, we present in Fig. 41 the frequency-dependent conductivity of the organic metal (NMeH)(I)(TCNQ) from Tanner et al. (1979). This material has a room-temperature conductivity of order $20 \, \Omega^{-1} \, cm^{-1}$, and it is seen that $\sigma(\omega)$ connects to this value and displays a

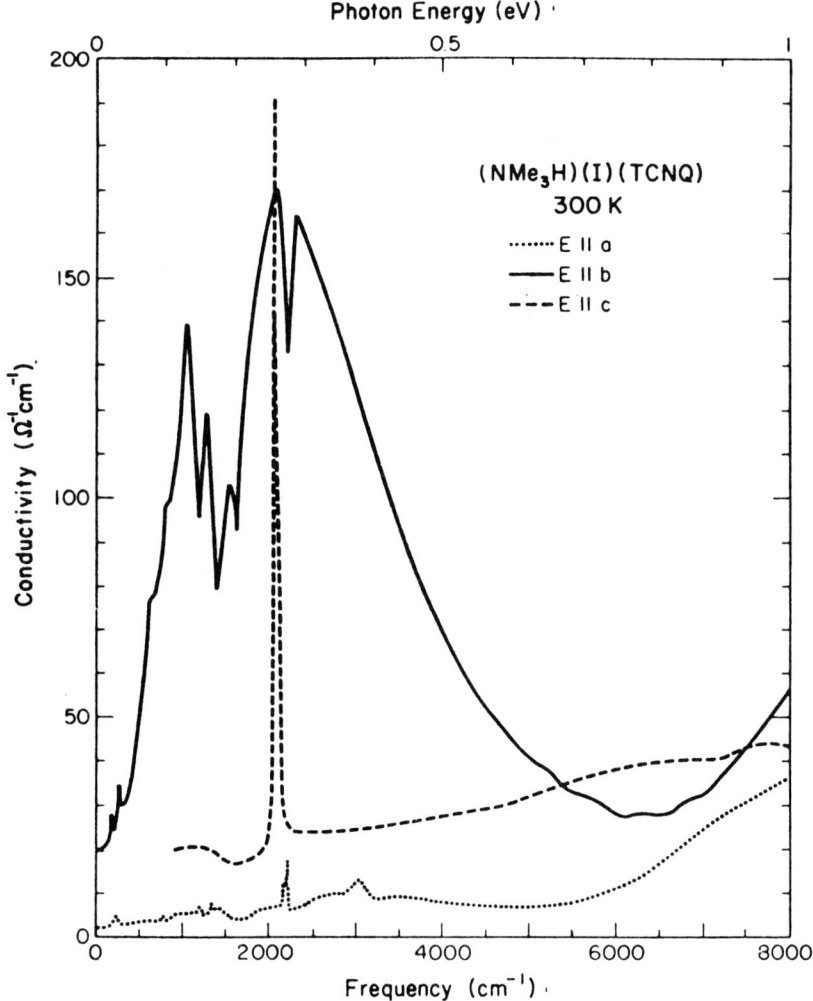

FIG. 41. Frequency-dependent conductivity determined by Kramers–Kronig analysis of the reflectance of (NMe$_3$H)(I)(TCNQ). The chain axis is b (solid lines). [From Tanner *et al.* (1979).]

pronounced broad maximum around 2000 cm^{-1} with superimposed vibronic structure. Although it is suggested that a structurally induced gap may explain the peak (Tanner *et al.*, 1979), such behavior seems to be rather general.

In Fig. 42 spectra are shown for four materials of the TTF-TCNQ family ($T = 300$ K, Jacobsen *et al.*, 1984). HMTSF-TNAP, TTF-TCNQ, and DBTTF-TCNQCl$_2$ are incommensurate conductors with slightly more than

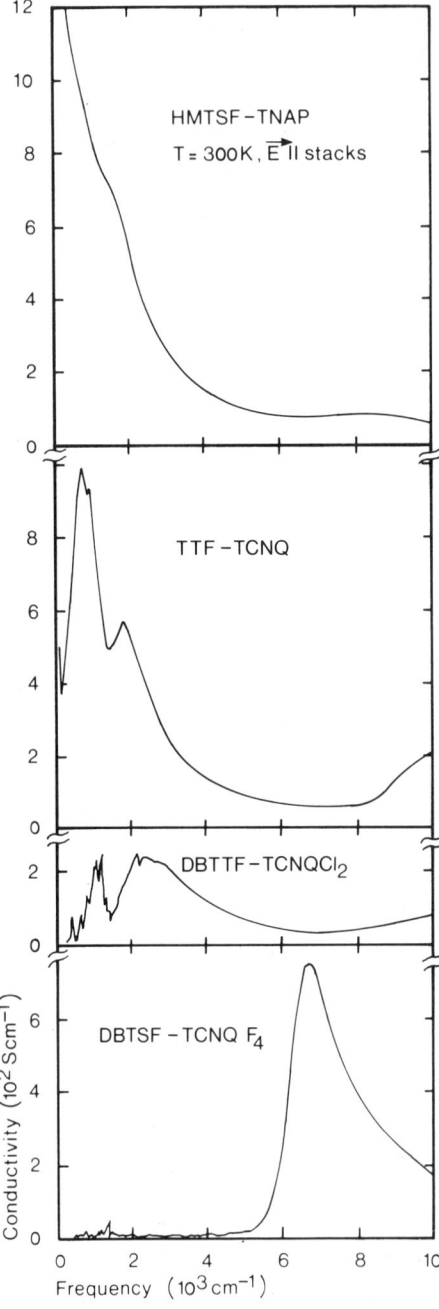

FIG. 42. Frequency-dependent conductivities of four double-stack conductors, $T = 300$ K, $\vec{E} \parallel$ stacks. [From Jacobsen, et al. (1984).]

quarter-filled one-electron bands and room-temperature conductivities of 2000, 600, and 40 Ω^{-1} cm^{-1}, respectively, while DBTSF-TCNQF$_4$ has half-filled bands and is a Mott-insulator (Lerstrup et al., 1983).

Thus the optical spectrum of the latter compound is easily understood as being correlation dominated, the gap being a rough measure of U. The three other spectra simply show the generality of the relation between dc conductivity and distribution of infrared oscillator strength: The lower the dc conductivity, the higher is the frequency around which the oscillator strength is centered, and the more distinct the vibrational structure.

There are a number of theoretical models, that lead to an infrared peak in $\sigma(\omega)$, and we review them briefly here.

Hinkelmann (1976) has discussed the influence of the dynamical structure factor of the lattice for weak electron-phonon coupling. He indeed finds a maximum in $\sigma(\omega)$ due to a frequency dependence in the relaxation rate. However, this dependence is linked to the existence of a Kohn anomaly at $2k_F$ and should therefore, in general, show correlation with the intensity of diffuse x-ray scattering. However, systems with no detectable scattering also have their peak in $\sigma(\omega)$.

Hinkelmann and Reik (1975) and Benoit et al. (1979) have treated aspects of small polaron formation. They attribute the maximum in $\sigma(\omega)$ in TTF-TCNQ to small polaron hopping (shake off of the polarization). These ideas are within the framework of strong electron-phonon coupling, and a continuous transition to a fluctuating Peierls system could be imagined.

Sadovskii (1974), Wonneberger and Lautenschläger (1976), Wonneberger (1977), and Blunck and Reik (1977 and 1979) have studied the infrared properties of a fluctuating Peierls system. The main result is a broad peak in $\sigma(\omega)$ at the gap position. Its sharpness depends on temperature, but its position is relatively temperature independent.

Formation of charge-density waves at wavevector $4k_F$ are frequently observed in organic conductors and are usually taken to imply strong electron-electron interactions. In the limit $U \to \infty$ the electrons act like spinless fermions in the usual tight-binding band, thus the density-of-states is halved and the band is filled to $\pm 2k_F$. The Peierls instability may then take place at $4k_F$. The infrared properties of the $4k_F$ Peierls system are completely analogous to those of the $2k_F$ system, if proper regard is taken to the smaller density-of-states. A rather detailed model for DBTTF-TCNQCl$_2$ based on this approach has been worked out (Jacobsen and Bechgaard, 1985). Strong $4k_F$ charge-density waves are indeed observed at all temperatures. One salient feature is that the model accounts for the intensity distribution between vibrational features and main electronic absorption band: The reduced density-of-states is crucial.

Similar big-U models have been used for a number of materials in which the lattice has $4k_F$ periodicity. Thus, in addition to possible $4k_F$ charge-density waves, the lattice itself might create a gap. Such materials include $Qn(TCNQ)_2$ (McCall *et al.*, 1985) and $TMTTF_2X$ (Bozio *et al.*, 1982).

A number of authors have considered the influence of static and dynamic (thermal) disorder on the infrared properties (Berezinskii, 1973; Cohen *et al.*, 1975; Bush, 1976; Gogolin and Melnikov, 1978; Gogolin, 1979; Freedman, 1980). Typically, a crossover from a ω^2 dependence to a ω^{-2} dependence in $\sigma(\omega)$ is found. The crossover frequency is basically $1/\tau_i$, where τ_i is the mean free time (Gogolin and Melnikov, 1978). These models may have some validity for systems with strong static disorder and for systems with no static disorder at high temperatures.

Finally, Mazumdar and Dixit (1986) attribute the infrared absorption maximum to correlation effects. They adopt a near-neighbor interaction, $V = U/3$ (cf., Section 7) and assume that the main part of the oscillator strength goes into charge-transfer bands measuring V and $U - 2V$. In less conducting materials, where $\sigma(\omega)$ typically peaks at 2000–3000 cm^{-1}, this is consistent with other estimates of U and V. The theory accounts for the $4k_F$ charge-density waves as formed in the Wigner lattice sense (Section 7), while the strong vibrational structure seen in the spectra is assumed to be a secondary effect. However, in the phase-phonon model, the emv coupling strongly adds to the stability of the charge-density wave, thus the electron–phonon coupling can hardly be neglected.

None of the above theories seem to be generally applicable, but the present state-of-the-art experimentally, as well as theoretically, strongly suggest that a proper theory should simultaneously treat electron–electron and electron–phonon (including emv) interactions.

V. Conclusion

We have seen how studies of the optical properties òf organic conductors from the very far infrared to the ultraviolet give a wealth of information on phase transitions, electron–phonon coupling, degree of charge transfer, bandwidths, electron–electron interaction, and even more.

While many separate phenomena are well understood, several important issues remain to be settled, especially with respect to the infrared excitation spectrum.

However, with the increasing efforts in the area, these questions are expected to be resolved, making studies of optical properties an even more powerful tool in the investigation and development of organic conductors.

Acknowledgments

The author wants to acknowledge many illuminating discussions with colleagues over the years, especially those with R. Bozio, D. B. Tanner, and J. B. Torrance.

He would also like to thank Margot Wisborg for her efficient typing of the manuscript.

The work has been supported by the Danish Natural Science Research Council and by the Royal Danish Academy of Sciences and Letters through a Niels Bohr Fellowship.

List of Symbols and Conversion Factors

All equations have been written in SI units. To convert to cgs, replace the vacuum permittivity, ε_0, with $1/4\pi$.

The favorite frequency measure of the spectroscopist is the wave number $\bar{\nu} = 1/\lambda$, where λ is the wavelength in vacuum. Wave numbers are converted to cyclic frequency and photon energy by the relation

$$E = (1.240 \times 10^{-4} \text{ eV cm}) \times \bar{\nu}$$

$$\omega = (1.89 \times 10^{+11} \text{ cm sec}^{-1}) \times \bar{\nu}.$$

The frequency scale is conventionally divided into the following ranges:

	$\bar{\nu}$ (cm^{-1})	E (eV)
Far infrared	10–400	0.001–0.05
Mid infrared	400–4000	0.05–0.5
Near infrared	4000–13000	0.5–1.6
Visible	13000–25000	1.6–3
Near ultraviolet	25000–50000	3–6
Vacuum ultraviolet	50000–10^6	6–100

Other symbols and notation:

- **a**: vector
- $\tilde{a}$: complex quantity or operator
- $\hat{a}$: unit vector
- $\tilde{\bar{a}}$: tensor

References

Aharon-Shalom, E., Weger, M., Agranat, I., and Wiener-Avnear, E. (1977). *Solid State Commun.* **23**, 53.
Allen, P. B. (1971). *Phys. Rev. B* **3**, 305.
Allender, D., Bray, J. W., and Bardeen, J. (1974). *Phys. Rev. B* **9**, 119.
Andrieux, A., Jérome, D., and Bechgaard, K. (1981). *J. de Physique-Lettres* **42**, L-87.
Atkins, P. W. (1970). "Molecular Quantum Mechanics," p. 397. Clarendon Press, Oxford.
Baeriswyl, D., Carmelo, J., and Luther, A. (1986). *Phys. Rev. B* **33**, 7247.
Banik, N. C. (1981). *Phys. Rev. B* **24**, 3564.
Bates, F. E., Eldridge, J. E., and Bryce, M. R. (1981). *Can. J. Phys.* **59**, 339.
Bechgaard, K., Jacobsen, C. S., and Andersen, N. H. (1978). *Solid State Commun.* **25**, 875.
Bechgaard, K., Jacobsen, C. S., Mortensen, K., Pedersen, H. J., and Thorup, N. (1980). *Solid State Commun.* **33**, 1119.

Bechgaard, K., Carneiro, K., Olsen, M., Rasmussen, F. B., and Jacobsen, C. S. (1981). *Phys. Rev. Lett.* **46**, 852.
Benoit, C., Galtier, M., Montaner, A., Deumie, J., Robert, H., and Fabre, J. M. (1976). *Solid State Commun.* **20**, 257.
Benoit, C., Galtier, M., and Montaner, A. (1979). *Phys. Stat. Sol. (b)* **91**, 269.
Berezinskii, V. L. (1973). *Zh. Eksp. Teor. Fiz.* **65**, 1251. [*Sov. Phys.—JETP* **38**, 620 (1974).]
Bernasconi, J., Brüesch, P., Kuse, D., and Zeller, H. R. (1974). *J. Phys. Chem. Solids* **35**, 145.
Bloch, A. N. (1977). *Lecture Notes in Physics* **65**, 317.
Blunck, M., and Reik, H. G. (1977). *Solid State Commun.* **21**, 141 and 797.
Blunck, M., and Reik, H. G. (1979). *Z Physik B* **32**, 147.
Boyd, R. H., and Phillips, W. D. (1965). *J. Chem. Phys.* **43**, 2927.
Bozio, R., Zanon, I., Girlando, A., and Pecile, C. (1978). *J. Chem. Soc. Faraday Trans II* **74**, 235.
Bozio, R., Zanon, I., Girlando, A., and Pecile, C. (1979). *J. Chem. Phys.* **71**, 2282.
Bozio, R., Pecile, C., Bechgaard, K., Wudl, F., and Nalewajek, D. (1982a). *Solid State Commun.* **41**, 905.
Bozio, R., Meneghetti, M., and Pecile, C. (1982b). *J. Chem. Phys.* **76**, 5785.
Bozio, R., Pecile, C., Scott, J. C., and Engler, E. M. (1985). *Mol. Cryst. Liq. Cryst.* **119**, 211.
Bozio, R., and Pecile, C. (1980a). In "The Physics and Chemistry of Low Dimensional Solids" (L. Alcácer, ed.), p. 165. Reidel, Dordrecht.
Bozio, R., and Pecile, C. (1980b). *J. Phys. C: Solid State Phys.* **13**, 6205.
Bozio, R., and Pecile, C. (1981). *Solid State Commun.* **37**, 193.
Brau, A., Brüesch, P., Farges, J. P., Hinz, W., and Kuse, D. (1974). *Phys. Stat. Sol. (b)* **62**, 615.
Brazovskii, S. A., and Finkel'stein, A. M. (1981). *Solid State Commun.* **38**, 745.
Bright, A. A., Garito, A. F., and Heeger, A. J. (1973). *Solid State Commun.* **13**, 943.
Bright, A. A., Garito, A. F., and Heeger, A. J. (1974). *Phys. Rev. B* **10**, 1328.
Brosens, F., Devreese, J. T., Kahn, L. M., and Ruvalds, J. (1982). *Phys. Stat. Sol. (b)* **111**, 95.
Brüesch, P. (1973). *Solid State Commun.* **13**, 13.
Bulaevskii, L. N., and Kukharenko, Y. A. (1972). *Fiz. Tverd. Tela* **14**, 2401. [*Sov. Phys. Solid State* **14**, 2076 (1973).]
Bush, R. L. (1976). *Phys. Rev. B* **13**, 805.
Buzdin, A. I., and Bulaevskii, L. N. (1977). *Pis'ma Zh. Eksp. Teor. Fiz.* **26**, 388. (*JETP Lett.* **26**, 266.)
Campos, V. B., Hipólito, O., and Lobo, R. (1977). *Phys. Stat. Sol. (b)* **81**, 657.
Cao, Y., Yakushi, K., and Kuroda, H. (1980a). *Solid State Commun.* **35**, 601.
Cao, Y., Yakushi, K., and Kuroda, H. (1980b). *Solid State Commun.* **35**, 739.
Chaikin, P. M., Grüner, G., Shchegolev, I. F., and Yagubskii, E. B. (1979). *Solid State Commun.* **32**, 1211.
Chaikin, P. M., Haen, P., Engler, E. M., and Greene, R. L. (1981). *Phys. Rev. B* **24**, 7155.
Chappell, J. S., Bloch, A. N., Bryden, W. A., Maxfield, M., Poehler, T. O., and Cowan, D. O. (1981). *J. Am. Chem. Soc.* **103**, 2442.
Chaudhari, P., Scott, B. A., Laibowitz, R. B., Tomkiewicz, Y., and Torrance, J. B. (1974). *Appl. Phys. Lett.* **24**, 439.
Cohen, M. J., Coleman, L. B., Garito, A. F., and Heeger, A. J. (1974). *Phys. Rev. B* **10**, 1298.
Cohen, M. H., Hertz, J. A., Horn, P. M., Madhukar, A., and Shante, V. K. (1975). *Bull. Am. Phys. Soc.* **20**, 415.
Cohen, M. J., and Heeger, A. J. (1977). *Phys. Rev. B* **16**, 688.
Coleman, L. B., Fincher, C. R., Garito, A. F., and Heeger, A. J. (1976). *Phys. Stat. Sol. (b)* **75**, 239.
Cooper, J. R., Ivezić, T., and Zorić, I. (1982). *J. Phys. C: Solid State Phys.* **30**, L397.

Cummings, K. D., Tanner, D. B., and Miller, J. S. (1981). *Phys. Rev. B* **24**, 4142.
Davydov, A. S. (1981). "Theory of Molecular Excitons". Plenum Press, New York.
Debray, J. D., Millet, R., Jérome, D., Barisic, A., Fabre, J. M., and Giral, L. (1977). *J. Physique Lettres* **38**, L-227.
Delhaes, P., Coulon, C., Amiell, J., Flandrois, S., Toreilles, E., Fabre, J. M., and Giral, L. (1979). *Mol. Cryst. Liq. Cryst.* **50**, 43.
Delhaes, P., Coulon, C., Flandrois, S., Hilti, B., Mayer, C. W., Rihs, G., and Rivory, J. (1980). *J. Chem. Phys.* **73**, 1452.
Denner, W., Schönfeld, B., and von Baltz, R. (1974). *Physics Lett.* **48A**, 313.
Duke, C. B. (1978). *Ann. NY Acad. Sci.* **313**, 166.
Duke, C. B., Lipari, N. O., and Pietronero, L. (1975). *Chem. Phys. Lett.* **30**, 415.
Dumke, W. P. (1961). *Phys. Rev.* **124**, 1813.
Ehrenreich, H., and Cohen, M. H. (1959). *Phys. Rev.* **115**, 786.
Eldridge, J. E. (1976). *Solid State Commun.* **19**, 607.
Eldridge, J. E. (1977). *Solid State Commun.* **21**, 737.
Eldridge, J. E. (1978). *Solid State Commun.* **26**, 243.
Eldridge, J. E. (1985). *Phys. Rev. B* **31**, 5465.
Eldridge, J. E., Homes, C. C., Bates, F. E., and Bates, G. S. (1985). *Phys. Rev. B* **32**, 5156.
Eldridge, J. E., and Bates, F. E. (1979). *Solid State Commun.* **30**, 195.
Eldridge, J. E., and Bates, F. E. (1982). *Phys. Rev. B* **26**, 1590.
Eldridge, J. E., and Bates, F. E. (1983). *Phys. Rev. B* **28**, 6972.
Eldridge, J. E., and Bates, G. S. (1985). *Mol. Cryst. Liq. Cryst.* **119**, 183.
Eldridge, J. E., and Bates, G. S. (1986). *Physica* **143B**, 428.
Epstein, A. J., and Miller, J. S. (1978). *Solid State Commun.* **27**, 325.
Etemad, S. (1976). *Phys. Rev. B* **13**, 2254.
Etemad, S. (1981). *Phys. Rev. B* **24**, 4959.
Fano, U. (1961). *Phys. Rev.* **124**, 1866.
Fenton, E. W., and Aers, G. C. (1985). *Mol. Cryst. Liq. Cryst.* **119**, 201.
Fenton, E. W., and Psaltakis, G. C. (1983). *Solid State Commun.* **47**, 767.
Ferguson, E. E., and Matsen, F. A. (1958). *J. Chem. Phys.* **29**, 105.
Freedman, R. (1980). *Solid State Commun.* **33**, 565.
Friedel, J., and Jérome, D. (1982). *Contemporary Physics*, in press.
Fritchie, C. J. (1966). *Acta Cryst.* **20**, 892.
Fritchie, C. J., and Arthur, P. (1966). *Acta Cryst.* **21**, 139.
Fröhlich, H. (1954). *Proc. Royal Soc. A* **223**, 296.
Fukuyama, H. (1976). *J. Phys. Soc. Jpn.* **41**, 513.
Garrigou-Lagrange, C., Graja, A., Coulon, C., and Delhaes, P. (1984). *J. Phys. C: Solid State Phys.* **17**, 5437.
Gasser, W., and Höfling, R. (1979). *Phys. Stat. Sol. (b)* **92**, 91.
Girlando, A., and Pecile, C. (1973). *Spectrochim. Acta* **29A**, 1859.
Gogolin, A. A. (1979). *Zh. Eksp. Teor. Fiz.* **76**, 1759. (*Sov. Phys.—JETP* **49**, 895.)
Gogolin, A. A., and Melnikov, V. I. (1978). *Phys. Stat. Sol. (b)* **88**, 377.
Gor'kov, L. P., and Rashba, E. I. (1978). *Solid State Commun.* **27**, 1211.
Gorshunov, B. P., Kozlov, G. V., Volkov, A. A., Zelezný, V., Petzelt, J., and Jacobsen, C. S. (1986). *Solid State Commun.* **60**, 681.
Graja, A., Huong, P. V., and Cornut, J.-C. (1981). *Solid State Commun.* **39**, 929
Grant, P. M. (1982). *Phys. Rev. B* **26**, 6888.
Grant, P. M., Greene, R. L., Wrighton, G. C., and Castro, G. (1973). *Phys. Rev. Lett.* **31**, 1311.
Greene, R. L., Haen, P., Huang, S. Z., Engler, E. M., Choi, M. Y., and Chaikin, P. M. (1982). *Mol. Cryst. Liq. Cryst.* **79**, 183.

Gunning, W. J., Khanna, S. K., Garito, A. F., and Heeger, A. J. (1977). *Solid State Commun.* **21**, 765.
Gunning, W. J., and Heeger, A. J. (1979). *Solid State Commun.* **29**, 585.
Gutfreund, H., Horovitz, B., and Weger, M. (1974a). *J. Phys. C: Solid State Phys.* **7**, 383.
Gutfreund, H., Horovitz, B., and Weger, M. (1974b). *Solid State Commun.* **15**, 849.
Harris, A. B., and Lange, R. V. (1967). *Phys. Rev.* **157**, 295.
Heeger, A. J. (1979). *Comments Solid State Phys.* **9**, 65.
Helberg, H. W. (1976). *Phys. Stat. Sol. (a)* **33**, 453.
Helberg, H. W. (1985). *Mol. Cryst. Liq. Cryst.* **119**, 179.
Herman, F. (1977). *Physica Scripta* **16**, 303.
Herman, F., Salahub, D. R., and Messmer, R. P. (1977). *Phys. Rev. B* **16**, 2453.
Hinkelmann, H. (1976). *Solid State Commun.* **18**, 957; *Z. Physik B* **25**, 147.
Hinkelmann, H., and Reik, H. G. (1975). *Solid State Commun.* **16**, 567.
Hiroma, S., Kuroda, H., and Akamatu, H. (1970). *Bull. Chem. Soc. Jpn.* **43**, 3626.
Hoekstra, A., Spoelder, T., and Vos, A. (1972). *Acta Cryst. B* **28**, 14.
Holstein, T. (1954). *Phys. Rev.* **96**, 535.
Holstein, T. (1964). *Ann. Phys. (N.Y.)* **29**, 410.
Hopfield, J. J. (1965). *Phys. Rev.* **139**, A419.
Hopfield, J. J. (1970). *Comments Solid State Physics* **3**, 48.
Horovitz, B., Gutfreund, H., and Weger, M. (1978). *Phys. Rev. B* **17**, 2796.
Hubbard, J. (1963). *Proc. Roy. Soc. (London)* **A276**, 238.
Hubbard, J. (1978). *Phys. Rev. B* **17**, 494.
Ikemoto, I., Sugano, T., and Kuroda, H. (1977). *Chem. Phys. Lett.* **49**, 45.
Iwahana, K., Kuzmany, H., Wudl, F., and Aharon-Shalom, E. (1982). *Mol. Cryst. Liq. Cryst.* **79**, 39.
Jacobsen, C. S. (1979). *Lecture Notes in Physics* **95**, 223.
Jacobsen, C. S. (1985). *Mat. Fys. Medd. Dan. Vidensk. Selsk.* **41**, 251.
Jacobsen, C. S. (1986). *J. Phys. C: Solid State Phys.* **19**, 5643.
Jacobsen, C. S., Tanner, D. B., Garito, A. F., and Heeger, A. J. (1974). *Phys. Rev. Lett.* **33**, 1559.
Jacobsen, C. S., Bechgaard, K., and Andersen, J. R. (1977). *Lecture Notes in Physics* **65**, 349.
Jacobsen, C. S., Mortensen, K., Andersen, J. R., and Bechgaard, K. (1978). *Phys. Rev. B* **18**, 905.
Jacobsen, C. S., Pedersen, H. J., Mortensen, K., and Bechgaard, K. (1980). *J. Physics C: Solid State Phys* **13**, 3411.
Jacobsen, C. S., Tanner, D. B., and Bechgaard, K. (1981). *Phys. Rev. Lett.* **46**, 1142.
Jacobsen, C. S., Pedersen, H. J., Mortensen, K., Rindorf, G., Thorup, N., Torrance, J. B., and Bechgaard, K. (1982). *J. Phys. C: Solid State Phys.* **15**, 2651.
Jacobsen, C. S., Tanner, D. B., and Bechgaard, K. (1983). *Phys. Rev. B* **28**, 7019.
Jacobsen, C. S., Johannsen, Ib., and Bechgaard, K. (1984). *Phys. Rev. Lett.* **53**, 194.
Jacobsen, C. S., Williams, J. M., and Wang, H. H. (1985). *Solid State Commun.* **54**, 937. (Note that the polarizations, ∥ and ⊥, in that paper should be interchanged.)
Jacobsen, C. S., Tanner, D. B., Williams, J. M., and Wang, H. H. (1987). *Synthetic Metals* **19**, 125.
Jacobsen, C. S., and Bechgaard, K. (1985). *Mol. Cryst. Liq. Cryst.* **120**, 71.
Javadi, H. H., Sridar, S., Grüner, G., Chiang, L., and Wudl, F. (1985). *Phys. Rev. Lett.* **55**, 1216.
Jérome, D. (1980). *In* "The Physics and Chemistry of Low Dimensional Solids''' (L. Alcácer, ed.), p. 123. Reidel, Dordrecht.
Jérome, D. (1981). *Chemica Scripta* **17**, 13.

Jordan, M. E. (1981). *Diss. Abstr. Int. B* **41**, 3459.
Kahn, L. M., Ruvalds, J., and Hastings, R. (1978). *Phys. Rev. B* **17**, 4600.
Kamarás, K., Ritvay-Emandity, K., Mihály, G., and Grüner, G. (1977). *Solid State Commun.* **24**, 93.
Kamarás, K., Grüner, G., and Sawatzky, G. A. (1978). *Solid State Commun.* **27**, 1171.
Kamarás, K., Holczer, K., and Jánossy, A. (1980). *Phys. Stat. Sol. (b)* **102**, 467.
Kamarás, K., and Grüner, G. (1979). *Solid State Commun.* **30**, 277.
Kapachina, L. M., Kaplunov, M. G., Kotov, A. I., Yagubskii, E. B., and Borodko, Y. G. (1978). *Opt. Spektrosk* **45**, 82. [*Opt. Spectrosc. (USSR)* **45**, 44.]
Kaplunov, M. G., Panova, T. P., and Borodko, Y. G. (1972). *Phys. Stat. Sol. (a)* **13**, K67.
Kaplunov, M. G., Pokhoduya, K. I., Kotov, A. I., Yagubskii, E. B., Kitaeva, T. A., and Borodko, Yu. G. (1977). *Phys. Stat. Sol. (a)* **43**, K73.
Kaplunov, M. G., Yagubskii, E. B., Rosenberg, L. P., and Borodko, Yu. G. (1985). *Phys. Stat. Sol. (a)* **89**, 509.
Kaplunov, M. G., and Lyubovskaya (1980). *Fiz. Tverd. Tela (Leningrad)* **22**, 3362. (*Sov. Phys. Solid State* **22**, 1968.)
Keller, H. J., Nöthe, D., Pritzkow, H., Dehe, D., Werner, M., Harms, R. H., Koch, P., and Schweitzer, D. (1981). *Chemica Scripta* **17**, 101.
Khanna, S. K., Pouget, J. P., Comès, R., Garito, A. F., and Heeger, A. J. (1977). *Phys. Rev. B* **16**, 1468.
Kikuchi, K., Ikemoto, Y., Yakushi, K., Kuroda, H., and Kobayashi, K. (1982). *Solid State Commun.* **42**, 433.
Koch, B., Geserich, H. P., Ruppel, W., Schweitzer, D., Dietz, K. H., and Keller, H. J. (1985). *Mol. Cryst. Liq. Cryst.* **119**, 343.
Král, K. (1976a). *Czech. J. Phys. B* **26**, 226.
Král, K. (1976b). *Czech. J. Phys. B* **26**, 660.
Krauzman, M., Poulet, H., and Pick, R. M. (1986). *Phys. Rev. B* **33**, 99.
Kurihara, S. (1976). *J. Phys. Soc. Jpn.* **41**, 1488.
Kuroda, H., Yakushi, K., and Cao, Y. (1982). *Mol. Cryst. Liq. Cryst.* **85**, 325.
Kuzmany, H. (1978). *Phys. Stat. Sol. (b)* **89**, K139.
Kuzmany, H., and Elbert, M. (1980). *Solid State Commun.* **35**, 597.
Kuzmany, H., and Stolz, H. J. (1977). *J. Phys. C: Solid State Phys.* **10**, 2241.
Kuzmany, H., Kundu, B., and Stolz, H. J. (1978). In "Proc. Int. Conf. Lattice Dynamics" (M. Balkanski, ed.), p. 584. Flammarion, Paris.
Kwak, J. F. (1982). *Phys. Rev. B* **26**, 4789.
Landau, L. D., and Lifshitz, E. M. (1960). "Electrodynamics of Continuous Media," p. 324. Pergamon Press, Oxford.
Lee, P. A., Rice, T. M., and Anderson, P. W. (1974). *Solid State Commun.* **14**, 703.
Lerstrup, K., Lee, M., Wiygul, F. M., Kistenmacher, T. J., and Cowan, D. O. (1983). *J. Chem. Soc. Chem. Commun.* 294.
Lieb, E. H., and Wu, F. Y. (1968). *Phys. Rev. Lett.* **20**, 1445.
Liebmann, R., Lemke, P., and Appel, J. (1977). *Phys. Rev. B* **16**, 4230.
Lipari, N. O., Duke, C. B., Bozio, R., Girlando, A., Pecile, C., and Pavda, A. (1976). *Chem. Phys. Lett.* **44**, 236.
Lipari, N. O., Rice, M. J., Duke, C. B., Bozio, R., Girlando, A., and Pecile, C. (1977). *Int. J. Quantum Chem. Symp.* **11**, 583.
Lyo, S. K. (1978). *Phys. Rev. B* **18**, 1854.
Lyo, S. K., and Gallinar, J.-P. (1977). *J. Phys. C: Solid State Phys.* **10**, 1696.
Maldague, P. F. (1977). *Phys. Rev. B* **16**, 2437.
Malek, J., Drchal, V., Hejda, B., and Záliš, S. (1984). *Chemical Physics* **89**, 361.

Marianer, S., Kaveh, M., and Weger, M. (1982). *Phys. Rev. B* **25**, 5197.
Martinsen, J., Palmer, S. M., Tanaka, J., Greene, R. C., and Hoffman, B. M. (1984). *Phys. Rev. B* **30**, 6269.
Matsuzaki, S., Kuwata, R., and Toyoda, K. (1980a). *Solid State Commun.* **33**, 403.
Matsuzaki, S., Moriyama, T., and Toyoda, K. (1980b). *Solid State Commun.* **34**, 857.
Matsuzaki, S., Onomichi, M., Tomura, H., Yoshida, S., and Toyoda, K. (1985). *Mol. Cryst. Liq. Cryst.* **120**, , 93.
Mattis, D. C., and Bardeen, J. (1958). *Phys. Rev.* **111**, 412.
Mazumdar, S., and Dixit, S. N. (1986). *Phys. Rev. B* **34**, 3683.
Mazumdar, S., and Soos, Z. (1981). *Phys. Rev. B* **23**, 2810.
McCall, R. P., Tanner, D. B., Miller, J. S., Epstein, A. J., Howard, I. A., and Conwell, E. M. (1985). *Synthetic Metals* **11**, 231.
Megtert, S., Pouget, J. P., Comès, R., Garito, A. F., Bechgaard, K., Fabre, J. M., and Giral, L. (1978). *J. de Physique—Lettres* **39**, L-118.
Megtert, S., Comès, R., Vettier, C., Pynn, R., and Garito, A. F. (1979). *Solid State Commun.* **31**, 977.
Meneghetti, M., Bozio, R., Zanon, I., Pecile, C., Ricotta, C., and Zanetti, M. (1984). *J. Chem. Phys.* **80**, 6210.
Meneghetti, M., Bozio, R., and Pecile, C. (1986). *J. Physique (Paris)* **47**, 1377.
Miller, J. S., and Epstein, A. J. (1978). *J. Am. Chem. Soc.* **100**, 1639.
Morawitz, H. (1981). *Chemica Scripta* **17**, 75.
Mortensen, K. (1982). *Solid State Commun.* **44**, 643.
Mortensen, K., Tomkiewicz, Y., Schultz, T. D., and Engler, E. M. (1981). *Phys. Rev. Lett.* **46**, 1234.
Mortensen, K., Jacobsen, C. S., Lindegaard-Andersen, A., and Bechgaard, K. (1983). *J. Physique (Paris) Colloq.* **44**-C3, 1349.
Ng, H. K., Timusk, T., and Bechgaard, K. (1983). *J. Physique (Paris) Colloq.* **44**-C3, 867.
Ng, H. K., Timusk, T., and Bechgaard, K. (1984). *Phys. Rev. B* **30**, 5842.
Ng, H. K., Timusk, T., and Bechgaard, K. (1985a). *Mol. Cryst. Liq. Cryst.* **119**, 191.
Ng, N. K., Timusk, T., and Bechgaard, K. (1985b). *Phys. Rev. B* **32**, 8041.
Nobile, A., and Tosatti, E. (1980). *J. Phys. C: Solid St. Phys.* **13**, 589.
Nücker, N., Fink, J., Schweitzer, D., and Keller, H. J. (1986). *Physica* **143B**, 482.
Ovchinnikov, A. A. (1969). *Zh. Eksp. Teor. Fiz.* **57**, 2137. [*Sov. Phys.—JETP* **30**, 1160 (1970).]
Ovchinnikov, A. A., and Ukrainskii, I. I. (1974). *Fiz. Tverd. Tela* **16**, 3239. [*Sov. Phys. Solid State* **16**, 2107 (1975).]
Painelli, A., Girlando, A., and Pecile, C. (1984). *Solid State Commun.* **52**, 801.
Painelli, A., and Girlando, A. (1986). *J. Chem. Phys.* **84**, 5655.
Perov, P. I., and Fischer, J. E. (1974). *Phys. Rev. Lett.* **33**, 521.
Phillips, T. E., Kistenmacher, T. J., Bloch, A. N., and Cowan, D. O. (1976). *J. Chem. Soc. Chem. Commun.* **1976**, 334.
Poehler, T. O., Bloch, A. N., Ferraris, J. P., and Cowan, D. O. (1974). *Solid State Commun.* **15**, 337.
Pouget, J. P. (1981). *Chemica Scripta* **17**, 85.
Pouget, J. P., Megtert, S., Comès, R., and Epstein, A. J. (1980). *Phys. Rev. B* **21**, 486.
Pouget, J. P., Comès, R., Epstein, A. J., and Miller, J. S. (1982). *Mol. Cryst. Liq. Cryst.* **85**, 203.
Rice, M. J. (1976). *Phys. Rev. Lett.* **37**, 36.
Rice, M. J. (1978). *Solid State Commun.* **25**, 1083.
Rice, M. J. (1979). *Solid State Commun.* **31**, 93.
Rice, M. J., Duke, C. B., and Lipari, N. O. (1975a). *Solid State Commun.* **17**, 1089.

Rice, M. J., Strässler, S., and Schneider, W. R. (1975b). *Lecture Notes in Physics* **34**, 282.
Rice, M. J., Pietronero, L., and Brüesch, P. (1977a). *Solid State Commun.* **21**, 757.
Rice, M. J., Lipari, N. O., and Strässler, S. (1977b). *Phys. Rev. Lett.* **39**, 1359.
Rice, M. J., Yartsev, V. M., and Jacobsen, C. S. (1980). *Phys. Rev. B* **21**, 3437.
Ritsko, J. J., Sandman, D. J., Epstein, A. J., Gibbons, P. C., Schnatterly, S. E., and Fields, J. (1975). *Phys. Rev. Lett.* **34**, 1330.
Ritsko, J. J., Brillson, L. J., and Sandman, D. J. (1977). *Solid State Commun.* **24**, 109.
Rosencwaig, A. (1975). *Physics Today*, September, p. 23.
Sadovskii, M. V. (1974). *Fiz. Tverd. Tela* **16**, 2504. (*Sov. Phys. Solid State* **16**, 1632.)
Salahub, D. R., Messmer, R. P., and Herman, F. (1976). *Phys. Rev. B* **13**, 4252.
Schultz, A. J., Stucky, G. D., Blessing, R. H., and Coppens, P. (1976). *J. Am. Chem. Soc.* **98**, 3194.
Schultz, T. D., and Craven, R. A. (1979). In "Highly Conducting One-Dimensional Solids" (J. T. Devreese, R. P. Evrard, and V. E. van Doren, eds.), p. 147. Plenum Press, New York.
Schuster, H. G. (1975). *Phys. Rev. B* **11**, 613.
Scott, J. C. (1982). *Mol. Cryst. Liq. Cryst.* **79**, 49.
Scott, B. A., La Placa, S. J., Torrance, J. B., Silverman, B. D., and Welber, B. (1977). *J. Am. Chem. Soc.* **99**, 6631.
Seiden, P. E., and Cabib, D. (1976). *Phys. Rev. B* **13**, 1846.
Shibaeva, R. P., Kaminskii, V. F., and Yagubskii, E. B. (1985). *Mol. Cryst. Liq. Cryst.* **119**, 373.
Siemons, W. J., Bierstedt, P. E., and Kepler, R. G. (1963). *J. Chem. Phys.* **39**, 3523.
Simonsen, M. G., and Coleman, R. V. (1973). *Phys. Rev. B* **8**, 5875.
Somoano, R. B., Gupta, A., Hadek, V., Novotny, M., Jones, M., Datta, T., Deck, R., and Herman, A. M. (1977). *Phys. Rev. B* **15**, , 595.
Somoano, R. B., Yen, S. P., Hadek, V., Khanna, S. K., Novotny, M., Datta, T., Hermann, A. M., and Woolam, J. A. (1978). *Phys. Rev. B* **17**, 2853.
Strzelecka, H., Weyl, C., and Rivory, J. (1979). *Lecture Notes in Physics* **96**, 348.
Strzelecka, H., and Rivory, J. (1981). *Chemica Scripta* **17**, 95.
Sugano, T., Yakushi, K., and Kuroda, H. (1978). *Bull. Chem. Soc. Jpn.* **51**, 1041.
Sugano, T., Yamada, K., Saito, G., and Kinoshita, M. (1985). *Solid State Commun.* **55**, 137.
Sugano, T., and Kuroda, H. (1977). *Chem. Phys. Lett.* **47**, 92.
Tajima, H., Yakushi, K., Kuroda, H., Saito, G., and Inokuchi, H. (1984). *Solid State Commun.* **49**, 769.
Tajima, H., Yakushi, K., Kuroda, H., and Saito, G. (1985). *Solid State Commun.* **56**, 159.
Tajima, H., Kanbara, H., Yakuski, K., Kuroda, H., and Saito, G. (1986). *Solid State Commun.* **57**, 911.
Tanaka, J., Tanaka, M., Kawai, T., Takabe, T., and Maki, O. (1976). *Bull. Chem. Soc. Jpn.* **49**, 2358.
Tanaka, J., Tanaka, M., Tanaka, C., Ohno, T., Takabe, T., and Anzai, H. (1978). *Ann. NY Acad. Sci.* **313**, 256.
Tanner, D. B. (1982). In "Extended Linear Chain Compounds" (J. S. Miller, ed.), Vol. 2, p. 205. Plenum, New York.
Tanner, D. B., Jacobsen, C. S., Garito, A. F., and Heeger, A. J. (1974). *Phys. Rev. Lett.* **32**, 1301.
Tanner, D. B., Jacobsen, C. S., Garito, A. F., and Heeger, A. J. (1976). *Phys. Rev. B* **13**, 3381.
Tanner, D. B., Jacobsen, C. S., Bright, A. A., and Heeger, A. J. (1977). *Phys. Rev. B* **16**, 3283.
Tanner, D. B., Deis, J. E., Epstein, A. J., and Miller, J. S. (1979). *Solid State Commun.* **31**, 671.
Tanner, D. B., Miller, J. S., Rice, M. J., and Ritsko, J. J. (1980). *Phys. Rev. B* **21**, 5835.
Tanner, D. B., Cummings, K. D., and Jacobsen, C. S. (1981). *Phys. Rev. Lett.* **47**, 597.

Tanner, D. B., Jacobsen, C. S., Williams, J. M., and Wang, H. H. (1987). *Synthetic Metals* **19**, 197.
Tanner, D. B., and Jacobsen, C. S. (1982). *Mol. Cryst. Liq. Cryst.* **85**, 137.
Temkin, H., and Fitchen, D. B. (1978). *In* "Proc. Int. Conf. Lattice Dynamics" (M. Balkanski, ed.), p. 587. Flammarion, Paris.
Thomas, G. A., Moncton, D. E., Wudl, F., Kaplan, M., and Lee, P. A. (1978). *Phys. Rev. Lett.* **41**, 486.
Timusk, T. (1987). *In* "Low-Dimensional Conductors and Superconductors" (L. Caron and D. Jerome, eds.), p. 275. Plenum, New York.
Tokumoto, M., Koshizuka, N., Anzai, H., and Ishiguro, T. (1982). *J. Phys. Soc. Jpn.* **51**, 332.
Tomkiewicz, Y., Welber, B., Seiden, P. E., and Schumaker, R. (1977). *Solid State Commun.* **23**, 471.
Torrance, J. B., Scott, B. A., and Kaufman, F. B. (1975a). *Solid State Commun.* **17**, 1369.
Torrance, J. B., Simonyi, E. E., and Bloch, A. N. (1975b). *Bull. Am. Phys. Soc.* **20**, 497.
Torrance, J. B., Scott, B. A., Welber, B., Kaufman, F. B., and Seiden, P. E. (1979). *Phys. Rev. B* **19**, 730.
Torrance, J. B., Mayerlee, J. J., Lee, V. Y., Bozio, R., and Pecile, C. (1981). *Solid State Commun.* **38**, 1165.
Torrance, J. B., and Nicoli, D. F. (1974). *Bull. Am. Phys. Soc.* **19**, 336.
Vlasova, R. M., Gutman, A. I., Kartenko, N. F., Rozenshtein, L. D., Agroskin, L. S., Papayan, G. V., Rautian, L. P., and Sherle, A. I. (1975). *Fiz. Tverd. Tela* **17**, 3529. [*Sov. Phys. Solid State* **17**, 2301 (1976).]
Vlasova, R. M., Ivanova, E. A., and Semkin, V. N. (1985). *Fiz. Tverd. Tela (Leningrad)* **27**, 530. (*Sov. Phys. Solid State* **27**, 326.)
Warmack, R. J., Callcott, T. A., and Watson, C. R. (1975). *Phys. Rev B* **12**, 3336.
Warmack, R. J., and Callcott, T. A. (1976). *Phys. Rev. B* **14**, 3238.
Weinstein, B. A., Slade, M. L., Epstein, A. J., and Miller, J. S. (1981). *Solid State Commun.* **37**, 643.
Welber, B., Seiden, P. E., and Grant, P. M. (1978). *Phys. Rev. B* **18**, 2692.
Weyl, C., Engler, E. M., Bechgaard, K., Jehanno, G., and Etemad, S. (1976). *Solid State Commun.* **19**, 925.
Wilckens, R., Geserich, H. P., Ruppel, W., Koch, P., Schweitzer, D., and Keller, H. J. (1982). *Solid State Commun.* **41**, 615.
Williams, P. F., Butler, M. A., Rousseau, D. L., and Bloch, A. N. (1974). *Phys. Rev. B* **10**, 1109.
Williams, P. F., and Bloch, A. N. (1974). *Phys. Rev. B* **10**, 1097.
Williams, P. F., and Bloch, A. N. (1976). *Phys. Rev. Lett.* **36**, 64.
Wonneberger, W. (1977). *J. Phys. C: Solid State Phys.* **10**, 1073.
Wonneberger, W., and Lautenschläger (1976). *J. Phys. C: Solid State Phys.* **9**, 2865.
Wooten, F. (1972). "Optical Properties of Solid." Academic Press, New York.
Wozniak, W. T., Depasquali, G., Klein, M. V., Sweany, R. L., and Brown, T. L. (1975). *Chem. Phys. Lett.* **33**, 33.
Yakushi, K., Kusaka, T., and Kuroda, H. (1979). *Chem. Phys. Lett.* **68**, 139.
Yakushi, K., Iguchi, M., Katagiri, G., Kusaka, T., Ohta, T., and Kuroda, H. (1981). *Bull. Chem. Soc. Jpn.* **54**, 348.
Yartsev, V. M. (1982). *Phys. Stat. Sol. (b)* **112**, 279.
Yartsev, V. M. (1984). *Phys. Stat. Sol. (b)* **126**, 501.
Yartsev, V. M., and Jacobsen, C. S. (1981). *Phys. Rev. B* **24**, 6167.
Young, R. H., and Walker, E. I. P. (1977). *Phys. Rev. B* **15**, 631.
Zuppiroli, L., Jacobsen, C. S., and Bechgaard, K. (1985). *J. Physique (Paris)* **46**, 799.

CHAPTER 6

Magnetic Properties

J. C. Scott

IBM RESEARCH
ALMADEN RESEARCH CENTER
SAN JOSE, CALIFORNIA

I.	MAGNETIC MEASUREMENTS	386
	1. *Introduction*	386
	2. *Magnetic Susceptibility*	386
	3. *ESR*	389
	4. *NMR*	392
II.	THE METALLIC STATE	398
	5. *Spin Susceptibility*	398
	6. *Decomposition of Susceptibility by g Factor*	402
	7. *Decomposition of Susceptibility by Knight Shift*	403
	8. *ESR Linewidth*	404
	9. *Nuclear Magnetic Relaxation*	408
	10. *The Semimetallic State of HMTSF-TCNQ*	411
III.	THE PEIERLS TRANSITION AND THE CHARGE-DENSITY WAVE STATE	412
	11. *Peierls Transition*	412
	12. *Critical Behavior*	414
	13. *Energy Gaps in the Charge-Density Wave State*	415
	14. *Impurities and Defects*	416
	15. *Anion Ordering*	417
IV.	THE SPIN-DENSITY WAVE STATE	418
	16. *Spin-Density Wave Transition*	418
	17. *Susceptibility*	421
	18. *Nuclear Magnetic Resonance*	421
	19. *Antiferromagnetic Resonance*	422
V.	SUMMARY BY MATERIAL	425
	20. *TTF-TCNQ*	425
	21. *TSF-TCNQ*	426
	22. *Alloys of TTF-TCNQ and TSF-TCNQ*	427
	23. *Other Analogs of TTF-TCNQ*	427
	24. *The (TMTSF)$_2$X Family*	428
	25. *The (TMTTF)$_2$X Family*	429
	26. *Other Materials*	430
	ACKNOWLEDGMENTS AND APOLOGIA	431
	REFERENCES	431

I. Magnetic Measurements

1. Introduction

The magnetism of highly conducting organic charge-transfer salts has attracted considerable experimental and theoretical effort, and has contributed much to the overall understanding of these rather complicated solids. This chapter contains a summary of magnetic susceptibility, electron spin resonance, and nuclear magnetic resonance data, and discusses the various theoretical interpretations that have been given to the experimental results. It will soon become apparent to the reader that a simplistic view of magnetism in metals in terms of, for example, Pauli susceptibility, spin–phonon scattering, and Korringa relaxation is inadequate to account for any but the grossest features of the data. This is not surprising since, as demonstrated frequently in other chapters of this volume, many-body effects are paramount in the physics of one-dimensional conductors, and magnetic properties are particularly sensitive to the interactions of the Fermi gas.

In discussing the magnetic properties of organic conductors it is important to separate the molecular contributions from those phenomena that arise as a result of interactions in the solid state. For example, core diamagnetism, g tensors, and hyperfine coupling constants are largely molecular properties, whereas conduction electron paramagnetism, ESR linewidths, and NMR relaxation rates are solid-state properties. In many cases it is important to have a knowledge of the molecular parameters before attempting to interpret measurements made on the charge-transfer salt of interest. It is therefore a secondary goal of this review to collect all the available data on molecular magnetic parameters and to point out where such information is missing.

Since many of the points of this introductory section are standard results of magnetism in metals, reference is made to suitable secondary sources. Where the work is specific to organic solids, primary references are given.

2. Magnetic Susceptibility

Together with dc conductivity, static susceptibility has traditionally been one of the first measurements to be carried out on newly synthesized organic salts. The motivation is that in a relatively straightforward experiment on a polycrystalline sample, one can easily distinguish among the nonmagnetic state of an insulator or wide-gap semiconductor, the Curie-like paramagnetism associated with localized spins, and the Pauli-like susceptibility of a degenerate Fermi system. In addition, one gains information concerning phase transitions that occur as a function of temperature.

The most frequently used technique has been the Faraday balance. This has the advantage of considerable precision and ease of use over a

6. MAGNETIC PROPERTIES

temperature range from above ambient down to liquid helium. This method measures, perhaps unfortunately, the total susceptibility, which is conventionally subdivided as (e.g., White, 1970)

$$\chi_T = \chi_D + \chi_S + \chi_{VV} + \chi_L, \tag{1}$$

where the terms on the right refer, respectively, to the core diamagnetism, the spin-paramagnetism (e.g., Pauli or Curie–Weiss), the Van Vleck susceptibility and the orbital Landau contribution. In general, since the uninteresting core contribution is difficult to measure or estimate accurately for the molecular *ion*, but is frequently as large or larger than the spin term, some of the precision of the Faraday method is lost. Other, less sensitive, methods have been used to determine χ_S directly, namely, calibrated microwave ESR intensity measurements and the Schumacher–Slichter (1956) technique (see below).

Core diamagnetism is basically a molecular property, reflecting the response of inner atomic orbitals and filled valence states to an applied magnetic field. Although it is, in principle, possible to calculate χ_D within molecular orbital theory, in practice, an empirical approach is universally used. In several cases the susceptibilities of the neutral constituents have been measured. Other values have been calculated from a tabulation of Pascal's constants (e.g., Konig, 1966), which include additive contributions for each atomic species, for certain bonding arrangements, for aromaticity, etc. Note that neither of these methods takes into account the effect of charge on the molecular anion or cation, which will have valence and core orbitals perturbed relative to the neutral species. Hence, any such estimates should be considered accurate to only about 10%. Other methods of determination include extrapolation to zero temperature in salts that have a nonmagnetic ground state and measurement of the difference between the total susceptibility and spin contribution measured by a resonance technique. Table I contains value of core diamagnetic contributions determined both experimentally and from Pascal's constants.

In metals, the degeneracy of the Fermi gas leads to a (molar) spin susceptibility given, to a first approximation, by the Pauli expression (Ashcroft and Mermin, 1976)

$$\chi_S^0 = \tfrac{1}{2} g^2 \mu_B^2 N_A \rho(\varepsilon_F), \tag{2}$$

where N_A is Avogadro's number, μ_B is the Bohr magneton and $\rho(\varepsilon_F)$ is the Fermi energy density of states, per molecule per spin. Measurement of the spin susceptibility therefore permits an estimate of the density of states which for a tight-binding band is given by

$$\rho(\varepsilon_F) = \left[2\pi t \sin\left(\frac{\pi \nu}{2}\right)\right]^{-1}, \tag{3}$$

TABLE I

Core Diamagnetism of Constituent Molecules of Organic Metals

Molecule	Pascal's Consts[a] $\times -10^{-6}$ cm^3 mole^{-1}	Neutral Molecule	TCNQ salt ($T = 0$)
TTF	91	99[b]	205[b,c]
DMTTF	115	—	197[b]
TMTTF	139	—	257[b]
HMTTF	151	—	—
TSF	123	127[d]	210[d]
TMTSF	171	149[e]	287[e]
HMTSF	183	177[f]	—
TCNQ	79	121[b]	—

[a] Konig (1966)
[b] Scott et al. (1974)
[c] Herman et al (1974)
[d] Scott et al. (1978)
[e] Bloch et al. (1977)
[f] Soda et al. (1976b)

where t is the transfer integral and v is the charge transfer. However, the interactions of the electron gas affect the susceptibility directly, leading to an enhancement, which is given by the Stoner factor

$$\chi_S = \frac{\chi_S^0}{1 - U_{\text{eff}}\rho(\varepsilon_F)} = \chi_S^0(1 - \alpha)^{-1} \tag{4}$$

Here U_{eff} is an effective Coulomb repulsion energy, reduced from the bare Hubbard (1963, 1964) U by correlation effects. (In the Hartree-Fock approximation the susceptibility is given by the same expression, but with $U_{\text{eff}} = U$.)

As we shall see below, the susceptibility of the one-dimensional electron gas does not conform to this simple expression. Indeed, the paramagnetism shows a temperature dependence that remains the subject of debate.

The Van Vleck susceptibility in metals arises because of admixture in the presence of the magnetic field of higher bands into the conduction band. This admixture involves matrix elements of the orbital angular momentum and, as shown by Boriack (1980), leads to zero Van Vleck susceptibility for a one-dimensional band structure, even when the energy denominator is the small gap arising from a CDW distortion, χ_{VV} has therefore been neglected in all discussions of the magnetism of organic metals.

The Landau term in Eq. (1) arises from orbital effects, which can usually be ignored in quasi-one-dimensional systems where interchain motion is accomplished by hopping and destroys the coherence of the electronic wavefunctions. There is, however, at least one case, namely HMTSF–TCNQ (see Part II, Section 10) in which orbital magnetism is found to be important at low temperature. The Landau diamagnetic susceptibility depends on the cyclotron effective mass (m^*)

$$\chi_L = -\frac{1}{3}\chi_S \frac{m_e}{m^*} \quad (5)$$

and is therefore, in general, anisotropic.

3. ESR

In contrast to three-dimensional metals, highly anisotropic organic conductors exhibit an ESR signal that is often easily observable up to room temperature. As discussed by Bloch (1977), the narrow line can be attributed to the reduced phase space available for spin-flip scattering from phonons and to the vanishing, via time reversal invariance, of the matrix element for $2k_F$ spin-flip scattering. ESR yields three experimental parameters that may be amenable to theoretical interpretation: intensity, g tensor, and linewidth.

The Kramers–Kronig relation (Slichter, 1978, Chapter 2) gives the spin susceptibility in terms of the integrated intensity of the ESR absorption signal:

$$\chi_S = \frac{2}{\pi H_0} \int_0^\infty \chi''(H)\, dH, \quad (6)$$

where H_0 is the resonance field, which is assumed to be much greater than the linewidth. χ'' is the imaginary part of the complex susceptibility and is proportional to the microwave absorption. In practice, it is rather difficult to measure absolute absorption, and susceptibilities are generally measured relative to some standard sample which is, ideally, placed in the microwave cavity adjacent to the material under study. The Schumacher–Slichter technique involves using, for this "standard sample," nuclei (in organics, usually the protons) of the compound itself. The same rf spectrometer is then used to compare the NMR absorption (at several kiloGauss) and the ESR absorption (at several Gauss). Since nuclear susceptibilities have the well-known Curie temperature dependence, the electronic susceptibility is easily determined. A great advantage of this technique is that skin-depth problems are considerably reduced relative to measurements in the microwave range. The major limitation is that, because of the low resonance-field for electrons, the method can only be used for narrow ESR lines.

The g tensor is a property that depends, first, upon the constituent molecules of the salt. Although there are methods for computation of g tensors in molecular radicals (Dalgard and Linderberg, 1975, 1976), calculations are not yet available for species of the complexity of those in organic metals. It has been universal practice to rely upon experimental data from model compounds. Results are collected in Table II. The principal directions of the tensor with respect to the molecular axes are also known. For the TTF and TSF derivatives g_{MAX} is obtained with external field parallel to the long axis of the molecule, g_{INT} for the short axis, and g_{MIN} perpendicular to the molecular plane. In the case of TCNQ, g_{MAX} is associated with the short axis and g_{INT} with the long axis.

Several features of the g tensor data are worthy of comment. The negative g shift perpendicular to the molecular plane of the fulvalene donors permits a quick way of identifying the stacking axis of their salts in the absence of detailed structural information. Substitution of protons with alkyl groups has little effect on g values, whereas substitution of sulphur by selenium increases the g shift by up to an order of magnitude, due to the much larger spin-orbit coupling in the heavier atom. The range of values given in Table II implies that the g tensor is slightly dependent on the molecular environment of the species, presumably reflecting solid-state perturbations of the molecular orbitals and amounting to about a 10 to 20% effect in the g shift.

The ESR signal of organic metals consists, in general, of a single Lorentzian line. This is true even in the case of conduction on separate donor and acceptor stacks and even when there are several inequivalent molecules per unit cell, because electronic spin dynamics both parallel and transverse to the highly conducting axis are sufficiently fast to average the different molecular environments.

TABLE II

g VALUES OF MOLECULAR IONS

Molecule	g_{AV}	g_{MAX}	g_{INT}	g_{MIN}
TTF	2.00838[a]	2.013–2.015[b,c,d]	2.007–2.008[b,c,d]	2.0020–2.0021[b,c,d]
TMTTF	—	2.0103–2.0115[b,e,f]	2.0086–2.0097[b,e,f]	2.0019–2.0029[b,e,f]
HMTTF	2.0080[g,h]	2.017–2.019[g,h]	2.0064–2.0072[g,h]	2.001[g,h]
TSF	2.027[i]	2.055–2.064[b]	2.019–2.021[b]	1.9954–1.9956[b]
TMTSF	—	2.042–2.049[j,k,l,m]	2.027–2.040[j,k,l,m]	1.984–1.994[j,k,l,m]
TTT	2.0077[n]	—	—	—
TCNQ	2.0025[o]	2.0035–2.0038[b,p]	2.0025–2.0029[b,p]	2.0023–2.0024[b,p]

[a] Wudl et al (1970)
[b] Walsh et al. (1980a)
[c] Wudl et al. (1977)
[d] Bray et al. (1975)
[e] Delhaes et al. (1979a)
[f] Brun et al. (1977)
[g] Tomkiewicz et al. (1977a)
[h] Tomkiewicz et al. (1980)
[i] Tomkiewicz et al. (1975)
[j] Pedersen et al. (1980)
[k] Pedersen et al. (1981)
[l] Scott et al. (1982a)
[m] Flandrois et al. (1982)
[n] Bramwell et al. (1978)
[o] Kinoshita and Akamatu (1965)
[p] Clark et al. (1979)

6. MAGNETIC PROPERTIES

In compounds containing both donor (D) and acceptor (A) conducting stacks, such averaging leads to a g value intermediate between the values of the separate species:

$$g = \frac{g_A \chi_S^A + g_D \chi_S^D}{\chi_S^A + \chi_S^D} \tag{7}$$

If $g_{A,D}$, the respective g values for the appropriate field direction, are known then Eq. (7) permits the evaluation of the susceptibility contributions, χ_S^A and χ_S^D ($\chi_S^A + \chi_S^D = \chi_S$). Such a decomposition of the susceptibility has been performed for many two-chain compounds by Tomkiewicz and coworkers (see Sections 6 and 11).

The third parameter available from ESR spectra is the linewidth. In metals, motional averaging of g value and hyperfine inhomogeneity yields a homogeneous Lorentzian line of width $\Gamma = \hbar(g\mu_B T_{2e})^{-1}$, where T_{2e} is the transverse relaxation rate. In isotropic metals, $T_{2e} = T_{1e}$ (the spin-lattice or longitudinal relaxation time), but the equality does not hold in one dimension (Yafet, 1963). However, as discussed below, both relaxation times depend on the same scattering processes, and most of the discussion in the literature has focused on T_{1e} as the quantity of interest for theoretical computation.

The dominant contribution to the linewidth of organic metals is due to spin–phonon scattering, as described for conventional metals by Elliot (1954). Lattice vibrations modulate the spin–orbit interaction, and hence the relaxation rates are greatest for metals with large spin–orbit couplings. Since the g shift is governed by the same interaction, the following relationship is easily derived:

$$T_{1e}^{-1} = \beta \left(\frac{\delta g}{g}\right)^2 \tau^{-1}, \tag{8}$$

where τ^{-1} is an electron–phonon scattering rate and β is of order unity. In isotropic metals, where the same scattering processes relax both momentum and spin distributions, τ is proportional to the conductivity. In contrast, for 1D metals the conductivity is relaxed primarily by backward scattering across the Fermi surface, a process that cannot, by time reversal symmetry, simultaneously flip a spin. Spin relaxation occurs primarily by interchain scattering events, and hence τ is more closely related to the transverse hopping time.

In spite of the considerable insight that has been gained into relaxation processes in organic conductors, a detailed microscopic theory for anisotropic metals is still lacking.

4. NMR

Since the molecules of which organic conductors are composed typically contain carbon, hydrogen, sulphur, selenium, and/or nitrogen, all of which have isotopes with a nuclear magnetic moment, NMR has proved to be a valuable technique in the study of organic metals, providing a rich variety of information about electron distribution and dynamics. As in the case of conventional metals, two of the primary quantities of experimental interest are the Knight shift (K) and the spin-lattice relaxation rate (T_1^{-1}), which arise as a result of hyperfine interactions with the conduction electrons. The simple theory of Knight shift and relaxation via the isotropic contact term in the hyperfine interaction fails for several reasons: The dipolar electron–nuclear interaction is found to be non-negligible, leading to an anisotropic Knight shift and an additional contribution to the relaxation rate; Coulomb interactions among the electrons enhance the relaxation rate in a manner that is quantitatively different from the 3D case due to the nature of the one-dimensional Fermi surface and the divergence of the response function at wavevector $2k_F$; anisotropic diffusion alters the spectral distribution of spin-density fluctuations and yields a frequency-dependent relaxation rate. In this section these various features of the NMR behavior will be discussed in turn.

Recall, first, the simple theory of the hyperfine interaction in metals (Slichter, 1978, Chapter 4), based on the scalar contact interaction between one nucleus, n, spin $\mathbf{I}$, at $r = 0$ and electrons i, spin $\mathbf{S}_i$

$$H_n = \frac{8\pi}{3} \hbar \gamma_n g \mu_B \sum_i [I_z S_{iz} + \tfrac{1}{2}(I_+ S_{i-} + I_- S_{i+})]\delta(r). \tag{9}$$

(Here and elsewhere the notation $g\mu_B$ is used for the electron moment, while $\hbar\gamma$ describes nuclear moments.) In a magnetic field the electrons gain a net polarization, $\langle S_{iz}\rangle = \chi_S H_0/N_A g\mu_B$. There is, therefore, an effective additional field at the nucleus such that the applied field for resonance is shifted to lower fields that that in diamagnetic materials by an amount

$$K = \frac{\Delta H}{H_0} = \frac{H_{\text{hyp}}\chi_S}{N_A g\mu_B}, \tag{10}$$

where $H_{\text{hyp}} = (8\pi/3)g\mu_B \langle |u(0)|^2\rangle_F$ is the hyperfine field at the nucleus, found by averaging the wavefunctions over the Fermi surface.

Spin-lattice relaxation through the conduction electrons occurs because of fluctuations in the hyperfine field and therefore probes the electron dynamics. The $(I_+ S_- + I_- S_+)$ term in the Hamiltonian induces transitions of the nucleus, which, by the Golden Rule, leads to a spin lattice relaxation rate

$$T_1^{-1} = \pi\hbar H_{\text{hyp}}^2 \gamma_n^2 \rho^2(\varepsilon_F)kT. \tag{11}$$

For noninteracting electrons the density of states yields directly the susceptibility (see Section 2) and hence the Knight shift. Therefore Eq. (11) can be modified to yield the Korringa relation

$$(K^2 T_1 T)^{-1} = \frac{4\pi k}{\hbar} \left(\frac{\hbar \gamma_n}{g\mu_B}\right)^2. \tag{12}$$

For interacting electrons the susceptibility is enhanced, as discussed above, and therefore the Knight shift and relaxation rate are also enhanced. However, it was shown by Moriya (1963) that, although the hyperfine field is enhanced, the fluctuation spectrum is altered such that the relaxation rate does not change proportionately. He found that the Korringa relation has to be modified according to:

$$(K^2 T_1 T)^{-1} = \frac{4\pi k}{\hbar} \left(\frac{\hbar \gamma_n}{g\mu_B}\right)^2 \eta(\alpha), \tag{13}$$

where

$$\eta(\alpha) = \left\langle \frac{(1-\alpha)^2}{[1-\alpha F(x)]^2} \right\rangle_F \tag{14}$$

$F(q/2k_F)$ is the Lindhard function, and the average is over the Fermi surface.

Now let us turn to the differences associated with the molecular structure and one dimensionality of organic conductors alluded to in the introduction of this section.

a. Dipolar Electron–Nuclear Coupling

In organic metals the nature of the molecular π orbitals that form the conduction bands leads to a dipolar hyperfine field, which may be comparable to the contact contribution described above (Devreux et al., 1979; Avalos et al., 1978). The various terms in the dipolar Hamiltonian modify the Knight shift and relaxation rate in different ways. The Hamiltonian for the interaction of one nucleus with a single electron may be written (Slichter, 1978, p. 58)

$$H_{\text{dip}} = \frac{g\mu_B \hbar \gamma_n}{r^3} (A + B + C + D + E + F), \tag{15}$$

where

$$A = I_z S_z (1 - 3\cos^2\theta) \tag{15a}$$

$$B = -(\tfrac{1}{4})(I_+ S_- + I_- S_+)(1 - 3\cos^2\theta) \tag{15b}$$

$$C = -(\tfrac{3}{2})(I_+ S_z + I_z S_+) \sin\theta \cos\theta\, e^{-i\phi} \tag{15c}$$

$$D = -(\tfrac{3}{2})(I_- S_z + I_z S_-) \sin\theta \cos\theta\, e^{i\phi} \tag{15d}$$

$$E = -(\tfrac{3}{4})I_+ S_+ \sin^2\theta\, e^{-2i\phi} \tag{15e}$$

$$F = -(\tfrac{3}{4})I_- S_- \sin^2\theta\, e^{2i\phi}. \tag{15f}$$

The term A produces an anisotropic Knight shift or, for a powder sample, an additional broadening of the resonance linewidth, but no extra average shift. The term B gives a contribution to the spin-lattice relaxation rate, but since it contains a different numerical coefficient than the scalar expression, Eq. (9), there is no longer the simple proportionality of T_1^{-1} and K^2, even for a single crystal sample. A further complication arises because there may be several inequivalent nuclei of the same type, e.g., the ring carbons, the methanide carbons, and the nitrile carbons of TCNQ. Depending on the strength of the mutual interactions among ineqivalent nuclei, the nuclear relaxation rates and the Knight shifts may or may not be averaged out. It is advantageous for the analysis of experimental data to perform measurements on samples that have been selectively isotopically enriched.

The terms C and D flip the nuclear spin system without affecting the electronic moment and therefore contribute to the relaxation rate. As we will see in more detail below, this contribution is proportional to the spectral density of electron fluctuations at the nuclear Larmor frequency.

b. One-Dimensional Diffusion

The derivation of the Korringa relation implicitly contains the assumption that the spectral density of spin fluctuations is "white" up to a frequency much greater than both the nuclear and electronic Larmor frequencies. In three-dimensional metals such a spectrum is indeed found, with the cutoff being the Fermi energy $\varepsilon_F/\hbar$. In one-dimension, however, the assumption is no longer necessarily valid, and one must reexamine the derivation of the relaxation rate.

The planarity of the one-dimensional Fermi surface and the degeneracy of the electron gas ($kT \ll \varepsilon_F$) imply that the important spin-density fluctuations occur with wavevectors $q \approx 0$ and $q \approx 2k_F$. The long-wavelength modes are diffusive in nature, and hence the spin-lattice relaxation reflects the characteristic property of one-dimensional diffusion, namely, that the spectral density diverges as $\omega^{-1/2}$. The nature of the fluctuation spectrum for $q \approx 2k_F$ has been a subject of debate. It is not clear whether the relevant wavevector is q itself, in which case $q^{-1} \ll \Lambda$ (the electronic mean free path) and the dynamics are coherent, or whether the small difference $q - 2k_F$ is important, in which case the motion is again diffusive. These points are discussed in a comprehensive paper by Soda *et al.* (1977a), who derive the Korringa products for the two cases.

The calculation of the relaxation rate now requires more general treatment of time-dependent perturbation theory. Since the contact hyperfine interaction and the B term of the dipole interaction involve simultaneous flip of the electron and nuclear spin, their contribution to the relaxation rate is proportional to the spectral density of fluctuations at the difference frequency: $f(\omega_e - \omega_n) \approx f(\omega_e)$. The C and D parts of the dipolar interaction flip only the nuclear spin and hence lead to a contribution proportional to $f(\omega_n)$. The general form of the relaxation rate is a sum of these contributions (Devreux et al., 1974)

$$T_1^{-1} = \Omega_z f(\omega_n) + \Omega_+ f(\omega_e), \tag{16}$$

where, for a powder sample, the coupling constants are given by

$$\Omega_z = (\tfrac{3}{5})\gamma_n^2 \overline{H_{\text{dip}}^2} \tag{17}$$

$$\Omega_+ = \gamma_n^2 [(\tfrac{7}{5})\overline{H_{\text{dip}}^2} + \overline{H_{\text{con}}^2})] \tag{18}$$

$\overline{H_{\text{dip}}^2}$ and $\overline{H_{\text{con}}^2}$ are the mean square dipole and contact hyperfine fields. The averages are carried out over all electrons and all nuclei that are sufficiently well coupled to relax with a common rate.

For one-dimensional diffusion, the spectral density is given by

$$f_{1D}(\omega) = \frac{kT\chi_S}{N_A g^2 \mu_B^2} (2D_1 \omega)^{-1/2}. \tag{19}$$

The first factor yields the fraction of spins at temperature T, which are effective in producing relaxation. The second factor contains the diffusion rate, D_1, for motion along the chain. At sufficiently low frequency, i.e., at long times, the electron dynamics reflect the possibility of interchain hopping. In this regime the spectral density exhibits behavior characteristic of two- or three-dimensional diffusion: $f_{2D} \sim \ln(\omega)$ or f_{3D} = constant. Determination of the crossover frequency or frequencies can provide quantitative information about the dimensionality of the electron motion.

Each nondiffusive mode (e.g., $q = 2k_F$) contributes a frequency-independent term to the spectral density, which may be conveniently written for one-dimensional systems (Devreux and Nechtschein, 1979)

$$f_{\text{ND}} = 2\pi\hbar k T \left(\frac{\chi_S}{N_A g^2 \mu_B^2}\right)^2. \tag{20}$$

This form is exactly equivalent to the simple Korringa expression, consisting of the product of the fraction of spins contributing to the relaxation ($\sim kT\chi_S$) and inverse cutoff frequency associated with band motion ($\hbar/\varepsilon_F \sim \chi_S$).

c. Enhancement

Electron–electron Coulomb interactions modify the nuclear relaxation rate in one-dimensional metals in a manner that is quantitatively different from that of conventional metals in two ways. As we have seen, the one-dimensional dynamics are diffusive, at least at long wavelengths. Coulomb interactions reduce the diffusion rate for spin excitations relative to that for charge motion (Fulde and Luther, 1968)

$$D_1 = D_1^0(1 - \alpha), \tag{21}$$

where $D_1^0 = v_F^2 \tau_\parallel$ is the diffusion constant associated with charge transport parallel to the chains. Secondly, the incipient divergence of the Lindhard function $F(q/2k_F)$ at $q = 2k_F$ leads to additional enhancement of the $2k_F$ fluctuations regardless of whether or not they are diffusive. It becomes necessary to separate the $q = 0$ and $q = 2k_F$ enhancements

$$\eta_0 = \frac{(1-\alpha)^2}{[1 - \alpha F(0)]^2} = 1 \tag{22a}$$

$$\eta_{2k_F} = \frac{(1-\alpha)^2}{[1 - \alpha F(1)]^2}. \tag{22b}$$

η_{2k_F} is implicitly temperature dependent through the temperature dependence of the Lindhart function, where the divergence, at $q/2k_F = 1$ and as $T \to 0$ in a strictly one-dimensional system, is removed by interchain interactions and nonzero temperature.

In order to write a general expression for the spin-lattice relaxation rate, it is necessary to sum all products of the form $\Omega f(\omega)\eta$, for (1) contact and dipolar hyperfine interactions that depend on the coupling, Ω, and frequency ω; (2) diffusive or nondiffusive dynamics and dimensionality that determine the spectral density, f; (3) enhancement $\eta_{0,2k_F}$. Conversely, in order to decide which terms in such a sum are contributing in a given experiment, one must, in principle, determine T_1 as a function of field (i.e., of Larmor frequencies ω_n and ω_e) and of temperature. The reward for such an exhaustive sequence of measurements is information on the important microscopic parameters relating to dimensionality and to the role of Coulomb interactions.

Knowledge of the hyperfine coupling constants is useful for deconvolution of NMR data. To the extent that the conduction bands derive from the orbitals occupied by the unpaired electron in the molecular radical ion, the isotropic coupling constants can be measured via solution ESR of the appropriate species. The hyperfine interaction results in a splitting in the ESR spectrum that is used to give the coupling, conventionally in field units. This is the effective field at the electron due to each nucleus and is related to the hyperfine field discussed above by the ratio of electron to nuclear moments

$$a_H = \left(\frac{\hbar\gamma}{g\mu_B}\right) H_{hyp}. \tag{23}$$

Such data are available for several of the sulphur-containing donors and for TCNQ and are given in Table III. Unfortunately, the solution ESR spectra of the selenium donors are severely broadened and do not exhibit resolvable hyperfine structure. In several cases, hyperfine contact interactions have been calculated, using molecular orbital theory, from the electron densities at the nuclei and the known hyperfine fields of the atomic orbitals (Rieger and Fraenkel, 1962; Lowitz, 1967; Jonkman et al., 1974; Ladik et al., 1975). Again, some modification of the molecular parameters may be expected in the solid state and the values of Table III should be considered only a first approximation.

This section has discussed only the conduction electron effects on NMR in organic metals. As in other organic materials, molecular dynamics may contribute to the spin-lattice relaxation rate. A detailed discussion of such phenomena is beyond the scope of this chapter. Suffice it to say that important motional relaxation effects have been observed in organic metals, especially due to methyl rotation in TMTTF and TMTSF salts.

TABLE III

EXPERIMENTAL HYPERFINE COUPLING CONSTANTS OF MOLECULAR IONS

Molecule	Nucleus	Scalar (gauss)	Dipolar (gauss)
TTF	^{1}H	1.26[a]	0.67[b]
	^{33}S	4.2[c]	
TTT	^{1}H (both types)	0.55[c]	
	^{33}S	3.30[c]	
TCNQ	^{1}H	$(-)1.44 \pm 0.01$[d]	0.58[b,e]
		-1.393 ± 0.005[f]	
	^{14}N	1.02 ± 0.01[d]	
		0.998 ± 0.005[f]	
	^{13}C (methanide)	7.18[d]	
	^{13}C (carbon-1)	$(-)4.40$[d]	
	^{13}C (carbon-2)	0.62[d]	
	^{13}C (cyano)	$(-)6.38$[d]	
		-6.966 ± 0.005[f]	

[a] Wudl et al. (1970)
[b] Devreux et al. (1979)
[c] Bramwell et al. (1978)
[d] Fischer and McDowell (1963)
[e] Avalos et al. (1978)
[f] Haustein et al. (1971)

II. The Metallic State

5. SPIN SUSCEPTIBILITY

The static susceptibility as a function of temperature has been measured for virtually every highly conducting organic metal. In several cases data are available from different laboratories, and it is heartening to note that there are no substantial discrepancies among the results of different groups. The interpretation of the data, however, is a very different matter, with half a dozen separate theories having been proposed to explain the observed departure from temperature-independent Pauli paramagnetism.

The experimental data for several representative materials are shown in Fig. 1. The earliest measurements of the susceptibility of TTF–TCNQ were reported by Perlstein et al. (1973). Subsequent studies were made by Tomkiewicz et al. (1974) and by Scott et al. (1974), who found, by comparison of the static and ESR susceptibilities, that the temperature-dependent part is due entirely to the spin contribution, χ_S. By careful consideration of the core diamagnetism, Scott et al. (1974) determined that χ_S reached a maximum of 6.2×10^{-4} cm^3 mole at 340 K and fell to 2.5×10^{-4} cm^3 mole at 60 K, just above the series of metal–insulator transitions. As can be seen

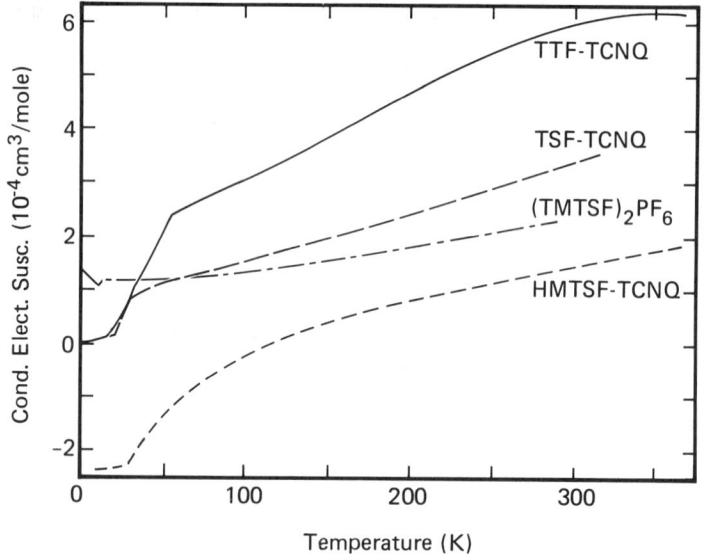

FIG. 1. Conduction electron contribution to the susceptibility for several organic metals: TTF–TCNQ [from Scott et al. (1974); Herman et al. (1976)]; TSF–TCNQ [from Scott et al. (1978)]; TMTSF$_2$PF$_6$, powder susceptibility at high field [from Scott et al. (1980)]; HMTSF–TCNQ [from Bloch et al. (1977)].

from the data presented in Fig. 1, this temperature dependence is typical of all organic charge-transfer conductors in the TTF-TCNQ and $(TMTSF)_2X$ families. Similar behavior has been found in TMTTF-TCNQ and DMTTF-TCNQ (Scott *et al.*, 1974), $TTT-TCNQ_2$ (Buravov *et al.*, 1974), TSF-TCNQ (Buravov *et al.*, 1977; Scott *et al.*, 1978), HMTSF-TCNQ (Soda *et al.*, 1976b; Bloch *et al.*, 1977), TMTSF-TCNQ (Bloch *et al.*, 1977), HMTTF-TCNQ (Tomkiewicz *et al.*, 1977a, 1977b; Delhaes *et al.*, 1977; Torrance *et al.*, 1980), and TMTSF-DMTCNQ (Tomkiewicz, 1979; Hardebusch *et al.*, 1979).

Among the single-cation chain metals there is the same temperature dependence in the metallic state susceptibility. Data are available for $TTF-I_x$, $TTF-(SCN)_x$ and $TTF-(SeCN)_x$ (Wudl *et al.*, 1977; Somoano *et al.*, 1977), $TTF-Br_{0.86}$ (Tomkiewicz and Taranko, 1978), $TTT_2-I_{3+\delta}$ (Kaminskii *et al.*, 1977; Miljak *et al.*, 1977; Isett and Perez-Albuerne, 1977; Delhaes *et al.*, 1979b), $(TMTTF)_2X$ with $X = Br, BF_4, ClO_4, PF_6, NO_3$ (Delhaes *et al.*, 1979a; Coulon *et al.*, 1982a), $(TMTSF)_2PF_2$ (Scott *et al.*, 1980; Wudl *et al.*, 1981), $(TMTSF)_2X$ with $X = PF_6, AsF_6, SbF_6, NO_3, BF_4, ClO_4, ReO_4$ (Pedersen *et al.*, 1981, 1982).

It is clear that the overall behavior in the metallic regime of all the compounds mentioned in the preceding two paragraphs is basically the same. Differences arise in the absolute magnitude of the spin susceptibilities, ranging from 6.0×10^{-4} cm^2 mole for TTF-TCNQ at room temperature to about 0.5×10^{-4} cm^3 mole for TTF pseudohalides, and in the extent of the temperature dependence. The ratio of the susceptibility just above the metal-insulator transition to that at room-temperature ranges from 0.3 for TSF-TCNQ to close to 1 for some of the $(TMTSF)_2X$ salts.

In addition to the temperature dependence of the susceptibility that is commonly measured for organic metals, pressure dependences have been obtained for a few compounds. Berthier *et al.* (1976b) present 40 MHz ESR data for the spin susceptibility of TTF-TCNQ up to 4 kbar for temperatures between 170 K and 340 K, and of TMTTF-TCNQ up to 10 kbar at room temperature. They find that the susceptibillity decreases by 7% kbar for TTF-TCNQ and 5.6% kbar for TMTTF-TCNQ. Hardebusch *et al.* (1979) measured the susceptibility of TMTSF-DMTCNQ at a pressure of 12 kbar, using a Faraday apparatus in which the entire clamp is suspended from the balance. The results are compared to the ambient pressure data on HMTSF-TCNQ.

The temperature dependence of the susceptibility in the metallic phase indicates a clear departure from the simple picture of Pauli paramagnetism. During the 10 years that have elapsed since this was first noticed in TTF-TCNQ, many explanations have been proposed. Several have fallen out of favor, but no single mechanism has yet evoked universal agreement.

For the sake of completeness, all the explanations that have appeared in print will be enumerated, along with the reasons why some have been eliminated.

a. Antiferromagnetism

One of the earliest suggestions, by Tomkiewicz *et al.* (1974), was that the donor and acceptor chains were fundamentally different in that one was metallic and the other insulating with localized spins. Thus the measured susceptibility would be the sum of a temperature-independent Pauli term and the Bonner–Fisher (1964) behavior characteristic of a one-dimensional antiferromagnet. Subsequent measurements, including g-value decomposition by Tomkiewicz herself and coworkers (see below), have shown that both chains are metallic and that the susceptibility variation occurs on both donor and acceptor stacks. Moreover, such a model cannot explain the temperature variation seen in the single-chain conductors such as the TMTSF family.

b. Interchain Hybridization Gap

Tight-binding band structure calculations by Bernstein *et al.* (1975) showed that nonzero interchain transfer integrals produced a deep minimum in the density of states near the Fermi level. If this gap were comparable in width to typical thermal energies, then a temperature-dependent susceptibility could be obtained. However, these authors were unable to obtain a satisfactory fit to the observed susceptibility. Again, the similar temperature dependence seen in the one-chain conductors, where interchain mixing does not produce a gap at the Fermi surface, rules out this as the mechanism responsible.

c. Peierls Pseudogap

In the one-dimensional fluctuation regime above the Peierls transition temperature, it is predicted (Lee *et al.*, 1973) that the density of states near the Fermi surface will be depressed. This mechanism was discussed in the context of the experimental data by Scott *et al.* (1974). Rice and Strassler (1973) and Bjelis and Barasic (1975) showed that the susceptibility is proportional to the inverse coherence length and therefore linear in temperature, qualitatively as observed. No fully satisfactory agreement between theory and experiment has been obtained in this picture, but, since it is known from the observation of diffuse x-ray scattering that fluctuations persist up to 150 K in TTF–TCNQ, the presence of a pseudogap must be considered as a contributing factor.

d. Thermal Contraction

The thermal expansion coefficient of organic metals is large compared to that of conventional metals. Since overlap integrals depend exponentially on lattice parameters, considerable increase in bandwidth, and hence reduction in susceptibility, is expected to occur upon cooling. In collaboration with the Orsay group, Caron (Caron *et al.*, 1978; Jérome and Caron, 1979) showed that the contraction occuring in TTF-TCNQ between 300 and 60 K was equivalent to the application of 5 kbar pressure. By analyzing both the pressure and temperature dependences of the susceptibility, they were able to extract a constant volume susceptibility, which shows a less pronounced temperature dependence than the constant pressure data. The remaining temperature dependence was attributed to the onset of a pseudogap.

e. Coulomb Interactions

It was pointed out by Torrance (Torrance *et al.*, 1977; Torrance, 1978) that the room-temperature susceptibility approaches that for noninteracting *localized* spins and that, therefore, there must be considerable enhancement over the value for noninteracting electrons in a tight-binding band of the width implied by thermopower and plasma frequency measurements. This enhancement was attributed to the effect of a Coulomb interaction $U \gtrsim 4t$ ("large U"). The temperature dependence was compared to the Bonner-Fisher (1964) behavior, with exchange interactions given by the Hubbard (1963, 1964) models. Since the Hubbard model is not solved for intermediate values of $U/4t$, quantitative agreement was neither expected nor obtained.

Lee, Rice, and Klemm (1977) used renormalization group techniques to analyze a model with δ function Coulomb interactions between electrons on the same chain. They showed that interchain Coulomb interactions renormalized the enhancement factor for the susceptibility [see Eq. (4)], reducing it at lower temperatures. However, they were still unable to obtain fully quantitative agreement with experiment and attributed the discrepancy to pseudogap effects.

The role of Coulomb interactions and the relative magnitudes of U and the bandwidth $4t$ will be discussed further in Section 9, which deals with NMR relaxation.

f. Dynamical Band Narrowing

Following a suggestion by Gutfreund *et al.* (1979) that electron–phonon interactions lead to motional narrowing of the conduction band, Entin-Wohlman *et al.* (1982) performed a calculation based on small polaron theory. They find that for strong electron–phonon interactions the density of states is enhanced by a factor that varies approximately as $\exp(-\Lambda/a)$, where Λ is the mean free path and a the lattice constant. Thus as phonon scattering becomes more dominant at higher temperature, the susceptibility increases. An experimentally observable prediction of this theory is an isotope effect, which has indeed been observed by Cooper and Korin (1979). Unfortunately, the real systems with which this theory can be quantitatively compared are in the regime of parameters for which the calculational approximations begin to break down. Hence again, agreement with experiment is only semiquantitative.

It should be emphasized that the mechanisms discussed above are, for the most part, not mutually exclusive and may therefore each contribute, to a greater or lesser extent, to the temperature dependence of the susceptibilities of the different organic metals. It has been recognized since the first comparisons of susceptibility with other measures of the bandwidth (Scott et al., 1974) that Coulomb enhancement occurs in TTF–TCNQ. The observation of the Peierls transition and the associated one-dimensional diffuse scattering indicates that electron–phonon interactions are significant and hence that polaronic effects and pseudogap formation are likely to occur. In the TMTSF and TMTTF families the observation of a spin-density wave indicates that the repulsive Coulomb interactions dominate the electron–phonon interaction, but the latter must be assumed to be present to some degree.

It might be argued that since so many mechanisms contribute to the thermal variation of the susceptibility, little fundamental understanding of the physics can be achieved by comparing theory with experiment. The counterargument is that, since no single mechanism has provided detailed agreement with the data, there has been a great impetus to develop new models. Though no one of them accounts entirely for the susceptibility, many of them do make testable predictions concerning other physical properties of organic metals and provide a framework for the analysis of data from other experiments, such as the NMR work to be discussed below.

6. Decomposition of Susceptibility by g Factor

The method of decomposing the susceptibility of two-chain conductors into the individual contributions, according to Eq. (7), has been applied to many organic metals, particularly by Tomkiewicz and her coworkers. In TTF–TCNQ it is found that in the metallic state ($T > 60$ K) the g tensor is independent of temperature, implying constant fractions of the susceptibility on the donor (54%) and acceptor (46%) chains (Tomkiewicz et al., 1976a, 1980). Work by the same group on TMTTF–TCNQ found it to have a two-line ESR spectrum (Tomkiewicz et al., 1976b) attributed to two different donor-acceptor pairs in the unit cell. Each line has a temperature independent

g value corresponding to a partitioning of susceptibilities similar to that in TTF-TCNQ. However, Ehrenfreund *et al.* (1977) observed only a single line in crystals of TMTTF-TCNQ, which were proven to have a crystal structure isomorphous with the unsubstituted parent compound. It is not clear what phase of TMTTF-TCNQ was studied by the IBM group. HMTTF-TCNQ has a g shift which changes by a factor of two between 50 and 300 K (Tomkiewicz *et al.*, 1977b). Decomposition according to Eq. (7) leads to the result that the susceptibility on the donor chain is temperature independent, while that on the acceptor chain increases linearly (from zero at 50 K) as the temperature is raised.

The method of g-value decomposition has been criticized by Conwell (1980) on the grounds that small changes in g value can lead to relatively large differences in the assigned susceptibilities. Such changes occur due to differences in molecular environment between different materials, and between solution and solid, and even as a function of temperature in the same material. Tomkiewicz *et al.* (1980) defend the method and present studies which show that the donor acceptor susceptibility ratio is independent of the orientation chosen for the g-value measurement. In spite of any remaining controversies that may exist concerning the *detailed* values of the susceptibilities obtained, it is clear that the method yields rather good semiquantitative results, in rough agreement with determinations made by independent techniques such as Knight-shift measurements (see below).

Due to the much larger ESR linewidths of TSF-TCNQ and its derivatives, compared to the sulphur-based compounds, g-value decomposition has been much less successful. Results are reported for TSF-TCNQ only at room temperature: namely $\chi_S(TCNQ) \approx 3.5\chi_S(TSF)$ (Tomkiewicz *et al.*, 1977b). The same group has analyzed the temperature dependence of the g values of the alloy series $TSF_xTTF_{1-x}TCNQ$ with x in the range 0.01–0.7 (Tomkiewicz *et al.*, 1976c, 1977d). It is found that the induced disorder on the donor chain introduces a change in the behavior of the g value, which tends towards the TCNQ value as the temperature is lowered. This implies increasingly more susceptibility on the acceptor chain as the Peierls transition is approached, at variance with simple ideas of a fluctuation-induced pseudogap on the undisordered metallic chain. However, it is not clear what other effects, for example, a disorder-induced mobility gap or changes in interchain coupling, arise in the alloy system.

7. DECOMPOSITION OF SUSCEPTIBILITY BY KNIGHT SHIFT

The localized nature of NMR as a probe of the hyperfine field permits Knight-shift measurements on selectivity isotopically enriched samples to be used to determine local susceptibilities. Rybaczewski *et al.* (1976) performed such an experiment on TTF-TCNQ enriched with ^{13}C at the cyano group of the TCNQ molecules. The Knight shift therefore probes the susceptibility on the acceptor stack

$$\chi_S^A = \chi_S - \chi_S^D = \frac{g\mu_B K^C}{a_H^C}. \qquad (24)$$

Above about 80 K, the ^{13}C Knight shift (K^C) varies approximately linearly with the total susceptibility, indicating that the temperature variation of the local susceptibilities is the same on both donor and acceptor chains. This is in agreement with the results of g-value decomposition. Since the carbon hyperfine coupling constant (a_H^C) is not known for the TCNQ molecular anion, Eq. (24) cannot be used directly to determine the ratio χ_S^A/χ_S^D. However, once the ratio is known at any one temperature, (a_H^C) is known and the data can be analyzed for all temperatures. By assuming equal enhancement of the spin-lattice relaxation rate (see below) for protons on the two chains, Rybaczewski used T_1 data to evaluate the individual susceptibilities at 300 K. They found 66% on the donor and 34% on the acceptor chain, slightly different from the partition given by g-value analysis.

To my knowledge, TTF-TCNQ is the only organic conductor for which such Knight-shift measurements have been carried out. We shall return to these data again when we discuss the metal insulator transition and semiconducting states in Parts III and IV.

8. ESR Linewidth

The relatively narrow linewidths of one-dimensional organic conductors, as compared to conventional three-dimensional metals, has facilitated the collection of a large amount of experimental data on the temperature dependence of the linewidth of a large number of compounds. Several representative examples are shown in Fig. 2, where it is immediately obvious that there are some striking differences from material to material.

First, systems in which TTF or a derivative is the donor have lines narrower by two orders of magnitude than their TSF analogs. Tomkiewicz and her colleagues, who first published the linewidths of TTF-TCNQ (Tomkiewicz et al., 1974) and the TSF$_x$TTF$_{1-x}$TCNQ alloys (Tomkiewicz et al., 1975), presented this as evidence for the dominance of the spin–phonon scattering mechanism, since the spin–orbit interaction depends on the atomic number of the elements contributing to the conduction electron wavefunction [cf., Eq. (8)]. Second, several materials have a linewidth that decreases with temperature in the metallic state, while others decrease. This was remarked upon by Tomkiewicz (1979), who compared the decreasing temperature dependence of TTF-TCNQ, TSF-TCNQ, and HMTTF-TCNQ with

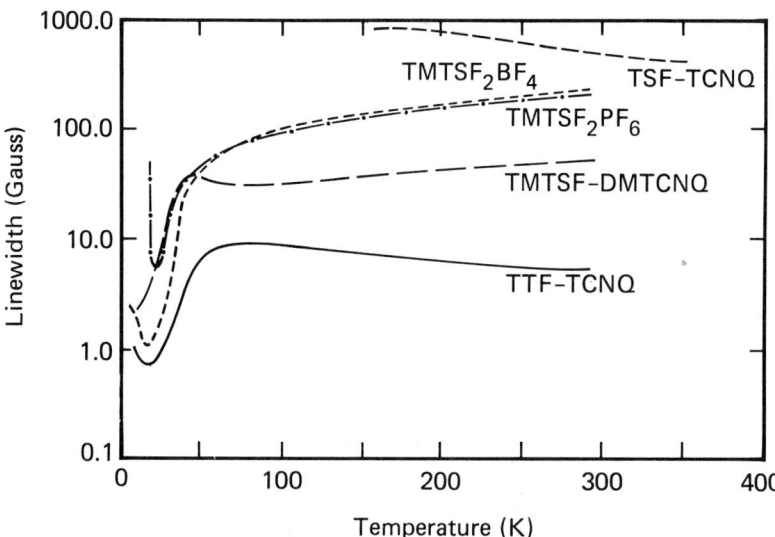

FIG. 2. ESR linewidth of several organic metals: TTF–TCNQ [from Tomkiewicz et al. (1974)]; TSF–TCNQ [from Tomkiewicz (1979)]; TMTSF–DMTCNQ [from Tomkiewicz (1979)]; TMTSF$_2$PF$_6$ [from Pedersen et al. (1980)]; TMTSF$_2$BF$_4$ [from Pedersen et al. (1981)].

the opposite behavior in TMTTF–TCNQ and TMTSF–DMTCNQ. The situation with TMTTF–TCNQ is complicated by the occurrence of nonstoichiometric phases (Ehrenfreund et al., 1977). These authors report a linewidth decreasing with temperature in the 1:1 phase that is similar in structure to TTF–TCNQ. Thus, of the two-chain conductors, it seems that only TMTSF–DMTCNQ is anomalous. In the metallic phase of the single-chain cation conductors (TMTSF)$_2$X (Pedersen et al., 1980, 1981; Scott, 1982a; Flandrois et al., 1982) and (TMTTF)$_2$X (Delhaes et al., 1979a; Flandrois et al., 1982; Coulon et al., 1982a), the linewidth is an increasing function of temperature.

No detailed theory for the temperature dependence of the ESR linewidth has yet been formulated for organic metals, because several different broadening processes may be contributing. In a one-dimensional system time-reversal invariance reduces the phase space available for spin–flip electron–phonon scattering (Bloch. 1977). Hence interchain interactions directly affect the linewidth. The relaxation time, τ, that enters Eq. (8) depends on the phonon density and therefore on temperature. However, not all phonons are equally efficient in causing spin relaxation, and in the fluctuation regime of Peierls systems $2k_F$ phonons are softening. Hence the temperature dependence is not trivially calculated from phonon occupation numbers. Finally, even in the absence of phonons there are contributions to the linewidth due to, for example, hyperfine, dipolar, and spin-other-orbit contributions. A full decomposition of these contributions to the linewidth has not been achieved.

Tomkiewicz (1979) discussed in qualitative terms the effect of Peierls fluctuations on the interchain scattering rate. Two factors may be expected to have opposing behavior: The softening of $2k_F$ phonons will lead to an increased occupation at lower temperature and increase the scattering rate; the formation of a pseudogap and the associated reduced density of states available for scattering will tend to decrease the scattering rate. The balance between these competing effects was used to explain the differences in behavior between TMTTF-DMTCNQ and other Peierls systems.

The temperature dependence of the linewidths of $(TMTTF)_2X$ (Delhaes et al., 1979a; Flandrois et al., 1982; Coulon et al., 1982a) and $(TMTSF)_2X$ (Pedersen et al., 1980, 1981: Scott, 1982a) is more typical of a normal metal in which the dominant factor is the density of phonons. However, once again the detailed temperature dependence is not fully understood.

In a recent paper, Adrian (1982) considers explicitly the matrix elements of the spin-orbit interaction between molecular orbitals of neighboring chains and obtains room-temperature linewidths in agreement with the experimental values for TTF-TCNQ and TSF-TCNQ. In his model the temperature dependence of the linewidth is the same as the parallel conductivity, in rough agreement with observation in most two-chain conductors, but not with the TMTTF and TMTSF salts.

The effect of interchain interactions can be deduced in the following way. Since the scattering time τ is predominantly an interchain hopping time (Bloch, 1977; Weger, 1978), its relative size among similar materials can be estimated by taking the ratio $\Gamma/(\delta g)^2$. There is not a common method of taking this ratio so that intercomparison of the data from different groups is rather difficult. Tomkiewicz and coworkers (Tomkiewicz et al., 1965, 1978; Tomkiewicz, 1979) analyzed the data for the TTF-TCNQ/TSF-TCNQ alloy series using for δg the value for the fulvalene chain, as measured below the metal-insulator transition temperature, where all of the susceptibility is on the donor molecules. The ratio increases with high TSF concentration, implying a larger interchain coupling due to the size of the selenium atomic orbitals. Flandrois et al. (1982) used the orientational average value of $(\delta g)^2$, as measured at room temperature, to determine the ratio for the $(TMTTF)_2X$ and $(TMTSF)_2X$ salts. The results correlate well with the ratio of interchain S-S or Se-Se distances to their van der Waals radii.

In spite of these successes for applicability of Eq. (8) to the dimensionality of isostructural materials, $\Gamma/(\delta g)^2$ is not a unique determinant of transverse bandwidth for compounds that have different structures, since the additional contributions to the linewidth may be expected to enter to different extents.

Nevertheless, qualitative comparisons may be made, and in Table IV are collected the available data on some compounds of interest. From these results one may conclude that, in general, selenium compounds have higher dimensionality than their sulphur analogs and that there is a trend to greater dimensionality from unsubstituted donor to tetramethyl to hexamethylene derivatives. These conclusions are in substantial agreement with those from crystallographic evidence (Bloch et al., 1977; Kistenmacher, 1978). From Table IV it is apparent that the (TMTTF)$_2$X and (TMTSF)$_2$X families are anomalous in that the dimensionality predicted from the ESR data is considerably less than that deduced from the structures (see, e.g., Thorup et al., 1981). It is not clear whether the origin of this discrepancy is related to

TABLE IV

Linewidths (Γ) and g Shifts (δg) at Room Temperature of Organic Metals

Material	Orientation	Γ (gauss)	δg	$\Gamma/(\delta g)^2$ (10^5 gauss)	$\langle\Gamma\rangle/\langle(\delta g)^2\rangle$ (10^5 gauss)
TTF-TCNQ[a]	b ($\parallel$)	5.6	0.0003	600	
	a ($\perp$)	6	0.0042	3.4	3.1
	c^* ($\perp$)	6	0.0063	1.5	
TMTTF-TCNQ[b,c]	a ($\perp$)	5	0.0037	3.6	
HMTTF-TCNQ[d]	b ($\parallel$)	8	−0.0006	200	
	a ($\perp$)	10	0.0042	5.7	13
	c ($\perp$)	8	0.0030	8.9	
TSF-TCNQ[e,f]	b ($\parallel$)	—	0.008	—	
	a ($\perp$)	—	0.010	—	
	c^* ($\perp$)	480	0.013	28	
TMTSF-TCNQ[c]	($\perp$)	100	0.009	12	
HMTSF-TCNQ[c]	—	>4000	~0.01?	≥400	
TMTSF-DMTCNQ[f,g]	($\perp$)	54	0.009	6.7	
(TMTTF)$_2$PF$_6$[h]	b ($\parallel$)	2.9	−0.0003	320	
	a ($\perp$)	3.6	0.0063	0.9	1.0
	c ($\perp$)	4.6	0.0084	0.7	
(TMTSF)$_2$PF$_6$[i]	a ($\parallel$)	160	0.013	9.5	
	b' ($\perp$)	210	0.025	3.4	2.5
	c^* ($\perp$)	240	0.041	1.4	

[a] Tomkiewicz et al. (1974)
[b] Ehrenfreund et al. (1977)
[c] Bloch et al. (1977)
[d] Delhaes et al. (1977)
[e] Walsh et al. (1980a)
[f] Tomkiewicz (1979)
[g] Tomkiewicz et al. (1978)
[h] Delhaes et al. (1979a)
[i] Pedersen et al. (1980, 1981)

the tendency of the single-chain conductors to exhibit spin-density wave instabilities, as compared to the charge-density wave transitions that prevail among the two-chain conductors.

The linewidths of organic conductors are invariably anisotropic. Unfortunately, it has not been uniform practice to report in the literature the details of the anisotropy, and in several cases linewidths are reported without reference to the direction in which the external magnetic field was applied. An additional complication is that the anisotropies, as well as the linewidths themselves, are temperature dependent. From the limited amount of information available, a few generalizations may be made: Anisotropies in the plane transverse to the highly conducting axis are typically in the range of 10–20%, while the ratio of the parallel to transverse linewidth ranges from 1 to 2. Data are available for TTF–TCNQ (Tomkiewicz et al., 1974; Tomkiewicz, 1979), for TSF–TCNQ (Tomkiewicz, 1979), for the $(TMTSF)_2X$ series (Pedersen et al., 1981), and for the $(TMTTF)_2X$ series (Delhaes et al., 1979a) and are given, for room temperature, in Table IV. It is clear that the linewidth anisotropy does not reflect the g anisotropy in the simple way suggested by Eq. (8). The reason for this was discussed by Pedersen et al. (1981) in terms of a phenomenological model in which the interchain scattering event, by changing the orbital motion of the electron, affected the spin–orbit coupling to produce an effective field acting on the spin. They concluded that the ratio of parallel to perpendicular linewidths should be of order 1.5, with deviations that reflect the anisotropy of the g value and scattering rates in the transverse plane.

9. Nuclear Magnetic Relaxation

Proton spin-lattice relaxation has been studied in detail for only a few compounds, but the relatively small amount of available information has contributed much to our knowledge of the microscopic parameters. The earliest measurements of proton relaxations rates in TTF–TCNQ were performed by the Pennsylvania group (Rybaczewski et al., 1975). Using selectively deuterated samples in order to probe separately the donor and acceptor chains, and working at a frequency of 38 MHz, they found that the relaxation rates followed a Korringa-like behavior in the metallic regime. From this it was deduced that the enhancement factor η [see Eq. (13)] did not reflect the temperature dependence expected for a divergent $2k_F$ response function and that, therefore, Coulomb correlations were of minor importance. This work, however, did not consider the diffusive contribution to the relaxation rate and the consequent frequency dependence.

A more extensive study, including frequency and pressure dependent measurements, was performed by Soda, Jérome, Weger, and coworkers and published in a series of articles (Soda et al., 1976a,b, 1977b; Jérome and Giral, 1977; Jérome and Weger, 1977; Jérome et al., 1977), culminating in comprehensive review (Soda et al., 1977a). These workers analyzed their data according to a form that took into account diffusive motion for the $q \approx 0$ component and coherent motion for the $q \approx 2k_F$ component of electronic spin dynamics

$$\left[T_1 T \chi_S^2 \left(\frac{\hbar \gamma a_H}{g \mu_B} \right) \right]^{-1} = \frac{2\pi k}{\hbar g^2 \mu_B^2} \left[\left(\frac{\tau_\perp}{\tau_\parallel} \right)^{1/2} g(\omega_e) \eta_0(\alpha) + \eta_{2k_F}(\alpha) \right]. \quad (25)$$

The first term on the right-hand side gives the contribution of the diffusive component, and when interchain hopping is taken into account as a cutoff to the one-dimensional divergence of the spectral density, the function $g(\omega_e)$ can be written

$$g(\omega_e) = \left[\frac{1 + (1 + \omega_e^2 \tau_\perp^2)^{1/2}}{2(1 + \omega_e^2 \tau_\perp^2)} \right]^{1/2}. \quad (26)$$

The scattering time for motion parallel to the chain, $\tau_\parallel$, is proportional to the conductivity for small and intermediate values of the Coulomb interaction, U. In the case of strong correlations $\tau_\parallel$ would be proportional to the exchange interaction, given in the Hubbard model by $t_\parallel^2/U$. The functions, η, are the enhancement factors defined in Part I, Section 3. Contributions due to the dipolar part of the hyperfine interaction are ignored, an approximation that is justified a posteriori by Soda et al. on the grounds that the temperature dependence of the high field-relaxation rate (i.e., the part dominated by diffusion) is relatively weak (but see below). The data presented by Soda et al. (1977a) show the expected $T_1^{-1} \sim H^{-1/2}$ dependence at high fields with a crossover to field-independent behavior for fields less than about 5-10 kG. Analysis of the data permits the determination of the temperature and pressure dependence of the parameters $\tau_\perp$, $\tau_\parallel$, and η_{2k_F}. The interchain hopping rate is found to be of order 1×10^{11} sec^{-1} at 300 K and 1 atm, increasing by a factor of 2 as the temperature is lowered to 100 K, or by a factor of 3-4 as the pressure is raised to 8 kbar. These variations are in good agreement with the temperature and pressure dependence of the transverse conductivity. The parallel scattering rate, $\tau_\parallel^{-1} \approx 5\text{-}10 \times 10^{-15}$ sec, is also in agreement with values obtained from optical and dc conductivities. The low-frequency limit of the data yields the $2k_F$ Korringa enhancement factor η_{2k_F} for both the TTF and TCNQ chains. For the TTF chain at room temperature and ambient pressure, the value is ~ 10, increasing to 20 as the temperature goes to 100 K; for TCNQ the $\eta_{2k_F} \approx 20$, increasing to 30 as the temperature is lowered. All values increase by about 25% with the application of 8 kbar pressure. From these values of η, and taking into account the temperature dependence of the $2k_F$ Lindhard function, Soda *et al.* arrive at a value of α [$\equiv U_{\text{eff}} \rho(\varepsilon_F)$] ≈ 0.6–0.7 for both donor and acceptor chains. This corresponds to a Coulomb interaction $U_{\text{eff}}/4t_\parallel \approx 0.8$–$1$.

In light of the frequency dependence revealed by Soda's measurements, Ehrenfreund and Heeger (1977) reanalyzed the data of Rybaczewski *et al.* (1975) and obtained values for $\alpha \sim 0.3$, somewhat smaller than the Orsay group.

Other frequency-dependent measurements have been made by Devreux and Nechtschein (1979), extending the field range to 90 kG, which corresponds to a proton Larmor frequency $\omega_n = 2.1 \times 10^9 \text{ sec}^{-1}$ and electron Larmor frequency $\omega_e = 1.6 \times 10^{12} \text{ sec}^{-1}$. The analysis of their data followed that of Soda *et al.* (1977a), except that several additional effects were taken into account: interchain coupling via dipolar interactions, such that electrons on the TTF chain contribute to the relaxation of protons on the TCNQ chain and *vice versa*; incomplete deuteration such that the measured relaxation rate does not arise solely from a single chain; and the possibility of large-U behavior for the spin dynamics and a consequent relaxation rate similar to that of an antiferromagnetic chain. The corrections introduced for incomplete deuteration and interchain dipolar effects make the data analysis less decisive than that of Soda *et al.* The corrected relaxation rate of the TCNQ protons has the logarithmic frequency dependence expected for two-dimensional diffusion, whereas that of the TTF protons is one dimensional. The results can be analyzed, with internal consistency, in terms of either an electron gas with Coulomb enhancements (the "small-u" approach) or an antiferromagnetic chain with spin–wave excitations ("large-U"). However, this work does not consider temperature dependence, in contrast to Soda *et al.* (1977a), who showed that the thermal variation of the enhancement factor, η_{2k_F}, follows that expected for the Lindhard function in a tight-binding metal, with small u.

Thus it seems clear that, in spite of differences in detail between the data analyses of the several groups, the relaxation rates for protons in TTF-TCNQ are best described in terms of an electron gas with small to intermediate Coulomb interactions.;

Proton relaxation has been measured in several other organic metals, though not in such detail as for TTF-TCNQ. The pressure and temperature dependence of the donor chain proton relaxation rate in TMTTF-TCNQ was measured by Berthier *et al.* (1976a,b). They found a behavior essentially similar to that of the parent compound, but with an additional contribution evident as maxima in T_1^{-1} at temperatures below 100 K, associated with the rotational motion of the methyl groups. Data are also available for HMTSF-TCNQ (Soda *et al.*, 1976b; Jérome *et al.*, 1977), which exhibits a frequency-independent rate up to 276 MHz. This imposes a lower limit on the interchain bandwidth of $t_\perp > 20$ meV. The pressure dependence of the proton relaxation rate is relatively weak and not much different from that expected on the basis of pressure-induced band broadening. From this comparison, Jérome et al. (1977) conclude that Coulomb interactions in HMTSF–TCNQ are half as large as in TTF–TCNQ.

In $TTT_2-I_{3+\delta}$, Schaffhauser et al. (1981a,b) found a proton relaxation rate that was almost frequency independent at room temperature, suggesting rapid transverse diffusion. By examining aligned crystals, these authors were able to conclude that the high-temperature relaxation mechanism was dominated by the scalar part of the hyperfine interaction. The temperature dependence in the metallic state was attributed to an increasing diffusion rate [cf., Eq. (19)] at lower temperature.

The frequency dependence of the relaxation rate in $(TMTSF)_2PF_6$ in the pressure-induced low-temperature metallic phase shows the logarithmic behavior characteristic of two-dimensional diffusion (Azevedo et al., 1982), with no evidence of crossover to one-dimensional dynamics at frequencies up to $\omega_e = 3 \times 10^{12}$ sec^{-1}. These results impose a lower limit on the larger interchain transfer integral of order 2 meV. The crossover to three-dimensional frequency-independent behavior occurs at 12 kOe or 2×10^{11} sec^{-1}, yielding a value for interchain overlap in the third direction of 0.1 meV.

10. The Semimetallic State of HMTSF–TCNQ

It is evident from the conduction electron susceptibility (Fig. 1) of HMTSF–TCNQ that there is a fundamental difference in its low-temperature state. No metal–insulator transition is observed down to helium temperature, although in the low-temperature regime there is some controversy concerning the behavior of the resistivity, apparently related to sample quality. Two factors in the magnetic properties, as well as evidence from other types of experiments, point to increased dimensionality in HMTSF–TCNQ: There is no observable ESR line (Bloch et al., 1977), implying a linewidth increased by the more effective spin–phonon scattering in a system with larger interchain coupling (see Table IV); and the proton relaxation rate, at room temperature, is frequency independent, implying three-dimensional spin dynamics (Soda et al., 1976a). It is thus not surprising that the Peierls transition should be suppressed in this material.

At temperatures sufficiently low for transverse motion to become coherent, as opposed to diffusive, the properties are determined by the band structure, which reflects the coupling of electrons on the TCNQ chain and holes on HMTSF (Weger, 1976). In this regime HMTSF–TCNQ is a two-dimensional semimetal with small electron and hole pockets, and correspondingly light effect masses. It is now possible for the carriers to undergo orbital motion in response to a magnetic field, leading to Landau diamagnetism (Soda et al., 1976b). For fields in the c direction (the axis of the cylindrical carrier pockets), the ratio of low-temperature Landau and Pauli susceptibilities is given by

$$\frac{\chi_L^{LT}}{\chi_P^{LT}} = -\frac{1}{3}\frac{m_e^2}{m_a^* m_b^*}, \qquad (27)$$

where m_a^* and m_b^* are the band effective masses for motion in the a and b directions, respectively. In a powdered sample those crystallites with the c axis perpendicular to the field do not contribute to the Landau susceptibility. Soda et al. (1976b) derive the following values for the high- and low-temperature Pauli and Landau contributions:

$$\chi_P^{HT} = 1.46, \quad \chi_L^{HT} = 0, \quad \chi_P^{LT} = 0.4, \quad \chi_L^{LT} = -2.32 \times 10^{-4} \text{ cm}^3 \text{ mole}^{-1}.$$

III. The Peierls Transition and the Charge-Density Wave State

11. Peierls Transition

The magnetic properties of TTF–TCNQ and its analogs show clearly the occurrence of phase transitions in the 20–60 K range, but the magnetism alone does not unambiguously identify these transitions as being due to the Peierls mechanism. The opening of a gap at the Fermi energy causes a reduction in the spin susceptibility and a crossover to an activated temperature dependence, producing a characteristic "knee" in the magnetic susceptibility at or just above the transition temperature and corresponding changes in those properties that also depend on the density of states, for example, Knight shift and spin–lattice relaxation rate. The condensation of soft $2k_F$ phonons into a static-lattice distortion removes the major mechanism of ESR line broadening, which was active in the metallic state. The ESR linewidth, therefore, falls to a small residual value dictated by spin scattering from acoustic phonons and hyperfine interactions.

Once it is established from other techniques, such as conductivity and diffraction, that the low temperature phase is semiconducting due to the onset of a charge density wave, closer examination of the magnetic properties, both at and below the transition temperature, can be used to gain useful information about the microscopic aspects of the phase change. Primary among these is the determination of the "driving-chain" by examination of the separate donor and acceptor susceptibilities using either the g value of Knight-shift decomposition techniques discussed in Parts I and II.

The Peierls transition at 54 K in TTF–TCNQ appears clearly in all measurements (Tomkiewicz et al., 1974; Scott et al., 1974; Gulley and

Weiher, 1975) of the temperature dependence of its susceptibility as a crossover from exponentially activated behavior at low temperature, to a much weaker dependence in the metallic phase. High-precision susceptibility measurements on single crystals of TTF–TCNQ (Herman *et al.*, 1976) reveal another phase transition at 38 K, and the observation of hysteresis (Herman *et al.*, 1977) demonstrates that it is first order. However, it is not possible to decide on the basis of static susceptibility alone the relationship of the 38 and 54 K transitions. Moreover, no anomaly is evident in the magnetic susceptibility at 48 K, where neutron diffraction data indicate the onset of a distortion on the TTF stacks.

The g-value decomposition was carried out for TTF–TCNQ by Tomkiewicz *et al.* (1976a, 1977c), who found that the susceptibilities of two chains behave differently: The susceptibility on the acceptor (TCNQ) chain begins to disappear as the temperature is lowered below 54 K; that of the donor (TTF) chain does not start dropping till about 45 K and then falls more precipitously below 38 K. This was interpreted as evidence that the higher transition involved a distortion on the acceptor chain and that the entire sequence of transitions was primarily driven by this instability.

A similar g-value decomposition was reported by Tomkiewicz *et al.* (1976b) for TMTTF–TCNQ, but since there is some controversy regarding the existence of several phases of this compound (Ehrenfreund *et al.*, 1977; Berthier *et al.*, 1976), the results will not be discussed here. For HMTTF–TCNQ (Tomkiewicz *et al.*, 1977a,b), g-value decomposition reveals that the acceptor-chain susceptibility has virtually disappeared on cooling to the Peierls transition temperature of 50 K. Below the transition, the donor-chain susceptibility, which is temperature independent in the metallic state, becomes thermally activated. These results are interpreted by the authors to imply a broad thermal range of Peierls fluctuations on the TCNQ stack, which finally achieve sufficient coherence to induce a distortion on the HMTTF stacks at the three-dimensional ordering temperature. However, it should be noted that some of these conclusions have been the subject of controversy (Conwell, 1980; Tomkiewicz *et al.*, 1980), since the small negative thermopower at 50 K implies comparable numbers of electrons and holes, in contrast to what would be expected for a well-developed pseudogap on the acceptor chain.

Knight-shift studies on TTF–TCNQ enriched with ^{13}C in the TCNQ nitrile groups (Rybaczewski *et al.*, 1976) demonstrate a susceptibility partition similar to the g-value results; The shift, which is proportional to the total-spin susceptibility in the metallic phase, drops much faster below 54 K, indicating that the susceptibility is removed primarily from the acceptor chain. Only below 49 K does the TTF susceptibility begin to fall, indicating the opening of a gap on the donor chains.

12. Critical Behavior

Relatively little attention has been devoted to the study of the Peierls transition as a critical phenomenon. Horn *et al.* (1977) analyze the exponents associated with the critical behavior of the derivative $d\chi/dT$. They argue, since the susceptibility reflects fluctuations in the mean-square order parameter, $\langle|\Delta|^2\rangle$, which in turn mirrors the energy associated with ordering, that the exponent is identical to the specific heat exponent, α.

$$\frac{d\chi}{dT} = \frac{|T - T_c|^{\alpha}}{T_c}; \quad T < T_c \tag{28a}$$

$$\frac{d\chi}{dT} = \frac{|T - T_c|^{\alpha'}}{T_c}; \quad T > T_c. \tag{28b}$$

Fitting the experimental data over a range of approximately $10^{-3} < |T - T_c|/T_c < 10^{-2}$ yields $\alpha = \alpha' = 0.5$. This is consistent with the classical Ornstein–Zernike (mean-field) approximation, $\alpha = 2 - d/2$, for dimensionality $d = 3$, implying significant interchain coupling. The absence of any observable deviations from mean-field behavior indicates a very narrow temperature range, given by an appropriate Ginsberg–Landau criterion in which the fluctuations become nonclassical.

Since the Peierls transition is nonmagnetic in nature, order parameter fluctuations have no direct coupling via the hyperfine interaction to the nuclear moments. Thus, in contrast to the spin-density-wave case, to be discussed below, there is no evidence of critical behavior in the nuclear relaxation rates (Rybaczewski *et al.*, 1975, 1976). Removal of spin susceptibility merely causes a drop in the rate to a value dictated by other relaxation mechanisms, such as molecular dynamics and coupling to paramagnetic defects.

The low-temperature regime was examined in somewhat more detail by Schaffhauser *et al.* (1981a,b) for the material $TTT_2\text{-}I_{3+\delta}$. It was found that the proton spin–lattice relaxation rate passes through a minimum near the metal–insulator transition and develops a pronounced anisotropy. They attribute this change in behavior to a reduction in the electronic diffusion rate such that the dipolar term [Eqs. (16, 17)] becomes dominant. The authors were aware of the large paramagnetic spin susceptibility (see Section 14) but concluded that paramagnetic defects were not responsible for the low-temperature relaxation. However, it was implicit in their analysis that such defects would be localized on a molecular site, and it is possible that the more extended states that result from disorder in one dimension could account for both the relaxation rate and the susceptibility.

13. Energy Gaps in the Charge-Density Wave State

As discussed in Section 11, the nonmagnetic nature of the CDW state implies that there is nothing dramatic in the magnetic properties. However, a certain amount of information concerning the parameters of the semiconducting state may be obtained from analysis of magnetic data. The magnetic excitations of such a system consist of electron-hole pairs, requiring a creation energy, in the absence of correlation, equal to the semiconducting gap. Thus any property that depends on the density of magnetic species, such as the susceptibility, the Knight shift, and nuclear spin-lattice relaxation rates, is exponentially activated.

By fitting the low-temperature static susceptibility data of TTF-TCNQ to the form $\chi T \sim \exp(-\Delta/T)$, Scott et al. (1974) obtain a gap, $2\Delta = 170$ K. Herman et al. (1976) analyzed their data in two ways. First, assuming two different gaps, one of which (on the TCNQ chain) opens at 54 K, while the other (on the TTF chain) opens at 38 K, they obtain $2\Delta_{TCNQ} = 510$ K and $2\Delta_{TTF} = 280$ K. Then fitting to a single gap over the temperature range up to 55 K, but ignoring the region near the 38 K anomaly, yields a value of $2\Delta = 300$ K. Bloch (1977) analyzed the susceptibility data according to $\chi T^{1/2} \sim \exp(-\Delta/T)$ and obtained $2\Delta = 208$ K. The origin of the different pre-exponential factor in the expression for χ is due a better accounting of the density of states of a one-dimensional semiconductor, and this difference in analysis explains the different values of Δ quoted by Scott and by Bloch. However, the discrepency with the result of Herman remains unexplained.

Activation gaps have also been obtained from the susceptibility data of TSF-TCNQ and its alloys. For TSF-TCNQ, Buravov et al. (1977) obtain $2\Delta = 190$ K, while Scott et al. (1978) quote 260 K. Similar values are given for the alloys $TTF_{1-x}TSF_xTCNQ$ (Scott et al., 1978): 2Δ ($x = 0.05$) $= 300$ K; 2Δ ($x = 0.25$) $= 260$ K.

Although the susceptibility of many other Peierls systems has been reported, and all show qualitatively the same temperature dependence, analysis in terms of Arrhenius behavior has not, in general, been given. This may be attributed, at least in part, to the difficulty of extracting the intrinsic low-temperature behavior from the dominant Curie-like susceptibility, due to impurities and other defects. We shall return to this in the following section.

There have been several suggestions in the literature (Tomkiewicz et al., 1977c; Scott et al., 1978) that the magnetic activation energy of TTF-TCNQ is not the same as the conductivity activation energy. Such a difference has been denied by Bloch (1977), and, in view of the difficulty of obtaining accurate magnetic activation energies and the discrepancies between different groups, the data are best described as ambiguous. However, if the difference should turn out to be real, there are at least two explanations waiting in the wings. The first involves the role of Coulomb correlations in separating the spin and charge degrees of freedom (Torrance et al., 1977; Torrance, 1978), thereby leading to a lower activation energy for the susceptibility. The second explanation invokes the possibility of phase solitons as excitations from the CDW ground state, carrying spin but no charge (Krumhansl et al., 1980). It is not appropriate in an article of this type to elaborate further on such ideas, with their tentative basis in experimental data.

14. Impurities and Defects

In virtually every highly conducting organic charge-transfer salt, especially those grown by crystallization from solution, as opposed to electrochemically, a rather pronounced "Curie-tail" dominates the low-temperature magnetism. The magnitude of this contribution to the susceptibility corresponds typically to a spin concentration of order 0.1%, which is a rather high value considering the painstaking chemical purification that is routinely done on the neutral donor and acceptor starting materials. It is well known (Gogolin et al., 1975) that the slightest disorder in one-dimensional systems leads to localization of states, and it has been suggested (Cooper et al., 1981) that the Curie signal arises from such states in the narrow gap of the Peierls semiconductor. Thus the relative high concentration of Curie spins is not *per se* a measure of impurities, but rather arises from strains and defects in the molecular lattice. Such a view is supported by the observation that the low-temperature g tensors (Tomkiewicz et al., 1976a; Walsh et al., 1980a; Wudl et al., 1977; Pedersen et al., 1981) are very similar to those of the constituent molecules. Thus it is clear that the "impurities" are intimately associated with the molecular stack themselves and are not due to foreign species embedded in the lattice.

There has been a considerable amount of work on materials into which defects have been deliberately introduced by irradiation or by alloying. There are many parallels with organic materials having inherent disorder as a consequence of asymmetry in the donor, such as the TCNQ salts of quinoline, N-methylphenazine, and N-methylacridinium–TCNQ (Miljak et al., 1980).

A frequent characteristic of disorder in one-dimensional magnetic systems is a low-temperature susceptibility contribution that varies as $T^{-\alpha}$, where α is approximately 0.8. Such a temperature dependence arises in many analyses of models where spins interact with a random exchange potential and was originally discussed by Bulaevskii et al. (1972). A detailed analysis of the one-dimensional disordered Hubbard model was presented by Theodorou and Cohen (1976) and by Theodorou (1977a, 1977b, 1977c), who derived the relationship between the form of the exchange distribution function and the exponent, α. More recently, renormalization group techniques (Hirsh and

Jose, 1980; Ma et al., 1979; Dasgupta and Ma, 1980) and computer simulation methods (Bondeson and Soos, 1980) have been applied to the problem. All give essentially the same predictions for the behavior of the susceptibility.

In $TTT_2-I_{3+\delta}$ disorder occurs because of the lack of iodine stoichiometry. Below about 20 K the susceptibility follows the $T^{-\alpha}$ form with $\alpha = 0.76$ (Miljak et al., 1977), suggesting that the effect of disorder is to localize the electronic states in such a way that there is a random exchange coupling between them. It was shown by Kaminskii et al. (1977) that the temperature of the metal-insulator transition in $TTT_2-I_{3+\delta}$, as measured by susceptibility as well as conductivity, depends on the degree of nonstoichiometry, increasing as the iodine content is decreased.

The susceptibility of HMTSF-TNAP (Miljak et al., 1980; Cooper et al., 1981) also shows a $T^{-\alpha}$ dependence at low temperature. This is attributed to the disorder induced by random orientation of the asymmetric TNAP acceptor molecule.

In stoichiometric organic conductors with symmetric molecules, disorder can be induced progressively by irradiation. This was done in TTF-TCNQ using a beam of deuterons (Chiang et al., 1977). The magnetic susceptibility was used to determine the concentration of induced defects, under the assumption that the number of spins was equal (or at least proportional) to the number of defects. Such an assumption has been validated by a more detailed analysis for TMTSF-DMTCNQ by Zuppiroli et al. (1982).

In a study of irradiated TTF-TCNQ, Miljak et al. (1980) found that the low-temperature susceptibility follows a Curie law (i.e., $\alpha = 1$) before irradiation but a $T^{0.74}$ dependence after the introduction of about 1% paramagnetic defects by neutron bombardment.

Gunning et al. (1979) applied the g-decomposition method to neutron-irradiated TTF-TCNQ and reported a smearing of the transition associated with the TCNQ chains, while the TTF chain appears never to develop a complete gap. These results, in conjunction with conductivity measurements on the same samples, were used as evidence that series of transitions at 54, 47, and 38 K are due to the three-dimensional ordering of fluctuating one-dimensional charge-density waves having a much higher mean-field temperature. Briefly, the argument is that defects limit the growth of the one-dimensional coherence length and therefore suppress the interchain coupling which leads to three-dimensional ordering.

15. ANION ORDERING

In the one-chain conductors, $(TMTSF)_2X$ and $(TMTTF)_2X$, where X is a noncentrosymmetric anion such as BF_4^-, ReO_4^-, NO_3^-, or ClO_4^-, the anions sit at inversion centers of the high-temperature lattice structure and are therefore orientationally disordered. At low-temperature order may appear such that the unit cell is doubled along one or more directions. If the superlattice wavevector has a component along the chain direction, then a gap will be opened at the Fermi surface and the low-temperature state is a semiconductor. This is not necessarily a Peierls transition, since the distortion may not be driven by electron–phonon interactions working to remove the electronic Fermi-energy degeneracy. Rather, the anions interact among themselves, by a mechanism that is not yet clear, to lower their lattice energy, and the charge density oscillation of the electronic system is merely a slave of the new periodic potential.

The resulting band structure has an energy gap large compared to the transition temperature. Thus the magnetic susceptibility falls much more quickly below the transition than for the Peierls systems. This behavior is seen in $(TMTSF)_2BF_4$ (Pedersen et al., 1981) and $(TMTSF)_2ReO_4$ (Pedersen et al., 1982; Jacobsen et al., 1982). In contrast to the spin-density wave transition (see Part IV), the ESR linewidth decreases below the anion ordering temperature, and there is no shift in the resonance field. Thus, in most respects the magnetic properties of anion-ordered systems resemble those of the Peierls semiconductors.

IV. The Spin-Density Wave State

16. Spin-Density Wave Transition

In contrast to the Peierls transition, which has no magnetic character and therefore only indirect effects on magnetic properties, the appearance of a spin-density wave (SDW) has profound consequences for magnetism. Indeed, it is by examining the magnetic properties of the $(TMTSF)_2X$ family of materials that one is able to deduce that the metal-semiconductor transition is driven by the appearance of a SDW. It had long been expected, on theoretical grounds (for a review of the relevant theory, see Solyom, 1979; also Overhauser, 1960, 1962; des Cloiseaux, 1959), that the one-dimensional electron gas with repulsive interactions should exhibit a $2k_F$ magnetic distortion. Briefly, the instability of the one-dimensional Fermi surface is manifested by the development of a spin-density, rather than charge-density, modulation. The exchange potential associated with Coulomb interactions opens a gap at the Fermi energy, and the system becomes semiconducting. It is only relatively recently that such a ground state has been discovered in several members of the two families of quasi-one-dimensional organic conductors: $(TMTSF)_2X$ and $(TMTTF)_2X$.

6. MAGNETIC PROPERTIES

The ground state of a one-dimensional metal with a SDW is almost indistinguishable from an antiferromagnetic semiconductor, although the details of the transition itself are somewhat different. The properties have been described in many places, starting with the pioneering work of Van Vleck (1941). The static susceptibility is anisotropic at low fields, tending to zero in one direction (the "easy" or parallel direction, along which the sublattice magnetization is aligned), but not in the other two. When a field applied parallel to the easy direction exceeds a critical value, the magnetization rotates into the direction of intermediate anisotropy energy; this is the "spin–flop" transition. The dynamic response of the system is no longer at the free-electron ESR frequency but is shifted by exchange and anisotropy energies into two, in general, antiferromagnetic resonances, which are the collective modes of the system. The presence of a net local magnetization produces a hyperfine field that is different in its magnitude and fluctuation spectrum from both the metallic and CDW states. Hence, there are characteristic signatures in all magnetic properties at the onset of the SDW.

A SDW was discovered first in $(TMTSF)_2PF_6$. The earliest susceptibility data (Bechgaard et al., 1980) showed no evidence of metal-insulator transition, in contrast to the abrupt disappearance of the ESR intensity (Pedersen et al., 1980). More accurate susceptibility measurements were made by Scott et al. (1980), who showed that there was a weak anomaly at the phase transition but that, indeed, the integrated ESR intensity did not agree with the static susceptibility. Additional clues that the metal-semiconductor transition is different from the Peierls transitions described in Part II came from study of the ESR spectrum of $TMTSF_2PF_6$ (Pedersen et al., 1980). The line broadens dramatically on cooling through the transition and shifts in field from the $g = 2$ position. Both features reflect the development of internal magnetic fields associated with the appearance of a magnetic moment on the donor molecules. It was also found (Scott et al., 1980) that the static susceptibility in the low-temperature state is field dependent in a manner suggestive of spin–flop behavior. Subsequently, Mortensen et al. (1981) measured the field-dependent susceptibility of several crystals of $(TMTSF)_2PF_6$ and demonstrated the anisotropy expected of an antiferromagnet.

The existence of a SDW ground state in $(TMTSF)_2PF_6$, and also in $(TMTSF)_2AsF_6$, was postulated independently by Walsh et al. (1980b) on the basis of the nonlinear response to dc and microwave electric fields. It was found that sufficiently high fields and/or frequencies restored the conductivity to its metallic value and "resurrected" the ESR signal. Depinning of a CDW could account for the nonlinear electrical response, but not for the reappearance of the susceptibility. By contrast, a sliding SDW would be expected to average out the anisotropy fields, resulting in a resonance at the free-electron g value.

Further evidence for the magnetic nature of the ground state of $(TMTSF)_2PF_6$ is found in NMR experiments (Andrieux et al., 1981; Scott et al., 1981). In the metallic state the proton linewidth is given predominantly by nuclear dipole–dipole interactions. At the metal-insulator transition, an additional local field arises due to the electronic magnetization. Since the conduction electron wavefunction is located in the center of the molecule, the hyperfine field is dipolar in nature, and since the easiest magnetic axis is oriented at an angle to the molecular axes, the field is different for all 24 protons in the unit cell. Hence, the SDW produces an additional inhomogeneous broadening mechanism for the protons. The linewidth is found to increase by several gauss, the exact amount depending on the strength of the static field and upon orientation.

The SDW state has also been found in $(TMTSF)_2SbF_6$ (Pedersen et al., 1981) and $(TMTSF)_2NbF_6$ (Pedersen et al., 1982) on the basis of ESR intensity and linewidth, which show analogous behavior to the other hexafluorides.

More recently, the existence of a SDW ground state has been discovered in two salts of TMTTF, the sulphur analog of TMTSF. In $(TMTTF)_2Br$, this was established by the observation of antiferromagnetic resonance (Parkin et al., 1982) and by NMR studies (Creuzet et al., 1982), which reveal inhomogeneous broadening of the proton resonance, as well as an additional contribution to the relaxation rate. In $(TMTTF)_2SCN$ the SDW state was discovered by the observation of anisotropy and spin–flop in the susceptibility (Coulon et al., 1982b).

The perchlorate of TMTSF deserves special mention. Early reports of its low-temperature magnetic properties contained many reports of irreproducible behavior. The details of the ESR spectrum (Scott, 1982a) were found to depend upon the manner in which the sample was handled, and the observation of antiferromagnetic resonance (Walsh et al., 1982), seemed to depend critically on sample history. The apparent irreproducibility of the magnetic properties is now understood in terms of an anion order-disorder transition at 25 K and the possibility of retaining the disordered high-temperature state by rapid quenching of the sample. This behavior was demonstrated in NMR experiments by Takahashi et al. (1982), who found that with fast cooling the ^{77}Se resonance broadened and disappeared at about 4 K, as expected for a SDW transition, but that with slow cooling the SDW behavior was suppressed. Subsequent ESR studies by Tomic et al. (1982) and by Kajimura et al. (1982) have confirmed this behavior and clarified the conditions necessary to achieve, respectively, the quenched or annealed state. In the fully annealed state anion ordering opens a gap at the Fermi energy and the SDW transition is completely suppressed.

17. Susceptibility

The static susceptibility of $(TMTSF)_2PF_6$ was first reported by Bechgaard et al. (1980), who found no anomaly associated with the metal-insulator transition. Subsequent measurements by Scott et al. (1980) and by Wudl et al. (1981) on samples with fewer paramagnetic defects show clearly that the low-field susceptibility falls sharply below the transition temperature of 12 K. In addition, it was also found that for fields greater than about 4 to 6 kOe, the low temperature susceptibility rose again to a value higher than that just above the transition, reminiscent of spin–flop behavior. More detailed measurements on a partially aligned sample of $(TMTSF)_2PF_6$ (Mortensen et al., 1981) and on single crystals of $(TMTSF)_2AsF_6$ (Mortensen, 1982; Mortensen et al., 1982) demonstrated the anisotropy of the susceptibility and confirmed the spin–flop transition for fields parallel to the easy (b') axis of the hexafluoroarsenate. The spin–flop field is 4.5 kOe, independent of temperature.

Among the other salts of TMTSF, the static susceptibility has been measured only for the perchlorate. As discussed above, it is now known that in this material the low-temperature behavior is complicated by the existence of an anion-ordering transition at 25 K (Pouget et al., 1983) and that the low-temperature properties depend on whether the sample was cooled slowly through this transition, to give the "relaxed" state, which exhibits no SDW, or quickly, yielding the "quenched" state, which does have a SDW transition near 5 K (Takahashi et al., 1982; Tomic et al., 1982). Thus the susceptibility measurements that have been made on this material must be viewed with caution. The drop in the conduction electron susceptibility (Scott, 1982a) and in the ESR intensity (Bechgaard et al., 1981) indicates that both samples were in the quenched state, the high-temperature anion disorder having been frozen in.

The existence of a spin-density wave has also been confirmed in two salts of the TMTTF family, the bromide and the thiocyanate. The earliest static susceptibility measurements on $(TMTTF)_2Br$ by Delhaes et al. (1979a) revealed a weak temperature dependence in the metallic state and a sharp upturn below about 20 K. Single-crystal measurements by Parkin et al. (1983) show an anisotropy developing below the SDW transition temperature (13 K) and a spin–flop transition at 4.2 kOe when the field is applied along the easy (b') axis. The susceptibility anistropy of $(TMTTF)_2SCN$ has been measured directly, using a torque technique, by Coulon et al. (1982b). The results show an antiferromagnetic ordering temperature of 7 K, with the easy axis again close to b' and a spin–flop field of approximately 9 kOe.

18. Nuclear Magnetic Resonance

Proton NMR provides a means for probing the internal local fields associated with the SDW state. As the SDW develops, each molecule acquires a magnetic moment that couples *via* a dipolar interaction with the proton nuclear moments on both the same molecule and the neighboring molecules. Because of the low-symmetry triclinic structure of the TMTSF salts, all 24 protons in the unit cell are inequivalent and experience different magnitudes and directions of the internal dipole field. Therefore, in an NMR experiment the resonance is inhomogeneously broadened by an amount that depends on the amplitude of the SDW.

This effect was exploited by Andrieux *et al.* (1981) and by Scott *et al.* (1981) to estimate the amplitude of the SDW in $(TMTSF)_2PF_6$ at a value of order 0.01–0.1 Bohr Magnetons. A more careful analysis of the dipole sums involved by Morawitz and Scott (1982) and by Scott (1982b) place the value closer to the upper end of this range. However, an accurate determination is difficult to obtain because of partial sample orientation, the uncertainty of the spin-density distribution throughout the molecule, and an apparent dependence of the magnitude of the broadening on the applied resonance field. Unfortunately, in the absence of direct neutron diffraction measurements, these NMR results give the only estimate of the SDW amplitude.

Similar broadening of the proton line is seen in $(TMTTF)_2Br$ (Creuzet *et al.*, 1982) and is attributed to an inhomogeneous local field of order 4 Oe. The authors do not convert this number into a SDW amplitude, but, given the similar crystal structures of $(TMTSF)_2PF_6$ and $(TMTTF)_2Br$ and the slightly smaller unit cell of the latter, it is clear that the value is again of order a few percent of a Bohr Magneton.

The methyl protons are, in a sense, in an ideal position to probe the microscopic aspects of the SDW state, since they are strongly enough coupled to it to show measureable effects and yet not so strongly coupled as to make the resonance unobservable. By contrast, the hyperfine field at the ^{77}Se nuclei of the TMTSF salts is considerably larger such that the NMR linewidth below the transition is so broad that a signal cannot be detected. Although no quantitative information can be obtained from Se NMR, the disappearance of the line has been presented as qualitative evidence for the SDW state in $(TMTSF)_2PF_6$ (Andrieux *et al.*, 1981) and was used to demonstrate the sensitivity to cooling rate of the SDW transition in $(TMTSF)_2ClO_4$ (Takahashi *et al.*, 1982).

19. Antiferromagnetic Resonance

Antiferromagnetic resonance (AFMR) is observed using experimental techniques very similar to electron-spin resonance, but instead of the microwave field exciting single-spin flips, the absorption is due to the collective modes (zero-wavenumber spin waves) of the ordered spin system. The theory of AFMR was developed many years ago (Yosida, 1951, 1952a,b;

Nagamiya, 1951, 1954; Keffer and Kittel, 1952; Ubbink, 1953a,b; Ubbink et al., 1953; Nagamiya et al., 1955; Gerritsen, 1955; Gerritsen et al., 1955) and has been reviewed by Foner (1963). Therefore, only a brief description is given here.

In the absence of a magnetic field, the AFMR mode frequencies depend on the magnetic anisotropy, which arises from single-ion crystal-field effect, from anisotropic exchange interactions, and from dipolar interactions between the localized moments. In the general low-symmetry case for a simple two-sublattice antiferromagnet, there are two modes at zero field having frequencies Ω_- and Ω_+. When a magnetic field is applied, the frequencies of these modes change in a manner that depends sensitively on the orientation of the field with respect to the magnetic axes of the sample. Since the experiment is done in a microwave cavity at fixed frequency, the field must be swept such that the mode frequencies pass that of the cavity. The dependence of the resonance frequencies for fields along the three principal directions and in the three corresponding planes is illustrated in Fig. 3. The observed AFMR fields and rotation patterns can be seen to depend critically on whether the microwave frequency is less than both zero-field mode frequencies, between them or above them.

The first observation of AFMR in organic SDW systems was made by Torrance et al. (1982) on $(TMTSF)_2AsF_6$. Their measurement frequency of 35 GHz lay just above the upper zero-field mode, which could therefore be observed by applying a magnetic field in the hard direction. For fields along the intermediate direction, the lower mode is pulled up in frequency and is observed at 11.5 kOe, just less than the ESR field for 35 GHz, and moves rapidly to lower field with rotation towards the easy direction. The easy axis spin-flop mode, which lies just above the ESR field, was not observed. The data are quantitatively described by two parameters, namely, the zero-field zero-temperature frequencies, $\Omega_+ = 35.3$ GHz, and $\Omega_- = 12.3$ GHz. The magnetic axes are b', easy, a, intermediate, and c^*, hard.

In their AFMR study of $(TMTSF)_2ClO_4$, Walsh et al. (1982) used microwave frequencies of 12 and 17 GHz, both of which lie between the two zero-field modes. Thus only the lower mode was seen for fields approaching the easy-intermediate plane. As the field is rotated towards the hard axis, the resonance field diverges as $\cos^{-1}$. Again, this is the behavior predicted by standard AFMR theory, and fitting the data yields $\Omega_- = 8.7$ GHz. (The upper-mode frequency cannot be determined from these results.) The magnetic axes are not aligned with any principle crystallographic directions but lie as follows: easy—approximately 25° from b'; intermediate—approximately 25° from a; hard—near c^*.

Antiferromagnetic resonance was observed in $(TMTTF)_2Br$ by Parkin et al. (1982). Their microwave frequency of 9.3 GHz is below the lower

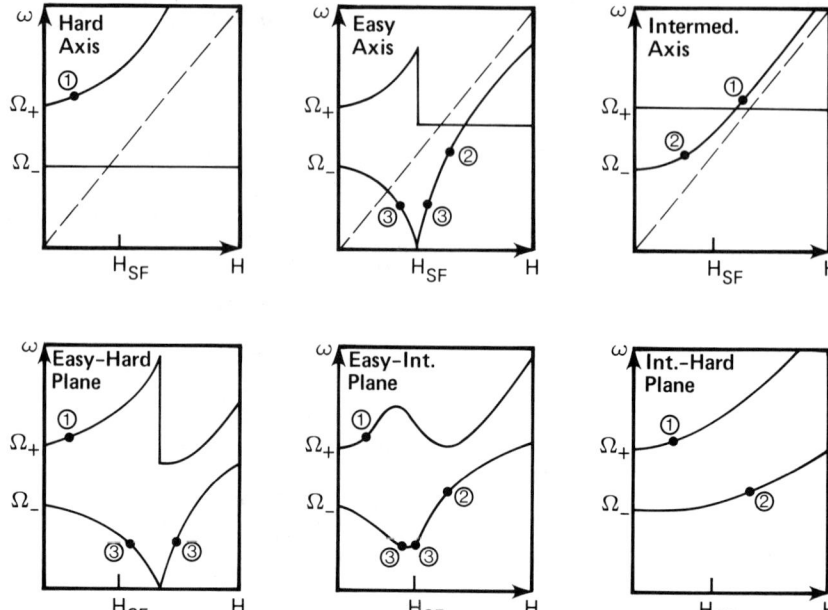

FIG. 3. Antiferromagnetic resonance frequencies as a function of magnetic field in the principal directions and for arbitrary orientations in the principal planes. The numbered points mark the observed AFMR modes in organic SDW systems: 1—TMTSF$_2$AsF$_6$ (Torrance et al., 1982); 2—TMTSF$_2$ClO$_4$ (Walsh et al., 1982); 3—TMTTF$_2$Br (Parkin et al., 1982).

zero-field mode in this material, which permits the observation of resonances both below and above the spin–flop field for orientations near the easy (b') axis. As the field is rotated in towards the intermediate axis (c^*), the two resonances approach each other until, beyond some angle, neither is observed. Near the hard (a) axis the resonant fields diverge as an inverse cosine. The detailed fitting of this orientation dependence allows the authors to obtain both zero-field modes (at 4.2 K): $\Omega_+ = 18.2$ GHz and $\Omega_- = 11.7$ GHz.

The standard theories of antiferromagnetic resonance give the relationship between the experimental parameters Ω_+ and Ω_-, and the parameters of the phenomenological Hamiltonian

$$H = J \sum_i \mathbf{S}_i \cdot \mathbf{S}_{i+1} - D \sum_i S_{iz}^2 + E \sum_i (S_{ix}^2 - S_{iy}^2). \tag{29}$$

Here each of the parameters, J (the exchange interaction) and $D > E$ (the anisotropies), are positive such that the ground state is antiferromagnetic with easy, intermediate, and hard axes along z, y, and x, respectively. One obtains

$$\hbar\Omega_\pm = [2J(D \pm E)]^{1/2}. \tag{30}$$

The spin–flop field is given by $\mu_B H_{SF} = \hbar\Omega_-$. The exchange interaction (J) can be obtained from the perpendicular static susceptibility

$$\chi_\perp = \frac{Ng^2\mu_B^2}{2J} \qquad (31)$$

and is of order 300–600 K in all the materials discussed in this part. When the amplitude of the SDW is μ, less than a full Bohr Magneton, it can be shown (Scott, 1982a) that the zero-field modes and the spin–flop field are both reduced by the ratio μ/μ_B. The susceptibility, however, and hence the derived exchange interaction, is independent of the SDW amplitude in this phenomenological description. Using the numbers given above, one obtains values for the anisotropy constants, D and E, of order 0.1 K. This is of the correct order of magnitude for a predominantly dipolar origin of the anisotropy, but a simple description of dipole coupling alone cannot account for either the signs (i.e., the fact that b' is the easy axis) of the anisotropy constants or the differences among the three materials. Thus the details of the low-temperature Hamiltonian remain a problem.

V. Summary by Material

In the final part of the chapter, the magnetic properties of each material are summarized briefly. In the spirit of an easily read synopsis of experimental results, no references are given here. The reader is directed to examine the relevant section above.

20. TTF-TCNQ

As the first of the highly conducting one-dimensional organic metals, TTF–TCNQ retains a special place in the literature of the field. It is the prototypical Peierls system, and yet, because of its two-chain structure, it exhibits an unexpected richness of behavior as it undergoes several phase transitions in its approach to the semiconducting ground state. The magnetic properties of TTF–TCNQ have provided valuable information concerning the nature of the metallic and insulating states and the steps in between.

In the metallic phase, between 54 K and 340 K, the conduction-electron susceptibility of TTF–TCNQ increases by 150%. Such departure from the temperature independence of simple Pauli paramagnetism is a common feature of all organic metals studied to date and remains the subject of debate. At least part of the temperature dependence can be attributed to the effects of thermal contraction, which is relatively large in organic metals, on a system with a narrow tight-binding band structure. However, comparison with the pressure dependence shows that contraction cannot account for the entire change. Since diffuse x-ray diffraction reveals $2k_F$ fluctuations extending to 150 K, a pseudogap in the density of states is likely to contribute also. Furthermore, it is clear from comparison with the density of states derived from, for example, the plasma frequency and thermoelectric power, that the susceptibility is enhanced by a factor of two or three. In the one-dimensional interacting electron gas, calculations show that a temperature-dependent enhancement results from interchain coupling. Finally, there is the recent proposal that the susceptibility enhancement and its temperature dependence are due to dynamical band narrowing associated with the short, mean free path of the conduction electrons. Whatever the origin of the temperature-dependent susceptibility in the metallic state, it must affect both the donor and acceptor chains to the same extent, because both ESR g-value decomposition and NMR Knight-shift analysis shows that the ratio of the two separate susceptibilities remains constant.

The conduction electron susceptibility drops as TTF–TCNQ enters the semiconducting regime. There is a shoulder at the 54 K transition and a sharp cusp signaling the first-order transition at 38 K. The transition at 47 K does not appear in the susceptibility data. The g tensor and Knight shift both show that the upper transition affects primarily the susceptibility of the acceptor (TCNQ) chain, while the lower one results in a gap in the donor density-of-states. In the low temperature regime, the susceptibility is exponentially activated as expected for a semiconductor. There is some indication that the magnetic gap is less than the single-particle gap measured by conductivity, but there is not yet a consensus on this point. Such a separation of the spin and charge excitations can arise as a result of Coulomb interactions, and it has also been proposed that the solitons are responsible for the low-temperature magnetic response.

TTF–TCNQ remains the only organic metals for which extensive frequency-, temperature-, and pressure-dependent NMR relaxation measurements have been performed. The results confirm the dimensionality determined from conductivity anisotropy, namely, that the interchain hopping rate is of order 10^{11} sec^{-1} at ambient conditions and increases as the temperature is lowered or the pressure is raised. The Korringa enhancement of the relaxation rate implies a Coulomb interaction that is comparable to, but a little smaller than, the bandwidth.

21. TSF–TCNQ

In many ways TSF–TCNQ is quite similar to its structural analog, TTF–TCNQ, but there are also some differences. The static susceptibility has a similar temperature dependence, but the overall magnitude is lower by a factor of two, due primarily to the larger bandwidth of the selenium-based system. The room temperature g value implies that the susceptibility on the acceptor chain is 3.5 times that on the donor, but the decomposition has not been carried out at lower temperature due to the difficulties associated with the larger ESR linewidth.

The metal-insulator transition in TSF-TCNQ occurs at the single temperature of 29 K, below which the susceptibility is temperature activated with a gap equal to that of the conductivity. The ESR linewidth implies a larger interchain interaction than in the sulphur analog, accounting for the fact that both donor and acceptor chains order at a single transition temperature.

22. Alloys of TTF-TCNQ and TSF-TCNQ

The isostructure of the TTF-TCNQ and TSF-TCNQ lattices permits a continuous range of solid solutions of the two complexes. The static susceptibilities and the ESR spectra of several such alloys have been investigated. The most striking result is that with increasing concentration of TSF in the donor stack, the g value develops a temperature dependence, shifting towards the TCNQ value as the temperature is lowered. This was interpreted as evidence against Peierls fluctuations, because they would be expected to be reduced on the donor chain by the disorder, and since the acceptor chain would then have the larger pseudogap, the g value should shift in the opposite direction.

The ESR linewidth increases with increasing TSF concentration, as does the g shift. This implies that the primary line-broadening mechanism is spin-phonon scattering mediated by the spin-orbit interaction. The reduced linewidth, $\Gamma/\delta g^2$, also increases with TSF concentration, showing that interchain intractions (i.e., effective dimensionality) increase with the addition of the larger-radius selenium atoms.

23. Other Analogs of TTF-TCNQ

Each molecule of the TTF-TCNQ complex, especially TTF, has been modified by substitution of alkyls, alkoxys or halogens for protons, selenium for sulphur, and by the extension of the aromaticity (e.g., the fusing of benzoid rings on the ends of the TTF molecule). The resulting binary donor-acceptor complexes form an alphabetic menagerie, of which relatively few members possess the high conductivity resulting from uniform segregated chains, with fractional charge transfer. Those that have been most thoroughly studied to date are DMTTF-TCNQ, TMTTF-TCNQ, HMTTF-TCNQ, TMTSF-TCNQ, HMTSF-TCNQ, and TMTSF-DMTCNQ.

The static susceptibility of each one of these materials, in its metallic state, has a temperature dependence similar to that of TTF-TCNQ, though the absolute magnitudes differ by about a factor of three, being larger for the sulphur-containing donors, which tend to have smaller bandwidths than their selenium counterparts. With the exception of HMTSF-TCNQ, all of these organic metals undergo a Peierls transition to a semiconducting ground state. In the cases where the static susceptibility, or equivalently the ESR intensity, has been measured, the transition appears as a shoulder, where temperature dependence crosses over to an exponentially activated form.

HMTSF-TCNQ is unique among the TTF-TCNQ analogs in that its conduction electron susceptibility does not vanish at zero temperature but becomes negative, reflecting a Landau-Peierls orbital contribution. The absence of a Peierls transition and the coherent transverse electron motion implied by orbital diamagnetism both point to larger interchain coupling than in the other one-dimensional metals. The frequency independence of the proton spin-lattice relaxation rate confirms this interpretation.

24. The $(TMTSF)_2X$ Family

Electrochemical oxidation of the organic donor, TMTSF, and the simultaneous growth of high-quality single crystals led to the discovery of several new phenomena in these so-called Bechgaard salts. First, there was the observation of superconductivity under modest pressure in $(TMTSF)_2PF_6$ and the other hexafluoride salts. Then came the realization that the ambient pressure metal-insulator transition in these same materials was driven not by the Peierls distortion common to the TTF-TCNQ analogs, but rather by the onset of a spin-density wave. Finally, the crucial role of the anion symmetry was recognized in the effects of anion ordering.

The spin-density wave ground state has been demonstrated conclusively by susceptibility nonlinearity and anisotropy, by observation of inhomogeneous NMR linewidths, and/or by antiferromagnetic resonance in the PF_6, AsF_6 salts and in the quenched low-temperature state of the ClO_4 salt. The ESR properties of $(TMTSF)_2NbF_6$ strongly suggests that it, too, has a SDW transition. $(TMTSF)_2TaF_6$ has not been examined by magnetic measurement, but it has a metal-insulator transition at a temperature similar to the other hexafluorides, and it is reasonable to assume that it is in the same category. All MF_6 salts have SDW transition temperatures in the range of 11-14 K, and all require a pressure of about 10 kbar to stabilize the metallic state down to the superconducting transition temperature, which is about 1.4 K in all cases. The perchlorate has a SDW transition temperature near 5 K, depending on the quench rate. The maximum superconducting transition temperature is 1.4 K and is suppressed when, in quenched samples, the SDW state occurs first.

The anion lies at a center of inversion in the room-temperature crystal structure of the Bechgaard salts. Therefore, tetrahedral and lower symmetry anions, such as ClO_4, BF_4, ReO_4, FSO_3, and NO_3, must be orientationally disordered. As the temperature is lowered the anions order, and, depending on the new superlattice periodicity, the resulting phase may or may not have a gap at the Fermi surface. The structural aspects are beyond the scope of this chapter, but the magnetic, as well as the transport, properties show clearly that an insulating state occurs at 180 K in $(TMTSF)_2ReO_4$ and at 40 K in the BF_4 salt. The structural transitions in the nitrate (at 40 K) and the perchlorate (25 K) leave the material in a metallic state, and there is no observable effect on the magnetic properties. The susceptibility of the FSO_3 salt has not been reported. Of the five salts with noncentrosymmetric anions, three become superconducting (ClO_4 without pressure, ReO_4 and FSO_3 under pressure) and two (NO_3 and BF_4) apparently do not.

There is a metal-insulator transition in $(TMTSF)_2NO_3$ at 12 K, considerably below the observed anion ordering at 40 K. It is still not clear what is the nature of the ground state, since its magnetic properties are not fully characterized. ESR studies in the neighborhood of the transition temperature show none of the g shifts and line broadening seen in the SDW materials. This suggests, but does not prove, that a SDW does not occur. It is possible that there is a second anion rearrangement, yielding a new periodicity that does open a gap in the density of states.

25. THE $(TMTTF)_2X$ FAMILY

The sulphur analogs were actually first prepared before the TMTSF series, but their detailed study came only after the exciting properties of the latter were discovered. Sample quality again benefits from the application of electrochemical synthesis. Although the TMTTF family has a structure identical to that of the TMTSF salts, there are important differences in the physical properties. In particular, the maxima in the conductivities of each member of the family occur at or above about 100 K, relatively far above the temperatures of the phase transitions detected in the logarithmic derivative or by other methods.

The susceptibilities of the ClO_4 and BF_4 show anomalies that might be associated with phase transitions at 75 and 41 K, respectively. For the perchlorate, this transition agrees with that signaled by a discontinuity in the resistivity and by an anomaly in the specific heat. The high value of the resistance of $(TMTTF)_2BF_4$ in the vicinity of 41 K has limited its sensitivity to the phase transition, but there is a small specific-heat anomaly, which confirms the magnetic observation. These transitions are caused by the ordering of the tetrahedral anions, and there is, at least for ClO_4, direct structural evidence. Less magnetic data are available for the perrhenate, but in this case x-ray measurements confirm the role of the anions in the metal-insulator transition seen at 160 K.

A magnetic anomaly in $(TMTTF)_2PF_6$ occurs at 15 K, a temperature where the resistance behaves like that of a semiconductor. Thus the observation of a $2k_F$ lattice distortion in conjunction with a nonmagnetic ground state has been attributed to a spin–Peierls distortion.

In the case of $(TMTTF)_2Br$ magnetic and transport measurements, both indicate a phase transition below 20 K. However, the ESR intensity vanishes at a slightly lower temperature (13 K) than the peak in the logarithmic derivative of the resistance (19 K). Antiferromagnetic resonance, susceptibility anisotropy, and a spin–flop transition all indicate that the ground state is a spin-density wave.

In $(TMTTF)_2SCN$ a metal-insulator transition occurs at 160 K, but there is no anomalous magnetic behavior in this region. Rather, a magnetic transition is seen at 7 K, associated with the onset of an antiferromagnetic ground state. Below this temperature the susceptibility is anisotropic and has a spin–flop transition at approximately 9 kOe. Since there is already a gap in the electronic spectrum below 160 K, the origin of the magnetic instability must be, at least quantitatively, different than for the other SDW Bechgard phases.

26. Other Materials

Probably several hundred organic charge-transfer salts have been synthesized during the last 20 years, but of these only a couple of dozen meet the "high conductivity" criterion for inclusion in this book. Most fall into the two general categories discussed in the previous 6 sections (20 to 25): derivatives of TTF-TCNQ and the 2:1 Bechgaard salts. There are, however, several other highly conducting organic materials that are not so nicely categorized.

In terms of its magnetic properties, if not strictly its chemistry, TTT-$(TCNQ)_2$ belongs with the TTF-TCNQ analogs. It has the static susceptibility typical of Peierls systems, the transition occurring at about 50 K. HMTSF-TNAP and $TTT_2-I_{3+\delta}$ behave in a similar fashion at high temperature, but show evidence of disorder in the low-temperature susceptibility, which varies as $T^{-\alpha}$. The source of disorder in the former material is presumed to arise from the asymmetry of TNAP molecule, while in the latter it results from the nonstoichiometry.

There is another whole class of material that have not been treated in this review: the salts of TCNQ with asymmetric donors such as Qn and NMP. There are many fascinating features of their magnetism, particularly related to the effects of disorder at low temperature, but there is not yet, in my opinion, sufficient consensus about their overall properties for inclusion in the present review, which has concentrated on the metallic state and on the instabilities that destroy the Fermi surface.

New materials continue to be synthesized, and magnetic measurements will play a key role in their characterization. It has been shown that the combined application of susceptometry, ESR, and NMR is crucial in obtaining a detailed description of organic metals, providing information on the density of states; on the importance of interchain coupling and of Coulomb interactions; on the nature of the ground-state, be it CDW, SDW, or semimetal; and on the critical behavior near the metal-insulator transition.

ACKNOWLEDGMENTS AND APOLOGIA

Collaboration, discussion, and even arguments with many colleagues, spanning the last 10 years, are reflected in the contents of this review article. There are too many names to mention them all, but I do wish to give special recognition to Prof. Alan Heeger with whom I worked at the University of Pennsylvania, to Dr. Hans Pedersen, my colleague at Cornell University, and to my present colleagues at IBM, Drs. Rick Greene, Jerry Torrance, Stuart Parkin, Paul Grant, and Ed Engler.

I have attempted to give as completely as possible a descriptive literature up to the end of 1982, but I suppose it is inevitable that a few articles may have slipped through my search procedure. For this I apologize in advance, both to the authors whose works were omitted and to the reader. I request that any such omissions be brought to my attention.

REFERENCES

Adrian, F. J. (1982). *Phys. Rev. B* **26**, 2682.
Andrieux, A., Jérome, D., and Bechgaard, K. (1981). *J. Phys. (Paris) Lett.* **42**, L-87.
Ashcroft, N. W., and Mermin, N. D. (1976). "Solid State Physics." Holt Rinehart, and Winston, New York.
Avalos, J., Devreux, F., Guglielmi, M., and Nechtschein, M. (1978); *Molec. Phys.* **36**, 669.
Azevedo, L. J., Schirber, J. E., and Scott, J. C. (1982). *Phys. Rev. Lett.* **49**, 826.
Bechgaard, K., Jacobsen, C. S., Mortensen, K., Pedersen, H. J., and Thorup, N. (1980). *Solid State Commun.* **33**, 1119.
Bechgaard, K., Carneiro, K., Rasmussen, F. B., Rindorf, G., Jacobsen, C. S., Pedersen, H. J., and Scott, J. C. (1981). *J. Amer. Chem. Soc.* **103**, 2440.
Bernstein, U., Chaikin, P. M., and Pincus, P. (1975). *Phys. Rev. Lett.* **34**, 271.
Berthier, C., Jérome, D., Soda, G., Weyl, C., Zuppiroli, L., Fabre, J. M., and Giral, L. (1976a). *Mol. Cryst. Liq. Cryst.* **32**, 261.
Berthier, C., Cooper, J. R., Jérome, D., Soda, G., Weyl, C., Fabre, J. M., and Giral, L. (1976b). *Mol. Cryst. Liq. Cryst.* **32**, 267.
Bjelis, A., and Barisic, S. (1975). *J. Phys. (Paris) Lett.* **36**, L-169.
Bloch, A. N. (1977). *In* "Organic Conductors and Semiconductors," Lecture Notes in Physics, Vol. 65 (L. Pal, G. Gruner, A. Janossy, and J. Solyom, eds.), p. 317. Springer-Verlag, Berlin.
Bloch, A. N., Carruthers, T. F., Poehler, T. O., and Cowan, D. O. (1977). *In* "Chemistry and Physics of One-Dimensional Metals," NATO–ASI Series Vol. B25 (H. J. Keller, ed.), p. 47. Plenum, New York.
Bondeson, S. R., and Soos, Z. G. (1980). *Phys. Rev. B* **33**, 1793.
Bonner, J. C., and Fisher, M. E. (1964). *Phys. Rev.* **135**, A640.

Boriack, M. L. (1980). *Phys. Rev. B* **21**, 4478.
Bramwell, F. B., Haddon, R. C., Wudl, F., Kaplan, M. L., and Marshall, J. H. (1978). *J. Amer. Chem. Soc.* **100**, 4612.
Bray, J. W., Hart, H. R., Interrante, L. V., Jacobs, I. S., Kasper, J. S., Watkins, G. D., Wee, S. H., and Bonner, J. C. (1975). *Phys. Rev. Lett.* **35**, 744.
Brun, G., Peytavin, S., Liautard, B., Maurin, M., Toreilles, E., Fabre, J. M., and Giral, L. (1977). *Comptes Rendus Acad. Sc. Paris, Serie C* **284**, 211.
Bulaevski, L. N., Zvarykina, A. V., Karimov., Yu. S., Lyubovskii, R. B., and Shchegolev, I. F. (1972). *Sov. Phys. JETP* **35**, 384.
Buravov, L. I., Eremenko, O. N., Lyubovskii, R. B., Rozenberg, L. P., Khidekel, M. L., Shibaeva, R. P., Shchegolev, I. F., and Yagubskii, E. B. (1974). *JETP Lett.* **20**, 208.
Buravov, L. I., Lyubovskaya, R. N., Lyubovskii, R. B., and Khidekel, M. L. (1977). *Sov. Phys. JETP* **43**, 1033.
Caron, L. G., Miljak, M., and Jérome, D. (1978). *J. Phys. (Paris)* **39**, 1355.
Chiang, C. K., Cohen, M. J., Newman, P. R., and Heeger, A. J. (1977). *Phys. Rev. B* **16**, 5163.
Clark, W. G., Hammann, J., Sanny, J., and Tippie, L. C. (1979). In "Quasi One-Dimensional Conductors II," Lecture Notes in Physics, Vol. 96 (S. Barisic, A. Bjelis, J. R. Cooper, and B. Leontic, eds.), p. 255. Springer-Verlag, Berlin.
des Cloiseaux, J. (1959). *J. Phys. et Rad.* **20**, 607.
Conwell, E. (1980). *Phys. Rev. B* **22**, 3107.
Cooper, J. R., and Korin, B. (1979). In "Quasi One-Dimensional Conductors I," Lecture Notes in Physics, Vol. 95 (S. Barasic, A. Bjelis, J. R. Cooper, and B. Leontic, eds.), p. 181. Springer-Verlag, Berlin.
Cooper, J. R., Miljak, M., and Korin, B. (1981). *Chemica Scripta* **17**, 79.
Coulon, C., Delhaes, P., Flandrois, S., Lagnier, R., Bonjour, E., and Fabre, J. M. (1982a). *J. Phys. (Paris)* **43**, 1059.
Coulon, C., Maaroufi, A., Amiell, J., Dupart, E., Flandrios, S., Delhaes, P., Moret, R., Pouget, J. P., and Morand, J. P. (1982b). *Phys. Rev. B* **26**, 6322.
Creuzet, F., Takahashi, T., Jérome, D., and Fabre, J. M. (1982). *J. Phys. (Paris) Lett.* **43**, L-755.
Dalgard, E., and Linderberg, J. (1975). *Int. J. Quant. Chem. Symp.* **9**, 269.
Dalgard, E., and Linderberg, J. (1976). *J. Chem. Phys.* **65**, 692.
Dasgupta, C., and Ma, S. K. (1980). *Phys. Rev. B* **22**, 1305.
Delhaes, P., Flandrois, S., Amiell, J., Keryer, G., Toreilles, E., Fabre, J. M., Giral, L., Jacobsen, C. G., and Bechgaard, K. (1977). *J. Phys. (Paris) Lett.* **38**, L-233.
Delhaes, P., Coulon, C., Amiell, J., Flandrois, S., Toreilles, E., Fabre, J. M., and Giral, L. (1979a). *Molec. Cryst. Liq. Cryst.* **50**, 43.
Delhaes, P., Manceau, J.-P., Coulon, C., Flandrois, S., Hilti, B., and Mayer, C. W. (1979b). In "Quasi One-Dimensional Conductors II," Lecture Notes in Physics, Vol. 96 (S. Barisic, A. Bjelis, J. R. Cooper, and B. Leontic, eds.), p. 324. Springer-Verlag, Berlin.
Devreux, F., and Nechtschein, M. (1979). In "Quasi One-Dimensional Conductors I," Lecture Notes in Physics, Vol. 95 (S. Barisic, A. Bjelis, J. R. Cooper, and B. Leontic, eds.), p. 145. Springer-Verlag, Berlin.
Devreux, F., Boucher, J.-P., and Nechtschein, M. (1974). *J. Phys. (Paris)* **35**, 271.
Devreux, F., Jeandey, C., Nechtschein, M., Fabre, J. M., and Giral, L. (1979). *J. Phys. (Paris)* **40**, 671.
Ehrenfreund, E., and Heeger, A. J. (1977). *Phys. Rev. B* **16**, 3830.
Ehrenfreund, E., Khanna, S. K., Garito, A. F., and Heeger, A. J. (1977). *Solid State Commun.* **22**, 139.
Elliott, R. J. (1954). *Phys. Rev.* **96**, 280.

6. MAGNETIC PROPERTIES

Entin-Wohlman, O., Gutfreund, H., and Weger, M. (1982). *J. Phys. C* **15**, 5763.
Fischer, P. H., and McDowell, C. A. (1963). *J. Amer. Chem. Soc.* **85**, 2694.
Flandrois, S., Coulon, C., Delhaes, P., Chasseau, D., Hauw, C., Gaultier, J., Fabre, J. M., and Giral, L. (1982). *Mol. Cryst. Liq. Cryst.* **79**, 307.
Foner, S. (1963). *In* "Magnetism" (G. T. Rado and H. Suhl, eds.), Vol. I, p. 383. Academic Press, New York.
Fulde, P., and Luther, A. (1968). *Phys. Rev.* **170**, 570.
Gerritsen, H. J. (1955). *Physica* **21**, 639.
Gerritsen, H. J., Okkes, R., Bolger, B., and Gorter, C. J. (1955). *Physica* **21**, 629.
Gogolin, A. A., Melnikov, V. I., and Rashba, E. I. (1975). *Sov. Phys. JETP* **42**, 168.
Gulley, J. E., and Weiher, J. F. (1975). *Phys. Rev. Lett.* **34**, 1061.
Gunning, W. J., Chiang, C. K., Heeger, A. J., and Epstein, A. J. (1979). *In* "Quasi One-Dimensional Conductors I," Lecture Notes in Physics, Vol. 95 (S. Barisic, A. Bjelis, J. R. Cooper, and B. Leontic, eds.), p. 246. Springer-Verlag, Berlin.
Gutfreund, H., Kaveh, M., and Weger, M. (1979). *In* "Quasi One-Dimensional Conductors I," Lecture Notes in Physics, Vol. 95 (S. Barisic, A. Bjelis, J. R. Cooper, and B. Leontic, eds.), p. 105. Springer-Verlag, Berlin.
Hardebusch, U., Gerhardt, W., Schilling, J. S., Bechgaard, K., Weger, M., Miljak, M., and Cooper, J. R. (1979). *Solid State Commun.* **32**, 1151.
Haustein, H., Dinse, K. P., and Mobius, K. (1971). *Z. Naturforsch.* **26a**, 1230.
Herman, R. M., Salamon, M. B., DePasquali, G., and Stucky, G. (1976). *Solid State Commun.* **19**, 137.
Herman, R. M., Salamon, M. B., DePasquali, G., and Stucky, G. (1977). *In* "Organic Conductors and Semiconductors," Lecture Notes in Physics, Vol. 65 (L. Pal, G. Gruner, A. Janossy, and J. Solyom, eds.), p. 481. Springer-Verlag, Berlin.
Hirsch, J. E., and Jose, J. V. (1980). *Phys. Rev. B* **22**, 5339.
Horn, P. M., Herman, R. M., and Salamon, M. B. (1977). *Phys. Rev. B* **16**, 5012.
Hubbard, J. (1963). *Proc. Roy. Soc. A* **276**, 238.
Hubbard, J. (1964). *Proc. Roy. Soc. A* **281**, 401.
Isett, L. C., and Perez-Albuerne, E. A. (1977). *Solid State Commun.* **21**, 433.
Jacobsen, C. S., Pedersen, H. J., Mortensen, K., Rindorf, G., Thorup, N., Torrance, J. B., and Bechgaard, K. (1982). *J. Phys. C* **15**, 2651.
Jérome, D., and Giral, L. (1977). *In* "Organic Conductors and Semiconductors," Lecture Notes in Physics, Vol. 65 (L. Pal, G. Gruner, A. Janossy, and J. Solyom, eds.), p. 381. Springer-Verlag, Berlin.
Jérome, D., and Caron, L. G. (1979). *In* "Quasi One-Dimensional Conductors I," Lecture Notes in Physics, Vol. 95 (S. Barisic, A. Bjelis, J. R. Cooper, and B. Leontic, eds.), p. 91. Springer-Verlag, Berlin.
Jérome, D., and Weger, M. (1977). *In* "Chemistry and Physics of One-Dimensional Metals." NATO-ASI Series, Vol. B25 (H. J. Keller, ed.), p. 341. Plenum, New York.
Jérome, D., Soda, G., Cooper, J. R., Fabre, J. M., and Giral, L. (1977). *Solid State Commun.* **22**, 319.
Jonkman, H. T., Van Der Velde, G. A., and Nieuwpoort, W. C. (1974). *Chem. Phys. Lett.* **25**, 62.
Kajimura, K., Tokumoto, H., Tokumoto, M., Murata, K., Ukachi, T., Anzai, H., Ishiguro, T., and Saito, G. (1982). *Solid State Commun.* **44**, 1573.
Kaminskii, V. F., Khidekel, M. L., Lyubovskii, R. B., Shchegolev, I. F., Shibaeva, R. P., Yagubskii, E. B., Zvarykina, A. V., and Zvereva, G. L. (1977). *Phys. Stat. Sol. (a)* **44**, 77.
Keffer, F., and Kittel, C. (1952). *Phys. Rev.* **85**, 329.
Kinoshita, M., and Akamatu, H. (1965). *Nature (London)* **207**, 291.

Kistenmacher, T. J. (1978). *Ann. NY Acad. Sci.* **313**, 333.
Konig, E. (1966). *In* "Numerical Data and Functional Relationships in Science and Technology," Group II, Vol. 2. Landolt-Bornstein Series (K. H. Hellwege, ed.). Springer-Verlag, Berlin.
Krumhansl, J. A., Horowitz, B., and Heeger, A. J. (1980). *Solid State Commun.* **34**, 945.
Ladik, J., Karpfen, A., Stollhoff, G., and Fulde, P. (1975). *Chem. Phys.* **7**, 267.
Lee, P. A., Rice, T. M., and Anderson, P. W. (1973). *Phys. Rev. Lett.* **31**, 462.
Lee, P. A., Rice, T. M., and Klemm, R. A. (1977). *Phys. Rev. B* **15**, 2984.
Lowitz, D. A. (1967). *J. Chem. Phys.* **46**, 4698.
Ma, S., Dasgupta, C., and Hu, C. K. (1979). *Phys. Rev. Lett.* **43**, 1434.
Miljak, M., Korin, B., Cooper, J. R., and Gruner, G. (1977). *Commun. on Phys.* **2**, 193.
Miljak, M., Korin, B., Cooper, J. R., Holczer, K., and Janossy, A. (1980). *J. Phys. (Paris)* **41**, 639.
Morawitz, H., and Scott, J. C. (1982). *Mol. Cryst. Liq. Cryst.* **85**, 305.
Moriya, T. (1963). *J. Phys. Soc. Japan* **18**, 516.
Mortensen, K. (1982). *Physica Scripta* **25**, 854.
Mortensen, K., Tomkiewicz, Y., Schultz, T. D., and Engler, E. M. (1981). *Phys. Rev. Lett.* **46**, 1238.
Mortensen, K., Tomkiewicz, Y., and Bechgaard, K. (1982). *Phys. Rev. B* **25**, 3319.
Nagamiya, T. (1951). *Prog. Theor. Phys.* **6**, 342.
Nagamiya, T. (1954). *Prog. Theor. Phys.* **11**, 309.
Nagamiya, T., Yosida, K., and Kubo, R. (1955). *Adv. Phys.* **4**, 1.
Overhauser, A. W. (1960). *Phys. Rev. Lett.* **4**, 607.
Overhauser, A. W. (1962). *Phys. Rev.* **128**, 1437.
Parkin, S. S., Scott, J. C., Torrance, J. B., and Engler, E. M. (1982). *Phys. Rev. B.* **26**, 6319.
Parkin, S. S., Scott, J. C., Torrance, J. B., and Engler, E. M. (1983). *J. Phys. (Paris) Colloq.* **44**, C3-1111.
Pedersen, H. J., Scott, J. C., and Bechgaard, K. (1980). *Solid State Commun.* **35**, 207.
Pedersen, H. J., Scott, J. C., and Bechgaard, K. (1981). *Phys. Rev. B* **24**, 5014.
Pedersen, H. J., Scott, J. C., and Bechgaard, K. (1982). *Physics Scripta* **25**, 849.
Perlstein, J. H., Ferraris, J. P., and Candela, G. A. (1973). *In* "Magnetism and Magnetic Materials—1972," *A.I.P. Conf. Proc.* **10**, part 2, 1494.
Rice, M. J., and Strassler, S. (1973). *Solid State Commun.* **13**, 1389.
Rieger, P. H., and Fraenkel, G. K. (1962). *J. Chem. Phys.* **37**, 2795.
Rybaczewski, E. F., Garito, A. F., Heeger, A. J., and Ehrenfreund, E. (1975). *Phys. Rev. Lett.* **34**, 524.
Rybaczewski, E. F., Smith, L. S., Garito, A. F., Heeger, A. J., and Silbernagel, B. G. (1976). *Phys. Rev. B* **14**, 2746.
Schaffhauser, T., Ernst, R. R., Hilti, B., and Mayer, C. W. (1981a). *Phys. Rev. B* **24**, 76.
Schaffhauser, T., Ernst, R. R., Hilti, B., and Mayer, C. W. (1981b). *Chemica Scripta* **17**, 27.
Schumacher, R. T., and Slichter, C. P. (1956). *Phys. Rev.* **101**, 58.
Scott, J. C. (1982a). *Molec. Cryst. Liq. Cryst.* **79**, 49.
Scott, J. C. (1982b). *J. Appl. Phys.* **53**, 1845.
Scott, J. C., Garito, A. F., and Heeger, A. J. (1974). *Phys. Rev. B* **10**, 3131.
Scott, J. C., Etemad, S., and Engler, E. M. (1978). *Phys. Rev. B* **17**, 2269.
Scott, J. C., Pedersen, H. J., and Bechgaard, K. (1980). *Phys. Rev. Lett.* **45**, 2125.
Scott, J. C., Pedersen, H. J., and Bechgaard, K. (1981). *Phys. Rev. B* **24**, 475.
Slichter, C. P. (1978). "Principles of Magnetic Resonance," 2nd edition. Springer-Verlag, Berlin.
Soda, G., Jérome, D., Weger, M. Fabre, J. M., and Giral, L. (1976a). *Solid State Commun.* **18**, 1417.

Soda, G., Jérome, D., Weger, M., Bechgaard, K., and Pedersen, E. (1976b). *Solid State Commun.* **20**, 107.
Soda, G., Jérome, D., Weger, M., Alizon, J., Gallice, J., Robert, H., Fabre, J. M., and Giral, L. (1977a). *J. Phys. (Paris)* **38**, 931.
Soda, G., Jérome, D., Weger, M., Fabre, J. M., Giral, L., and Bechgaard, K. (1977b). In "Organic Conductors and Semiconductors," Lecture Notes in Physics, Vol. 65 (L. Pal, G. Gruner, A. Janossy, and J. Solyom, eds.), p. 371. Springer-Verlag, Berlin.
Solyom, J. (1979). *Adv. Phys.* **28**, 201.
Somoano, R. B., Gupta, A., Hadek, V., Novotny, M., Jones, M., Datta, T., Deck, R., and Hermann, A. M. (1977). *Phys. Rev. B* **15**, 595.
Takahashi, T., Jérome, D., and Bechgaard, K. (1982). *J. Phys. (Paris) Lett.* **43**, L-565.
Theodorou, G. (1977a). *Phys. Rev. B* **16**, 2254.
Theodorou, G. (1977b). *Phys. Rev. B* **16**, 2264.
Theodorou, G. (1977c). *Phys. Rev. B* **16**, 2273.
Theodorou, G., and Cohen, M. H. (1976). *Phys. Rev. Lett.* **37**, 1014..
Thorup, N., Rindorf, G., Soling, H., and Bechgaard, K. (1981). *Acta Cryst. B* **37**, 1236.
Tomic, S., Jérome, D., Monod, P., and Bechgaard, K. (1982). *J. Phys. (Paris) Lett.* **43**, L-839.
Tomkiewicz, Y. (1979). *Phys. Rev. B* **19**, 4038.
Tomkiewicz, Y., and Taranko, A. R. (1978). *Phys. Rev. B* **18**, 733.
Tomkiewicz, Y., Scott, B. A., Tao, L. J., and Title, R. S. (1974). *Phys. Rev. Lett.* **32**, 1363.
Tomkiewicz, Y., Engler, E. M., and Schultz, T. D. (1975). *Phys. Rev. Lett.* **35**, 456.
Tomkiewicz, Y., Taranko, A. R., and Torrance, J. B. (1976a). *Phys. Rev. Lett.* **36**, 751.
Tomkiewicz, Y., Taranko, A. R., and Green, D. C. (1976b). *Solid State Commun.* **20**, 767.
Tomkiewicz, Y., Taranko, A. R., and Engler, E. M. (1976c). *Phys. Rev. Lett.* **37**, 1705.
Tomkiewicz, Y., Taranko, A. R., and Schumaker, R. (1977a). *Phys. Rev. B* **16**, 1380.
Tomkiewicz, Y., Welber, B., Seiden, P. E., and Schumaker, R. (1977b). *Solid State Commun.* **23**, 471.
Tomkiewicz, Y., Taranko, A. R., and Torrance, J. B. (1977c). *Phys. Rev. B* **15**, 1017.
Tomkiewicz, Y., Craven, R. A., Schultz, T. D., Engler, E. M., and Taranko, A. R. (1977d). *Phys. Rev. B* **15**, 3643.
Tomkiewicz, Y., Anderson, J. R., and Taranko, A. R. (1978). *Phys. Rev. B* **17**, 1579.
Tomkiewicz, Y., Taranko, A. R., and Torrance, J. B. (1980). *Phys. Rev. B* **22**, 3113.
Torrance, J. B. (1978). In "Chemistry and Physics of One-Dimensional Metals," NATO-ASI Series, Vol. B25 (H. J. Keller, ed.), p. 137. Plenum, New York.
Torrance, J. B., Tomkiewicz, Y., and Silverman, B. D. (1977). *Phys. Rev. B* **15**, 4738.
Torrance, J. B., Mayerle, J. J., Bechgaard, K., Silverman, B. D., and Tomkiewicz, Y. (1980). *Phys. Rev. B* **22**, 4960.
Torrance, J. B., Pedersen, H. J., and Bechgaard, K. (1982). *Phys. Rev. Lett.* **49**, 881.
Ubbink, J. (1953). *Physica* **19**, 9.
Ubbink, J. (1953). *Physica* **19**, 919.
Ubbink, J., Poulis, J. A., Gerritsen, H. J., and Gorter, C. J. (1953). *Physica* **19**, 928.
Van Vleck, J. H. (1941). *J. Chem. Phys.* **9**, 85.
Walsh, W. M., Rupp, L. W., Wudl, F., Kaplan, M. L., Schafer, D. E., Thomas, G. A., and Gemmer, R. (1980a). *Solid State Commun.* **33**, 413.
Walsh, W. M., Wudl, F., Thomas, G. A., Nalewajek, D., Hauser, J. J., Lee, P. A., and Poehler, T. (1980b). *Phys. Rev. Lett.* **45**, 829.
Walsh, W. M., Wudl, F., Aharon-Shalom, E., Rupp, L. W., Vandenberg, J. M., Andres, K., and Torrance, J. B. (1982). *Phys. Rev. Lett.* **49**, 885.
Weger, M. (1976). *Solid State Commun.* **19**, 1149.
Weger, M. (1978). *J. Phys. (Paris) Suppl.* **39**, C6-1456.

White, R. M. (1970). "Quantum Theory of Magnetism." McGraw-Hill, New York.
Wudl, F., Smith, G. M., and Hufnagel, E. J. (1970). *Chem. Commun.* **1970**, 1453.
Wudl, F., Schafer, D. E., Walsh, W. M., Rupp, L. W., DiSalvo, F. J., Waszczak, J. V., Kaplan, M. L., and Thomas, G. A. (1977). *J. Chem. Phys.* **66**, 377.
Wudl, F., Andres, K., McWhan, D. B., Thomas, G. A., Nalewajek, D., Walsh, W. M., Rupp, L. W., DiSalvo, F. J., Wazczak, J. V., and Stevens, A. L. (1981). *Chemica Scripta* **17**, 19.
Yafet, Y. (1963). *In* "Solid State Physics" (F. Seitz and D. R. Turnbull, eds.), Vol. 14, p. 1. Academic Press, New York.
Yosida, K. (1951). *Prog. Theor. Phys.* **5**, 691.
Yosida, K. (1952a). *Prog. Theor. Phys.* **6**, 25.
Yosida, K. (1952b). *Prog. Theor. Phys.* **7**, 425.
Zuppiroli, L., Delhaes, P., and Amiell, J. (1982). *J. Phys. (Paris)* **43**, 1233.

CHAPTER 7

Irradiation Effects: Perfect Crystals and Real Crystals

L. Zuppiroli

LABORATOIRE DES SOLIDES IRRADIES
ECOLE POLYTECHNIQUE, PALAISEAU 91128, FRANCE

I.	INTRODUCTION	437
II.	EXPERIMENTAL RESULTS	439
	1. *Transport Properties.*	439
	2. *Phase Transitions.*	452
	3. *Structural Properties of the Charge-Density Waves when Disordered by Irradiation.*	453
	4. *Collective Modes*	455
	5. *The Low-Temperature Magnetic Properties of Irradiated Organic Conductors.*	460
III.	DISCUSSION	460
	6. *Introduction: A Qualitative Picture of Irradiation Effects.* .	460
	7. *A Charge-Density-Wave Mosaic*	462
	8. *Impurity States in the Gaps of Irradiated Insulating Phases of Organic Conductors.*	465
	9. *Localization Concentration Range and the Metallic Particle Picture*	467
IV.	IRRADIATION BY IONIZING PARTICLES: LOCAL CHEMISTRY OF EXCITED MOLECULES	473
	10. *Introduction: The Mechanisms of Damage Production* . .	473
	11. *Experimental Determinations of Defect Concentrations.* .	475
V.	CONCLUSION: PERFECT CRYSTALS AND REAL CRYSTALS	477
	REFERENCES.	479

I. Introduction

Disorder has always been an active field of solid-state physics. Because of the new theories of localization (Abrahams *et al.*, 1979) and the scientific and technological impact of granular materials (Abeles, 1976), the interest in the role of disorder on the electronic properties of real solids has become even greater in the last few years.

Even when moderate, disorder changes the low-temperature scattering times of the electrons of a metal, introduces states in the gap of a semiconductor, and changes the position of the Fermi level of a semimetal.

When disorder is strong, it causes the conducting electrons to become localized. Amorphous semiconductors and very "dirty" semimetals or metals, where the mean free paths are of the order of one atomic distance, are the subject of recent interest related to theories of localization.

Research in the field of low-dimensional conductors has become so active in the last 10 years that all the main fields of solid-state electronics have been extended to low dimensions. In particular, the role of disorder on the electronic properties of low-dimensional conductors and semiconductors has been investigated in great detail. In the years 1975 to 1980, a large effort was made by solid-state chemists to produce organic alloys, and the structural and electronic properties of the quasi-one-dimensional conductors disordered by chemical doping have been reviewed several times since then (Jacobsen et al., 1978; Schultz and Craven, 1979; Cooper et al., 1981).

Irradiation by ionizing particles was used for the first time in organic conductors around 1976 (Chiang et al., 1977; Zuppiroli et al., 1978) and is now accepted as an excellent method for introducing disorder in a controlled way. Doping and irradiation are complementary, because the chemical method produces a large number of weak random potentials, while irradiation creates strong perturbing potentials in concentrations as low as needed. Typical examples of systems produced in these two ways and studied recently, are, on the one hand, the alloy $(TTF)_{0.8}(TSF)_{0.2}(TCNQ)$ [(tetrathiofulvalene)$_{0.8}$(tetraselenafulvalene)$_{0.2}$(7,7,8,8-tetracyano-p-quinodimethane)] (Engler et al., 1977) and, on the other hand, the organic superconductor $(TMTSF)_2PF_6$ (ditetramethyltetraselenafulvalene hexafluorophosphate), containing a concentration of a few times 10^{-5} molecular fraction of irradiation defects (Bouffard et al., 1982). Thus the greatest advantages of irradiation are (1) it permits *in situ* studies and (2) at low irradiation doses it is possible to follow physical properties as a function of defect concentrations, which are always proportional to irradiation time. To know precisely the absolute value of defect concentrations is more difficult and requires, in each case, an accurate study of the radiation damage process. In the first years of investigation of the irradiated samples transport properties, knowledge about concentration levels and damage mechanisms was very poor, but the experimental results were still rich. Because of the efforts of the Fontenay-aux-Roses group in concentration scales determination and the mechanisms of damage production (Mihály and Zuppiroli, 1982; Zuppiroli, 1982), the knowledge of what are irradiation defects is presently more extended. A few elementary ideas on what irradiation does in organic conductors will be given at the end of the present review.

The first part will be devoted to the most relevant experimental results in the field of irradiated organic conductors. Very roughly speaking, they proceed from three different groups in the world. The group of the University of Pennsylvania has studied the transport in irradiated (TTF)(TCNQ), including nonlinear and ac. The accent was put on the changes in phase transitions and on the deviation from Mathiessen's rule. La Section d'Etude des Solides Irradiés in Fontenay-aux-Roses has focused on the transport, magnetic, and structural properties of the TMTSF family of organic conductors. Most of their results concern the metallic, insulating (charge-density waves, CDW, and spin-density waves, SDW), and superconducting phases of these systems. Finally, extensive research was conducted jointly by two groups in Budapest (transport properties) and in Zagreb (magnetic properties) on "pure" and irradiated $Qn(TCNQ)_2$ (Qn = quinolinium). This classification is, of course, very crude and excludes several other contributions that will appear in the course of the review. The influence of irradiation defects on the transport and magnetic properties of organic conductors will be examined in the metallic, insulating, and superconducting phases separately.

In the second part, the author has attempted to classify the experimental results in order to check the principal theoretical ideas in the field. Answers or elements of answers proceeding from experiments are proposed to many questions, including the following ones:

1. What is the role of localization in disordered organic conductors and to what extent are the quasi-one-dimensional processes different from the three dimensional?
2. Is there evidence of the existence of states in the gap of a disordered organic semiconductor?
3. What does a charge-density wave mosaïc look like?
4. What are the consequences of disordering charge-density waves or spin-density waves on single-electron transport, collective modes, and phase transitions?

In the last part of the review, irradiation will be demonstrated to be the local chemistry of excited molecules within an organic solid. Recent results, giving, in particular cases, more precise information on the chemical (or crystallographical) nature of irradiation defects, will be presented briefly.

II. Experimental Results

1. TRANSPORT PROPERTIES

a. Introduction

Most of the work on organic conductors is based on dc conductivity measurements. Irradiation experiments are not an exception to this general tendency. Before introducing disorder in organic conductors, it is important to remember what a typical curve for a "good," "pure" sample looks like. I have chosen, for this purpose, the resistivity vs. temperature curves of (TMTSF)(DMTCNQ) (tetramethyltetraselenafulvalene) (dimethyltetracyanoquinodimethane). They are presented in Fig. 1 in three perpendicular directions a, b, c, where a is the high conductivity longitudinal axis. In the lower part of the figure the resistivity ρ is plotted vs. temperature. This part, intended for metal physics researchers, shows the large metallic temperature range ($d\rho/dT > 0$) in which the anisotropies ρ_b/ρ_a and ρ_c/ρ_a are rather independent of temperature and are on the order of 100 to 1000. In the upper part of the figure, the same curves have been reproduced in the log of conductivity vs. reciprocal temperature scale, familiar to semiconductor physicists. They give a better view of the ill-defined "semiconducting" phase below the sharp phase transition at 42 K ($1/T = 0.024$).

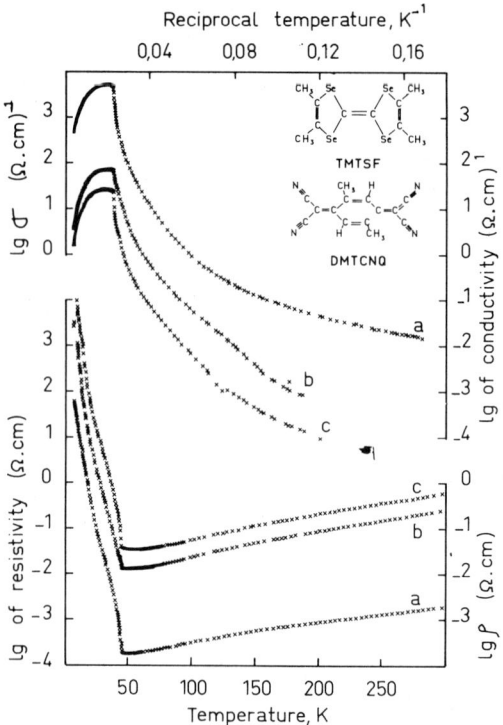

FIG. 1. Conductivity vs. temperature curves of (TMTSF)(DMTCNQ) measured in three perpendicular directions, a, b, and c, close to the crystal axes. In the lower part of the figure, the logarithm of the resistivity is plotted vs. temperature, while in the upper part, the logarithm of conductivity is plotted vs. reciprocal temperature. [Measurements from Mihály et al. (1980).]

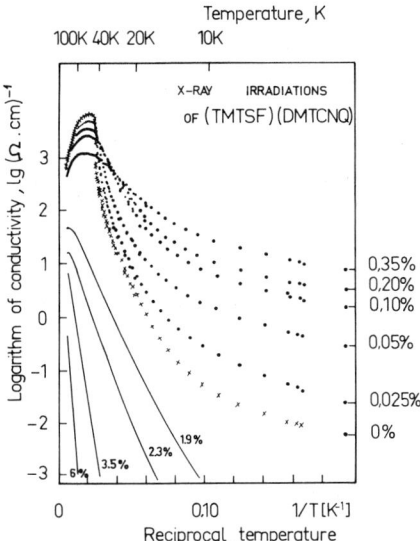

FIG. 2. Conductivity vs. temperature curves of (TMTSF)(DMTCNQ) submitted to x-ray irradiations. On each curve the defect concentrations are expressed in mole %. At low doses, the low-temperature conductivity increases by several orders of magnitude, while it increases only moderately in the metallic state. At higher doses (continuous curves), the curves become singly activated in the whole temperature range. [Measurements from Mihály et al. (1980) and Forró et al. (1982).]

Figure 2 shows one of the most complete sets of conductivity vs. temperature data concerning disordered organic conductors. It has been recorded using single crystals of (TMTSF)(DMTCNQ) irradiated with x-rays. The measurements are due to Forró et al. (1982) for the low-dose part and to Mihály et al. (1980) for the higher dose part. The principal effects of radiation-induced defects on the conductivity appear in this set of curves.

1. A low dose of irradiation corresponding to a molecular fraction of radiation-induced defects of 0.05% (5×10^{-4}) is enough to increase the conductivity of the sample by two orders of magnitude in the "insulating" phase, while it produces a negligible change at room temperature in the metallic phase.

2. After a strong low-temperature increase, the conductivity decreases at any temperature above a concentration of about 0.5% mole fraction.

3. During this disordering process, the phase transition has been completely smeared out of the conductivity curve.

4. At about 2% and more, the metallic regime ($d\rho/dT > 0$) disappears and the conductivity curve becomes singly activated in the whole temperature range, with an activation energy increasing with the irradiation dose.

The description of this first set of results strongly suggests that two different mechanisms are acting at low and high temperatures. Thus it is better to present separately in the present review high irradiation doses and high temperatures, on the one hand, and low doses and low temperatures, on the other hand. Before ending this introduction, it is important to stress that the tendencies that have been seen in this set of curves are not particular to (TMTSF)(DMTCNQ). They are quite general, as will be shown further on. In particular, at high doses, on the order of a few mole percent, the conductivity vs. temperature of all organic conductors that have been irradiated becomes singly activated over a large temperature range. A few examples are shown in Figs. 3 and 7.

b. Resistivity and Electron-Spin-Resonance Linewidth in the Metallic State

In ordinary metals, resistivity and electron-spin-resonance (ESR) linewidth are two transport properties related to each other by Eliott's mechanism (Eliott, 1954). The resistivity ρ is proportional to the electron-scattering frequency $1/\tau$, the ESR linewidth ΔH is proportional to the frequency $1/\tau_s$ of collisions that reverse the spin, and the quantities ΔH and ρ are connected

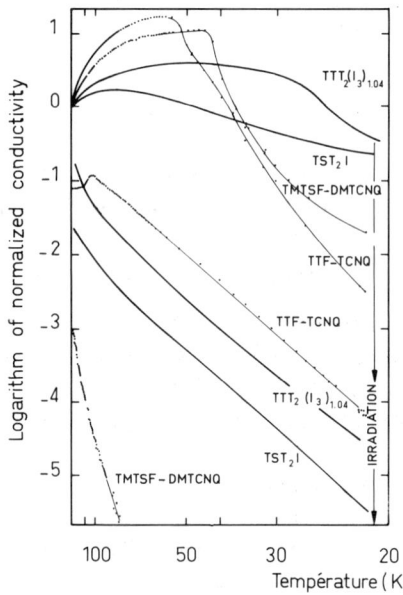

FIG. 3. Conductivity vs. reciprocal temperature curves of different organic conductors before and after a neutron irradiation introducing several mole % in defects. After irradiation the curves are single activated on the whole temperature range. [From Zuppiroli et al. (1980).]

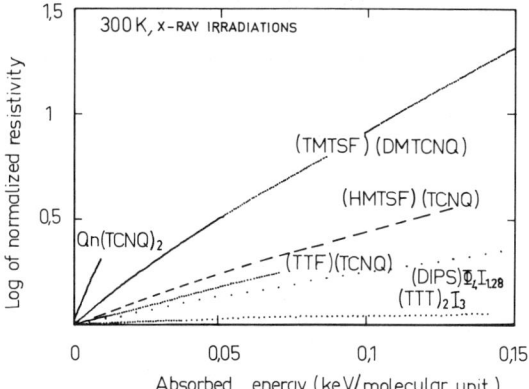

FIG. 4. 300 K resistivity vs. irradiation dose for six different organic conductors. The resistivity is normalized to the value before irradiation. The absorbed energy from the incident ionizing radiation is expressed in keV/molecular unit; 0.1 keV/molecular unit corresponds to a concentration of 1.4 mole % in TMTSF-DMTCNQ, 0.7% in TTF-TCNQ and about 1% in Qn(TCNQ)$_2$. Experiments are from G. Mihály and S. Bouffard. The samples of (TMTSF)(DMTCNQ), (HMTSF)(TCNQ), and (TTF)(TCNQ) have been provided by K. Bechgaard, those of TTT$_2$I$_3$ by B. Hilti, and those of (DIPS)ϕ_4I$_{1.28}$ (tetraphenyldithiopyranilidene iodide) by N. Strzelecka.

by Eliott's relation. Usually when defects are introduced in a metal, both the resistivity and the ESR linewidth increase in the same proportion (Beuneu and Monod, 1976); in the metallic phase of irradiated organic conductors, this is not at all the case, because resistivity increases exponentially with irradiation dose, while ESR linewidth decreases exponentially with dose. These results are shown for a few examples* in Figs. 4, 5 and 6. The resistivity at 300 K is shown to increase rather exponentially with the dose in six different organic conductors submitted to a x-ray irradiation (Fig. 4). The same result is shown on Fig. 5 for (TMTSF)$_2$ClO$_4$ and (TMTSF)(DMTCNQ) submitted to an electron irradiation up to a dose corresponding to about 10% of destroyed molecules. The increase in resistivity is roughly exponential over 5 or 6 orders of magnitude. Figure 6 deals with ESR linewidths measured at room temperature by Sanquer *et al.* (1985). It decreases by a factor of

* In these three figures the irradiation dose has been deliberately expressed in three different units, usual in the field of irradiation. When, for a particular irradiated organic conductor, the concentration scale has been determined experimentally (see Section IV.2 or Zuppiroli, 1982), it is more convenient to express directly the concentration dose in defect concentration (mole %). When it has not yet been determined for all the compounds in the figure, the more elaborate and convenient way of expressing the dose is to convert it into energy absorbed irreversibly by the sample during its exposure to the ionizing radiation (keV/molecular unit of Mrad; 1 rad is equivalent to 100 ergs of absorbed energy per gram of material).

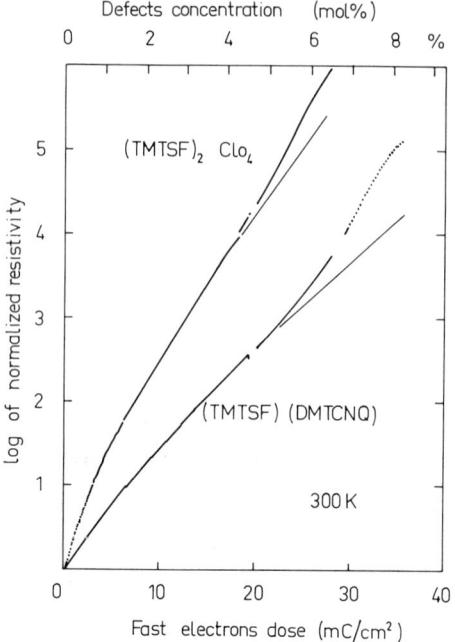

FIG. 5. 300 K resistivity vs. dose curves of $(TMTSF)_2ClO_4$ and $(TMTSF)(DMTCNQ)$ irradiated with 2 MeV electrons. The dose is expressed in mC/cm^2. The concentration scale has been checked for $(TMTSF)(DMTCNQ)$, and we assume that it is roughly the same for $(TMTSF)_2ClO_4$. The resistivity increase is rather exponential over several orders of magnitude. The continuous curve represents a fit to a two-parameter model presented later on.

2 to 4, reaches a minimum corresponding to a concentration of defects of about 5%, and starts increasing at higher doses. $Qn(TCNQ)_2$ is an exception. The linewidth starts increasing from the beginning of the irradiation.

The deviation from Eliott's rule is not the only strange manifestation of defects in the metallic phase of organic conductors. In the early work of Chiang *et al.* (1977), a strong deviation from Mathiessen's rule was emphasized by the authors: the higher the temperature in the metallic state the larger is the resistivity due to a given amount of defects.

c. *The Transport Properties of Irradiated Quinolinium* $(TCNQ)_2$

Although $Qn(TCNQ)_2$ was one of the first organic conductors studied from the point of view of the transport properties, the mechanism of electronic conduction in this compound has been the subject of large controversies, which are not yet completely resolved. The conductivity vs. temperature curve was considered by Bloch *et al.* (1972) as the signature of electronic

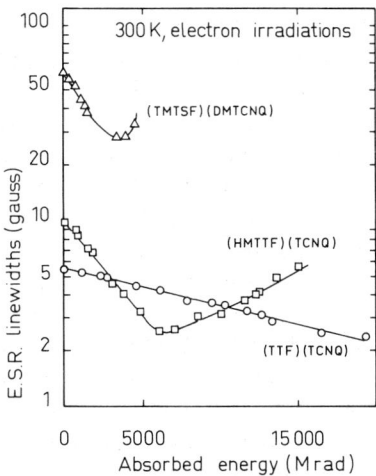

FIG. 6. 300 K ESR linewidth for three different organic conductors irradiated with 100 keV electrons. The irradiation dose has been expressed in megarads (1 rad represents an absorbed energy of 100 ergs g). In (TMTSF)(DMTCNQ) 1500 Mrad represent 1 mole % of radiation-induced defects. The experiments are from Sanquer et al. (1985). Similar results have been published in Forró et al. (1983). (Courtesy of S. Bouffard.)

properties determined principally by disorder. Several other authors refuted the dominant role of disorder. For example, Epstein et al. (1977) attributed an important role to a large strongly temperature-dependent mobility. Three years later, the disordered character of $Qn(TCNQ)_2$ was called upon more by Grüner (1981), who presented a model to explain ac and field-dependent transport. Strictly one dimensional, this model is based on the following observations and assumptions. $Qn(TCNQ)_2$ is composed of independent chains with random barriers. At high temperatures, larger than the average barrier height, the conductivity reflects the intrinsic resistance within the segments, the barriers do not play any significant role, and the chain is metallic, thus the conductivity is frequency independent, the dielectric constant tends to zero, and current is linear with fields up to 2 V/cm at least, as observed. On the other hand, at low temperatures, smaller than the average barrier height, phonon-assisted hopping over barriers leads to strongly decreased conductivity. Charge carriers accumulate at the barriers, thus the dielectric constant is positive and large and strongly depends on the frequency, like the conductivity. In this limit, when the random potentials are important, nonlinear conductivity is expected and indeed occurs. In this limit the material is a one-dimensional dielectric.

Irradiation experiments have brought new elements for the solution of this controversy. It is important to present here the main results concerning irradiated Qn(TCNQ)$_2$. Figure 7 represents the conductivity vs. temperature curves of nominally pure and neutron-irradiated Qn(TCNQ)$_2$. It is interesting to compare this set of curves with those of irradiated (TMTSF) (DMTCNQ) represented in Fig. 2. The results of Qn(TCNQ)$_2$ are from (Holczer *et al.*, 1979). More recently Jánossy *et al.* (1982) have measured at 9.1 GHz the dielectric constant of the same kind of neutron-irradiated Qn(TCNQ)$_2$ samples and demonstrated that ε is proportional to c^{-2}, where c is the molar concentration of defects (Fig. 8). Finally, Grüner and Khanna (1979) have measured the thermopower S of Qn(TCNQ)$_2$, nominally pure and irradiated. They attributed the variation of S shown in Fig. 9 to changes in the orbital entropy of localized electrons due to the presence of defect sites. In this way, the TEP measurement can be considered an elegant determination of the concentration level (0.2 mole % per hour of neutron irradiation). It will be shown in the discussion that these three experiments, together with the study of low-temperature magnetic properties and x-ray diffuse scattering experiments, support the idea that Qn(TCNQ)$_2$ may be viewed as a collection of one-dimensional strands with discrete energy-level spacings, even if these strands are not necessarily "metallic" and if Coulomb interactions play an important role in this conductor as well as a $4k_F$ distortion (Conwell and Howard, 1986).

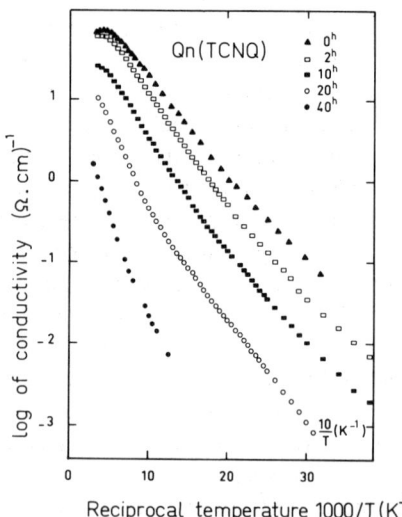

FIG. 7. dc conductivity of samples of quinolinium (TCNQ)$_2$ irradiated with different doses expressed in hours of neutron irradiations. A rough estimate of the defects induced is 0.2% disintegrated molecules per hour of irradiation. [From Holczer *et al.* (1979); courtesy of G. Grüner and *Solid State Communications*.]

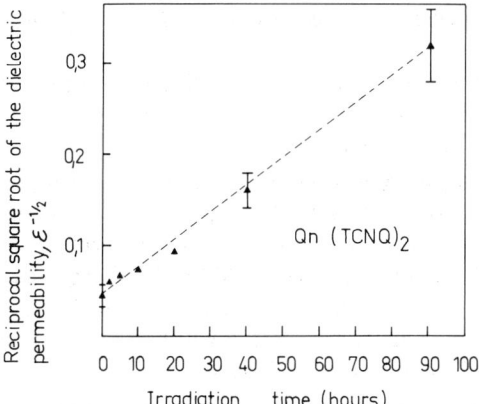

FIG. 8. Dielectric constant ε of neutron-irradiated quinolinium $(TCNQ)_2$ as a function of the dose. More precisely, $\varepsilon^{-1/2}$ has been plotted vs. the irradiation time in order to demonstrate the c^{-2} dependence of ε with the concentration (mole %). $\varepsilon \simeq 0.45(c_0 + c)^{-2}$ with $c_0 \simeq 3\%$. (c_0 is the extrapolated concentration in the nominally pure sample.) The measurements have been performed at 9.1×10^9 Hz by Jánossy et al. (1982). (Courtesy of A. Jánossy and Solid State Communications.)

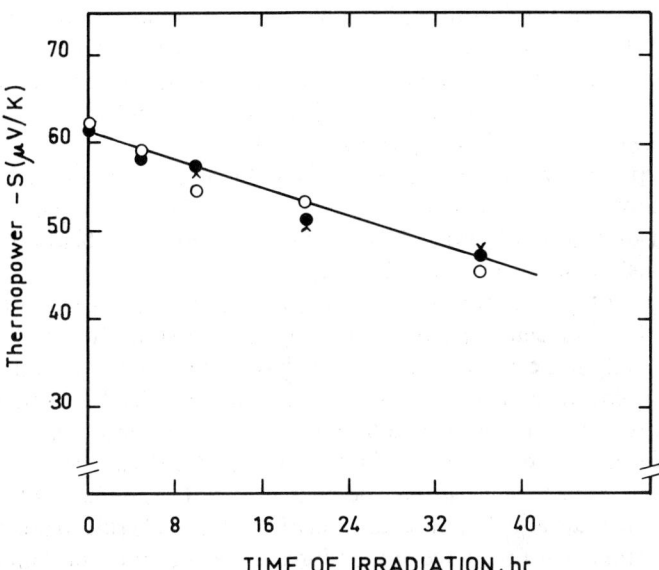

FIG. 9. Concentration dependence of the room-temperature thermoelectric power for neutron irradiated $Qn(TCNQ)_2$. Each hour of irradiation induces 0.2% damaged molecules. ○, ●, and × represent different samples. [From Grüner and Khanna (1979); courtesy of G. Grüner and Solid State Communications.]

d. Transport Properties in the Insulating Phases of Organic Conductors

The existence of low-temperature insulating phases in organic conductors is generally related to instabilities in the quasi-one-dimensional electron gas in the presence of electron–electron or electron–phonon interactions. The ordering of charge-density waves (CDW) or spin-density waves (SDW) leads to the opening of small gaps on the Fermi surface from a few meV to a few tens of meV. In addition to these small-gap semiconductors like (TTF)(TCNQ) (charge-density waves) or $(TMTSF)_2PF_6$ (spin-density waves), there also exist plenty of true organic semiconductors with gaps from a few tenths of eV to a few eV. Our first example will be one of these semiconductors, n-methyl-3,5-methyl-pyridine $(TCNQ)_2$, the conductivity vs. temperature curves of which are shown in Fig. 10. The two slopes in the Arrhenius plot of the conductivity of the nominally pure sample have been interpreted by Przybyski and Graja (1981) in terms of a classical model for compensated semiconductors with shallow donor and acceptor in the gap. Figure 11 illustrates the behavior of this semiconductor submitted to electron irradiation. The conductivity starts increasing with the defect concentration; it reaches about one order of magnitude more than the initial value, and then it decreases with further irradiation. In Fig. 10, where the temperature dependence of several irradiated samples has been reported, one can see the two activation energies at the zero dose, a conductivity plateau at 3 mC/cm^2 (exhaustion range), and finally a single activation energy on the whole temperature range, increasing with dose at higher defect concentrations (Przybylski *et al.*, 1983; Zuppiroli *et al.*, 1985).

The type of curve of Fig. 11 with a conductivity maximum can also be obtained by irradiation of charge-density wave insulators such as (TTF)(TCNQ) or (TMTSF)(DMTCNQ) at low temperatures (below the phase transitions). This is shown in Fig. 12.

The case of spin-density wave insulators is less clear. $(TMTSF)_2PF_6$ has been studied independently by Chaikin *et al.* (1982) on the one hand, and Forró (1982) on the other. Both authors have found many resistance jumps during cooldown and were unable to measure the absolute values of the resistances as a function of irradiation dose. Nevertheless, their magnetoresistance measurements in the SDW state, at 4.2 K as a function of the amount of damage, suggest that the main effect of the defects is to increase the scattering rate and that the extension of Mathiessen's rule to the magnetoresistance (magnetoresistance proportional to the square of the damage) does apply to the SDW phase of $(TMFSF)_2PF_6$ (Chaikin *et al.*, 1982).

Several organic conductors exhibit smooth and structurally ill-defined phase transitions. This is the case for $Qn(TCNQ)_2$, where charge-density waves are present but do not order at low temperatures (Pouget, 1981).

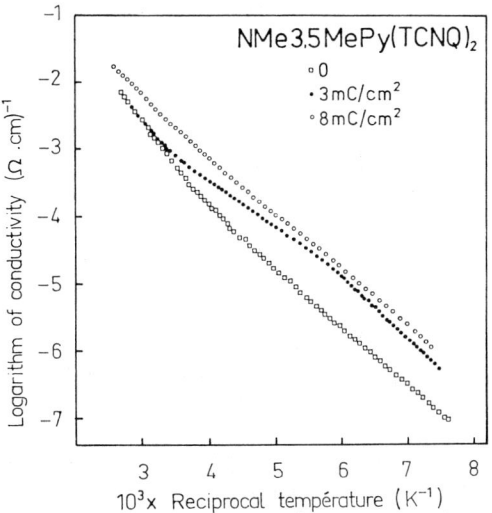

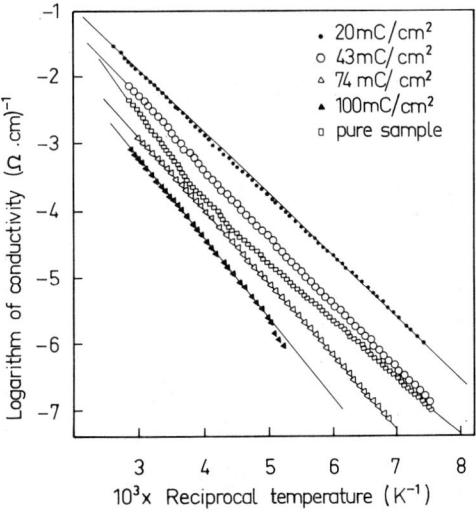

FIG. 10. Temperature dependences of the long axis conductivity of N-Me-3,5-MePy(TCNQ)$_2$ measured on pure and electron-irradiated crystals. The nominally pure sample exhibits two different activation energies of 0.18 and 0.28 eV. The intrinsic range is pushed to higher temperature by a 3 mC/cm^2 irradiation and a plateau appears (upper part of the figure). At higher doses the intrinsic regime disappears completely and a single activation energy, increasing with dose, is observed in the entire temperature range (lower part of the figure). The doses are expressed in mC/cm^2. A rough estimate of the defects induced by irradiation is 0.015% per mC/cm^2. [From Przybylski et al. (1983) and Zuppiroli et al. (1985).]

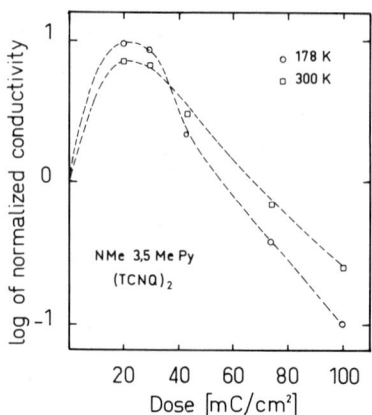

FIG. 11. Dose dependences at 178 K and 300 K of the long axis conductivity of N-Me-3,5-MePy(TCNQ)$_2$ irradiated with 2.5 MeV electrons. The dose of 100 mC/cm^2 corresponds roughly to a concentration of radiation-induced defects of 1.5 mole %. [From Przybylski et al. (1983).]

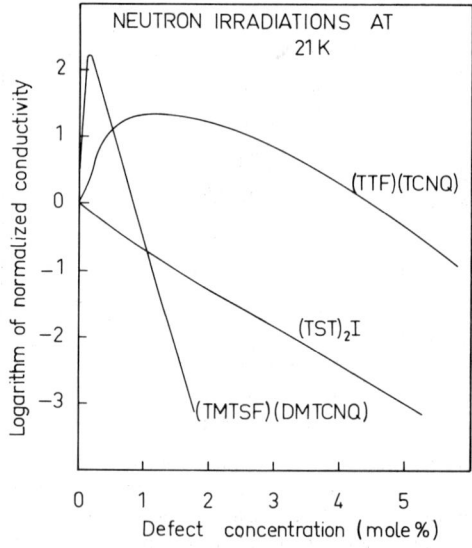

FIG. 12. Neutron irradiations of three quasi-one-dimensional conductors at 21 K. Their conductivities are plotted vs. the molecular fraction of radiation-induced defects. After an initial increase related to the existence of a well-defined Peierls gap, the conductivities decrease exponentially over three to six orders of magnitude. [Data from Zuppiroli et al. (1980).]

This is the case for (HMTSF)(TCNQ) (HMTSF = hexamethylenetetraselenafulvalene) (Cooper et al., 1976) for TTT_2I_3 or TST_2I (TTT = tetrathiotetracene; TST = tetraselenatetracene) (Delhaes et al., 1978), and for many other more or less "obscure" organic conductors. In these cases, irradiation is not able to produce any increase of the conductivity at any temperature, thus the conductivity decreases rather exponentially with the defect concentration, as shown in Fig. 12, for TST_2I.

Hall effect measurements in pure and irradiated organic conductors have been shown to complement very nicely the conductivity data. The metal to insulator phase transition is reflected in a large increase in the Hall coefficient of the pure sample at low temperature with respect to the small high-temperature value, typical of a metallic state. A small defect concentration of the order of 0.1 mole % decreases the low temperature Hall coefficient by several orders of magnitude; then the Hall coefficient reaches values of the order of the room-temperature values. This effect has been demonstrated by Forró et al. (1982) and by Forró (1982). It is illustrated in Fig. 13, where the two examples of a charge-density wave insulator (TMTSF)(DMTCNQ) and a spin-density wave insulator $(TMTSF)_2PF_6$ are presented. The latter is especially interesting, because the sign of the Hall coefficient changes when the magnetic field is moved from one of the transverse directions to the other, due to some peculiar anisotropy in the band structure or in the electronic

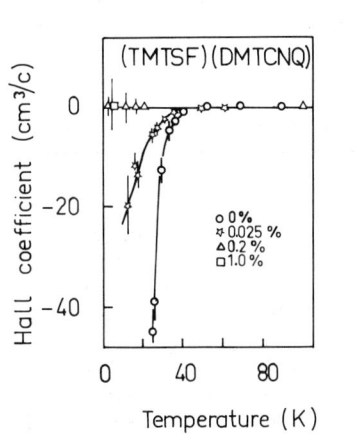
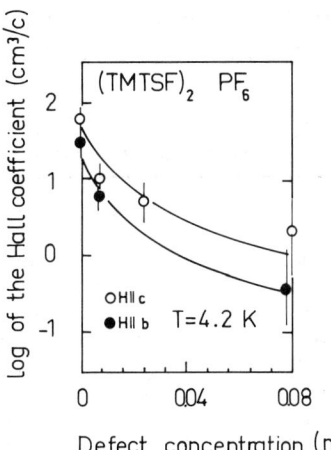

FIG. 13. Hall coefficients of the pure and irradiated CDW insulator (TMTSF)(DMTCNQ) and SDW insulator $(TMTSF)_2PF_6$. The defect concentrations are expressed in mole %. In the case of $(TMTSF)_2PF_6$, the Hall coefficient is negative when $H \parallel c$ and positive when $H \parallel b$ (b and c are the two transverse directions); thus only the absolute values have been reported on the figure. [Experiments from Forró et al.(1982) and Forró (1982).]

hopping (Forró, 1982; Conwell, 1983); when the crystal is irradiated, the absolute value of the Hall coefficient is changed in both directions in the same proportion.

A few measurements of the thermopower of irradiated (TMTSF) (DMTCNQ), performed by Forró et al. (1982), have confirmed the Hall coefficient results previously mentioned.

2. Phase Transitions

The electron gas in one dimension is well known to be highly unstable, and three kinds of low-temperature electron condensations have been observed: charge-density waves, spin-density waves, and superconductivity. It is important to mention that a phase transition is achieved only when a three-dimensional ordering occurs. Generally, precursor one-dimensional fluctuations occur well above the phase-transition temperature, and their ordering in three dimensions produces the phase transition.

The effects of irradiation defects on the phase transitions have been studied in many organic conductors. A transition of any kind has been found to shift downwards at an initial rate of the order of 100 K per percent defect. Table I summarizes the experimental values of shift rates.

One of the most precise irradiation studies in the field of phase transitions remains the early one of Chiang et al. (1977). Figure 14 summarizes these experimental results.

It is worth mentioning that the shifts in the phase transitions and their disappearance from the conductivity or susceptibility vs. temperature curves

TABLE I

Phase Transition Shifts in Irradiated Quasi-One-Dimensional Conductors

Compound	Transition temperature (K)	Nature of the transition[a]	$-dT/dC$[b] (K)	Reference
(TTF)(TCNQ)	54	CDW	15,000	Chiang et al. (1977)
(TTF)(TCNQ)	38	CDW	20,000	Chiang et al. (1977)
(TMTSF)(DMTCNQ)	42	CDW	11,000	Forró et al. (1982)
(TMTSF)$_2$PF$_6$	14	SDW	6,500	Forró and Beuneu (1982)
(TMTSF)$_2$PF$_6$	0.9	SC	100,000	Bouffard et al. (1982)
n-propylquinolinium (TCNQ)$_2$	220	?	1,600	Ero-Gecs et al. (1979)
TaS$_3$	160	CDW	12,000	Unpublished results

[a] CDW = charge-density wave; SDW = spin-density wave; SC = superconducting.
[b] dT/dC = 15,000 K means that the transition shifts downwards with an initial rate of 150 K per percent defects.

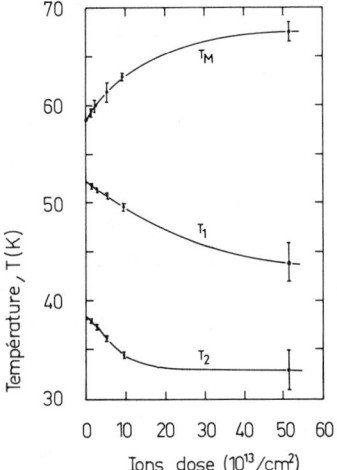

FIG. 14. 8 MeV deuteron irradiation of (TTF)(TCNQ). Characteristic temperatures as a function of incident ion dose. T_M is the temperature of the conductivity maximum, T_1 is the first transition associated with the onset of long-range order, and T_2 is the 38 K transition associated with the locking of CDWs. The dose of 5.10^{14} corresponds roughly to a concentration of 0.1 mole %. (Courtesy of A. Heeger and American Physical Society.)

are not independent of the transport properties of the insulating phases when submitted to irradiation (Section II.1.d). We shall see further that the changes of all these properties have the same origin: the pinning of charge-density waves or spin-density waves to irradiation defects.

3. Structural Properties of the Charge-Density Waves When Disordered by Irradiation

A few authors have referred to the pinning of charge-density waves as a likely reason for the changes in the transport properties of the Peierls phases due to irradiation (Zuppiroli *et al.*, 1982). It was conjectured that each irradiation-induced defect behaves as a high foreign potential, defining rigidly the phase of a piece of charge-density waves in a given volume of the sample. The low-temperature transport properties of irradiated samples were analyzed phenomenologically either as the consequence of a random distribution of conducting volumes created in an insulating Peierls matrix or as the consequences of a discommensuration network, where new carriers were created at phase defects of the charge-density waves. But these intuitions were suggested mainly from transport experiments. A more direct investigation was necessary to check the spatial distribution of charge-density waves in the presence of disorder and was performed recently, in the particular case of irradiated (TMTSF)(DMTCNQ), by Forró *et al.* (1983) and, in the case of (TTF)(TCNQ), by Forró *et al.* (1984). Both these studies are based on x-ray diffuse scattering experiments on irradiated samples.

The results are summarized on the schematic picture of Fig. 15. Even in heavily irradiated samples (~3 mole %) $2k_F$ and $4k_F$ diffuse scattering are still present in the whole temperature range, even when the phase transitions resulting from their condensation into superstructure have completely disappeared and the transport properties have changed so markedly (see Figs. 2 and 13). The pinning of the charge-density waves to the defects destroys the three-dimensional long-range order even at 10 K, and anisotropically broadened reflections appear in the irradiated samples instead of the

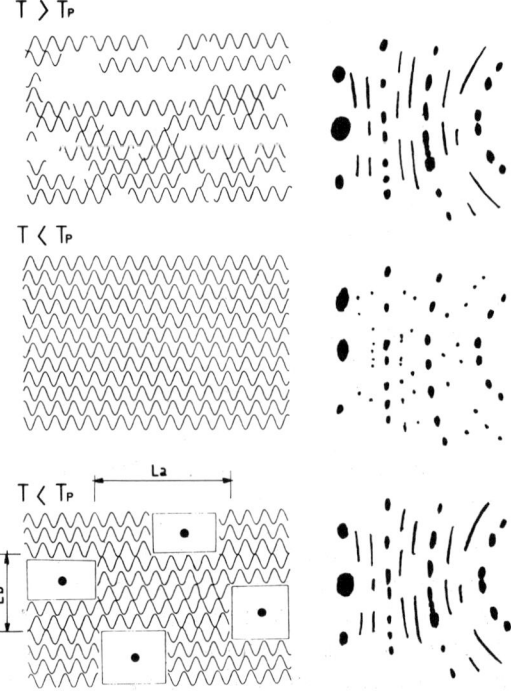

FIG. 15. The coherence of the charge-density waves in irradiated TMTSF–DMTCNQ: A schematic picture inspired by the experimental results of Forró *et al.* (1983). Above the temperature T_P of the Peierls transition, the disordered fluctuating charge-density waves give rise to diffuse lines on the x-ray scattering patterns of pure (TMTSF)(DMTCNQ). Below T_P the charge-density waves are ordered; then the diffuse lines condense into superstructure spots in the pure sample. In irradiated samples, strong pinning centers disturb the coherence of charge-density waves and give rise to the diffuse lines below T_c. L_a and L_b denote the dimensions of the unperturbed volume that diffuse x-rays coherently. Their averages are related to the widths of the diffuse lines and proportional to the coherence lengths ξ_a and ξ_b plotted in Fig. 16.

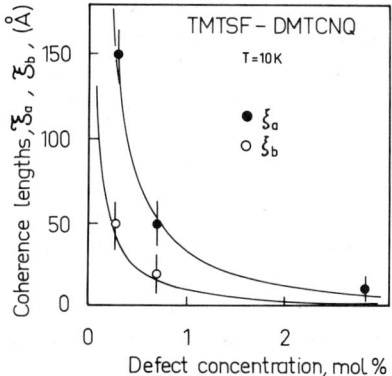

FIG. 16. Correlation lengths ξ_a, ξ_b of the charge-density waves measured at 10 K in irradiated TMTSF-DMTCNQ (a is the longitudinal direction and b the lowest anisotropy transverse direction) and plotted vs. defect concentration. The full lines represent the fit to a three-parameter model in which each defect pins the CDW in a given volume of the crystal. The model was developed in the paper by Forró et al. (1983).

superstructure reflections in the pure sample. The values of the longitudinal ξ_a as well as the transverse ξ_b correlation lengths have been measured as a function of the temperature and the irradiation dose; they are plotted at 10 K on Fig. 16.

It is also interesting to report on the issue of the thermal $4k_F$ diffuse scattering under irradiation. This anomaly is generally attributed to lattice effects induced by substantial electron Coulomb interactions (Emery, 1976). At small defect concentration, the intensity of the thermal $4k_F$ anomaly observed in the pure (TMTSF)(DMTCNQ) decreases with dose and becomes almost undetectable at 0.3 mole %. At higher doses the $4k_F$ anomaly is resurrected by the defects and becomes static in the whole temperature range.

In the x-ray patterns of strongly irradiated (TMTSF)(DMTCNQ) or (TTF)(TCNQ) (2 or 3 mole %), the following features have been observed: a temperature-dependent $2k_F$ scattering, always keeping the one-dimensional character until 20 K at most; never condensing in superlattice spots; and coexisting with a $4k_F$ scattering that is almost temperature independent in width and intensity. It is striking to note that these are precisely the main features of the patterns of nominally pure Qn(TCNQ)$_2$ (Epstein et al., 1981).

4. Collective Modes

At each electron instability is associated a collective mode for the electron transport: The opening of a BCS gap leads to superconducting properties, and the opening of a Peierls gap leads to charge-density wave transport (Fröhlich, 1954). The present section demonstrates the sensitivity to irradiation disorder of these collective modes. Although tantalum trisulfide TaS$_3$ or niobium triselenide NbSe$_3$ are not organic conductors, but quasi-one-dimensional, the threshold conduction in these compounds will be discussed as a first clear example of the effects of disorder on a collective mode.

a. Charge-Density Wave Conduction in Irradiated TaS$_3$

The main interest in trichalcogenides for studying charge-density wave conduction is that the collective mode occurs when the CDWs are ordered, and it corresponds to a macroscopic motion of condensed electrons as a whole. The very peculiar nonlinear conductivity, the frequency-dependent conductivity, the narrow band noise, and current oscillations are recognized as manifestations of this motion (Grüner, 1983). As expected, this kind of conduction is very sensitive to the pinning of charge-density waves to strong defects, as demonstrated by the recent experiment of Mutka *et al.* (1984), who measured the nonlinear conductivity and narrow-band noise in electron-irradiated TaS$_3$. Figure 17 shows the nonlinear conductivity at different irradiation doses. The existence of a very sharp conduction threshold is usually attributed to lattice and impurity pinning of the CDWs, and indeed the threshold is displaced markedly by irradiation-induced strong pinning centers present at the level of one defect over one million tantalum atoms. At a concentration of the order of 10^{-4}, there is no longer any clear threshold conduction. The CDW has become so "polycrystalline" that the coherent motion is very restricted, and the remaining smooth nonlinear conductivity is related to the accommodation of the incoherent CDW to the defect centers.

Two less recent papers have reported the results of irradiation experiments and their effects on the threshold conduction in NbSe$_3$ (Fuller *et al.*, 1981; Monceau *et al.*, 1981). Surprisingly, in the latter, Monceau *et al.* have not found any important pinning effect of the threshold by electron-irradiation-induced defects. Their lowest dose is probably one order of magnitude too large to see any clear conduction threshold in their irradiated samples. In the former, Fuller *et al.* have irradiated NbSe$_3$ with 2.5 meV protons and found that the threshold field varies linearly with the defect concentration of more than one order of magnitude. Although the damage by protons in a 25 μm sample is not homogeneous and the concentration estimation of Fuller *et al.* seems to be wrong by an order of magnitude, their results seem to be more consistent with the recent experiments of Mutka *et al.* in TaS$_3$.

b. Fluctuating Charge-Density Wave Conductivity in (TTF)(TCNQ)

Is it possible to observe threshold conduction in the charge-density wave phases of (TTF)(TCNQ)? Below 38 K the coherent notion of the CDW is impeded by two pinning forces: commensurability pinning and Coulomb

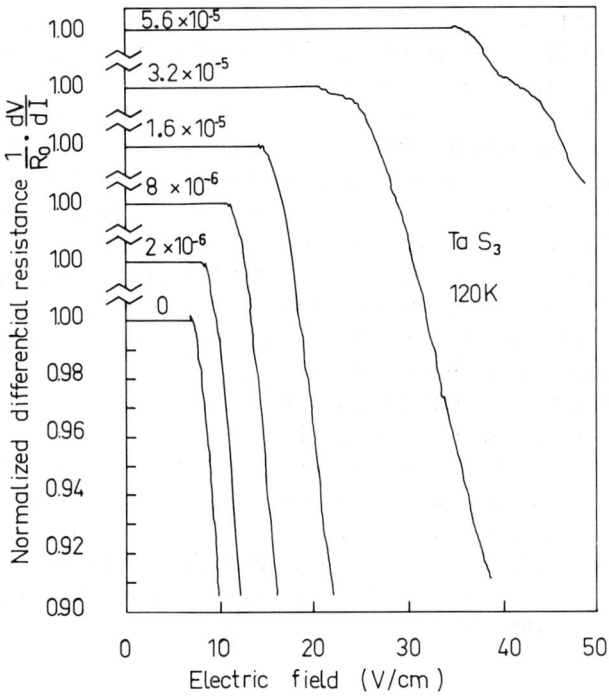

FIG. 17. Charge-density wave threshold conduction in electron irradiated TaS_3. Below the threshold field, current and voltage are proportional; thus the normalized differential resistance is constant and equal to 1. Above the threshold field corresponding to the depinning of charge-density waves, the conductivity increases markedly due to the coherent motion of condensed electrons. The concentrations of defects are expressed in atomic fraction of tantalum atoms irreversibly displaced from their normal position by the incident fast electrons. The conduction threshold is significantly increased by the introduction of new pinning centers at the level of 10^{-6} atom fractions. (Experiments from H. Mutka and S. Bouffard.)

interaction between oppositely charged chains of the charge-transfer complex. Between 38 K and 54 K, the CDW are developed and ordered in the TCNQ sublattice, but still fluctuating in the TTF sublattice. Nonlinear electrical transport at the onset of the Peierls transition has been observed recently in this temperature range (Lacoe et al., 1985), but the excess current is much smaller than in inorganic CDW conductors such as TaS_3 or blue bronzes.

A few years ago the Orsay group claimed to have discovered a charge-density wave collective conduction of another nature, in the metallic state near the conductivity maximum of (TTF)(TCNQ) (Andrieux et al., 1979). In fact they made two experimental observations: first, a hydrostatic pressure of 19 kbar makes the CDW commensurate with the lattice; second, around such a pressure there is a large decrease in the conductivity. The interpretation of the Orsay group was that the pinning of fluctuating charge-density waves to the lattice by the external pressure destroys the collective contribution to the conductivity due to the CDW gliding. Many other authors have refuted this interpretation for the explanation of these results. They have proposed that the pseudogap produced by the CDW fluctuations near the conductivity maximum was indeed sensitive to pressure through commensurability effects. Changes in the pseudogap were assumed to produce, in turn, changes in the number of single-particle carriers and, consequently, the changes in the conductivity recorded by the Orsay group. Unfortunately this last interpretation is not able to explain the strong anisotropy of the effect recorded by the Orsay group and it appears that the former interpretation is now generally considered the better.

Thus it is interesting to know what are the effects of defects in this particular collective contribution of CDWs. This was done by Bouffard et al. (1981) by using a coupled pressure and irradiation experiment on (TTF)(TCNQ). The result, illustrated by Fig. 18, is that an irradiation defect concentration of 0.2 mole % is enough to completely destroy the collective contribution to conduction by the fluctuating CDW, schematically represented on the upper picture of Fig. 15.

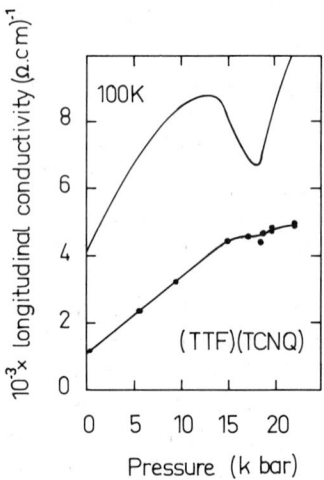

FIG. 18. 100 K conductivity of (TTF)(TCNQ) as a function of external hydrostatic pressure for a pure (upper curve) and an irradiated (lower curve) sample containing 0.2 mole % of x-ray irradiation, strong pinning centers. The CDW commensurability effect recorded on the pure sample at 19 kbar and attributed to the suppression of the CDW collective conductivity disappears in the irradiated sample. [From Bouffard et al. (1981); courtesy of S. Bouffard and Solid State Communications.]

c. Superconducting Transport in $(TMTSF)_2PF_6$

$(TMTSF)_2PF_6$ has been irradiated under pressure by Chaikin *et al.* (1982) and by Bouffard *et al.* (1982) independently. Both have found that a concentration of irradiation induced defects of the order of few 0.01 mole % (10^{-4}) is enough to completely destroy the superconductivity in this compound. In Fig. 19 the resistance vs. temperature curves of a sample containing about 2.10^{-4} irradiation defects have been plotted for different values of the magnetic field near the superconducting transition. The behavior of the sample when submitted to a magnetic field confirms the superconducting nature of the transition, but the resistivity never drops to zero, showing the inhomogeneous character of the superconductivity in the irradiated sample.

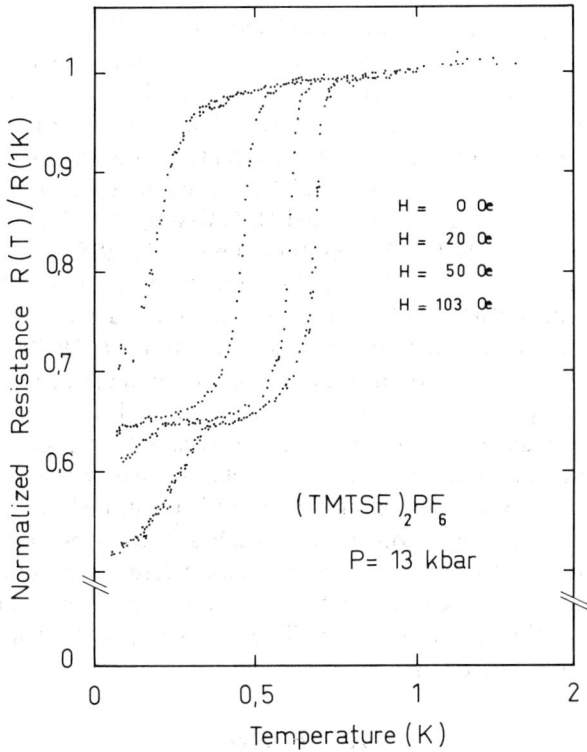

FIG. 19. Temperature dependence of the resistance of irradiated $(TMTSF)_2PF_6$ at 13 kbars near the superconducting transition. The resistance has been normalized to its value at 1 K in the earth's magnetic field. The four curves correspond to different applied magnetic fields (the higher the field, the lower the transition temperature). The concentration of irradiation defects was estimated to be 0.02 mole %. [From Bouffard *et al.* (1982); courtesy of S. Bouffard and *J. Phys. C: Solid State Physics*.]

5. Low-Temperature Magnetic Properties of Irradiated Organic Conductors

The low-temperature magnetic properties of disordered linear chain conductors have been the subject of recent and less recent interest. Shchegolev first demonstrated that the susceptibility of $Qn(TCNQ)_2$ shows a large low-temperature upturn similar to a Curie tail (Shchegolev, 1972). Measuring it down to 0.1 K, he demonstrated that the low-temperature variation was not T^{-1} but $T^{-\alpha}$ with $\alpha \simeq 0.7$. Similar results were obtained on a large variety of more or less disordered linear chain conductors; they are reviewed in the paper of Cooper et al. (1981). A few months later, Tippie and Clark (1981) measured the electron-spin-resonance susceptibility of $Qn(TCNQ)_2$ down to 30 mK and demonstrated unambiguously the $T^{-\alpha}$ character of the law with $\alpha \sim 0.8$.

From the beginning, theorists associated these $T^{-\alpha}$ laws to the magnetic interactions between spins created by disorder (Bulaevskii et al., 1972; Bondenson and Soos, 1980). In order to answer the question of the origin of these spins, their distributions on the chains, and the nature of their interaction, several authors have used irradiation as a way of introducing a controlled number of strong defects. They have measured the low-temperature magnetic properties of irradiated $Qn(TCNQ)_2$ (Ero-Gecs et al., 1979; Jánossy et al., 1982), (TTF)(TCNQ) (Gunning et al., 1979; Korin-Hamzić et al., 1982), and (TMTSF)(DMTCNQ) (Zuppiroli et al., 1982). The last two references (Korin et al., 1982; Zuppiroli et al., 1982) demonstrate definitely the results announced in the early papers of Chiang et al. (1977) and Gunning et al. (1979): Within a factor of two or three, the number of spins responsible for the low-temperature upturn in the susceptibility is equal to the number of defects introduced by irradiation counted as interruptions of the conducting chains or as pinning centers for the charge-density waves. The highest quality experiment demonstrating this result is illustrated in Fig. 20 (Korin et al., 1982). It is worth noticing that such a one-to-one correspondence between spins and defects is valid for irradiation defects but not for chemical impurities in alloys based on the previous molecules such as $(n-propyl)_{1-x}$ $(n-ethyl)_x$ quinolinium $(TCNQ)_2$ or $(TTF)_x(TSF)_{1-x}TCNQ$ (Jánossy et al., 1982; Cooper et al., 1981).

III. Discussion

6. Introduction: A Qualitative Picture of Irradiation Effects

Most of the qualitative conclusions concerning the role of irradiation disorder follow quite directly from the experimental results.

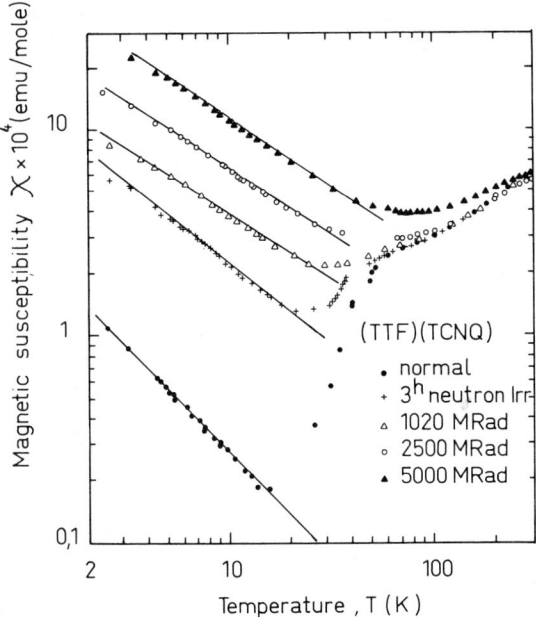

FIG. 20. Spin susceptibility of pure and irradiated (TTF)(TCNQ) vs. temperature on log–log scale. Solid lines are fits to $AT^{-\alpha}$ laws: the exponent $\alpha \simeq 1$ in the pure sample and $\alpha \simeq 0.7$ in irradiated ones. For the gamma-irradiated samples, the dose is expressed in absorbed energy (Mrads). 5000 Mrads correspond to 4% of spins. [From Korin-Hamzić et al. (1982); courtesy of J. R. Cooper and Mol. Cryst. Liq. Cryst.]

At very low concentrations, less than 0.1 mole %, the effects of defects are visible on the low-temperature condensed phases only. There, they change the coherence lengths of charge-density waves, spin-density waves, or superconductivity and, in turn, the transport properties of these phases. The collective modes, that need large coherence lengths, begin to be affected first at the level of a few 10^{-3} mole %; this was checked for charge-density wave conduction and superconductivity (Section II.4). Single-electron properties in the insulating phases are also sensitive to irradiation disorder at the level of a few 10^{-2} mole %, as shown by the conductivity and Hall-effect measurements on (TMTSF)(DMTCNQ) and (TMTSF)$_2$PF$_6$ reported in Figs. 2 and 13.

When the defect concentration reaches a few times 0.1%, the picture changes fundamentally. The low-temperature conductivity reaches a maximum (Figs. 11 and 12); the Hall coefficient stops decreasing and is saturated at some value close to the room-temperature metallic-state value (Fig. 13). Then irradiated organic conductors enter the localization range. The nature of this localization will be discussed later on. This concentration range is usually characterized by an exponential decrease of conductivity with defect concentration (Figs. 4 and 5), a more or less activated behavior of the conductivity over a large temperature range (especially above 2 mole %) (Figs. 2, 3, and 7), and a strong decrease of the electron-spin-resonance (ESR) linewidth with increasing concentration (Fig. 6). At these doses, of the order of 1%, there are no more signs of phase transitions on the conductivity vs. temperature curves (Fig. 14), and the disorder state characterized by a large number of carriers with low mobility related to some localization process extends over the whole temperature range.

It is striking to notice that, at the high defect concentration of a few percent, most of the individual, specific, electronic transport features that mark off each organic conductor species are completely lost in favor of a common macroscopic state: In their new radiation-induced poverty they begin to all look the same, at least from the point of view of their transport and magnetic properties.

This qualitative, synthetic view gives rise to more precise questions regarding the low-temperature insulating, as well as the high-temperature metallic, phases of irradiated organic conductors. In the following sections the discussion will focus successively on the mosaicity of charge-density waves, the possible existence of localized states in the gaps, and the localization concentration range.

7. A Charge-Density-Wave Mosaic

Because most of the low-temperature properties of organic conductors are related to the coherence lengths in the condensed phases, it was important to perform direct experimental investigations of these coherence lengths in pure and irradiated samples. Unfortunately, only the charge-density wave phases were investigated in this way. Structural studies of the spin-density waves are difficult (Neutron diffraction), and no serious study of the critical fields in irradiated organic superconductors is available at present. The coherence lengths of charge-density waves in irradiated (TMTSF) (DMTCNQ) and (TTF)(TCNQ) have been obtained in the works of Forró et al. (1983, 1984) mentioned in Section II.3. The loss of coherence is due to the pinning of the charge-density waves to irradiation defects. Each irradiation-induced pinning center creates an electronic perturbation—giant Friedel oscillation with period $2k_F$—as well as a mechanical perturbation—size effect—leading to an average change of lattice parameter of 1% per percent defect (Trouilloud et al., 1982). Thus the potential of the defect triggers and locally drives the Fermi surface instability. This is the reason why the phase of the CDW is imposed by the defect in some volume of the

7. IRRADIATION EFFECTS: PERFECT CRYSTALS AND REAL CRYSTALS   463

sample. One of the clear results of the structural study of Forró *et al.* (1983) is that it is impossible to understand the variations of the coherence lengths if the effect of the defect is limited to a single chain. Charge-density wave pinning is a tridimensional effect.

The need to accommodate phase mismatches between pieces of adjacent charge-density waves raises several questions. Are the phase defects localized in the form of steep walls (discommensurations) or delocalized over several periods of the CDW by elastically stretching the phase of the CDW? This should depend on the intensity and harmonicity of the charge-density wave, a strong inharmonicity favoring a wall with respect to a smooth change.

The study of Forró *et al.* (1983) provides a rather complete view of the CDW in the reciprocal space, but in the real space only the coherence lengths are known and not the details of the CDW structure. Nevertheless, in a particular charge-density wave system, direct images of the CDWs have been obtained by means of an electron microscope. Instead of the main Bragg spots of the structure, which are used to produce the conventional electron-microscopy images, superstructure spots have been used for the purpose of imaging the CDW. Unfortunately, this recent study (Mutka and Housseau, 1983), which clearly shows a granular CDW with sharp grain boundaries and microdomains closely correlated to the number of defects, has been performed on tantalum disulfide, which is neither organic nor one-dimensional but a charge-density-wave lamellar compound. Yet it is interesting to show, in Fig. 21, the evolution of the CDW structure with irradiation in TaS_2. This is the kind of CDW mosaic that perhaps could be expected in organic conductors as well.

The pinning of the CDWs is one of the central problems of charge-density-wave conduction in trichalcogenides. From the beginning of these studies the potential of undefined point defects was recognized as the source of a threshold in the CDW conduction. Indeed, this threshold increases with irradiation at very low doses (Fig. 17). However, in all the charge-density-wave conduction models developed up to now [see, e.g., Grüner (1983) and references therein], the pinning is introduced essentially as a phenomenological parameter. The phenomenological models, such as the "overdamped oscillator" or the more recent model of Bardeen (1980), are very simple and fit to several charge-density-wave-conduction experiments so well that one may wonder if it will be possible in the near and even distant future to extract some microscopic information about the CDW pinning from threshold conduction experiments. The same kind of problem arises in metallurgy when macroscopic plastic deformation data for a metal or an alloy are used to learn something microscopic about dislocation interactions with impurities or defects.

Why is threshold CDW conduction not observed (Lacoe *et al.*, 1985) in the low-temperature ordered CDW phase of organic conductors? The

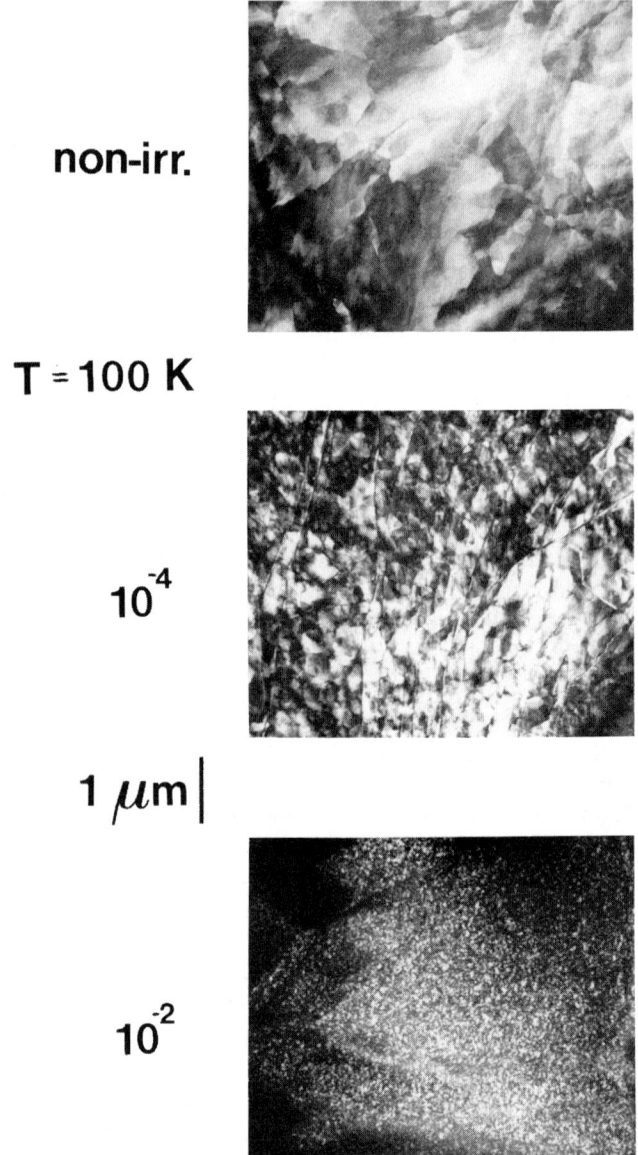

FIG. 21. Charge-density wave domains and walls in lamellar tantalum disulfide 1T–TaS$_2$. The temperature of 100 K corresponds to the commensurate CDW phase. Superstructure spots have been used to image the CDWs in the electron microscope. The existence of microdomains and walls is closely correlated to the number of defects in the lattice. [Work of Mutka and Housseau (1983); courtesy of N. Housseau and Philosophical Magazine.]

answer to this question is partially included in the experimental work of Tanner and Jacobsen (1982), who have measured the low-temperature polarized infrared reflectance of (TTF)(TCNQ). The pinning frequency of the CDW was found to be $k_P = 40\,\text{cm}^{-1}$. This is indeed strong pinning: Bardeen's theory of charge-density waves depinning (Bardeen, 1980) connects the depinning electric field E_0 for CDW motion to the pinning frequency ω_P.

$$E_0 = \frac{E_g}{2\xi_0 e} \frac{M_F}{m},$$

where E_g is the pinning gap ($E_g = \hbar\omega_P = \hbar c k_P$); M_F/m is the ratio of the Frölich mass to the band mass [~ 20 in TTF–TCNQ according to Tanner and Jacobsen (1982)]; and ξ_0 is the coherence length of the ordered CDW, ~ 100 molecular distances according to the diffuse x-ray scattering studies. With $k_P = 40\,\text{cm}^{-1}$, $M_F/m \sim 20$ and $\xi_0 \sim 100a$, we find $E_0 \simeq 2500\,\text{V/cm}$.

Nonlinearities of another nature occur in (TTF)(TCNQ) for much lower fields, on the order of 100 V/cm. Conwell and Banik (1981) have stressed, first, that these nonlinearities cannot be related to the CDW conduction. They proposed a more likely interpretation of single-particle-gap decrease due to injection across internal barriers between regions of different gaps (Conwell and Banik, 1982). In an actual quasi-one-dimensional conductor, defects can cause the gap to be nonuniform. Thus barrier heights of about 5% of the gap are expected. Electron and hole injection across these barriers will cause the conductivity σ to increase with field. At high enough fields, some of the larger gaps that are in a favorable position to receive large injection should become unstable and decrease, decreasing the corresponding barrier heights. This puts more voltage on the remaining barriers so that, with further increase in applied voltage, other large gaps will decrease, lowering more barriers. This mechanism could explain the very steep increases in σ observed for $E > 150\,\text{V/cm}$ in (TTF)(TCNQ) at 4 K.

At the end of the present section, where several aspects of the charge-density-wave mosaic inhomogeneity and pinning have been discussed, it is worth mentioning that in the presence of a very low concentration of irradiation-induced defects of 10^{-4}, the superconducting phase of $(\text{TMTSF})_2\text{PF}_6$ also becomes inhomogeneous, as shown on Fig. 19: a residual resistance appears due to parts of the sample in the normal state.

8. Impurity States in the Gaps of Irradiated Insulating Phases of Organic Conductors

The pinning of the CDW and SDW to defects causes the coherence length in the condensed phases to decrease, producing in turn the attenuation of the collective modes. However, single-electron properties are also affected by this pinning through changes in the CDW or SDW gap. How is it possible to describe the effects of irradiation-induced defects on such a gap? In usual semiconductors, it is expected that at low enough temperatures, the properties will be dominated by electrons and holes provided by defect levels in the gap. The application of this picture to TTF–TCNQ or $(TMTSF)_2PF_6$ is theoretically possible, even when the defect foreign potential is much larger than the gap, because the smaller the gap, the more effective is the screening of a defect, which scales with the gap through the dielectric constant value. The difficulty here lies in the fact that the opening of the gap occurs through a phase transition. It is related to the condensation of the electron gas and accompanied by the ordering of CDWs or SDWs. Now it has been shown that all these processes are affected by irradiation-induced defects.

Before discussing this point further, it is convenient to examine the case of organic semiconductors based on the TCNQ molecule-exhibiting stable gaps of the order of 0.5 eV. N-methyl derivatives of pyridinium with TCNQ, belonging to this class, were irradiated recently, and part of the results are presented in Figs. 10 and 11 (Przybylski *et al.*, 1983; Zuppiroli *et al.*, 1984). The conductivity and thermopower temperature variations in several nominally pure, as well as irradiated, N-methyl derivatives of pyridinium have been fitted sucessfully by using a model of compensated semiconductors in which irradiation defects are considered to provide new donor and acceptor levels in the gap. In the nominally pure sample, the Arrhenius plot of the conductivity exhibits two different activation energies: the high-temperature intrinsic regime and the low-temperature extrinsic one. Low-dose irradiation tends to push the extrinsic regime to higher temperatures, and an exhaustion range plateau corresponding to the situation where all donors and acceptors are ionized appears (Fig. 10). In a few words, the classical transport properties of doped semiconductors, as they appear in usual textbooks, do appear in the irradiated semiconductors as well and can be described in terms of states in the gap, even if this extended-state single-electron picture is somewhat questionable in a system in which electron mobilities are on the order of 1 cm^2/V sec and electron correlations are well known to play a very important role (magnetic properties). In fact, a more careful analysis of the localized states, which reveals a very strange and systematic electron-hole symmetry, has suggested that the main current-carrying excitations could be soliton-antisoliton pairs, the presence of which is triggered by the defects. These solitons do appear as states in the gap in the equivalent single-electron picture of the transport properties (Zuppiroli *et al.*, 1984).

Is it possible to extend the same kind of picture to the small CDW or SDW gaps? It probably is, at least at low doses. The Hall coefficient results in Fig. 13 on (TMTSF)(DMTCNQ) and $(TMTSF)_2PF_6$ show that a very low irradiation-defect concentration of 4.10^{-4} changes the Hall coefficient of the CDW as well as SDW system by an order of magnitude at 4 K. This corresponds approximately to the creation of one charge carrier per defect and may be consistent with a simple impurity picture. However, when the concentration of defects reaches 0.1 or 0.2%, the number of carriers reaches a value in the metallic state of the order of one per molecule; each defect is then the source of the order of 100 carriers. In this range of concentrations, transport properties may probably not be explained within a single impurity-band picture. In the previous discussion on CDW mosaicity, the possibility of setting up a network of phase defects of CDW structure of irradiated samples was mentioned. These charged discommensurations are probably the new sources of conduction in irradiated samples more than the defects themselves. These discommensurations states, sitting within the gap, are probably of the same nature as the soliton-like states considered by Conwell (1983) as a good explanation of the puzzling low-temperature transport properties of $(TMTSF)_2PF_6$. When the defects are numerous enough to restore a situation close to the metallic state, one can also speak in terms of a simple destruction of the gap due to disorder.

9. Localization Concentration Range and the Metallic Particle Picture

All the organic conductors and semiconductors containing irradiation defects at a level of a few percent display, without any exception, all the features of electron localization, in the sense that electrons are confined in more or less extended volumes of the sample. They explore these volumes mainly by elastic collisions, whereby they are forced to jump from volume to volume through an activated hopping, which is an inelastic process involving phonons.

In Figs. 11 and 12, showing the dose dependency of conductivity of several organic conductors, this localization range corresponds to the second part of the curves, in which conductivity decreases with dose, the first part corresponding to the situation in which defects act mainly as dopants.

In most organic conductors, the localization volume bounded by irradiation defects has been demonstrated to be a chain segment and the inelastic escape process a transverse short-range hopping (Zuppiroli *et al.*, 1980). In some, those in which the conductivity anisotropy is lower than 10 (methyl derivatives of pyridinium and TCNQ) or which are metallic at low temperatures [polysulfur nitride $(SN)_x$, niobium triselenide $NbSe_3$, (HMTSF)(TCNQ), (TMTSF)(DMTCNQ), or $(TMTSF)_2PF_6$ under pressure], the localization process is unquestionable, but its mechanisms are rather obscure. These two families of organic conductors will be examined successively in the two following sections.

a. An Assembly of Weakly Interacting Metallic Segments

It was stressed a few years ago that the energy-level structures within a metallic segment could give rise to very peculiar transport and magnetic effects (Rice and Bernasconi, 1972; Schegolev, 1972; Zuppiroli et al., 1980; Korin et al., 1982). In a segment of n molecules bounded by strong defects, the electronic-energy-level spacing is of the order of $\Delta E/n$ where ΔE is the longitudinal bandwidth. For very short segments (dimers, trimers, tetramers, etc.), energy splitting has been calculated by Herman et al. (1977). An average segment length of 100 molecules, corresponding to a defect concentration of 1 mole %, gives an energy splitting on the order of 10 meV in the TMTSF-based compounds. This is much larger than thermal effects ($k_B T$) in the whole usual temperature range, and larger than the usual magnetic excitation $\mu_B H$ for an electron-spin-resonance (ESR) or a classical magnetic susceptibility experiment. This energy splitting causes the transport and magnetic properties of the assembly of metallic chains to be changed to the electronic properties of an assembly of quasi-one-dimensional metallic particles.

The splitting was demonstrated to modify the transverse hopping frequency between neighboring chains τ^{-1} in the following manner (Zuppiroli et al., 1980).

$$\tau_\perp^{-1}(c) = \tau_\perp^{-1}(0) \exp\left(-\frac{\varepsilon c}{k_B T}\right), \quad (1)$$

where c is the defect molecular concentration and ε is on the order of the longitudinal bandwidth. This will, in turn, change the transverse resistivity in the lower anisotropy direction,

$$\rho_\perp(c) = \rho_\perp(0) \exp\left(\frac{\varepsilon c}{k_B T}\right), \quad (2)$$

and also the longitudinal resistivity because of the blocking effects of defects on the chains, which cause interchain series resistance to play an important role (Mihály et al., 1982).

$$\rho_\parallel(c) = \rho_\parallel(0) + \alpha c \exp\left(\frac{\varepsilon c}{k_B T}\right), \quad (3)$$

where α is a coefficient of the order of one. Because of the presence of metallic strands interrupted by defects, the microwave dielectric constant ε should vary as c^{-2} with the defect concentration (Rice and Bernasconi, 1972),

$$\varepsilon \sim c^{-2}. \quad (4)$$

The ESR linewidth is also a transport property related to the spin–flip rate in the sample. In quasi-one-dimensional conductors, even pure ones, the

ESR linewidth is much lower than the values predicted and observed in usual metals. The low spin relaxation rate is simply due to the confinement of the spins, obliged to diffuse in one dimension during the average time $\tau_\perp$, when there are no possibilities of spin relaxation in one-dimension. Weger (1978) has suggested that the spin–flip rate ΔH in a quasi-one-dimensional system is related to the escape time $\tau_\perp$ by the following relation

$$\Delta H \sim \frac{\Delta g^2}{\tau_\parallel} \frac{\tau_\parallel}{\tau_\perp},$$

where the first factor, containing the deviation Δg from the free electron value and the relaxation time $\tau_\parallel$ for the longitudinal coherent motion of the electron comes from the usual metal spin–flip theory of Eliott (1954); while the second term is the quasi-one-dimensional decrease factor of Weger.

When the chains of an organic conductor are transformed into metallic segments, the spins are even more confined than in the infinite chains and the linewidth is expected to decrease with irradiation because of the increase of the transverse time $\tau_\perp$ due to hopping. Following Weger (1978) and applying relation (1), an exponential decrease of the linewidth can be predicted.

$$\Delta H(c) \sim \Delta g^2 \tau_\perp^{-1}(c) = \Delta H(0) \exp\left(-\frac{\varepsilon c}{k_B T}\right). \tag{5}$$

These are the main features that can be predicted for the transport properties of organic conductors irradiated in their metallic state, and, indeed, the very simple relations (1) to (5) apply very well, as shown in Sections II.1.b and II.1.c and in Figs. 2–8.

As mentioned by Korin et al. (1982), most of the interpretations of the magnetic properties of irradiated organic conductors have focused on the idea that the molecular decomposition causes the chains to be broken into weakly interacting magnetic segments. Half of the segments, on average, enclose an unpaired spin and behave as an entity of spin $\frac{1}{2}$. These spin $\frac{1}{2}$ segments interact with each other to give a $T^{-\alpha}$ law at low temperatures. It is not surprising to find in these conditions that the number of spins measured in the low-temperature magnetic susceptibility experiments on irradiated organic conductors is indeed of the same order as the number of defects counted as interruptions in the conducting chains, as shown in Section II.5.

More precisely, it is not possible to localize spins on magnetic segments without sufficient Coulomb interactions preventing the spin carriers from tunneling and associating into nonmagnetic pairs. The models accounting for $T^{-\alpha}$ laws always contain some non-negligible on-site Coulomb interaction

U between electrons. Bondeson and Soos (1980) have developed a large U magnetic segment approach (random-exchange Heisenberg antiferromagnetic chain), which, in most cases, can only be considered a crude approximation. More realistic is the moderate U approach of Gorkov et al. (1978), giving a molecular concentration c of localized moments on the order of $(U/t)(a/L)$, where U is the on-site Coulomb interaction, t is the on-chain transfer integral, a the molecular stacking interspace, and L some localization length (or segment size).

Recently, the interrupted strand model with transverse hopping developed here was completed in order to include Coulomb effects (Zuppiroli, 1984).

b. Open Questions in the Field of Localization

In principle, in a one-dimensional system an infinitesimal random potential creates a localized state. All electronic eigenstates would therefore be localized, and electronic conduction could only occur via phonon-assisted hopping (Mott and Twose, 1961; Bush, 1972). This well-known principle has raised the idea in several minds that irradiation-induced defects, when present at the level of 1% of the molecules, should not substantially change the properties of a system in which all states are localized, even before irradiation. The first part of the present review has clearly shown that experiments contradict this idea. In practice, a large class of organic conductors exhibit in their metallic state a typical Drude behavior characteristic of the transport properties of travelling waves, and they are very sensitive to irradiation disorder at any temperature. A recent theoretical result of Bentosela et al. (1980) illustrates how much care an experimentalist needs to take to handle the theoretical results in the field of localization. First, these authors have proven, once again, that in one dimension all states are localized, that is to say, that the Hamiltonian $-(d^2/dx^2) + V(x)$ has a pure point spectrum with exponentially decaying eigenfunctions when $V(x)$ is any random potential describing any arbitrarily chosen small degree of disorder. Second, they have applied an electric field F and have shown that all states become extended, that is to say, the Hamiltonian $-(d^2/dx^2) + V(x)$ has a purely absolutely continuous spectrum for any arbitrarily small, constant external field F and for any arbitrarily high disorder $V(x)$. It is indeed difficult for an experimentalist working at zero temperature to measure a conductivity without applying an electric field! One may wonder to what extent some of the very stimulating results of the localization theories can aid in explaining experimental situations.

Of course, there are visible signs of localization in irradiated organic conductors, in the sense that electrons can be found in many situations to be confined within volumes of the sample that they can explore through elastic collisions and from which they can escape only by inelastic collisions involving phonons. In most organic conductors, the localization volume bounded by irradiation defects has been demonstrated to be a chain segment and the inelastic process a transverse short-range hopping, as shown in Section III.4.a. This is quite remarkable because there are very few localized systems in solid-state physics in which the nature of the localization, the shape of the localization volume, and the transport in the localized phase are understood so well. However, in several other irradiated organic conductors, even though the localization is evident, its mechanisms and the transport in the localized phase are not understood so well. This is the case each time the anisotropy of the organic conductor becomes small enough and/or the metallic phase is extended to low temperatures (by application of a hydrostatic pressure, for example), leading to a low-temperature, more or less dirty metal.

The interrupted-strands model with transverse hopping (Zuppiroli et al., 1980) is only applicable for sufficiently high anisotropies. When the conductivity anisotropy decreases below 10, the organic conductor becoming more quasi-two-dimensional than quasi-one-dimensional, the problem is different. It is easier to go around large random potentials in two dimensions than in one dimension. Thus a smoother decrease in conductivity is expected in 2D than the exponential decrease in 1D. This dimensionality effect on the type of localization has indeed been observed in the N-methyl derivatives of pyridinium with TCNQ, as shown on Fig. 22. However, the more 2D the system is, the less we understand the details of the transport processes in the localized state.

The more puzzling results concern low-temperature dirty metals such as irradiated $(SN)_x$, (TMTSF)(DMTCNQ), or (HMTSF)(TCNQ) under pressure and $NbSe_3$. Because the transport properties in the disordered phases of these compounds are not understood properly, it is interesting to look for correlations. Figure 23 demonstrates that at a given temperature (4.2 K), there is a rather general correlation between the temperature coefficient of resistivity $(d\rho/dT)_{4.2K}$ and the residual resistivity $\rho_{4.2K}$. At low irradiation doses the behavior of the three systems presented in Fig. 23 is still metallic. At doses corresponding to a resistivity on the order of 80 μOhm cm, the sign of $d\rho/dT$ changes. This value is approximately the same for the three compounds and seems to be some kind of maximum metallic resistivity at a given temperature.

This type of experimental correlation discovered by Mooij (1973) in a large variety of disordered systems is one of the open questions in the field of localization and dirty metals. Imry (1980) has expressed the hope of understanding this kind of correlation through the modern theories of incipient quantum localization effects. Girvin and Jonson (1980) attributed it to some

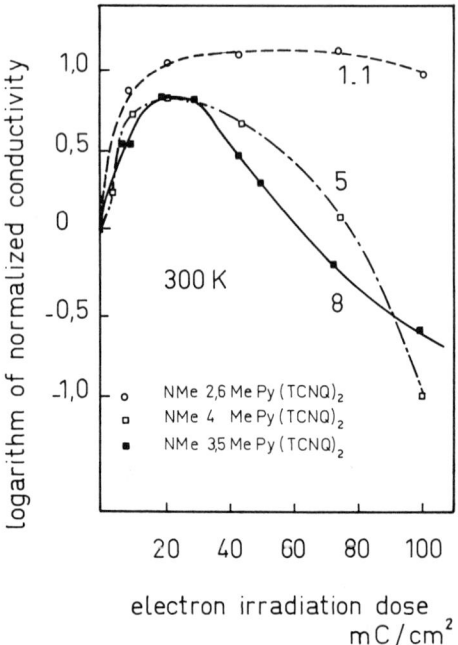

FIG. 22. Electron irradiations of low anisotropy N-methyl derivatives of pyridinium with TCNQ. Below the dose of 20 mC/cm², defects act mainly as dopants increasing the number of carriers; thereafter the behavior of the three compounds are very similar. With further irradiation the samples enter the localization range and conductivity begins to decrease. The manner in which it does so depends very much on the anisotropy. The quasi-two-dimensional-conductor N-Me-2,6-MePy(TCNQ)₂ exhibits a very smooth decrease. The higher the anisotropy, the faster the decrease. The numbers near the curves are the anisotropies of conductivity (1.1, 5, and 8). A quasi-one-dimensional-conductor with anistropy 100 would exhibit a strictly exponential decrease (see Figs. 11 and 12). [Experiments from Zuppiroli et al. (1985).]

general features of the electron–phonon interaction in disordered systems. Below some maximum metallic resistivity, which they calculated to be larger than 1000 μOhm cm, phonon scattering slows down the electron motion, as in usual metals, while above this value phonons could help the electronic motion. Finally, in a paper where a low-temperature upturn in resistivity proportional to the reciprocal square root of temperature ($T^{-1/2}$) was observed in nominally pure $(SN)_x$, Brimlow and Priestley (1983) called upon the interaction theory of Altshuler et al. (1980), according to which Coulomb interaction between electrons was found to interfere with elastic scattering by defects leading to a $T^{1/2}$ conductivity variation at low temperatures. Several new questions about localization in organic conductors are discussed more deeply in a paper of Zuppiroli (1987).

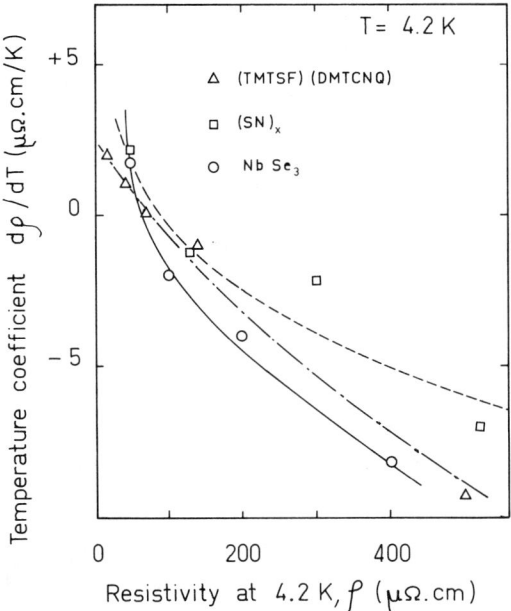

FIG. 23. Mooij correlation in irradiated low-dimensional conductors. The temperature coefficient of resistivity at 4.2 K has been plotted vs. the residual resistance at the same temperature. Although the three different compounds exhibit very few common features in their nominally pure state, the plots of $d\rho/dT$ vs. ρ of irradiated samples, at a given temperature, resemble each other very much. [The results concerning $(SN)_x$ are unpublished from Bandyopadhyay et al. (1984); those concerning NbSe$_3$ are from Fuller et al., (1981); and those concerning (TMTSF)(DMTCNQ) under pressure are from Bouffard et al. (1982).]

IV. Irradiation by Ionizing Particles: Local Chemistry of Excited Molecules

10. INTRODUCTION: THE MECHANISMS OF DAMAGE PRODUCTION

Having examined and discussed the physical properties of organic conductors disordered by irradiation, it is useful to present a few elements of and references to the mechanisms of damage production and to outline the actual knowledge about the chemical and crystallographical structure of radiation-induced defects.

Physicists have the tendency to forget that organic conductors are very interesting and rather complex chemical systems. Physics can be considered as a strong simplification of these systems into assemblies of metallic, weakly interacting conducting chains. Therefore, most of the richness of the chemical approach, with all the details of the different bondings and proximity effects, is lost in favor of a simpler scheme, containing only two or three parameters such as forward- and backward-scattering g_1, g_2, g_3 or longitudinal and transverse scattering times, etc. Irradiation at a sufficient dose has been shown to be a further simplification of the organic conductor properties, leading to universal magnetic and transport properties, independent of the details of the electronic structure. Fortunately, the production of irradiation defects is a molecular chemical process, and the conducting character of the crystal, as well as the anisotropy in the band electronic properties, will be shown to not play a very important role in this process, directed mainly by the local properties of the molecules.

The starting point of the defect production by ionizing particles is the local ionization or excitation of an atom of a molecule. Because of the high polarizability of the molecules composing organic conductors, this kind of excitation can travel very easily and quickly within the molecule. It can also travel quickly in the crystal, from one molecular species to another, through a particular form of the charge transfer, or along the chain, because of the large electronic overlaps in this direction. Most of the excited states will recombine either radiatively or nonradiatively through multiphonon processes. Only a small part of them will produce more or less stable free radicals. Most of these radicals will, in turn, recombine. There are no signs, in fact, of the presence of stable, charged free radicals in irradiated organic conductors, but a small part of the unstable radicals will remain in the structure long enough to participate in a local chemistry leading to the creation of a new local chemical species: the neutral stable defect. In conclusion, the production of stable defects by ionizing particles is a radiolytic process, resulting from the existence of a strongly excited state and the competition between all the escape and recombination processes, on the one hand, and the limited possibility of some local chemistry, on the other hand.

Some very important properties of irradiation in organic conductors are related simply to the radiolytic nature of the production process. Similar results can be found in irradiated ionic crystals (Hersh, 1966) or ordinary molecular crystals or polymers (Gaumann and Hoigne, 1968; Slifkin and Yakoumakis, 1976). The three principal properties of the radiolytic process are:

1. The concentration of defects, c, is proportional to the energy E absorbed by the crystal through electronic excitations. This energy can be easily calculated using only atomic considerations (Mihály and Zuppiroli, 1982). $c = E/\bar{E}$, where $\bar{E}$ is the average energy absorbed per molecular unit in order to produce a stable defect.

2. The damage production rate, being limited by multiphonon processes,

is expected to be temperature dependent. This applies to ionic crystals and normal molecular crystals and to organic conductors as well (Zuppiroli *et al.*, 1986).

3. There are only limited possibilities for a local chemistry of excited molecules to take place, and this leads to a relatively low number of defect centers.

The main quantitative results and discussions about organic conductor radiolysis can be found in a review paper (Zuppiroli. 1982). The average absorbed energy for production of a defect is shown to be on the order of a few tens of keV. This means that only one excitation event over a few hundred or even thousand leads to the production of a stable defect. These numbers, $\bar{E}$, are of the same order in ionic crystals but are usually one or two orders of magnitude lower in nonconducting molecular crystals, exept those in which molecules are very stable and where charges can travel rather quickly, such as in semiconducting phtallocyanines.

However, what are the chemical and crystallographical configurations of the defects? The only attempt to answer this question is contained in the recent paper of Mermilliod and Sellier (1983) on irradiated quinolinium (TCNQ)$_2$. The mass spectrometry of pure and irradiated samples has revealed the presence in the damaged samples of a new molecule formed by the association of a quinolinium and a neighboring TCNQ molecule.

Before concluding this very short discussion on the damage mechanisms in irradiated organic conductors, it is worth mentioning that everything that has been presented here applies only to fast-electron, x-ray, gamma-ray or fast-ion irradiation, when the main damage is due to electronic excitations. In the case of neutron irradiations, which have also been used extensively for irradiating organic conductors (Ero Gecs *et al.*, 1979; Grüner and Khanna, 1979; Holczer *et al.*, 1979; Zuppiroli *et al.*, 1980), the damage is mainly due to atomic collisions. Calculations of the displacements cross sections for these processes are given in Zuppiroli *et al.* (1980) and Mihály and Zuppiroli (1982).

11. EXPERIMENTAL DETERMINATIONS OF
    DEFECT CONCENTRATIONS

In their early work on (TTF)(TCNQ) Chiang *et al.* (1977) determined the number of defects by studying the low-temperature magnetic properties of their irradiated samples and by counting the paramagnetic centers that they found. Section II.5 and Fig. 20 of this review provide a justification of their magnetic method. A few years later, Mihály *et al.* (1981, 1982) proposed another experimental method based on the simultaneous measurement of the longitudinal and transverse resistance of the irradiated sample at room temperature. The justification of their procedure was based on the following

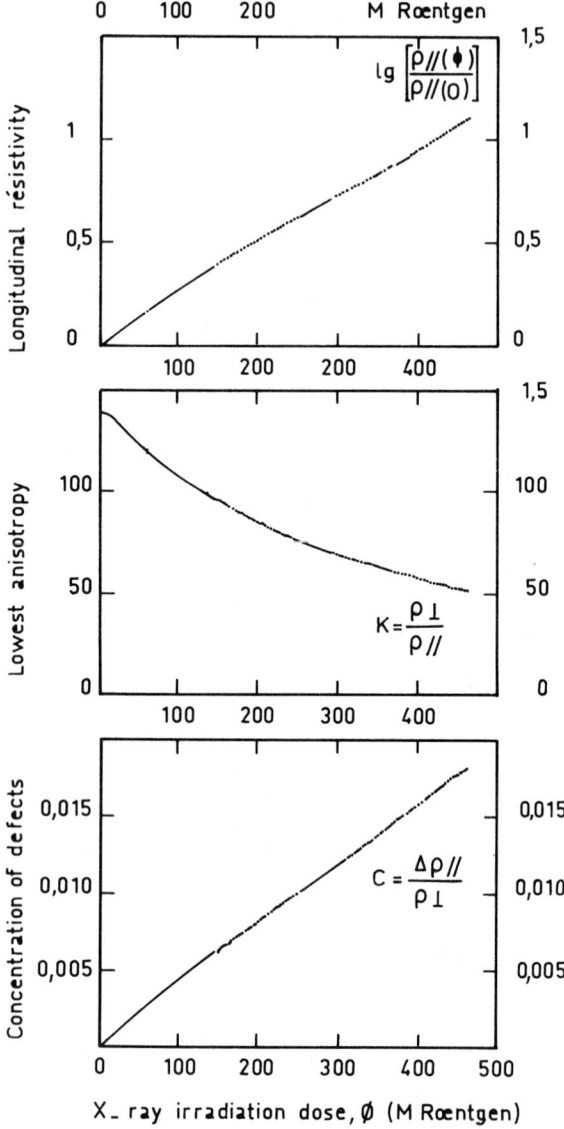

FIG. 24. Room temperature determination of the concentration of defects in irradiated (TMTSF)(DMTCNQ). The sample was irradiated in a conventional x-ray generator using a copper anode. The anisotropy $K = \rho_\perp/\rho_\parallel$, as well as the longitudinal resistivity, were measured by the Montgomery method. The defect concentration is a simple combination of the two quantities $c(\phi) = [1/K(\phi)][1 - \rho_\parallel(0)/\rho_\parallel(\phi)]$. The linear variation of the defect concentration with the dose ϕ, though $\rho_\parallel$ varies exponentially with ϕ, verifies the validity of the damage model.

argument: Each defect interrupts the conducting chain and the electrons are forced to jump to a neighboring chain. The chain direction resistivity increases by the fraction of transverse resistivity mixed into the longitudinal conduction path. This leads to a decrease in the anisotropy with increasing defect concentration. On the other hand, the chain segmentation effect also modifies the transverse conductivity, which varies exponentially with the defect concentration. These effects, described in detail in Section III.4.a, lead to the following formula for the fraction c of damaged molecules producing breaks in the conducting chains:

$$c \sim \frac{\Delta \rho_\parallel(\phi)}{\rho_\perp(\phi)},$$

where ϕ is the fluence of incident particles, $\Delta \rho_\parallel$ is the increase in longitudinal resistance, and $\rho_\perp$ is the value of the transverse resistance in the direction of lowest anisotropy. The applicability and the precision of the above model, especially for *in situ* dose-dependent measurements, is demonstrated in Fig. 24, in which the longitudinal resistance, the anisotropy, and the concentration c are plotted vs. irradiation dose in the case of a 8 keV x-ray irradiation of (TMTSF)(DMTCNQ).

It was interesting to compare in the same samples the number of spins deduced from the low-temperature magnetic properties with the number of defects determined by room-temperature conductivity measurements. This was done recently by Zuppiroli *et al.* (1982) and by Korin *et al.* (1982). In both cases, the number of spins and the number of defects are of the same order, but in both cases there is a significant difference between the two results (two or three times as many spins as interruptions in the conducting chains). This is probably due to the percolative character of the conduction in the irradiated sample, which the above model neglects. If we assume, in the case of (TTF)(TCNQ), for example, that there are twice as many defects produced in the TCNQ chains as in the TTF chains, the conducting path will contain preferentially long TTF segments and the number of interruptions counted along this path will be lower than the average.

V. Conclusion: Perfect Crystals and Real Crystals

Irradiation experiments on organic conductors have provided an important series of new results on the role of disorder in the different phases of these systems.

At low concentrations ($< 10^{-3}$), irradiation-induced defects change the coherence lengths of charge-density waves, spin-density waves, or superconductivity and, in turn, the transport properties of these phases. Our knowledge about these temperature and concentration ranges remains primarily experimental and the interpretations very qualitative. At higher doses, larger than 1%, most organic conductors lose a large part of their specificity, at least in their transport and magnetic properties. They become granular systems of metallic segments, the properties of which are mainly related to the energy separation between the electronic states in the metallic particles. This is quite remarkable, because the effects of the quantization of the motion of electrons in metallic small particles are not so easy to observe in 3D granular systems, although they have raised considerable interest since they were predicted by Frölich in 1937. In the 1D case, the transport and magnetic properties of such assemblies of metallic segments are described quite satisfactorily by the phenomenological models presented in the review. Organic conductors, which are more 2D systems than 1D, and/or which are low-temperature dirty metals, also enter the range of electron localization when irradiated at a sufficient dose; however, the nature of this localization and the dependences of the transport properties in these phases are not properly understood. One may wonder to what extent the modern theories of localization can help to find a solution. In the same vein it is interesting to notice that the motion of electrons in the semiconducting phases of organic conductors are surely strongly polaronic, and the more they are disordered by irradiation, the more they are polaronic. One may wonder if the proper application of polaronic theories will increase the present knowledge of these systems.

Irradiation is an artificial way of disordering solids. In organic conductors, the radiation-induced defects appear as strong electronic perturbations in the sense that they usually affect the electronic properties much more than isoelectronic foreign molecules of an alloy. This statement gives rise to a question: In the nominally pure crystals produced by organic or inorganic syntheses, are there defects similar to those created artificially by irradiation? I believe that most of the defects that limit the properties of organic conductors are indeed of the same nature as irradiation defects. They define an average chain length for electronic and magnetic properties similar to the chain length introduced for different purposes in polymer sciences. Quinolinium $(TCNQ)_2$ provides a good example of an organic conductor containing 2 or 3% of strong defects that are very similar to irradiation defects. It was already mentioned that the conductivity vs. temperature curve of nominally pure $Qn(TCNQ)_2$ strongly resembles that of irradiated quasi-one-dimensional conductors of the (TTF)(TCNQ) family (Zuppiroli and Bouffard, 1981). This is consistent with the similarities in the susceptibilities of irradiated (TTF)(TCNQ) and "pure" $Qn(TCNQ)_2$ noted recently (Korin et al., 1982). In their study of the dielectric constant of irradiated $Qn(TCNQ)_2$, Jánossy et al. (1982) noticed that the extrapolated concentration of impurities that can account for the properties of nominally pure material is 3% (see Fig. 8). Cooper *et al.* (1981) also needed 3% of "localized" spins to explain the intensity of the low-temperature tail in the magnetic susceptibility of the "pure" sample. Finally, in their study of the x-ray diffuse scattering by irradiated samples, Forró *et al.* (1982, 1983) have demonstrated the large resemblance between strongly irradiated organic conductors and the patterns of the so-called "pure" $Qn(TCNQ)_2$.

All these results converge on the idea that nominally pure $Qn(TCNQ)_2$, as well as other unirradiated organic conductors, contain strong defects similar to irradiation defects. Produced by some local chemistry of excited molecules in the irradiated samples, the strong defects can also be produced by some accident in the normal chemistry of the syntheses and crystal growths. This result does not mean, of course, that $Qn(TCNQ)_2$ is a simple disordered 1D-metal. Coulomb interactions as well as a $4k_F$ local distortion are known to play also a role in this conduction (Conwell and Howard, 1986).

Acknowledgments

I very much thank my colleagues of the organic conductor group in Fontenay-aux-Roses for their help and D. Poletti for the typing of the manuscript.

References

Abeles, B. (1976). *In* "Applied Solid State Science" (R. Wolfe, ed.), Vol. 6, Chap. 1. Academic Press, New York.
Abrahams, E., Anderson, P. W., Licciardello, D. C., and Ramakrishnan, T. V. (1979). *Phys. Rev. Lett.* **42**, 673.
Andrieux, A., Schulz, H. J., Jérome, D., and Bechgaard, K. (1979). *Phys. Rev. Lett.* **40**, 227.
Altshuler, B. L., Khmel'nitzkii, D., Larkin, A. I., and Lee, P. A. (1980). *Phys. Rev. B* **22**, 5142.
Bardeen, J. (1980). *Phys. Rev. Lett.* **45**, 1978.
Bentosela, F., Carmona, R., Duclos, P., Simon, B., Souillard, B., and Weder, R. (1983). *Commun. Math. Phys.* **88**, 387.
Beuneu, F., and Monod, P. (1976). *Phys. Rev. B* **13**, 3424.
Brimlow, G. M., and Priestley, M. G. (1983). *Solid State Commun.* **45**, 1063.
Bloch, A. N., Weisman, R. B., and Varma, C. M. (1972). *Phys. Rev. Lett.* **28**, 753.
Bondenson, S. R., and Soos, Z. G. (1980). *Phys. Rev. B* **22**, 1973.
Bouffard, S., Chipaux, R., Jérome, D., and Bechgaard, K. (1981). *Solid State Commun.* **37**, 405.
Bouffard, S., Ribault, M., Brusetti, R., Jérome, D., and Bechgaard, K. (1982). *J. Phys. C.: Solid State Phys.* **15**, 2951.
Bulaevskii, L. N., Zvarykina, A. V., Karimov, Y. S., Lyubovskii, R. B., and Shchegolev, I. F. (1972). *Zh. Eksp. Teor. Fiz.* **62**, 725; (*JETP* **35**, 384).
Bush, R. L. (1972). *Phys. Rev. B* **6**, 1182.
Chaikin, P. M., Mu-Yong Choi, Haen, P., Engler, E. M., and Green, R. L. (1982). *Mol. Cryst. Liq. Cryst.* **79**, 79.
Chiang, C. K., Cohen, M. J., Newman, P. R., and Heeger, A. J. (1977). *Phys. Rev. B* **16**, 5163.
Conwell, E. M. (1983). *Phys. Rev. B* **27**, 7654.

Conwell, E. M., and Banik, N. C. (1981). *J. de Physique* (Paris) Colloque C7, **42**, C7-315.
Conwell, E. M., and Banik, N. C. (1982). *Phys. Rev.* B **26**, 530
Conwell, E. M., and Howard, I. A. (1986). *Synthetic Metals* **13**, 71.
Cooper, J. R., Miljak, M., and Korin, B. (1981). *Chemical Scripta* **17**, 79.
Cooper, J. R., Weger, M., Jérome, D., Le Fur, D., Bechgaard, K., and Bloch, A. N. (1976). *Solid State Commun.* **19**, 1149.
Delhaes, P., Coulon, C., Manceau, J. P., Flandrois, S., Hilti, B., and Mayer, C. W. (1978). Proceedings of the International Conference on Quasi-one-Dimensional Conductors, Dubrovnik. Lecture Notes in Physics (J. Ehlers *et al.*, eds.), Vol. 96. Springer, New York.
Eliott, R. J. (1954). *Phys. Rev.* **96**, 266.
Emery, V. J. (1976). *Phys. Rev. Lett.* **37**, 107.
Engler, E. M., Scott, B. A., Etemad, S., Penny, T., and Patel, V. V. (1977). *J. Am. Chem. Soc.* **99**, 5909.
Epstein, A. J., Conwell, E. M., Sandman, D. J., and Miller, J. S. (1977). *Solid State Commun.* **23**, 355.
Epstein, A. J., Miller, J. S., Pouget, J. P., and Comès, R. (1981). *Phys. Rev. Lett.* **47**, 741.
Ero-Gecs, M., Forró, L., Vancsó, G., Holczer, K., Mihály, G., and Jánossy, A. (1979). *Solid State Commun.* **32**, 845.
Farges, J. P. (1980). *In* "The Physics and Chemistry of Low Dimensional Solids" (L. Alcacer, ed.), pp. 223–232. Reidel, Dordrecht.
Forró, L. (1982). *Mol. Cryst. Liq. Cryst.* **85**, 315.
Forró, L., Bouffard, S., and Zuppiroli, L. (1983). Proceedings of the Internation Conference on Low-dimensional Conductor (Les Arcs, 1982). *J. de Physique.* C **3**, 927.
Forró, L., Jánossy, A., Zuppiroli, L., and Bechgaard, K. (1982). *J. de Physique* **43**, 977.
Forró, L., Bouffard, S., and Pouget, J. P. (1984). *J. Physique Lett.* **45**, L-543.
Forró, L., Zuppiroli, L., Pouget, J. P., and Bechgaard, K. (1983). *Phys. Rev.* B **27**, 7600.
Fröhlich, H. (1937). *Physica* **6**, 406.
Fröhlich, H. (1954). *Proc. R. Soc. London* A **223**, 296.
Fuller, W. W., Grüner, G., and Chaikin, P. M. (1981). *Phys. Rev.* B **23**, 6259.
Gaumann, T., and Hoigne, J. (1968). "Aspects of Hydrocarbon Radialysis." Academic Press, New York.
Girvin, S. M., and Jonson, M. (1980). *Phys. Rev.* B **22**, 3583.
Grüner, G. (1981). *Chemica Scripta* **17**, 207.
Grüner, G. (1983). *Comments Solid State Phys.* **10**, 183.
Grüner, G., and Khanna, S. K. (1979). *Solid State Communications* **32**, 1233.
Gunning, W. J., Chiang, C. K., Heeger, A. J., and Epstein, A. J. (1979). *Phys. Stat. Solidi (b)* **96**, 145.
Herman, F., Saluhub, D. R., and Messmer, R. P. (1977). *Phys. Rev.* B **16**, 2453.
Holczer, K., Grüner, G., Mihály, G., and Jánossy, A. (1979). *Solid State Commun.* **31**, 145.
Imry, Y. (1980). *Phys. Rev. Lett.* **44**, 469.
Jacobsen, C. S., Mortensen, K., Andersen, J. R., and Bechgaard, K. (1978). *Phys. Rev.* B **18**, 905.
Jánossy, A., Holczer, K., Hsieh, P. L., Jackson, C. M., and Zettl, A. (1982). *Solid State Commun.* **43**, 507.
Jánossy, A., Mihály, G., Forró, L., Cooper, J. R., Miljak, M., and Korin-Hamzić, B. (1982). Proceedings of the International Conference on Low-Dimensional Conductors, Boulder, Colorado, August 1981. *Mol. Cryst. Liq. Cryst.* **85**, 233.
Kamaras, K., Holczer, K., and Jánossy, A. (1980). *Phys. Stat. Sol. (b)* **102**, 467.
Korin-Hamzic, B., Miljak, M., and Cooper, J. R. (1982). Proceedings of the International Conference on Low-Dimensional Conductors, Boulder, Colorado, August 1981. *Mol. Cryst. Liq. Cryst.* **85**, 177.

Mermilliod, N., and Sellier, N. (1983). International Conference on Quasi-One-Dimensional Conductors (Les Arcs, 1982). *J. de Physique.*
Lacoe, R. C., Schulz, H. J., Jérome, D., Bechgaard, K., and Johannsen, I. (1985). *Phys. Rev. Lett.* **55**, 2351.
Mihály, G., Bouffard, S., Zuppiroli, L., and Bechgaard, K. (1980). *J. de Physique* **41**, 1495.
Mihály, G., and Zuppiroli, L. (1982). *Philosophical Magazine A* **45**, 549.
Mooij, J. H. (1973). *Phys. Status Solidi (a)* **17**, 521.
Monceau, P., Richard, J., and Lannier, R. (1981). *J. Phys. C: Solid State Phys.* **14**, 2995.
Mott, N. F., and Twose, W. D. (1961). *Adv. Phys.* **10**, 107.
Mutka, H. (1983). *Phys. Rev. B* **28**, 2855.
Mutka, H., Bouffard, S., Mihály, G., and Mihály, L. (1984). *J. de Physique—Lettres* **45**, L-113.
Mutka, H., and Housseau, N. (1983). *Philosophical Magazine A* **47**, 797.
Poujet, J. P. (1982). Proceedings of the International Conference on Low-Dimensional Synthetic Metals. *Chemica Scripta* **17**, 85.
Przybylski, M., and Graja, A. (1981). *Physica* **104 B**, 278.
Przybylski, M., Pukacki, W., and Zuppiroli, L. (1983). Proceedings of the International Conference on Low-Dimensional Conductors (Les Arcs, 1982). *J. de Physique* **C3**, 1369.
Rice, M. J., and Bernasconi, J. (1972). *Phys. Rev. Lett.* **29**, 113.
Sanquer, M., Bouffard, S., and Forró, L. (1985). *Mol. Cryst. Liq. Cryst.* **120**, 183.
Schultz, T. D., and Craven, R. A. (1979). In "Highly Conducting One-Dimensional Solids" (J. T. Devreese, R. P. Evrard, and B. E. Van Doren, eds.). Plenum Press, New York.
Shchegolev, I. F. (1972). *Phys. Status Solidi (a)* **12**, 9.
Slifkin, M. A., and Yakoumakis, E. (1976). *Radiation Effects* **29**, 241.
Tanner, D. B., and Jacobsen, C. S. (1982). *Mol. Cryst. Liq. Cryst.* **85**, 137.
Tippie, L. C., and Clark, W. G. (1981). *Phys. Rev. B* **23**, 5846.
Trouilloud, P., Bouffard, S., Ardonceau, J., and Zuppiroli, L. (1982). *Philosophical Magazine B* **45**, 277.
Weger, M. (1978). *J. de Physique Colloque C6* **39**, C1456.
White, A. E., Tinkham, M., Skocpol, W. J., and Flanders, D. C. (1982). *Phys. Rev. Lett.* **42**, 1752.
Zuppiroli, L. (1982). *Radiation Effects* **62**, 53.
Zuppiroli, L. (1984). *Berichte der Bunsen-Gesellshaft für Physikalische Chemie* **88**, 304.
Zuppiroli, L. (1987). In NATO ASI Series B (D. Jérome and L. G. Caron, eds.). *Physics* **155**, 307.
Zuppiroli, L., Ardonceau, J., Weger, M., Bechgaard, K., and Weyl, C. (1978). *J. de Physique (Lettres)* **39**, L-170.
Zuppiroli, L., and Bouffard, S. (1981). *Chemica Scripta* **17**, 7.
Zuppiroli, L., Bouffard, S., Bechgaard, K., Hilti, B., and Mayer, C. W. (1980). *Phys. Rev. B* **22**, 6035.
Zuppiroli, L., Delhaës, P., and Amiell, J. (1982). *J. de Physique* **43**, 1233.
Zuppiroli, L., Housseau, N., Forró, L., Pelissier, J., and Guillot, J. P. (1986). *Ultramicroscopy* **22**, 325.
Zuppiroli, L., Mutka, H., and Bouffard, S. (1982). Proceedings of the International Conference on Low-Dimensional Conductors (Boulder, 1981). *Mol. Cryst. Liq. Cryst.* **85**, 1.
Zuppiroli, L., Przybylski, M., and Pukacki, W. (1985). *J. de Physique* **45**, 1925.

Index

A

Acceptor, 3, 88, 90
Acoustic phonons, 8, 222-223
   dispersion, 122, 124, 126
   electron coupling to, 15, 132, 224-226, 242, 244-245
   $2k_F$, 8
   pressure- and temperature-dependence, 252
   scattering by, 230-234, 246-247, 268
   soft phonons, 8, 9, 221
   structural instabilities, role in, 132
Ad(TCNQ)$_2$, 31, 305
Anion ordering, 8, 359, 417, 420, 429
Anisotropy
   electrical, 3
   optical, 15-16, 294, 329, 338-342

B

Band structure, 1D charge-transfer salts, 101, 218
   anisotropy, 338-342
   dispersion relation, 218-219
Bechgaard salts, 4-5, *see also* TMTSF salts; specific TMTSF and TMTTF compounds
(BEDT-TTF)$_2$I$_3$,
   α-form, 341-343, 370
   β-form, 5, 33, 323, 341-343
BEDT-TTF salts, 341, 356, 369-371
β-(BEDT-TTF)$_2$AuI$_2$, 370-372
BIPA(TCNQ)$_2$, 35, 333
Bolometry, 307, 361-362
Boltzmann equation, 229-230, 237, 249

C

Carrier concentration, 10, 243-244, *see also* Charge transfer
Charge-density waves, 6-8, 10, 99
   coherence (correlation) length, 110, 117, 122-123, 125, 140-142, 196-197, 462-464
   conduction by, 12-13, 19-20, 106, 114, 216, 281-282, 286-287, 456
   Coulomb interactions between, 20, 158
   3D coupling, 8, 156-161
   fluctuations, 8, 110-155
   phase oscillations (phonons), 10, 24, 352-353, 357
   pinned, infrared absorption due to, 23-24, 363, 366-368
   pinning due to
      commensurability, 106
      impurities, defects, 19-20, 453-454, 456
Charge transfer between donor and acceptor stacks, 93, 98, 101-102, 105
   determination by
      measuring $2k_F$ or $4k_F$, 98
      vibrational spectroscopy, 349-351
   effect on $2k_F$ and $4k_F$ instabilities, 153
   factors that determine, 98, 103-104
   lock-in at commensurate value, 104
   pressure, effect of, 104, 106
Collective transport, coupled electrons and phonons, 8, 12-13, 16-17, 216, *see also* Charge-density waves
Conductivity, d.c., 3, 4, *see also* Charge-density waves, conduction; Mobility; Nonlinear conduction; Relaxation times; (TMTSF)$_2$PF$_6$; TTF-TCNQ
   metallic range ($T > T_P$), 11-15, 215-254
   charge-density wave contribution, 12-13, 16-17, 19-20, 106, 114, 216, 458
   data, agreement with theory, 11-15, 242-253
   electron–electron scattering, 13, 216
   one-phonon vs two-phonon processes, 13-15, 225-229, 234-236, 238-240, 244-246

INDEX

phonon scattering, theory of, 224–240
polarization-fluctuation scattering, 228, 236
pressure-dependence, 14–15, 252–253
transverse, 3, 253–254
semiconducting range ($T < T_P$), 19–23, 266–289
  charge-density wave transport, 281–282
  data for low fields, agreement with theory, 274–281
  electric-field induced nonlinearity, 281–289
  hot electron range, 282–289
  impact ionization, 287
  ionized impurity scattering, 270–274
  phonon scattering, theory, 266–270
  spatially varying gap, effects of, 279–281, 288–289
  steep increase with field, mechanism for, 286–289
  warm electron range, 284–285
Conductivity frequency-dependent, 16–18, 23–24, 298, 300–301, 326, 352, 354, 362, 365, 371, 373–376
  infrared absorption peak, 372–376
Coulomb effects, *see* Electron–electron correlations
$(CS)_2(TCNQ)_3$, 41, 305, 325–328, 345, 350

D

DBTTF-TCNQCl$_2$, 44, 102, 104, 112, 135–136, 147–149, 333, 350, 373–375
Deformation potentials, 244–245, 249, *see also* Electron-phonon coupling
Density of states, 1D
  below, T_P, 134
  measurement through spin susceptibility, 387
  metallic range, 220
Dielectric function, 16, 18, 295–302, 309–312, 353, 366–367, 372
Dimer optics, 309–312, 316–318, 323–328
Dimethyltetracyanoquinodimethane, *see* DMTCNQ
DIPSφ$_4$I$_{2.28}$, 48, 333, 443
Disorder, 437–439, *see also* Irradiation effects
  effect on x-ray diffuse scattering, 117, 201

localization, 461
structural, 94–95
suppression of Peierls transition, 116–119, 416–417
Dispersion relation of electrons
  metallic range, 105, 218–219
  semiconducting range, 257, 264
DMeFc-TCNQ, 48, 325
DMTCNQ, 3, 91
DMTTF-TCNQ, 50, 399
Donor, 3, 88, 90
Drude behavior, 297–298, 302, 330
DSDTF-TCNQ, 51, 112
DSTSF-TCNQ, 102

E

Effective mass
  below T_P, 263–265, 276, 278
  metallic state, 219
Electron–electron interactions, 6, 100–110, 220–222
  band filling, dependence on, 7, 221
  Hubbard model, 6, 220
  $2k_F$, $4k_F$ distortions, effect on, 142–156
  magnetic properties, effect on, 221, 388, 395–397
  optical spectra, effect on, 312–315, 375
  screening, effect of, 7, 220–221
Electron–electron scattering, 13, 216
Electron energy-loss spectroscopy, 308, 343–345
Electron–molecular vibration coupling, *see* Electron–phonon coupling
Electron–phonon coupling, 8–9, 89, 110, 156, 203, *see also* Deformation potentials
  coupling constant from gap equation, 9, 259–261
  coupling due to modulation of transfer integral, 224–227, 261
  coupling due to polarization fluctuations, 228
  coupling to internal modes, 224, 227, 260, 350–356
  one and two-phonon processes, 15, 224–226, 244–246
  values of coupling constants for TCNQ, 242

F

Fano interference, 356
Free energy of electrons and lattice
　electronic free energy, 257, 260-261
　energy of lattice vibrations, 258-259
　fourth order terms, 182, 200
　Landau-Ginzburg development, 197-200
　second order terms, 177-178, 197-200
　umklapp processes, 153, 187-188, 190

G

Gap, 6-7, 9, *see also* Peierls transition
　carrier injection, effect of, 262-263
　contribution of internal modes and librons, 254-261
　determination of (optical), 360-362
　equation, for arbitrary band filling, 254-263
　high electric field, effect of, 288-289
　strain, effect of, 279-280
　spatial variation, consequences of, 279-281
　temperature-dependence, 19-20, 22, 24, 261-262

H

Hall coefficient, 12, 451, 461, 466-467
Hamiltonian, electron-phonon
　band of arbitrary filling, 255
　diagonalization, below T_P, 257
　Hubbard, 131, 313
　momentum representation, 225, 255
　site representation, 244
　spin-density wave case, 256-257
　wavefunctions and energies below transition temperature, 257-258
Hexamethylene-tetraselenafulvalene tetracyanoquinodimethane, *see* HMTSF-TCNQ
HMDSDTF-TCNQ, 350-351
HMTSF-TCNQ, 3, 56, 91-94, 102, 104, 112, 166, 120-121, 123, 125, 129, 132, 136-137, 139-144, 146, 149, 153, 160, 173-174, 190-193, 328-330, 333-334, 350-351, 388, 399, 407, 410-412, 428-429, 443, 451, 467, 471
HMTSF-TCNQF$_4$, 103, 221
HMTSF-TNAP, 56, 91-94, 102, 111-114, 129, 136, 139, 142, 149-150, 153, 160, 333, 373-375, 417, 430
HMTTF-TCNQ, 58, 91-94, 102, 112, 115, 125, 136-137, 139, 142, 151, 153, 157, 160, 172-174, 190-193, 219, 333-334, 350-351, 399, 403-404, 407, 413, 445
HMTTF-TCNQF$_4$, 103, 221

I

Interchain coupling, 110-111, 114-116, 156-161
Internal vibration modes, 10, 14-15, 130-131
　a_g (symmetric) modes, 223, 227, 350-351
　coupling of electrons to internal modes, 224, 227, 260, 350-356
　coupling of electrons to internal modes, experimental determination, 352-356
　infrared absorption due to a_g modes, 10, 24, 352-356
Ionized impurity scattering, 1D, 270-274
Irradiation effects, 437-481
　conductivity, on, 439-446, 448-451, 456-459, 462
　defect concentrations, 475-477
　dielectric constant, on, 447
　ESR linewidth, on, 442-447
　Hall effect, on, 451
　localization, 467-473, 478-479
　magnetic properties, on, 460
　mechanism of damage production, 473-475
　phase transitions, on, 441, 448, 452-453
　pinning of charge-density waves, 453-454, 458, 462-465
　superconductivity, on, 459

K

$2k_F$, $4k_F$ distortions of electron gas, 6-10, 17, 98-156, *see also* Charge-density waves
Coulomb interactions, effect of, 150-151
crossover temperature, 111, 113, 115
dimensionality of fluctuations, 111-116
electronic origin, 142-149
fluctuations, 110-155
irradiation effects, 453-454

large U, effect of, 6, 221
physical parameters controlling, 149–156
Kohn anomaly, 8, 122, 124, 126, 128, 200, 203, 375
$K_2Pt(CN)_4 0.3Br, \times H_2O$, 88, 126
Kramers–Kronig relations, 302, 306–307
K-TCNQ, 58, 103, 221, 305, 324–325, 350

L

Lattice vibrations, 222–223, 360, *see also* Acoustic phonons; Internal vibration modes; Librons
Librons, 152, 223
Localization, 461, 467–473
  interrupted strand model, 468–471

M

Madelung Energy, 93, 98
Magnetic Properties
  Curie spins, 416
  disorder and defect effects, 416–417
  electron–electron correlations, effect of, 221, 388, 395–397, 401–402
  ESR linewidth, 391, 404–408
  g-tensor, 389–391
  magnetic susceptibility, 6, 386–389
  Peierls transition, effects of, 412–413
  spin-density wave transition, effect of, 418–420
  spin-lattice relaxation, effect of, 1D diffusion, 394–395
  spin susceptibility, 389, 398–402
Matrix elements for electron–phonon scattering
  metallic range, 224–229
  semiconducting range, 266–267
Maximum metallic resistivity, 471–473
Mean-field theory
  critical temperature, 110, 200
  validity of, 8, 110–111, 134–136, 147–149, 152, 355
Mobility of carriers, 238, *see also* Conductivity, d.c.
  below Peierls transition, 20, 22, 266, 270
  dependence on size of Peierls gap, 22, 276–278
  metallic range, 15

Molecular vibrations, *see* Internal vibration modes
$MTPP(TCNQ)_2$, 62, 350

N

Na-TCNQ, 103, 350
$NbSe_3$, charge-density waves in, 19–20, 286, 467, 471, 473
Neutron scattering, 88, 122, 124, 128, 166
  theory, 201–203
Ni(Pc)I, 64, 332
$NMeAd(TCNQ)_2$, 66, 416
$(NMe_3H)I(TCNQ)$, 373–374
$(NMe\text{-}3,5\text{-}MeP_y)(TCNQ)_2$, 67, 448–450, 466–467
N-methyl derivatives of pyridinium with TCNQ, 67, 466, 471–472
N-methylphenazinium, *see* NMP-TCNQ
$(NMP)_x(Phen)_{1-x}TCNQ$, 94–95, 107–111, 116, 118–119, 129, 136, 147–149, 151, 153, 338–339, 416
NMP-TCNQ, 2, 68, 91–92, 94, 100–101, 107–110, 117, 129, 154, 305, 324–325, 350–351
Nonlinear conduction
  carrier heating, 20, 22, 282–289
  charge-density waves, 20, 281–282, 286–287, 456
  rapid rapid rise in conductivity at low temperatures, 288–289, 465
$(N\text{-propyl})_{1-x}(n\text{-ethyl})_xQn(TCNQ)_2$, 460
$NPQn(TCNQ)_2$, 69, 452

O

Optical properties, 15–19, 23–25, 293–384, *see also* Conductivity, frequency-dependent; Dielectric function
  absorption coefficient, 294–295
  Holstein model, 301–302
  absorption spectra, 298, 305, 316–319, 324, 358
  Drude parameters and transfer integral, 332–333
  infrared absorption maximum, 372–376
  molecular vibration absorption enhanced by charge-density wave coupling, 24–25, 350–366
  optical relaxation rate, 346–347

Peierls transition, effect of, 356–360, 364
reflectance, 306–307
reflectance spectra, 16, 298–300, 309, 319, 321–322, 327, 329, 336–341, 364
solution spectra, 316–317, 324

P

Peierls distortion, 7, see also Charge-density waves; Peierls transition; Periodic lattice distortion
Peierls transition, 8–10, 99, 215, see also Charge-density waves; Gap
commensurate, 9
internal mode coupling origin, 9
similarity to superconducting transition, 9
stacks involved in, in salts with two conducting chains, 156–157, 412–413
suppression by disorder, 16–119, 416–417, 452–458
vibrational spectra, effects on, 356–360, 364
Periodic lattice distortion, 99–100, 156, 175, see also Peierls transition
(Perylene)$_2$(PF$_6$)$_{1.1}$, 332
Phase transitions, 5–6, 156–194, see also Peierls transition
analysis within Landau theory, 175–194
interchain coupling mechanisms, 158–161
neutral-ionic, 88
phase diagram, 161–175
pressure-dependence, 166–170
spin-Peierls, 88
ϕPh(TCNQ)$_2$, 70, 333
Photoconductivity, 21, 307–308, 361–362
Plasma edge, 297, 330–342, 346–348
pressure dependence of, 336–338
temperature dependence of, 335–336, 340
Plasmons, 343–345, 348
Powder spectroscopy, 305–306, 317–318
Pressure effects, 96, on,
carrier concentration, 10, 104
d.c. conductivity, 13–14, 106, 217
gap, 22
incommensurate-to-commensurate change, 106
phase transitions, 166–170
plasma frequency, 337
superconductivity, 4
transfer integral, 97

Q

Qn(TCNQ)$_2$, 1, 10, 71, 92, 94, 110–111, 117–118, 129, 136–137, 147–148, 353, 416, 444–448, 455, 460, 475, 478–479
Quinoliniumtetracyanoquinodimethane, see Qn(TCNQ)$_2$

R

Raman spectroscopy, 308, 351
Rb-TCNQ, 103
Relaxation times for scattering
acoustic mode, above T_P, 230–234
acoustic mode, below T_P, 268
internal mode, 231–233
ionized impurity, 273–274
libron, 234
polarization fluctuation, 236
two-phonon processes, 234–236

S

Screening, 220, 228
treatment for 1D case, 271
(SN)$_x$, 467, 471, 473
Soft phonons, 221, see also Kohn Anomaly
Spin-density waves, 6–8, 153, 407–408, 418–425
antiferromagnetic resonances, 422–425
Spin-Peierls transition, 119, 135–137, 148
Stabilities of different arrays, 93
Stacking in charge-transfer crystals, 88–90
mixed, 88
segregated, 88–89
"skew", 90
slipped overlap, 89, 91
"zig-zag", 90
Structural fluctuations, see $2k_F$ and $4k_F$ distortions of electron gas
Sum rules, 303, 314
Superconducting fluctuations, 3, 4
Superconductivity, in quasi-one-dimensional conductors, 4, 428, 465
β-BEDT-(TTF)$_2$I$_3$
(TMTSF)$_2$PF$_6$, 4, 465
(TMTSF)$_2$ClO$_4$, 4
Susceptibility, charge-density wave, 110, 122, 131, 196–197
divergence, 137–140
3D, 158–161

T

TaS$_3$, 452, 456
TCNQ, 1-3, 6, 8, 305, 349, 355, 388, 390
TEA(TCNQ)$_2$, 73, 90, 134, 166, 305, 350, 353, 356
Tetracyanoquinodimethane, *see* TCNQ
Tetramethyltetraselenafulvalene, *see* TMTSF
Tetramethyltetrathiafulvalene, *see* TMTTF
Tetraselenafulvalene, *see* TSeF
Tetrathiafulvalene, *see* TTF
Tetrathiafulvalene tetracyanoquinodimethane, *see* TTF-TCNQ
Thermoelectric power, 12, 15, 238, 241-242, 446-447
TMA-I-TCNQ, 75, 90
TMTSF, 3, 76
(TMTSF)$_2$AsF$_6$, 4, 76, 322, 332, 370, 399, 421, 423, 428
(TMTSF)$_2$BF$_4$, 4, 76, 399, 405, 417-418, 428-429
(TMTSF)$_2$ClO$_4$, 4-5, 76, 370, 399, 417-418, 420-421, 423, 428-429, 443-444
TMTSF-DMTCNQ, 3, 76, 92-93, 95-96, 102, 104, 111-114, 117, 129, 134-137, 139, 142, 147, 149-151, 153-154, 174-175, 328, 333-334, 399, 405, 407, 417, 440-446, 448, 450-455, 460-462, 466-467, 471, 473, 476-477
(TMTSF)$_2$FSO$_3$, 428-429
(TMTSF)$_2$NbF$_6$, 420, 428
(TMTSF)$_2$NO$_3$, 4, 76, 399, 417-418, 428-429
(TMTSF)$_2$PF$_6$, 4, 76, 267, 332
  carrier heating, 20, 283, 286-288
  ESR linewidth, 405-407, 419
  gap, pressure-dependence of, 20, 279
  Hall mobility, 277-278
  irradiation effects, 4, 48, 451-452, 409, 466-467
  mobility, low-temperature, 20, 268-269, 278
  NMR, 411
  optical properties, 346, 369-370
  plasma frequency transverse to stacks, 4, 339-341
  spatially varying gap, consequences, 279-281, 465
  spin-density-wave state properties, 419-421, 428
  spin susceptibility, 399
  superconductivity, 4, 465
(TMTSF)$_2$ReO, 4, 76, 362, 399, 417-418, 428-429
TMTSF salts, *see also* specific salts
  anion ordering, 417-418
  band structure, 4
  conductivity, 4
  crystal structures, 90-94
  dimensionality, 4
  electron-electron scattering, 13
  ESR linewidth, 405-407
  external lattice modes, 360
  phase transitions, 5, 174-175
  spin-density-wave transitions, 418-420, 428-429
  spin susceptibility, 399
  superconducting transitions, 4
(TMTSF)$_2$SbF$_6$, 4, 76, 370, 399, 420
(TMTSF)$_2$TaF$_6$, 428
TMTSF-TCNQ, 3, 76, 91-94, 102, 111-114, 136, 139, 149, 153, 222, 333, 350-351, 399, 407
TMTTF, 4, 77, 149, 406-407, 417-420, 429-430
(TMTTF)$_2$BF$_4$, 77, 399, 429
(TMTTF)$_2$Br, 77, 399, 420-421, 423-424, 430
TMTTF-Bromanil, 91-93, 102, 104, 117, 136, 147-149, 157
(TMTTF)$_2$ClO$_4$, 77, 399, 429
TMTTF-DMTCNQ, 77, 92-93, 102, 112, 130, 136-137, 139-140, 146-147, 149, 151, 157, 254, 328, 333-334
(TMTTF)$_2$NO$_3$, 399
(TMTTF)$_2$PF$_6$, 77, 148, 332, 399, 407, 430
(TMTTF)$_2$ReO$_4$, 429
(TMTTF)$_2$SCN, 77, 420-421, 430
TMTTF-TCNQ, 78, 91-95, 102, 112, 136-137, 139, 147, 150-151, 153, 157, 174-175, 350-351, 399, 402-403, 405, 407, 410, 413
T_P, 8, 9
Transfer integral, 6, 89, 105, 131, 218, 224, 244-245, 331-342
Translon., *see* Acoustic phonon
Transport, 11-15, 19-20, 215-292, *see also* Conductivity, dc; Thermoelectric power

INDEX

TSeF–TCNQ, 3, 9–10, 13, 19, 22, 80, 91–93, 102, 106, 112, 115, 123, 125, 129, 132, 134, 136–137, 139–141, 146–147, 149–150, 152–153, 157, 166, 168, 170, 175–176, 181–183, 216, 219, 250–252, 266, 276, 305, 333–334, 350–351, 358–360, 362, 399, 403–404, 406–408, 415, 426–427
TSF–TCNQ, see TSeF–TCNQ
$(TSeT)_2Br$, 81, 322, 332
$(TSeT)_2I$, 81, 442, 450–451
TTF, 2, 3, 81, 149, 315–322, 349, 355, 388, 390, 397
$(TTF)Br_n$, $n = 0.7$, 24, 81, 317–321, 332, 334, 351, 399
TTF–chloranil, 88
TTF–$Cl_{0.80}$, 316–317, 351
TTF–$I_{0.711}$, 82, 320, 332, 335, 351, 399
TTF$(SCN)_{0.57}$, 82, 320–321, 332, 334, 357, 399
$(TTF)(SeCN)_{0.57}$, 82, 357, 399
TTF–TCNQ, 2, 3, 83, see also TCNQ, TTF
  acoustic phonons, 222–223
    dispersion, 122, 124, 126, 247
    electron coupling to, 132, 242, 244–245
    pressure and temperature-dependence, 252
    structural instabilities, role in, 132
  anisotropy, 3, 15, 16
  band structure, 100, 219
    bandwidths, 3, 9, 12, 219, 243, 249, 253
    dispersion relation above T_P, 105, 219
    dispersion relation below T_P, 257, 264
    effective mass below T_P, 264, 276
    effective mass in metallic state, 219–220
  carrier concentration below T_P, 276–277, see also charge transfer
  charge-density waves
    conduction by, 12–13, 16–17, 19–20, 106, 114, 216, 281–282, 456–458
    3-D coupling, 161
    fluctuations, 8, 122–130
    phase oscillations (phonons), 24
    pinned, infrared absorption due to, 23–24, 366–369
    pinning due to commensurability, 106
    pinning due to impurities, defects, 19–20, 454–458, 465
    susceptibility, 137, 139–140
  charge transfer, 10, 101–103, 106, 243–244, 350–351
  compressibility coefficients, 96, 252
  conductivity, d.c., in metallic range, 3, 11–15, 216–217
    charge-density wave contribution, see charge-density waves
    data, agreement with theory, 11–15, 242–253
    internal mode scattering, 14–15, 216, 248–249
    one-phonon vs two-phonon scattering, 13–15, 216–217, 242, 245–246
    pressure, dependence, 14–15, 217, 252–253
    transverse, 3, 253–254
  conductivity, d.c., in semiconducting range, 19–23
    charge-density wave transport, 281–282
    data for low fields, agreement with theory, 274–281
    electric-field induced nonlinearity, 281–289
    hot electrons, 282–289
    impact ionization, 287
    spatially varying gap, effects of, 279–281
    steep increase with field, mechanism for, 286–289
    warm electron range, 284–285
  conductivity, frequency-dependent, 16–18, 23–24, 365, 373–375
  crystal structure, 2, 90–93
  dielectric function, 16, 18–19, 24, 366
  elastic properties, 95–98, 252
  electron–electron interactions, 146, 220–222
  electron–phonon coupling
    external modes, 15, 242, 244–247
    molecular vibrations, 223, 355
  Hall coefficient, 12
  internal vibration modes
    a_g (symmetric) modes, 223, 355
    coupling of electrons to internal modes, 223, 355
  irradiation effects on
    charge-density wave transport, 458
    conductivity, 442–443, 448, 450, 477
    ESR linewidth, 445
    $2k_F$ and $4k_F$ distortions, 455
    magnetic properties, 460–461
    phase transitions, 452–453

$2k_F$ and $4k_F$ distortions, 9–10, 17–18, 112–115, 118, 120–121, 123, 125, 130, 133–134, 136, 146–147, 149–150, 156, 161–166, 221, 455
Kohn anomaly, 8, 12, 122, 124, 126–128, 133
lattice vibrations, 222–223, 360, *see also* acoustic phonons, internal vibration modes, librons
librons, 223
magnetic properties
 ESR linewidth, 404–408
 g value, 390
 irradiation effects, 417
 Peierls transition, effect of, 412–413, 426
 proton NMR, 408–411
 spin susceptibility, 398–399, 425–426
mobility, 15, 22, 266, 276–277
nonlinear conduction, 20, 281, 283
 charge-density waves, 12–13, 16–17, 19–20, 106, 114, 216, 281–282, 286–287, 456–458
 hot electrons, 20, 22, 283, 286–289
 rapid rise in conductivity at low temperatures, 288–289, 465
 warm electron range, 284–285
optical properties, 15–19, 23–25, 294, *see also* conductivity, frequency-dependent; dielectric function
 absorption spectra, 305, 316, 324, 358
 Drude parameters and transfer integral, 332–333
 infrared absorption maximum, 374–376
 molecular vibration absorption enhanced by charge-density wave coupling, 24–25, 350–351, 354–359
Peierls gap
 high electric field, effect of, 288–289
 size, 20, 22–23, 275–276, 361–362, 415
 strain, effect of, 279–280
 spatial variation, consequences of, 279–281
 variation with temperature, 19, 22, 24, 358–359
 Peierls transition, effect of, 24–25, 356–359, 364
 reflectance spectra, 15–16, 328–329, 334, 364

Peierls transition, 12, 20, 88, 216, 275, *see also* phase transitions sound velocity increase below T_P, 97
phase transitions
 3D ordering, 161–162
 phase diagram, 161
 pressure-dependence, 166–170
 theory of, 176–190
ϕ-particles, 21–22, 279, 282
photoconductivity, 21, 361
plasma edge
 pressure-dependence, 336–338
 temperature-dependence, 335–336
plasmons, 343–345
pressure effects on
 charge transfer, 10, 104
 d.c. conductivity, 14
 incommensurate to commensurate change, 106
 phase transitions, 166–170
 plasma frequency, 337
 relaxation time, 249
 internal mode scattering, 232
 polarization-fluctuation scattering vs acoustic phonon scattering, 236
screening, 220, 228
stacks involved in, 156–157, 412–413
suppression by disorder, 117, 417, 452, 454, 455, 458
vibrational spectra, effects on, 24–25, 354–359, 364
thermal expansion, 95, 252
thermoelectric power, 12, 15, 216, 238, 241–242, 246, 249–250, 253
transfer integral, 105, 331–338
transverse conductivity, 253–254
$(TTF)_x(TSeF)_{1-x}(TCNQ)$, 95, 101, 116–118, 123, 125, 130, 149, 170–172, 181–183, 189, 403, 406, 415, 426–427, 438, 460
$(TTT)_2I_{3+\delta}$, 85, 322, 332, 399, 411, 414, 417, 430, 442–443, 451
$TTT(TCNQ)_2$, 85, 399, 430

U

U, 6, 8, 131, 220, 244, 249, 470, *see also* Electron–electron interactions
 large U limit, 7–10, 99, 221, 312–313, 375–376, 401, 470

magnetic susceptibility enhancement, 221
optical properties, effect on, 313–315, 375–376

W

Wavefunctions below T_P, 254–258

X

X-ray damage, 95
X-ray diffuse scattering, 88, 111, 155, 221
   resolution corrections, 204–205
   theory, 195–201

Contents of Previous Volumes

Volume 1 Physics of III–V Compounds
C. Hilsum, Some Key Features of III–V Compounds
Franco Bassani, Methods of Band Calculations Applicable to III–V Compounds
E. O. Kane, The $k \cdot p$ Method
V. L. Bonch-Bruevich, Effect of Heavy Doping on the Semiconductor Band Structure
Donald Long, Energy Band Structures of Mixed Crystals of III–V Compounds
Laura M. Roth and Petros N. Argyres, Magnetic Quantum Effects
S. M. Puri and T. H. Geballe, Thermomagnetic Effects in the Quantum Region
W. M. Becker, Band Characteristics near Principal Minima from Magnetoresistance
E. H. Putley, Freeze-Out Effects, Hot Electron Effects, and Submillimeter Photoconductivity in InSb
H. Weiss, Magnetoresistance
Betsy Ancker-Johnson, Plasmas in Semiconductors and Semimetals

Volume 2 Physics of III–V Compounds
M. G. Holland, Thermal Conductivity
S. I. Novkova, Thermal Expansion
U. Piesbergen, Heat Capacity and Debye Temperatures
G. Giesecke, Lattice Constants
J. R. Drabble, Elastic Properties
A. U. Mac Rae and G. W. Gobeli, Low Energy Electron Diffraction Studies
Robert Lee Mieher, Nuclear Magnetic Resonance
Bernard Goldstein, Electron Paramagnetic Resonance
T. S. Moss, Photoconduction in III–V Compounds
E. Antončik and J. Tauc, Quantum Efficiency of the Internal Photoelectric Effect in InSb
G. W. Gobeli and F. G. Allen, Photoelectric Threshold and Work Function
P. S. Pershan, Nonlinear Optics in III–V Compounds
M. Gershenzon, Radiative Recombination in the III–V Compounds
Frank Stern, Stimulated Emission in Semiconductors

Volume 3 Optical of Properties III–V Compounds
Marvin Hass, Lattice Reflection
William G. Spitzer, Multiphonon Lattice Absorption
D. L. Stierwalt and R. F. Potter, Emittance Studies
H. R. Philipp and H. Ehrenreich, Ultraviolet Optical Properties
Manuel Cardona, Optical Absorption above the Fundamental Edge
Earnest J. Johnson, Absorption near the Fundamental Edge
John O. Dimmock, Introduction to the Theory of Exciton States in Semiconductors
B. Lax and J. G. Mavroides, Interband Magnetooptical Effects

H. Y. Fan, Effects of Free Carriers on Optical Properties
Edward D. Palik and George B. Wright, Free-Carrier Magnetooptical Effects
Richard H. Bube, Photoelectronic Analysis
B. O. Seraphin and H. E. Bennett, Optical Constants

Volume 4 Physics of III–V Compounds

N. A. Goryunova, A. S. Borschevskii, and D. N. Tretiakov, Hardness
N. N. Sirota, Heats of Formation and Temperatures and Heats of Fusion of Compounds $A^{III}B^{V}$
Don L. Kendall, Diffusion
A. G. Chynoweth, Charge Multiplication Phenomena
Robert W. Keyes, The Effects of Hydrostatic Pressure on the Properties of III–V Semiconductors
L. W. Aukerman, Radiation Effects
N. A. Goryunova, F. P. Kesamanly, and D. N. Nasledov, Phenomena in Solid Solutions
R. T. Bate, Electrical Properties of Nonuniform Crystals

Volume 5 Infrared Detectors

Henry Levinstein, Characterization of Infrared Detectors
Paul W. Kruse, Indium Antimonide Photoconductive and Photoelectromagnetic Detectors
M. B. Prince, Narrowband Self-Filtering Detectors
Ivars Melngailis and T. C. Harman, Single-Crystal Lead–Tin Chalcogenides
Donald Long and Joseph L. Schmit, Mercury–Cadmium Telluride and Closely Related Alloys
E. H. Putley, The Pyroelectric Detector
Norman B. Stevens, Radiation Thermopiles
R. J. Keyes and T. M. Quist, Low Level Coherent and Incoherent Detection in the Infrared
M. C. Teich, Coherent Detection in the Infrared
F. R. Arams, E. W. Sard, B. J. Peyton, and F. P. Pace, Infrared Heterodyne Detection with Gigahertz IF Response
H. S. Sommers, Jr., Microwave-Based Photoconductive Detector
Robert Sehr and Rainer Zuleeg, Imaging and Display

Volume 6 Injection Phenomena

Murray A. Lampert and Ronald B. Schilling, Current Injection in Solids: The Regional Approximation Method
Richard Williams, Injection by Internal Photoemission
Allen M. Barnett, Current Filament Formation
R. Baron and J. W. Mayer, Double Injection in Semiconductors
W. Ruppel, The Photoconductor–Metal Contact

Volume 7 Application and Devices
Part A

John A. Copeland and Stephen Knight, Applications Utilizing Bulk Negative Resistance
F. A. Padovani, The Voltage–Current Characteristics of Metal–Semiconductor Contacts
P. L. Hower, W. W. Hooper, B. R. Cairns, R. D. Fairman, and D. A. Tremere, The GaAs Field-Effect Transistor
Marvin H. White, MOS Transistors
G. R. Antell, Gallium Arsenide Transistors
T. L. Tansley, Heterojunction Properties

Part B

T. *Misawa*, IMPATT Diodes
H. C. *Okean*, Tunnel Diodes
Robert B. *Campbell and Hung-Chi Chang*, Silicon Carbide Junction Devices
R. E. *Enstrom, H. Kressel, and L. Krassner*, High-Temperature Power Rectifiers of $GaAs_{1-x}P_x$

Volume 8 Transport and Optical Phenomena

Richard J. *Stirn*, Band Structure and Galvanomagnetic Effects in III–V Compounds with Indirect Band Gaps
Roland W. *Ure, Jr.*, Thermoelectric Effects in III–V Compounds
Herbert *Piller*, Faraday Rotation
H. Barry *Bebb and E. W. Williams*, Photoluminescence I: Theory
E. W. *Williams and H. Barry Bebb*, Photoluminescence II: Gallium Arsenide

Volume 9 Modulation Techniques

B. O. *Seraphin*, Electroreflectance
R. L. *Aggarwal*, Modulated Interband Magnetooptics
Daniel F. *Blossey and Paul Handler*, Electroabsorption
Bruno *Batz*, Thermal and Wavelength Modulation Spectroscopy
Ivar *Balslev*, Piezooptical Effects
D. E. *Aspnes and N. Bottka*, Electric-Field Effects on the Dielectric Function of Semiconductors and Insulators

Volume 10 Transport Phenomena

R. L. *Rode*, Low-Field Electron Transport
J. D. *Wiley*, Mobility of Holes in III–V Compounds
C. M. *Wolfe and G. E. Stillman*, Apparent Mobility Enhancement in Inhomogeneous Crystals
Robert L. *Peterson*, The Magnetophonon Effect

Volume 11 Solar Cells

Harold J. *Hovel*, Introduction; Carrier Collection, Spectral Response, and Photocurrent; Solar Cell Electrical Characteristics; Efficiency; Thickness; Other Solar Cell Devices; Radiation Effects; Temperature and Intensity; Solar Cell Technology

Volume 12 Infrared Detectors (II)

W. L. *Eiseman, J. D. Merriam, and R. F. Potter*, Operational Characteristics of Infrared Photodetectors
Peter R. *Bratt*, Impurity Germanium and Silicon Infrared Detectors
E. H. *Putley*, InSb Submillimeter Photoconductive Detectors
G. E. *Stillman, C. M. Wolfe, and J. O. Dimmock*, Far-Infrared Photoconductivity in High Purity GaAs
G. E. *Stillman and C. M. Wolfe*, Avalanche Photodiodes
P. L. *Richards*, The Josephson Junction as a Detector of Microwave and Far-Infrared Radiation
E. H. *Putley*, The Pyroelectric Detector—An Update

Volume 13 Cadmium Telluride

Kenneth Zanio, Materials Preparation; Physics; Defects; Applications

Volume 14 Lasers, Junctions, Transport

N. Holonyak, Jr. and M. H. Lee, Photopumped III–V Semiconductor Lasers
Henry Kressel and Jerome K. Butler, Heterojunction Laser Diodes
A. Van der Ziel, Space-Charge-Limited Solid-State Diodes
Peter J. Price, Monte Carlo Calculation of Electron Transport in Solids

Volume 15 Contacts, Junctions, Emitters

B. L. Sharma, Ohmic Contacts to III–V Compound Semiconductors
Allen Nussbaum, The Theory of Semiconducting Junctions
John S. Escher, NEA Semiconductor Photoemitters

Volume 16 Defects, (HgCd)Se, (HgCd)Te

Henry Kressel, The Effect of Crystal Defects on Optoelectronic Devices
C. R. Whitsett, J. G. Broerman, and C. J. Summers, Crystal Growth and Properties of $Hg_{1-x}Cd_xSe$ Alloys
M. H. Weiler, Magnetooptical Properties of $Hg_{1-x}Cd_xTe$ Alloys
Paul W. Kruse and John G. Ready, Nonlinear Optical Effects in $Hg_{1-x}Cd_xTe$

Volume 17 CW Processing of Silicon and Other Semiconductors

James F. Gibbons, Beam Processing of Silicon
Arto Lietoila, Richard B. Gold, James F. Gibbons, and Lee A. Christel, Temperature Distributions and Solid Phase Reaction Rates Produced by Scanning CW Beams
Arto Lietoila and James F. Gibbons, Applications of CW Beam Processing to Ion Implanted Crystalline Silicon
N. M. Johnson, Electronic Defects in CW Transient Thermal Processed Silicon
K. F. Lee, T. J. Stultz, and James F. Gibbons, Beam Recrystallized Polycrystalline Silicon: Properties, Applications, and Techniques
T. Shibata, A. Wakita, T. W. Sigmon, and James F. Gibbons, Metal–Silicon Reactions and Silicide
Yves I. Nissim and James F. Gibbons, CW Beam Processing of Gallium Arsenide

Volume 18 Mercury Cadmium Telluride

Paul W. Kruse, The Emergence of $(Hg_{1-x}Cd_x)Te$ as a Modern Infrared Sensitive Material
H. E. Hirsch, S. C. Liang, and A. G. White, Preparation of High-Purity Cadmium, Mercury, and Tellurium
W. F. H. Micklethwaite, The Crystal Growth of Cadmium Mercury Telluride
Paul E. Petersen, Auger Recombination in Mercury Cadmium Telluride
R. M. Broudy and V. J. Mazurczyck, (HgCd)Te Photoconductive Detectors
M. B. Reine, A. K. Sood, and T. J. Tredwell, Photovoltaic Infrared Detectors
M. A. Kinch, Metal-Insulator-Semiconductor Infrared Detectors

Volume 19 Deep Levels, GaAs, Alloys, Photochemistry

G. F. Neumark and K. Kosai, Deep Levels in Wide Band-Gap III–V Semiconductors
David C. Look, The Electrical and Photoelectronic Properties of Semi-Insulating GaAs
R. F. Brebrick, Ching-Hua Su, and Pok-Kai Liao, Associated Solution Model for Ga–In–Sb and Hg–Cd–Te
Yu. Ya. Gurevich and Yu. V. Pleskov, Photoelectrochemistry of Semiconductors

Volume 20 Semi-Insulating GaAs

R. N. Thomas, H. M. Hobgood, G. W. Eldridge, D. L. Barrett, T. T. Braggins, L. B. Ta, and S. K. Wang, High-Purity LEC Growth and Direct Implantation of GaAs for Monolithic Microwave Circuits
C. A. Stolte, Ion Implantation and Materials for GaAs Integrated Circuits
C. G. Kirkpatrick, R. T. Chen, D. E. Holmes, P. M. Asbeck, K. R. Elliott, R. D. Fairman, and J. R. Oliver, LEC GaAs for Integrated Circuit Applications
J. S. Blakemore and S. Rahimi, Models for Mid-Gap Centers in Gallium Arsenide

Volume 21 Hydrogenated Amorphous Silicon
Part A

Jacques I. Pankove Introduction
Masataka Hirose, Glow Discharge; Chemical Vapor Deposition
Yoshiyuki Uchida, dc Glow Discharge
T. D. Moustakas, Sputtering
Isao Yamada, Ionized-Cluster Beam Deposition
Bruce A. Scott, Homogeneous Chemical Vapor Deposition
Frank J. Kampas, Chemical Reactions in Plasma Deposition
Paul A. Longeway, Plasma Kinetics
Herbert A. Weakliem, Diagnostics of Silane Glow Discharges Using Probes and Mass Spectroscopy
Lester Guttman, Relation between the Atomic and the Electronic Structures
A. Chenevas-Paule, Experimental Determination of Structure
S. Minomura, Pressure Effects on the Local Atomic Structure
David Adler, Defects and Density of Localized States

Part B

Jacques I. Pankove, Introduction
G. D. Cody, The Optical Absorption Edge of a-Si:H
Nabil M. Amer and Warren B. Jackson, Optical Properties of Defect States in a-Si:H
P. J. Zanzucchi, The Vibrational Spectra of a-Si:H
Yoshihiro Hamakawa, Electroreflectance and Electroabsorption
Jeffrey S. Lannin, Raman Scattering of Amorphous Si, Ge, and Their Alloys
R. A. Street, Luminescence in a-Si:H
Richard S. Crandall, Photoconductivity
J. Tauc, Time-Resolved Spectroscopy of Electronic Relaxation Processes
P. E. Vanier, IR-Induced Quenching and Enhancement of Photoconductivity and Photoluminescence
H. Schade, Irradiation-Induced Metastable Effects
L. Ley, Photoelectron Emission Studies

Part C

Jacques I. Pankove, Introduction
J. David Cohen, Density of States from Junction Measurements in Hydrogenated Amorphous Silicon
P. C. Taylor, Magnetic Resonance Measurements in a-Si:H
K. Morigaki, Optically Detected Magnetic Resonance
J. Dresner, Carrier Mobility in a-Si:H
T. Tiedje, Information about Band-Tail States from Time-of-Flight Experiments
Arnold R. Moore, Diffusion Length in Undoped a-Si:H
W. Beyer and H. Overhof, Doping Effects in a-Si:H
H. Fritzsche, Electronic Properties of Surfaces in a-Si:H
C. R. Wronski, The Staebler–Wronski Effect
R. J. Nemanich, Schottky Barriers on a-Si:H
B. Abeles and T. Tiedje, Amorphous Semiconductor Superlattices

Part D

Jacques I. Pankove, Introduction
D. E. Carlson, Solar Cells
G. A. Swartz, Closed-Form Solution of I–V Characteristic for a-Si:H Solar Cells
Isamu Shimizu, Electrophotography
Sachio Ishioka, Image Pickup Tubes
P. G. LeComber and W. E. Spear, The Development of the a-Si:H Field-Effect Transitor and Its Possible Applications
D. G. Ast, a-Si:H FET-Addressed LCD Panel
S. Kaneko, Solid-State Image Sensor
Masakiyo Matsumura, Charge-Coupled Devices
M. A. Bosch, Optical Recording
A. D'Amico and G. Fortunato, Ambient Sensors
Hiroshi Kukimoto, Amorphous Light-Emitting Devices
Robert J. Phelan, Jr., Fast Detectors and Modulators
Jacques I. Pankove, Hybrid Structures
P. G. LeComber, A. E. Owen, W. E. Spear, J. Hajto, and W. K. Choi, Electronic Switching in Amorphous Silicon Junction Devices

Volume 22 Lightwave Communications Technology
Part A

Kazuo Nakajima, The Liquid-Phase Epitaxial Growth of InGaAsP
W. T. Tsang, Molecular Beam Epitaxy for III–V Compound Semiconductors
G. B. Stringfellow, Organometallic Vapor-Phase Epitaxial Growth of III–V Semiconductors
G. Beuchet, Halide and Chloride Transport Vapor-Phase Deposition of InGaAsP and GaAs
Manijeh Razeghi, Low-Pressure Metallo-Organic Chemical Vapor Deposition of $Ga_xIn_{1-x}As_yP_{1-y}$ Alloys
P. M. Petroff, Defects in III–V Compound Semiconductors

Part B

J. P. van der Ziel, Mode Locking of Semiconductor Lasers
Kam Y. Lau and Amnon Yariv, High-Frequency Current Modulation of Semiconductor Injection Lasers
Charles H. Henry, Spectral Properties of Semiconductor Lasers
Yasuharu Suematsu, Katsumi Kishino, Shigehisa Arai, and Fumio Koyama, Dynamic Single-Mode Semiconductor Lasers with a Distributed Reflector
W. T. Tsang, The Cleaved-Coupled-Cavity (C^3) Laser

Part C

R. J. Nelson and N. K. Dutta, Review of InGaAsP/InP Laser Structures and Comparison of Their Performance
N. Chinone and M. Nakamura, Mode-Stabilized Semiconductor Lasers for 0.7–0.8- and 1.1–1.6-μm Regions
Yoshiji Horikoshi, Semiconductor Lasers with Wavelengths Exceeding 2 μm
B. A. Dean and M. Dixon, The Functional Reliability of Semiconductor Lasers as Optical Transmitters
R. H. Saul, T. P. Lee, and C. A. Burrus, Light-Emitting Device Design
C. L. Zipfel, Light-Emitting Diode Reliability
Tien Pei Lee and Tingye Li, LED-Based Multimode Lightwave Systems
Kinichiro Ogawa, Semiconductor Noise-Mode Partition Noise

Part D

Federico Capasso, The Physics of Avalanche Photodiodes
T. P. Pearsall and M. A. Pollack, Compound Semiconductor Photodiodes
Takao Kaneda, Silicon and Germanium Avalanche Photodiodes
S. R. Forrest, Sensitivity of Avalanche Photodetector Receivers for High-Bit-Rate Long-Wavelength Optical Communication Systems
J. C. Campbell, Phototransistors for Lightwave Communications

Part E

Shyh Wang, Principles and Characteristics of Integratable Active and Passive Optical Devices
Shlomo Margalit and Amnon Yariv, Integrated Electronic and Photonic Devices
Takaaki Mukai, Yoshihisa Yamamoto, and Tatsuya Kimura, Optical Amplification by Semiconductor Lasers

Volume 23 Pulsed Laser Processing of Semiconductors

R. F. Wood, C. W. White, and R. T. Young, Laser Processing of Semiconductors: An Overview
C. W. White, Segregation, Solute Trapping, and Supersaturated Alloys
G. E. Jellison, Jr., Optical and Electrical Properties of Pulsed Laser-Annealed Silicon
R. F. Wood and G. E. Jellison, Jr., Melting Model of Pulsed Laser Processing
R. F. Wood and F. W. Young, Jr., Nonequilibrium Solidification Following Pulsed Laser Melting
D. H. Lowndes and G. E. Jellison, Jr., Time-Resolved Measurements During Pulsed Laser Irradiation of Silicon
D. M. Zehner, Surface Studies of Pulsed Laser Irradiated Semiconductors
D. H. Lowndes, Pulsed Beam Processing of Gallium Arsenide
R. B. James, Pulsed CO_2 Laser Annealing of Semiconductors
R. T. Young and R. F. Wood, Applications of Pulsed Laser Processing

Volume 24 Applications of Multiquantum Wells, Selective Doping, and Superlattices

C. Weisbuch, Fundamental Properties of III–V Semiconductor Two-Dimensional Quantized Structures: The Basis for Optical and Electronic Device Applications

H. Morkoç and H. Unlu, Factors Affecting the Performance of (Al, Ga)As/GaAs and (Al, Ga)As/InGaAs Modulation-Doped Field-Effect Transistors: Microwave and Digital Applications

N. T. Linh, Two-Dimensional Electron Gas FETs: Microwave Applications

M. Abe et al., Ultra-High-Speed HEMT Integrated Circuits

D. S. Chemla, D. A. B. Miller, and P. W. Smith, Nonlinear Optical Properties of Multiple Quantum Well Structures for Optical Signal Processing

F. Capasso, Graded-Gap and Superlattice Devices by Band-gap Engineering

W. T. Tsang, Quantum Confinement Heterostructure Semiconductor Lasers

G. C. Osbourn et al., Principles and Applications of Semiconductor Strained-Layer Superlattices

Volume 25 Diluted Magnetic Semiconductors

W. Giriat and J. K. Furdyna, Crystal Structure, Composition, and Materials Preparation of Diluted Magnetic Semiconductors

W. M. Becker, Band Structure and Optical Properties of Wide-Gap $A_{1-x}^{II}Mn_xB^{VI}$ Alloys at Zero Magnetic Field

Saul Oseroff and Pieter H. Keesom, Magnetic Properties: Macroscopic Studies

T. Giebultowicz and T. M. Holden, Neutron Scattering Studies of the Magnetic Structure and Dynamics of Diluted Magnetic Semiconductors

J. Kossut, Band Structure and Quantum Transport Phenomena in Narrow-Gap Diluted Magnetic Semiconductors

C. Riqaux, Magnetooptics in Narrow Gap Diluted Magnetic Semiconductors

J. A. Gaj, Magnetooptical Properties of Large-Gap Diluted Magnetic Semiconductors

J. Mycielski, Shallow Acceptors in Diluted Magnetic Semiconductors: Splitting, Boil-off, Giant Negative Magnetoresistance

A. K. Ramdas and S. Rodriquez, Raman Scattering in Diluted Magnetic Semiconductors

P. A. Wolff, Theory of Bound Magnetic Polarons in Semimagnetic Semiconductors

Volume 26 III–V Compound Semiconductors and Semiconductor Properties of Superionic Materials

Zou Yuanxi, III–V Compounds

H. V. Winston, A. T. Hunter, H. Kimura, and R. E. Lee, InAs-Alloyed GaAs Substrates for Direct Implantation

P. K. Bhattacharya and S. Dhar, Deep Levels in III–V Compound Semiconductors Grown by MBE

Yu. Ya. Gurevich and A. K. Ivanov-Shits, Semiconductor Properties of Superionic Materials